Managing and Mining Graph Data

T0142335

ADVANCES IN DATABASE SYSTEMS
Volume 40

Series Editors

Ahmed K. Elmagarmid
Purdue University
West Lafayette, IN 47907

Amit P. Sheth
Wright State University
Dayton, OH 45435

For other titles published in this series, please visit www.springer.com/series/5573

Managing and Mining Graph Data

by

Charu C. Aggarwal
IBM T.J. Watson Research Center
Hawthorne, NY, USA

Haixun Wang
Microsoft Research Asia
Beijing, China

 Springer

Charu C. Aggarwal
IBM
Thomas J. Watson Research
Center
19 Skyline Drive
Hawthorne, NY10532
USA
charu@us.ibm.com

Haixun Wang
Microsoft Research Asia
49 Zhichun Road
100190 Beijing
5F Sigma Center
China, People's Republic
haixunw@microsoft.com

ISSN 1386-2944
ISBN 978-1-4614-2560-1   e-ISBN 978-1-4419-6045-0
DOI 10.1007/978-1-4419-6045-0
Springer New York Dordrecht Heidelberg London

© Springer Science+Business Media, LLC 2010
Softcover reprint of the hardcover 1st edition 2010
All rights reserved. This work may not be translated or copied in whole or in part without the written permission of the publisher (Springer Science+Business Media, LLC, 233 Spring Street, New York, NY 10013, USA), except for brief excerpts in connection with reviews or scholarly analysis. Use in connection with any form of information storage and retrieval, electronic adaptation, computer software, or by similar or dissimilar methodology now known or hereafter developed is forbidden.
The use in this publication of trade names, trademarks, service marks, and similar terms, even if they are not identified as such, is not to be taken as an expression of opinion as to whether or not they are subject to proprietary rights.

Printed on acid-free paper

Springer is part of Springer Science+Business Media (www.springer.com)

Contents

List of Figures

List of Tables

Preface

The field of graph mining has seen a rapid explosion in recent years because of new applications in computational biology, software bug localization, and social and communication networking. This book is designed for studying various applications in the context of managing and mining graphs. Graph mining has been studied by the theoretical community extensively in the context of numerous problems such as graph partitioning, node clustering, matching, and connectivity analysis. However the traditional work in the theoretical community cannot be directly used in practical applications because of the following reasons:

- The definitions of problems such as graph partitioning, matching and dimensionality reduction are too "clean" to be used with real applications. In real applications, the problem may have different variations such as a disk-resident case, a multi-graph case, or other constraints associated with the graphs. In many cases, problems such as frequent sub-graph mining and dense graph mining may have a variety of different flavors for different scenarios.

- The size of the applications in real scenarios are often very large. In such cases, the graphs may not be stored in main memory, but may be available only on disk. A classic example of this is the case of web and social network graphs, which may contain millions of nodes. As a result, it is often necessary to design specialized algorithms which are sensitive to disk access efficiency constraints. In some cases, the entire graph may not be available at one time, but may be available in the form of a continuous stream. This is the case in many applications such as social and telecommunication networks in which edges are received continuously.

The book will study the problem of managing and mining graphs from an applied point of view. It is assumed that the underlying graphs are massive and cannot be held in main memory. This change in assumption has a critical impact on the algorithms which are required to process such graphs. The problems studied in the book include algorithms for frequent pattern mining, graph matching, indexing, classification, clustering, and dense graph mining.In many cases, the problem of graph management and mining has been studied from the perspective of structured and XML data. Where possible, we have clarified the connections with the methods and algorithms designed by the XML data management community. We also provide a detailed discussion of the application of graph mining algorithms in a number of recent applications such as graph privacy, web and social networks.

Many of the graph algorithms are sensitive to the application scenario in which they are encountered. Therefore, we will study the usage of many of these techniques in real scenarios such as the web, social networks, and biological data. This provides a better understanding of how the algorithms in the book apply to different scenarios. Thus, the book provides a comprehensive summary both from an algorithmic and applied perspective.

Chapter 1

AN INTRODUCTION TO GRAPH DATA

Charu C. Aggarwal
IBM T. J. Watson Research Center
Hawthorne, NY 10532
charu@us.ibm.com

Haixun Wang
Microsoft Research Asia
Beijing, China 100190
haixunw@microsoft.com

Abstract   Graph mining and management has become an important topic of research recently because of numerous applications to a wide variety of data mining problems in computational biology, chemical data analysis, drug discovery and communication networking. Traditional data mining and management algorithms
such as clustering, classification, frequent pattern mining and indexing have now
been extended to the graph scenario. This book contains a number of chapters
which are carefully chosen in order to discuss the broad research issues in graph
management and mining. In addition, a number of important applications of
graph mining are also covered in the book. The purpose of this chapter is to
provide an overview of the different kinds of graph processing and mining techniques, and the coverage of these topics in this book.

Keywords: Graph Mining, Graph Management

1.   Introduction

This chapter will provide an introduction of the topic of graph management
and mining, and its relationship to the different chapters in the book. The
problem of graph management finds numerous applications in a wide variety
of application domains such as chemical data analysis, computational biology,

C.C. Aggarwal and H. Wang (eds.), *Managing and Mining Graph Data*,
Advances in Database Systems 40, DOI 10.1007/978-1-4419-6045-0_1,
© Springer Science+Business Media, LLC 2010

social networking, web link analysis, and computer networks. Different applications result in different kinds of graphs, and the corresponding challenges are also quite different. For example, chemical data graphs are relatively small but the labels on different nodes (which are drawn from a limited set of elements) may be repeated many times in a single molecule (graph). This results in issues involving graph isomorphism in mining and management applications. On the other hand, in many large scale domains [12, 21, 22] such as the web, computer networks, and social networks, the node labels (eg. URLs) are distinct, but there are a very large number of them. Such graphs are also challenging because the degree distributions of these graphs are highly skewed [10], and this leads to difficulty in characterizing such graphs succinctly. The massive size of computer network graphs is a considerable challenge for mining algorithms. In some cases, the graphs may be *dynamic* and *time-evolving*. This means that the structure of the graph may change rapidly over time. In such cases, the *temporal aspect* of network analysis is extremely interesting.

A closely related field is that of XML data. Complex and semi-structured data is often represented in the form of XML documents because of its natural expressive power. XML data is naturally represented in graphical form, in which the attributes along with their values are expressed as nodes, and the relationships among them are expressed as edges. The expressive power of graphs and XML data comes at a cost, since it is much more difficult to design mining and management operations for structured data. The design of management and mining algorithms for XML data also helps in the design of methods for graph data, since the two fields are closely related to one another.

The book is designed to survey different aspects of graph mining and management, and provide a compendium for other researchers in the field. The broad thrust of this book is divided into three areas:

- **Managing Graph Data:** Since graphs form a complex and expressive data type, we need methods for representing graphs in databases, manipulating and querying them. We study the problem of designing query languages for graphs [14], and show how to use such languages in order to retrieve structures from the underlying graphs [26]. We also explore the design of indexing and retrieval structures for graph data. In addition, a number of specialized queries such as matching, keyword search and reachability queries [4–7, 24] are studied in the book. We will see that the design of the index is much more sensitive to the underlying application in the case of structured data than in the case of multi-dimensional data. The problem of managing graph data is related to the widely studied field of managing XML data. Where possible, we will draw on the field of XML data, and show how some of these techniques may be used in order to manage graphs in different domains. We will also present some of the recently designed techniques for graph data.

- **Mining Graph Data:** As in the case of other data types such as multi-dimensional or text data, we can design mining problems for graph data. This includes techniques such as frequent pattern mining, clustering and classification [1, 11, 16, 18, 23, 25, 26, 28]. We note that these methods are much more challenging in the graph domain, because the structural nature of the data makes the intermediate representation and interpretability of the mining results much more challenging. This is of course related to the cost of the greater expressive power associated with graphs.

- **Graph Applications:** Many of the techniques discussed above are for the case of generic graphs under a number of specific assumptions. However, graph domains are extremely diverse, and this may result in a large number of differences in the algorithms which are designed for such cases. For example, the algorithms which are designed for the web or social networks need to be constructed for graphs with very large size, but with distinct node labels. On the other hand, the algorithms which are designed for chemical data need to take into account repetitions in node labels. Similarly many graphs may have additional information associated with nodes and edges. Such variations make different applications much more challenging. Furthermore, the generic techniques discussed above may need to be applied differently for different application domains. Therefore, we have included different chapters to handle these different cases. We will study applications relating to the web, social networks, software bug localization, chemical and biological data.

One of the goals of this book is to provide the reader with a comprehensive compendium of material in the area of graph management and mining. The book provides a number of introductory chapters in the beginning, and then discusses a variety of graph mining algorithms in more detail.

2. Graph Management and Mining Applications

In this section, we will discuss the organization of the different chapters in the book. We will discuss the different applications, and the chapters in which they are discussed. In the first two chapters, we provide an introduction to the area of graph mining an a general survey. This chapter (Chapter 1) provides a brief introduction to the area of graph mining and the organization of this book. Chapter 2 is a general survey which discusses the key problems and algorithms in each area. The aim of the first two chapters is to provide the reader with a general overview of the field without getting into too much detail. Subsequent chapters expand on the various areas of graph mining. We discuss these below.

Natural Properties of Real Graphs and Generators.   In order to understand the various management and mining techniques discussed in the book,
it is important to get a feel of what real graphs look like in practice. Graphs
which arise in many large scale applications such as the web and social networks satisfy many properties such as the power law distribution [10], sparsity,
and small diameters [19]. These properties play a key role in the design of effective management and mining algorithms for graphs. Therefore, we discuss
these properties at an early stage of the book. Furthermore, the evolution of
dynamic graphs such as social networks shows a number of interesting properties such as densification, and shrinking diameters [19]. Furthermore, since the
study of graph mining algorithms requires the design of effective graph generators, it is useful to study methods for constructing realistic generators [3].
Clearly, the understanding that we obtain from the study of the natural properties of graphs in real domains can be leveraged in order to design models
for effective generators. Chapter 3 studies the laws of real large-scale network
graphs and a number of techniques for synthetic generation of graphs.

Query Languages and Indexing for Graphs.   In order to effectively handle graph management applications, we need query languages which allow ex-
pressivity for management and manipulation of structural data. Furthermore,
such query languages also need to be efficiently implementable. In chapter 4,
a variety of query languages for graphs are presented.

A second issue is that of *efficient access* of the underlying information in
order to resolve the queries. Therefore, it is useful to study the design of index
structures for graphs. General techniques for efficiently indexing graphs are
presented in chapter 5. While chapter 5 is focussed exclusively on the graph
domain, we note that many of the indexing techniques for the XML domain can
also be useful for graphs. Chapter 2 explores some of the connections between
XML indexing and graph indexing. In addition to general queries such as
similarity search, which are typically designed on *multi-graph data sets*, graph
structures are naturally suited to the design of a number of different other kinds
of queries for a single massive graph. In such cases, we may have a single
graph, but we wish to determine important intra-node characteristics in the
graph. Such queries often arise in the context of social networks and the web.
Examples of such queries include reachability and distance based queries [2,
4–7, 24]. Such queries are based on the *intra-node distance behavior* in a large
network structure, and are often extremely challenging because the underlying
graph may be disk-resident. In chapter 6, the literature for reachability query
processing is reviewed.

Graph Matching.   Graph matching is a critical problem which arises in the
context of a number of different kinds of applications such as schema match-

ing, graph embedding and other business applications [9]. In the problem of graph matching, we have a pair of graphs, and we attempt to determine a mapping of nodes between the two graphs such that edge and/or label correspondence is preserved. Graph matching has traditionally been studied in the theoretical literature in the context of the *graph isomorphism* problem. However, in the context of practical applications, precise matching between two graphs may not be possible. Furthermore, many practical variations of the problem allow for partial knowledge about the matching between different nodes. Therefore, we also need to study inexact matching techniques which allow edits on the nodes and edges during the matching process. Chapter 7 studies exact and inexact matching techniques for graphs.

Keyword Search in Graphs.   In the problem of keyword search, we would like to determine small groups of link-connected nodes which are related to a particular keyword [15]. For example, a web graph or a social network may be considered a massive graph [21, 22], in which each node may contain a large amount of text data. Even though keyword search is defined with respect to the text inside the nodes, we note that the linkage structure also plays an important role in determining the appropriate set of nodes. The information in the text and linkage structure re-enforce each other, and this leads to higher quality results. Keyword search provides a simple but user-friendly interface for information retrieval on the web. It also proves to be an effective method for searching data of complex structures. Since many real life data sets are structured as tables, trees and graphs, keyword search over such data has become increasingly important and has attracted much research interest in both the database and the IR communities. It is important to design keyword search techniques which maintain query semantics, ranking accuracy, and query efficiency. Chapter 8 provides an exhaustive survey of keyword search techniques in graphs.

Graph Clustering and Dense Subgraph Extraction.   The problem of graph clustering arises in two different contexts:

- In the first case, we wish to determine dense node clusters in a *single large graph*. This problem arises in the context of a number of applications such as graph-partitioning and the minimum cut problem. The determination of dense regions in the graph is a critical problem from the perspective of a number of different applications in social networks, web graph clustering and summarization. In particular, most forms of graph summarization require the determination of dense regions in the underlying graphs. A number of techniques [11, 12, 23] have been designed in the literature for dense graph clustering.

- In the second case, we have multiple graphs, each of which may possibly be of modest size. In this case, we wish to cluster graphs as objects. The distance between graphs is defined based on a structural similarity function such as the edit distance. Alternatively, it may be based on other aggregate characteristics such as the membership of frequent patterns in graphs. Such techniques are particularly useful for graphs in the XML domain, which are naturally expressed as objects. A method for XML data clustering is discussed in [1].

In chapter 9, both the above methods for clustering graphs have been studied. A particularly closely related problem to clustering is of dense subgraph extraction. Whereas the problem of clustering is traditionally defined as a *strict partitioning of the nodes*, the problem of dense subgraph extraction is a relaxed variation of this problem in which dense subgraphs may have overlaps. Furthermore, many nodes may not be included in any dense component. The dense subgraph problem is often studied in the context of frequent pattern mining of multi-graph data sets. Other variations include the issue of repeated presence of subgraphs in a single graph or in multiple graphs. These problems are studied in chapter 10. The topics discussed in chapters 9 and 10 are closely related, and provide a good overview of the area.

Graph Classification.   As in the case of graph clustering, the problem of graph classification arises in two different contexts. The first context is that of vertex classification in which we attempt to label the nodes of a single graph based on training data. Such problems are based on that of determining *desired properties of nodes* with the use of training data. Examples of such methods may be found in [16, 18]. The second context is one in which we attempt to label entire graphs as objects. The first case arise in the context of massive graphs such as social networks, whereas the second case arises in many different contexts such as chemical or biological compound classification, or XML data [28]. Chapter 11 studies a number of different algorithms for graph classification.

Frequent Pattern Mining in Graphs.   The problem of frequent pattern mining is much more challenging in the case of graphs than in the case of standard transaction data. This is because not all frequent patterns are equally relevant in the case of graphs. In particular, patterns which are highly connected are much more relevant. As in the case of transactional data, a number of different measures may be defined in order to determine which graphs are the most significant. In the case of graphs, the structural constraints make the problem even more interesting. As in the case of the transactional data, many variations of graph pattern mining such as that of determining closed patterns or significant patterns [25, 26], provide different kinds of insights to the field.

The frequent pattern mining problem is particularly important for the graph domain, because the end-results of the algorithms provide an overview of the important structures in the underlying data set, which may be used for other applications such as indexing [27]. Chapter 12 provides an exhaustive survey of the different algorithms for frequent pattern mining in graphs.

Streaming Algorithms for Graphs. Many graph applications such as those in telecommunications and social networks create continuous streams of edges. Such applications create unique challenges, because the entire graph cannot be held either in main memory or on disk. This creates tremendous constraints for the underlying algorithms, since the standard one-pass constraint of streaming algorithms applies to this case. Furthermore, it is extremely difficult to explore the structural characteristics of the underlying graph, because a global view of the graph is hard to construct in the streaming case. Chapter 13 discusses a number of streaming applications for such edge streams. The chapter discusses how graph streams can be summarized in an application-specific way, so that important structural characteristics of the graph can be explored.

Privacy-Preserving Data Mining of Graphs. In many applications such as social networks, it is critical to preserve the privacy of the nodes in the underlying network. Simple de-identification of the nodes during the release of a network structure is not sufficient, because an adversary may use background information about known nodes in order to re-identify the other nodes [17]. Graph privacy is especially challenging, because background information about many structural characteristics such as the node degrees or structural distances can be used in order to mount identity-attacks on the nodes [17, 13]. A number of techniques have recently been proposed in the literature, which use node addition, deletion, or swapping in order to hide such structural characteristics for privacy-preservation purposes [20, 29]. The key in these techniques is to hide identifying structural characteristics, without losing the overall structural utility of the graph. Chapter 14 discusses the challenges of graph privacy, and a variety of algorithms which can be used for private processing of such graphs.

Web Applications. Since the web is naturally structured as a graph, numerous such applications require graph mining and management algorithms. A classic example is the case of social networks in which the linkage structure is defined in the form of a graph. Typical social networking applications require the determination of interesting regions in the graph such as the dense communities. Community detection is a direct application of the problem of clustering, since it requires the determination of dense regions of the underlying graph. Many other applications such as blog analysis, web graph analysis, and page rank analysis for search require the use of graph mining algorithms. Chapter 15 provides a comprehensive overview of graph mining techniques for web applications. Since social networking is an important area, which cannot be easily covered within the context of the single chapter on web applications, we devote a special chapter on social networking. Graph mining applications for social networking are discussed in chapter 16.

Software Bug Localization.   Software programs can be represented as graphs, in which the control flow is represented in the form of a graph. In many cases, the software bugs arise as a result of "typical" distortions in the underlying control flow. Such distortions can also be understood in the context of the graphical structure which represents this control flow. Therefore, software bug localization is a natural application is graph mining algorithms in which the structure of the control flow graph is studied in order to determine and isolate bugs in the underlying program. Chapter 17 provides a comprehensive survey of techniques for software bug localization.

Chemical and Biological Data.   Chemical compounds can be represented as graph structures in which the atoms represent the nodes, and the bonds represents the links. If desired, a higher level of representation can be used in which sub-units of the molecules represent the nodes and the bonds between them represent the links. For example, in the case of biological data, the amino-acids are represented as nodes, and the bonds between them are the links. Chemical and biological data are inherently different in the sense that the graphs corresponding to biological data are much larger and require different techniques which are more suitable to massive graphs. Therefore, we have devoted two separate chapters to the topic. In chapter 18, methods for mining biological compounds are presented. Techniques for mining chemical compounds are presented in chapter 19.

3.   Summary

This book provides an introduction to the problem of managing and mining graph data. We will present the key techniques for both management and mining of graph data sets. We will show that these techniques can be very useful in a wide variety of applications such as the web, social networks, biological data, chemical data and software bug localization. . The book also presents some of the latest trends for mining massive graphs and their applicability across different domains. A number of trends in graph mining are fertile areas of research for future applications:

- Scalability is the new frontier in graph mining applications. Applications such as the web and social networks are defined on *massive graphs* in which it is impossible to explicitly store the underlying edges in main memory and sometimes even on disk. While graph-theoretic algorithms have been studied extensively in the literature, these techniques implicitly assume that the graphs can be held in main memory and are therefore not very useful for the case of disk-resident. This is because disk access may result in random access to the underlying edges which is extremely inefficient in practice. This also leads to a lack of scalability of the underlying algorithms.

- Many communication and social networking applications create large sets of edges which arrive continuously over time. Such dynamic applications require quick responses to queries to a number of traditional applications such as the shortest path problem or connectivity queries. Such queries are an enormous challenge, since it is impossible to pre-store the massive volume of the data for future analysis. Therefore, effective techniques need to be designed to compress and store the graphical structures for future analysis.

- A number of recent data mining applications and advances such as privacy-preserving data mining and uncertain data need to be studied in the context of the graph domain. For example, social networks are structured as graphs, and privacy applications are particularly important in this context. Such applications are also very challenging since they are defined on a massive domain of nodes.

This book studies a number of important problems in the graph domain in the context of important graph and networking applications. We also introduce some of the recent trends for massive graph mining applications.

References

[1] C. Aggarwal, N. Ta, J. Feng, J. Wang, M. J. Zaki. XProj: A Framework for Projected Structural Clustering of XML Documents, *KDD Conference*, 2007.

[2] R. Agrawal, A. Borgida, H.V. Jagadish. Efficient Maintenance of transitive relationships in large data and knowledge bases, *ACM SIGMOD Conference*, 1989.

[3] D. Chakrabarti, Y. Zhan, C. Faloutsos R-MAT: A Recursive Model for Graph Mining. *SDM Conference*, 2004.

[4] J. Cheng, J. Xu Yu, X. Lin, H. Wang, and P. S. Yu, Fast Computing Reachability Labelings for Large Graphs with High Compression Rate, *EDBT Conference*, 2008.

[5] J. Cheng, J. Xu Yu, X. Lin, H. Wang, and P. S. Yu, Fast Computation of Reachability Labelings in Large Graphs, *EDBT Conference*, 2006.

[6] E. Cohen. Size-estimation framework with applications to transitive closure and reachability, *Journal of Computer and System Sciences*, v.55 n.3, p.441-453, Dec. 1997.

[7] E. Cohen, E. Halperin, H. Kaplan, and U. Zwick, Reachability and distance queries via 2-hop labels, *ACM Symposium on Discrete Algorithms*, 2002.

[8] D. Cook, L. Holder, Mining Graph Data, *John Wiley & Sons Inc*, 2007.

[9] D. Conte, P. Foggia, C. Sansone, and M. Vento. Thirty years of graph matching in pattern recognition. *Int. Journal of Pattern Recognition and Artificial Intelligence*, 18(3):265–298, 2004.

[10] M. Faloutsos, P. Faloutsos, C. Faloutsos, On Power Law Relationships of the Internet Topology. *SIGCOMM Conference*, 1999.

[11] G. Flake, R. Tarjan, M. Tsioutsiouliklis. Graph Clustering and Minimum Cut Trees, *Internet Mathematics*, 1(4), 385–408, 2003.

[12] D. Gibson, R. Kumar, A. Tomkins, Discovering Large Dense Subgraphs in Massive Graphs, *VLDB Conference*, 2005.

[13] M. Hay, G. Miklau, D. Jensen, D. Towsley, P. Weis. Resisting Structural Re-identification in Social Networks, *VLDB Conference*, 2008.

[14] H. He, A. K. Singh. Graphs-at-a-time: Query Language and Access Methods for Graph Databases. In *Proc. of SIGMOD '08*, pages 405–418, Vancouver, Canada, 2008.

[15] H. He, H. Wang, J. Yang, P. S. Yu. BLINKS: Ranked keyword searches on graphs. In *SIGMOD*, 2007.

[16] H. Kashima, K. Tsuda, A. Inokuchi. Marginalized Kernels between Labeled Graphs, *ICML*, 2003.

[17] L. Backstrom, C. Dwork, J. Kleinberg. Wherefore Art Thou R3579X? Anonymized Social Networks, Hidden Patterns, and Structural Steganography. *WWW Conference*, 2007.

[18] T. Kudo, E. Maeda, Y. Matsumoto. An Application of Boosting to Graph Classification, *NIPS Conf.* 2004.

[19] J. Leskovec, J. Kleinberg, C. Faloutsos. Graph Evolution: Densification and Shrinking Diameters. *ACM Transactions on Knowledge Discovery from Data (ACM TKDD)*, 1(1), 2007.

[20] K. Liu and E. Terzi. *Towards identity anonymization on graphs*. ACM SIGMOD Conference 2008.

[21] R. Kumar, P Raghavan, S. Rajagopalan, D. Sivakumar, A. Tomkins, E. Upfal. The Web as a Graph. *ACM PODS Conference*, 2000.

[22] S. Raghavan, H. Garcia-Molina. Representing web graphs. *ICDE Conference*, pages 405-416, 2003.

[23] M. Rattigan, M. Maier, D. Jensen: Graph Clustering with Network Sructure Indices. *ICML*, 2007.

[24] H. Wang, H. He, J. Yang, J. Xu-Yu, P. Yu. Dual Labeling: Answering Graph Reachability Queries in Constant Time. *ICDE Conference*, 2006.

[25] X. Yan, J. Han. CloseGraph: Mining Closed Frequent Graph Patterns, *ACM KDD Conference*, 2003.

[26] X. Yan, H. Cheng, J. Han, and P. S. Yu, Mining Significant Graph Patterns by Scalable Leap Search, *SIGMOD Conference*, 2008.

[27] X. Yan, P. S. Yu, and J. Han, Graph Indexing: A Frequent Structure-based Approach, *SIGMOD Conference*, 2004.

[28] M. J. Zaki, C. C. Aggarwal. XRules: An Effective Structural Classifier for XML Data, *KDD Conference*, 2003.

[29] B. Zhou, J. Pei. Preserving Privacy in Social Networks Against Neighborhood Attacks. *ICDE Conference*, pp. 506-515, 2008.

Chapter 2

GRAPH DATA MANAGEMENT AND MINING: A SURVEY OF ALGORITHMS AND APPLICATIONS

Charu C. Aggarwal
IBM T. J. Watson Research Center
Hawthorne, NY 10532, USA

charu@us.ibm.com

Haixun Wang
Microsoft Research Asia
Beijing, China 100190

haixunw@microsoft.com

Abstract   Graph mining and management has become a popular area of research in recent years because of its numerous applications in a wide variety of practical
fields, including computational biology, software bug localization and computer
networking. Different applications result in graphs of different sizes and complexities. Correspondingly, the applications have different requirements for the
underlying mining algorithms. In this chapter, we will provide a survey of different kinds of graph mining and management algorithms. We will also discuss
a number of applications, which are dependent upon graph representations. We
will discuss how the different graph mining algorithms can be adapted for different applications. Finally, we will discuss important avenues of future research
in the area.

Keywords: Graph Mining, Graph Management

1.   Introduction

Graph mining has been a popular area of research in recent years because
of numerous applications in computational biology, software bug localization
and computer networking. In addition, many new kinds of data such as semi-

C.C. Aggarwal and H. Wang (eds.), *Managing and Mining Graph Data*,
Advances in Database Systems 40, DOI 10.1007/978-1-4419-6045-0_2,
© Springer Science+Business Media, LLC 2010

structured data and XML [8] can typically be represented as graphs. A detailed discussion of various kinds of graph mining algorithms may be found in [58].

In the graph domain, the requirement of different applications is not very uniform. Thus, graph mining algorithms which work well in one domain may not work well in another. For example, let us consider the following domains of data:

- **Chemical Data:** Chemical data is often represented as graphs in which the nodes correspond to atoms, and the links correspond to bonds between the atoms. In some cases, substructures of the data may also be used as individual nodes. In this case, the individual graphs are quite small, though there are significant repetitions among the different nodes. This leads to isomorphism challenges in applications such as graph matching. The isomorphism challenge is that the nodes in a given pair of graphs may match in a variety of ways. The number of possible matches may be exponential in terms of the number of the nodes. In general, the problem of isomorphism is an issue in many applications such as frequent pattern mining, graph matching, and classification.

- **Biological Data:** Biological data is modeled in a similar way as chemical data. However, the individual graphs are typically much larger. Furthermore, the nodes are typically carefully designed portions of the biological models. A typical example of a node in a DNA application could be an amino-acid. A single biological network could easily contain thousands of nodes. The sizes of the overall database are also large enough for the underlying graphs to be disk-resident. The disk-resident nature of the data set often leads to unique issues which are not encountered in other scenarios. For example, the access order of the edges in the graph becomes much more critical in this case. Any algorithm which is designed to access the edges in random order will not work very effectively in this case.

- **Computer Networked and Web Data:** In the case of computer networks and the web, the number of nodes in the underlying graph may be massive. Since the number of nodes is massive, this can lead to a very large number of *distinct edges*. This is also referred to as the *massive domain issue* in networked data. In such cases, the number of distinct edges may be so large, that they may be hard to hold in the available storage space. Thus, techniques need to be designed to summarize and work with condensed representations of the graph data sets. In some of these applications, the edges in the underlying graph may arrive in the form of a data stream. In such cases, a second challenge arises from the fact that it may not be possible to store the incoming edges for future analysis. Therefore, the summarization techniques are especially essential for this case. The stream summaries may be leveraged for future processing of the underlying graphs.

- **XML data:** XML data is a natural form of graph data which is fairly general. We note that mining and management algorithms for XML data are also quite useful for graphs, since XML data can be viewed as labeled graphs. In addition, the attribute-value combinations associated with the nodes makes the problem much more challenging. However, the research in the field of XML data has often been quite independent of the research in the graph mining field. Therefore, we will make an attempt in this chapter to discuss the XML mining algorithms along with the graph mining and management algorithms. It is hoped that this will provide a more integrated view of the field.

It is clear that the design of a particular mining algorithm depends upon the application domain at hand. For example, a disk-resident data set requires careful algorithmic design in which the edges in the graph are not accessed randomly. Similarly, massive-domain networks require careful summarization of the underlying graphs in order to facilitate processing. On the other hand, a chemical molecule which contains a lot of repetitions of node-labels poses unique challenges to a variety of applications in the form of *graph isomorphism*.

In this chapter, we will discuss different kinds of graph management and mining applications, along with the corresponding applications. We note that the boundary between graph mining and management algorithms is often not very clear, since many kinds of algorithms can often be classified as both. The topics in this chapter can primarily be divided into three categories. These categories discuss the following:

- **Graph Management Algorithms:** This refers to the algorithms for managing and indexing large volumes of the graph data. We will present algorithms for indexing of graphs, as well as processing of graph queries. We will study other kinds of queries such as reachability queries as well. We will study algorithms for matching graphs and their applications.

- **Graph Mining Algorithms:** This refers to algorithms used to extract patterns, trends, classes, and clusters from graphs. In some cases, the algorithms may need to be applied to large collections of graphs on the disk. We will discuss methods for clustering, classification, and frequent pattern mining. We will also provide a detailed discussion of these algorithms in the literature.

- **Applications of Graph Data Management and Mining:** We will study various application domains in which graph data management and mining algorithms are required. This includes web data, social and computer networking, biological and chemical data, and software bug localization.

This chapter is organized as follows. In the next section, we will discuss a variety of graph data management algorithms. In section 3, we will discuss algorithms for mining graph data. A variety of application domains in which these algorithms are used is discussed in section 4. Section 5 discusses the conclusions and summary. Future research directions are discussed in the same section.

2.   Graph Data Management Algorithms

Data management of graphs has turned out to be much more challenging than that for multi-dimensional data. The structural representation of graphs has greater expressive power, but it comes at a cost. This cost is in terms of the complexity of data representation, access, and processing, because intermediate operations such as similarity computations, averaging, and distance computations cannot be naturally defined for structural data in as intuitive a way as is the case for multidimensional data. Furthermore, traditional relational databases can be efficiently accessed with the use of block read-writes; this is not as natural for structural data in which the edges may be accessed in arbitrary order. However, recent advances have been able to alleviate some of these concerns at least partially. In this section, we will provide a review of many of the recent graph management algorithms and applications.

2.1   Indexing and Query Processing Techniques

Existing database models and query languages, including the relational model and SQL, lack native support for advanced data structures such as trees and graphs. Recently, due to the wide adoption of XML as the de facto data exchange format, a number of new data models and query languages for tree-like structures have been proposed. More recently, a new wave of applications across various domains including web, ontology management, bioinformatics, etc., call for new data models, languages and systems for graph structured data.

Generally speaking, the task can be simple put as the following: For a query pattern (a tree or a graph), find graphs or trees in the database that contain or are similar to the query pattern. To accomplish this task elegantly and efficiently, we need to address several important issues: i) how to model the data and the query; ii) how to store the data; and iii) how to index the data for efficient query processing.

Query Processing of Tree Structured Data.   Much research has been done on XML query processing. On a high level, there are two approaches for modeling XML data. One approach is to leverage the existing relational model after mapping tree structured data into relational schema [169]. The other approach is to build a native XML database from scratch [106]. For instance, some works starts with creating a tree algebra and calculus for XML data [107]. The proposed tree algebra extends the relational algebra by defining new operators, such as node deletion and insertion, for tree structured data.

SQL is the standard access method for relational data. Much efforts have been made to design SQL's counterpart for tree structured data. The criteria are, first expressive power, which allows users the flexibility to express queries over tree structured data, and second declarativeness, which allows the system to optimize query processing. The wide adoption of XML has spurred standards body groups to expand the SQL specification to include XML processing functions. XQuery [26] extends XPath [52] by using a FLWOR[1] structure to express a query. The FLWOR structure is similar to SQL's SELECT-FROM-WHERE structure, with additional support for iteration and intermediary variable binding. With path expressions and the FLWOR construct, XQuery brings SQL-like query power to tree structured data, and has been recommended by the World Wide Web Consortium (W3C) as the query language for XML documents.

For XML data, the core of query processing lies in efficient tree pattern matching. Many XML indexing techniques have been proposed [85, 141, 132, 59, 51, 115] to support this operation. DataGuide [85], for example, provides a concise summary of the path structure in a tree-structured database. T-index [141], on the other hand, indexes a specific set of path expressions. Index Fabric [59] is conceptually similar to DataGuide in that it keeps all label paths starting from the root element. Index Fabric encodes each label path to each XML element with a data value as a string and inserts the encoded label path and data value into an index for strings such as the Patricia tree. APEX [51] uses data mining algorithms to find paths that appear frequently in query workload. While most techniques focused on simple path expressions, the F^+B Index [115] emphasizes on branching path expressions (twigs). Nevertheless, since a tree query is decomposed into node, path, or twig queries, joining intermediary results together has become a time consuming operation. Sequence-based XML indexing [185, 159, 186] makes tree patterns a first class citizen in XML query processing. It converts XML documents as well as queries to sequences and performs tree query processing by (non-contiguous) subsequence matching.

Query Processing of Graph Structured Data.   One of the common characteristics of a wide range of nascent applications including social networking, ontology management, biological network/pathways, etc., is that the data they are concerned with is all graph structured. As the data increases in size and complexity, it becomes important that it is managed by a database system.

There are several approaches to managing graphs in a database. One possibility is to extend a commercial RDBMS engine to support graph structured data. Another possibility is to use general purpose relational tables to store graphs. When these approaches fail to deliver needed performance, recent research has also embraced the challenges of designing a special purpose graph database. Oracle is currently the only commercial DBMS that provides internal support for graph data. Its new 10g database includes the Oracle Spatial network data model [3], which enables users to model and manipulate graph data. The network model contains logical information such as connectivity among nodes and links, directions of links, costs of nodes and links, etc. The logical model is mainly realized by two tables: a node table and a link table, which store the connectivity information of a graph. Still, many are concerned that the relational model is fundamentally inadequate for supporting graph structured data, for even the most basic operations, such as graph traversal, are costly to implement on relational DBMSs, especially when the graphs are large. Recent interest in Semantic Web has spurred increased attention to the Resource Description Framework (RDF) [139]. A *triplestore* is a special purpose database for the storage and retrieval of RDF data. Unlike a relational database, a triplestore is optimized for the storage and retrieval of a large number of short statements in the form of subject-predicate-object, which are called triples. Much work has been done to support efficient data access on the triplestore [14, 15, 19, 33, 91, 152, 182, 195, 38, 92, 194, 193]. Recently, the semantic web community has announced the billion triple challenge [4], which further highlights the need and urgency to support inferencing over massive RDF data.

A number of graph query languages have been proposed since early 1990s. For example, GraphLog [56], which has its roots in Datalog, performs inferencing on rules (possibly with negation) about graph paths represented by regular expressions. GOOD [89], which has its roots in object-oriented databases, defines a transformation language that contains five basic operations on graphs. GraphDB [88], another object-oriented data model and query language for graphs, performs queries in four steps, each carrying out operations on subgraphs specified by regular expressions. Unlike previous graph query languages that operate on nodes, edges, or paths, GraphQL [97] operates directly on graphs. In other words, graphs are used as the operand and return type of all operations. GraphQL extends the relational algebraic operators, including selection, Cartesian product, and set operations, to graph structures. For instance, the selection operator is generalized to graph pattern matching. GraphQL is relationally complete and the nonrecursive version of GraphQL is equivalent to the relational algebra. A detailed description of GraphQL and a comparison of GraphQL with other graph query languages can be found in [96].

With the rise of Semantic Web applications, the need to efficiently query RDF data has been propelled into the spotlight. The SPARQL query language [154] is designed for this purpose. As we mentioned before, a graph in the RDF format is described by a set of triples, each corresponding to an edge between two nodes. A SPARQL query, which is also SQL-like, may consist of triple patterns, conjunctions, disjunctions, and optional patterns. A triple pattern is syntactically close to an RDF triple except that each of the subject, predicate and object may be a variable. The SPARQL query processor will search for sets of triples that match the triple patterns, binding the variables in the query to the corresponding parts of each triple [154].

Another line of work in graph indexing uses important structural characteristics of the underlying graph in order to facilitate indexing and query processing. Such structural characteristics can be in the form of paths or frequent patterns in the underlying graphs. These can be used as *pre-processing filters*, which remove irrelevant graphs from the underlying data at an early stage. For example, the *GraphGrep* technique [83] uses the enumerated paths as index features which can be used in order to filter unmatched graphs. Similarly, the *GIndex* technique [201] uses discriminative frequent fragments as index features. A closely related technique [202] leverages on the substructures in the underlying graphs in order to facilitate indexing. Another way of indexing graphs is to use the tree structures [208] in the underlying graph in order to facilitate search and indexing.

The topic of query processing on graph data has been studied for many years, still, many challenges remain. On the one hand, data is becoming increasingly large. One possibility of handling such large data is through parallel processing, by using for example, the Map/Reduce framework. However, it is well known that many graph algorithms are very difficult to be parallelized. On the other hand, graph queries are becoming increasingly complicated. For example, queries against a complex ontology are often lengthy, no matter what graph query language is used to express the queries. Furthermore, when querying a complex graph (such as a complex ontology), users often have only a vague notion, rather than a clear understanding and definition, of what they query for. These call for alternative methods of expressing and processing graph queries. In other words, instead of explicitly expressing a query in the most exact terms, we might want to use keyword search to simplify queries [183], or using data mining methods to semi-automate query formation [134].

2.2 Reachability Queries

Graph reachability queries test whether there is a path from a node v to another node u in a large directed graph. Querying for reachability is a very basic operation that is important to many applications, including applications in semantic web, biology networks, XML query processing, etc.

Reachability queries can be answered by two obvious methods. In the first method, we traverse the graph starting from node v using breath- or depth-first search to see whether we can ever reach node u. The query time is $O(n + m)$, where n is the number of nodes and m is the number of edges in the graph. At the other extreme, we compute and store the edge transitive closure of the graph. With the transitive closure, which requires $O(n^2)$ storage, a reachability query can be answered in $O(1)$ time by simply checking whether (u, v) is in the transitive closure. However, for large graphs, neither of the two methods is feasible: the first method is too expensive at query time, and the second takes too much space.

Research in this area focuses on finding the best compromise between the $O(n + m)$ query time and the $O(n^2)$ storage cost. Intuitively, it tries to compress the reachability information in the transitive closure and answer queries using the compressed data.

Spanning tree based approaches.   Many approaches, for example [47, 176, 184], decompose a graph into two parts: i) a spanning tree, and ii) edges not on the spanning tree (non-tree edges). If there is a path on the spanning tree between u and v, reachability between u and v can be decidedly easily. This is done by assigning each node u an interval code (u_{start}, u_{end}), such that v is reachable from u if and only if $u_{start} \leq v_{start} \leq u_{end}$. The entire tree can be encoded by performing a simple depth-first traversal of the tree. With the encoding, reachability check can be done in $O(1)$ time.

If the two nodes are not connected by any path on the spanning tree, we need to check if there is a path that involves non-tree edges connecting the two nodes. In order to do this, we need to build index structures in addition to the interval code to speed up the reachability check. Chen et al. [47] and Trißl et al. [176] proposed index structures for this purpose, and both of their approaches achieve $O(m - n)$ query time. For instance, Chen et al.'s SSPI (Surrogate & Surplus Predecessor Index) maintains a predecessor list $PL(u)$ for each node u, which, together with the interval code, enables efficient reachability check. Wang et al. [184] made an observation that many large graphs in real applications are sparse, which means the number of non-tree edges is small. The algorithm proposed based on this assumption answers reachability queries in $O(1)$ time using a $O(n + t^2)$ size index structure, where t is the number of non-tree edges, and $t \ll n$.

Set covering based approaches.   Some approaches propose to use simpler data structures (e.g., trees, paths, etc) to "cover" the reachability information embodied by a graph structure. For example, if v can reach u, then v can reach any node in a tree rooted at u. Thus, if we include the tree in the index, we cover a large set of reachability in the graph. We then use multiple trees to cover an entire graph. Agrawal et al. [10]'s optimal tree cover achieves $O(\log n)$ query time, where n is the number of nodes in the graph. Instead of using trees, Jagadish et al. [105] proposes to decompose a graph into pairwise disjoint *chains*, and then use chains to cover the graph. The intuition of using a chain is similar to using a tree: if v can reach u on a chain, then v can reach any node that comes after u on that chain. The chain-cover approach achieves $O(nk)$ query time, where k is the number of chains in the graph. Cohen et al. [54] proposed a 2-hop cover for reachability queries. A node u is labeled by two sets of nodes, called $L_{in}(u)$ and $L_{out}(u)$, where $L_{in}(u)$ are the nodes that can reach u and $L_{out}(u)$ are the ones that u can reach. The 2-hop approach assigns the L_{in} and L_{out} labels to each node such that u can reach v if and only if $L_{out}(u) \cap L_{in}(v) \neq \emptyset$. The optimal 2-hop cover problem of finding the minimum size 2-hop cover is NP-hard. A greedy algorithm finds a 2-hop cover iteratively. In each iteration, it picks the node w that maximizes the value of $\frac{S(A_w, w, D_w) \cap TC'}{|A_w| + |D_w|}$, where $S(A_w, w, D_w) \cap TC'$ represents the new (uncovered) reachability that a 2-hop cluster centered at w can cover, and $|A_w| + |D_w|$ is the cost (size) of the 2-hop cluster centered at w. Several algorithms have been proposed to compute high quality 2-hop covers [54, 168, 49, 48] in a more efficient manner. Many extensions to existing set covering based approaches have been proposed. For example, Jin et al. [112] introduces a 3-hop cover approach that combines the chain cover and the 2-hop cover.

Extensions to the reachability problem.   Reachability queries are one of the most basic building blocks for many advanced graph operations, and some are directly related to reachability queries. One interesting problem is in the domain of labeled graphs. In many applications, edges are labeled to denote the relationships between the two nodes they connect. A new type of reachability query asks whether two nodes are connected by a path whose edges are constrained by a given set of labels [111]. In some other applications, we want to find the shortest path between two nodes. Similar to the simple reachability problem, the shortest path problem can be solved by brute force methods such as Dijkstra's algorithm, but such methods are not appropriate for online queries in large graphs. Cohen et al extended the 2-hop covering approach for this problem [54].

A detailed description of the strengths and weaknesses of various reachability approaches and a comparison of their query time, index size, and index construction time can be found in [204].

2.3   Graph Matching

The problem of graph matching is that of finding either an approximate or a one-to-one correspondence among the nodes of the two graphs. This correspondence is based on one or more of the following structural characteristics of the graph: (1) The labels on the nodes in the two graphs should be the same. (2) The existence of edges between corresponding nodes in the two graphs should match each other. (3) The labels on the edges in the two graphs should match each other.

These three characteristics may be used to define a matching between two graphs such that there is a one-to-one correspondence in the structures of the two graphs. Such problems often arise in the context of a number of different database applications such as schema matching, query matching, and vector space embedding. A detailed description of these different applications may be found in [161]. In *exact graph matching*, we attempt to determine a one-to-one correspondence between two graphs. Thus, if an edge exists between a pair of nodes in one graph, then that edge must also exist between the corresponding pair in the other graph. This may not be very practical in real applications in which *approximate matches may exist*, but an exact matching may not be feasible. Therefore, in many applications, it is possible to define an objective function which determines the similarity in the mapping between the two graphs. Fault tolerant mapping is a much more significant application in the graph domain, because common representations of graphs may have many missing nodes and edges. This problem is also referred to as *inexact graph matching*. Most variants of the graph matching problem are well known to be NP-hard. The most common method for graph matching is that of tree-based search techniques. In this technique, we start with a seed set of nodes which are matched, and iteratively expand the neighborhood defined by that set. Iterative expansion can be performed by adding nodes to the current node set, as long as no edge constraints are violated. If it turns out that the current node set cannot be expanded, then we initiate a backtracking procedure in which we undo the last set of matches. A number of algorithms which are based upon this broad idea are discussed in [60, 125, 180]. A survey of many of the classical algorithms for graph matching may be found in [57].

The problem of exact graph matching is closely related to that of graph isomorphism. In the case of the graph isomorphism problem, we attempt to find an exact one-to-one matching between nodes and edges of the two graphs. A generalization of this problem is that of finding the maximal common subgraph in which we attempt to match the maximum number of nodes between the two graphs. Note that the solution to the maximal common subgraph problem will also provide a solution to the problem of exact matching between two subgraphs, if such a solution exists. A number of similarity measures can be derived on the basis of the mapping behavior between two graphs. If the two graphs share a large number of nodes in common, then the similarity is more significant. A number of models and algorithms for quantifying and determining the common subgraphs between two graphs may be found in [34–37]. The broad idea in many of these methods is to define a distance metric based on the nature of the matching between the two graphs, and use this distance metric in order to guide the algorithms towards an effective solution.

Inexact graph matching is a much more practical model, because it accounts for the natural errors which may occur during the matching process. Clearly, a method is required in order to quantify these errors and the closeness between the different graphs. A common technique which may be used to quantify these errors is the use of a function such as the graph edit distance. The graph edit distance determines the distance between two graphs by measuring the *cost of the edits required* to transform one graph to the other. These edits may be node or edge insertions, deletions or substitutions. An *inexact graph matching* is one which allows for a matching between two graphs after a sequence of such edits. The quality of the matching is defined by the cost of the corresponding edits. We note that the concept of graph edit distance is closely related to that of finding a maximum common subgraph [34]. This is because it is possible to direct an edit-distance based algorithm to find the maximum common subgraph by defining an appropriate edit distance.

A particular variant of the problem is when we account for the values of the labels on the nodes and edges during the matching process. In this case, we need to compute the distance between the labels of the nodes and edges in order to define the cost of a label substitution. Clearly, the cost of the label substitution is *application-dependent*. In the case of numerical labels, it may be natural to define the distances based on numerical distance functions between the two graphs. In general, the cost of the edits is also application dependent, since different applications may use different notions of similarity. Thus, domain-specific techniques are often used in order to define the edit costs. In some cases, the edit costs may even be learned with the use of sample graphs [143, 144]. When we have cases in which the sample graphs have naturally defined distances between them, the edit costs may be determined as values for which the corresponding distances are as close to the sample values as possible.

The typical algorithms for inexact graph matching use combinatorial search over the space of possible edits in order to determine the optimal matching [35, 145]. The algorithm in [35] is relatively exhaustive in its approach, and can therefore be computationally intensive in practice. In order to solve this issue, the algorithms discussed in [145] explores local regions of the graph in order to define more focussed edits. In particular, the work in [145] proposes an important class of methods which are referred to as *kernel functions*. Such methods are extremely robust to structural errors, and are therefore a useful construct for solving graph matching problems. The broad idea is to incorporate the key ideas of the graph edit distance into kernel functions. Since kernel machines are known to be extremely powerful techniques for pattern recognition, it follows that these techniques can then be leveraged to the problem of graph matching. A variety of other kernel techniques for graph matching may be found in [94, 81, 119]. The key kernel methods include convolution kernels

[94], random walk kernels [81] and diffusion kernels [119]. In random walk kernels [81], we attempt to determine the number of random walks between the two graphs which have some labels in common. Diffusion kernels [119] can be considered a generalization of the standard gaussian kernel in Euclidian space.

The technique of *relaxation labeling* is another broad class of methods which is often used for graph matching. Note that in the case of the matching problem, we are really trying to assign labels to the nodes in a graph. The specific label for a node is drawn out of a discrete set of possibilities. This discrete set of possibilities correspond to the matching nodes in the other graph. The probability of matching is defined by Gaussian probability distributions. We start off with an initial labeling based on the structural characteristics of the underlying graph, and then successively improve the solution based on additional exploration of structural information. Detailed descriptions of techniques for relaxation labeling may be found in [76].

2.4   Keyword Search

In the problem of keyword search, we would like to determine small groups of link-connected nodes which are related to a particular keyword. For example, a web graph or a social network may be considered a massive graph, in which each node may contain a large amount of text data. Even though keyword search is defined with respect to the text inside the nodes, we note that the linkage structure also plays an important role in determining the appropriate set of nodes. It is well known the text in linked entities such as the web are related, when the corresponding objects are linked. Thus, by finding groups of closely connected nodes which share keywords, it is generally possible to determine the qualitatively effective nodes. Keyword search provides a simple but user-friendly interface for information retrieval on the Web. It also proves to be an effective method for accessing structured data. Since many real life data sets are structured as tables, trees and graphs, keyword search over such data has become increasingly important and has attracted much research interest in both the database and the IR communities.

Graph is a general structure and it can be used to model a variety of complex data, including relational data and XML data. Because the underlying data assumes a graph structure, keyword search becomes much more complex than traditional keyword search over documents. The challenges lie in three aspects:

- **Query semantics**. Keyword search over a set of text documents has very clear semantics: A document satisfies a keyword query if it contains every keyword in the query. In our case, the entire dataset is often considered as a single graph, so the algorithms must work on a finer granularity and return subgraphs as answers. We must decide what subgraphs are qualified as answers.

- **Ranking strategy**: For a given keyword query, it is likely that many subgraphs will satisfy the query, based on the query semantics in use. However, each subgraph has its own underlying graph structure, with subtle semantics that makes it different from other subgraphs that satisfy the query. Thus, we must take the graph structure into consideration and design ranking strategies that find most meaningful and relevant answers.

- **Query efficiency**: Many real life graphs are extremely large. A major challenge for keyword search over graph data is query efficiency, which, to a large extent, hinges on the semantics of the query and the ranking strategy.

Current approaches for keyword search can be classified into three categories based on the underlying structure of the data. In each category, we briefly discuss query semantics, ranking strategies, and representative algorithms.

Keyword search over XML data. XML data is mostly tree structured, where each node only has a single incoming path. This property has significant impact on query semantics and answer ranking, and it also provides great optimization opportunities in algorithm design [197].

Given a query, which contains a set of keywords, the search algorithm returns snippets of an XML document that are most relevant to the keywords. The interpretation of *relevant* varies, but the most common practice is to find smallest subtrees that contain the keywords.

It is straightforward to find subtrees that contain all the keywords. Let L_i be the set of nodes in the XML document that contain keyword k_i. If we pick one node n_i from each L_i, and form a subtree from these nodes, then the subtree will contain all the keywords. Thus, an answer to the query can be represented by $lca(n_1, \cdots, n_n)$, the lowest common ancestor of nodes $n_1, \cdots, n_n$ in the tree, where $n_i \in L_i$.

Most query semantics are only interested in *smallest* answers. There are different ways to interpret the notion of *smallest*. Several algorithms [197, 102, 196] are based on the SLCA (smallest lowest common ancestor) semantics, which requires that an answer (a least common ancestor of nodes that contain all the keywords) does not have any descendent that is also an answer. XRank [86] adopts a different query semantics for keyword search. In XRank, answers consist of substrees that contain at least one occurrence of all of the query keywords, after excluding the sub-nodes that already contain all of the query keywords. Thus, the set of answers based on the SLCA semantics is a subset of answers qualified for XRank.

A keyword query may find a large number of answers, but they are not all equal due to the differences in the way they are embedded in the nested XML structure. Many approaches for keyword search on XML data, including XRank [86] and XSEarch [55], present a ranking method. A ranking mechanism takes into consideration several factors. For instance, more specific answers should be ranked higher than less specific answers. Both SLCA and the semantics adopted by XRank signify this consideration. Furthermore, keywords in an answer should appear *close* to each other, and closeness is interpreted as the the semantic distance defined over the XML embedded structure.

Keyword search over relational data. SQL is the de-facto query language for accessing relational data. However, to use SQL, one must have knowledge about the schema of the relational data. This has become a hindrance for potential users to access tremendous amount of relational data.

Keyword search is a good alternative due to its ease of use. The challenges of applying keyword search on relational data come from the fact that in a relational database, information about a single entity is usually divided among several tables. This is resulted from the normalization principle, which is the design methodology of relational database schema.

Thus, to find entities that are relevant to a keyword query, the search algorithm has to join data from multiple tables. If we represent each table as a node, and each foreign key relationship as an edge between two nodes, then we obtain a graph, which allows us to convert the current problem to the problem of keyword search over graphs. However, there is the possibility of self-joins: that is, a table may contain a foreign key that references itself. More generally, there might be cycles in the graph, which means the size of the join is only limited by the size of the data. To avoid this problem, the search algorithm may adopt an upper bound to restrict the number of joins [103].

Two most well-known keyword search algorithm for relational data are DBXplorer [12] and DISCOVER [103]. They adopted new physical database design (including sophisticated indexing methods) to speed up keyword search over relational databases. Qin et al [155], instead, introduced a method that takes full advantage of the power of RDBMS and uses SQL to perform keyword search on relational data.

Keyword search over graph data. Keyword search over large, schema-free graphs faces the challenge of how to efficiently explore the graph structure and find subgraphs that contain all the keywords in the query. To measure the "goodness" of an answer, most approaches score each edge and node, and then aggregate the scores over the subgraph as a goodness measure [24, 113, 99].

Usually, an edge is scored by the strength of the connection, and a node is scored by its importance based on a PageRank like mechanism.

Graph keyword search algorithms can be classified into two categories. Algorithms in the first category finds matching subgraphs by exploring the graph link by link, without using any index of the graph. Representative algorithms in this category include BANKS [24] and the bidirectional search algorithm [113]. One drawback of these approaches is that they explore the graph blindly as they do not have a global picture of the graph structure, nor do they know the keyword distribution in the graph. Algorithms in the other category are index-based [99], and the index is used to control guide the graph exploration, and support forward-jumps in the search.

2.5   Synopsis Construction of Massive Graphs

A key challenge which arises in many of the applications discussed below is that the graphs they deal with are very large scale in nature. As a result, the graph may be available only on disk. Most of the traditional graph mining applications assume that the data is available in main memory. However, when the graph is available on disk, applications which access the edges in random order may be extremely expensive. For example, the problem of finding the minimum-cut between two nodes is extremely efficient with the use of memory resident algorithms, but it is extraordinarily expensive when the underlying graphs are available on disk [7]. As a result algorithms need to be carefully designed in order to reduce the disk-access costs. A typical technique which may often be used is to design a synopsis construction technique [7, 46, 142], which summarizes the graph in a much smaller space, but retains sufficient information in order to effectively respond to queries.

The synopsis construction is typically defined through either node or edge contractions. The key is to define a synopsis which retains the relevant structural property of the underlying graph. In [7], the algorithm in [177] is used in order to collapse the dense regions of the graph, and represent the summarized graph in terms of sparse regions. The resulting contracted graph still retains important structural properties such as the connectivity of the graph. In [46], a randomized summarization technique is used in order to determine frequent patterns in the underlying graph. A bound has been proposed in [46] for determining the false positives and false negatives with the use of this approach. Finally, the technique in [142] also compresses graphs by representing sets of nodes as super-nodes, and separately storing "edge corrections" in order to reconstruct the entire graph. A bound on the error has been proposed in [142] with the use of this approach.

A closely related problem is that of mining *graph streams*. In this case, the edges of the graph are received continuously over time. Such cases arise frequently in applications such as social networks, communication networks, and web log analysis. Graph streams are very challenging to mine, because the structure of the graph needs to be mined in real time. Therefore, a typical approach is to construct a synopsis from the graph stream, and leverage it for the purpose of structural analysis. It has been shown in [73] how to summarize the graph in such a way that the underlying distances are preserved. Therefore, this summarization can be used for distance-based applications such as the shortest path problem. A second application which has been studied in the context of graph streams is that of *graph matching* [140]. We note that this is a different version of the problem from our discussion in an earlier section. In this case, we attempt to find a set of edges in a single graph such that no two edges share an end point. We desire to find a maximum weight or maximum cardinality matching. The main idea in [140] is to always maintain a candidate matching and update it as new edges come in. When a new edge arrives, the process of inserting it may displace as many as two edges at its end points. We allow an incoming edge to displace the edges at its endpoints, if the weight of the incoming edge is a factor $(1 + \gamma)$ of the outgoing edges. It has been shown in [140] that this matching is within a factor $(3 + 2 \cdot \sqrt{2})$ of the optimal matching.

Recently, a number of techniques have also been designed to create synopses which can be used to estimate the aggregate structural properties of the underlying graphs. A technique has been proposed in [61] for estimating the statistics of the degrees in the underlying graph stream. The techniques proposed in [61] use a variety of techniques such as sketches, sampling, hashing and distinct counting. Methods have been proposed for determining the moments of the degrees, determining heavy hitter degrees, and determining range sums of degrees. In addition, techniques have been proposed in [18] to perform space-efficient reductions in data streams. This reduction has been used in order to count triangles in the data stream. A particularly useful application in graph streams is that of the problem of *PageRank*. In this problem, we attempt to determine significant pages in a collection with the use of the linkage structure of the underlying documents. Clearly, documents which are linked to by a larger number of documents are more significant [151]. In fact, the concept of page rank can be modeled as the probability that a node is visited by a random surfer on the world wide web. The algorithms designed in [151] are for static graphs. The problem becomes much more challenging when the graphs are dynamic, as is the case of social networks. A natural synopsis technique which can be used for such cases is the method of sampling. In [166], it has been shown how to use a sampling technique in order to estimate page rank for graph streams. The idea is to sample the nodes in the graph independently and perform random walks starting from these nodes. These random walks can be used in order to estimate the probability of the presence of a random surfer at a given node. This is essentially equal to the page rank.

3. Graph Mining Algorithms

Many of the traditional mining applications also apply to the case of graphs. As in the case of management applications, the mining applications are far more challenging to implement because of the additional constraints which arise from the structural nature of the underlying graph. In spite of these challenges, a number of techniques have been developed for traditional mining problems such as frequent pattern mining, clustering, and classification. In this section, we will provide a survey of many of the structural algorithms for graph mining.

3.1 Pattern Mining in Graphs

The problem of frequent pattern mining has been widely studied in the context of mining transactional data [11, 90]. Recently, the techniques for frequent pattern mining have also been extended to the case of graph data. The main difference in the case of graphs is that the process of determining support is quite different. The problem can be defined in different ways depending upon the application domain:

- In the first case, we have a group of graphs, and we wish to determine all patterns which support a fraction of the corresponding graphs [104, 123, 181].

- In the second case, we have a single large graph, and we wish to determine all patterns which are supported at least a certain number of times in this large graph [31, 75, 123].

In both cases, we need to account for the isomorphism issue in determining whether one graph is supported by another. However, the problem of defining the support is much more challenging, if overlaps are allowed between different embeddings. This is because if we allow such overlaps, then the anti-monotonicity property of most frequent pattern mining algorithms is violated.

For the first case, where we have a data set containing multiple graphs, most of the well known techniques for frequent pattern mining with transactional data can be easily extended. For example, *Apriori*-style algorithms can be extended to the case of graph data, by using a similar level-wise strategy of generating $(k + 1)$-candidates from k-patterns. The main difference is that we need to define the join process a little differently. Two graphs of size k can be joined, if they have a structure of size $(k - 1)$ in common. The *size of this structure* could be defined in terms of either nodes or edges. In the case of the AGM algorithm [104], this common structure is defined in terms of the number of common vertices. Thus, two graphs with k vertices are joined, only if they have a common subgraph with at least $(k-1)$ vertices. A second way of performing the mining is to join two graphs which have a subgraph containing at least $(k-1)$ edges in common. The FSG algorithm proposed in [123] can be used in order to perform edge-based joins. It is also possible to define the joins in terms of arbitrary structures. For example, it is possible to express the graphs in terms of edge-disjoint paths. In such cases, subgraphs with $(k+1)$-edge disjoint paths can be generated from two graphs which have k edge disjoint paths, of which $(k-1)$ must be common. An algorithm along these lines is proposed in [181]. Another strategy which is often used is that of *pattern growth techniques*, in which frequent graph patterns are extended with the use of additional edges [28, 200, 100]. As in the case of frequent pattern mining problem, we use lexicographic ordering among edges in order to structure the search process, so that a given pattern is encountered only once.

For the second case in which we have a single large graph, a number of different techniques may be used in order to define the support in presence of the overlaps. A common strategy is to use the size of the maximum independent set of the overlap graph to define the support. This is also referred to as the *maximum independent set support*. In [124], two algorithms HSIGRAM and VSIGRAM are proposed for determining the frequent subgraphs within a single large graph. In the former case, a breadth-first search approach is used in order to determine the frequent subgraphs, whereas a depth-first approach is used in the latter case. In [75], it has been shown that the maximum independent set measure continues to satisfy the anti-monotonicity property. The main problem with this measure is that it is extremely expensive to compute. Therefore, the technique in [31] defines a different measure in order to compute the support of a pattern. The idea is to compute a *minimum image based support* of a given pattern. For this case, we compute the number of unique nodes of the graph to which a node of the given pattern is mapped. This measure continues to satisfy the anti-monotonicity property, and can therefore be used in order to determine the underlying frequent patterns. An efficient algorithm with the use of this measure has been proposed in [31].

As in the case of standard frequent pattern mining, a number of variations are possible for the case of finding graph patterns, such as determining maximal patterns [100], closed patterns [198], or significant patterns [98, 157, 198]. We note that significant graph patterns can be defined in different ways depending upon the application. In [157], significant graphs are defined by transforming regions of the graphs into features and measuring the corresponding importance in terms of p-values. In [198], significant patterns are defined in terms of arbitrary objective functions. A meta-framework has been proposed in [198] to determine the significant patterns based on arbitrary objective functions. One interesting approach to discover significant patterns is to build a model-based search tree or MbT[71]. The idea is to use divide and conquer to mine the most significant patterns in a subspace of examples. It builds a decision tree that partitions the data onto different nodes. Then at each node, it directly discovers a discriminative pattern to further divide its examples into purer subsets. Since the number of examples towards leaf level is relatively small, this approach is able to examine patterns with extremely low global support that could not be enumerated on the whole data set. For some graph data sets which occur in drug discovery applications[71], it could mine significant graph patterns, which is very difficult for most other solutions. Since it uses the divide and conquer paradigm, the algorithm is almost linearly scalable with $1 - MinSupport$ and the number of examples[71]. The MbT technique is not limited to graphs, but also applicable to item sets and sequences, and mine pattern set is both small and significant.

One of the key challenges which arises in the context of all frequent pattern mining algorithms is the massive number of patterns which can be mined from the underlying database. This problem is particularly acute in the case of graphs since the size of the output can be extremely large. One solution for reducing the number of representative patterns is to report frequent patterns in terms of *orthogonality*. A model called *ORIGAMI* has been proposed in [93] which reports frequent graph patterns only if the similarity is below a threshold α. Such patterns are also referred to as α-*orthogonal patterns*. A pattern set P is said to be β-*representative*, if for every non-reported pattern g, at least one pattern can be found in P for which the underlying similarity to g is at least a threshold β. These two constraints address different aspects of the structural patterns. The method in [93] determines the set of all α-orthogonal and β-representative patterns. An efficient algorithm has been proposed in [93] in order to mine such patterns. The idea here is to reduce the redundancy in the underlying pattern set so as to provide a better understanding of the reported patterns.

Some particularly challenging variations of the problem arise in the context of either very large data sets or very large data graphs. Recently, a technique was proposed by [46], which uses randomized summarization in order to reduce the data set to a much smaller size. This summarization is then leveraged in order to determine the frequent subgraph patterns from the data. Bounds are derived in [46] on the false positives and false negatives with the use of such an approach. Another challenging variation is when the frequent patterns are overlaid on a very large graph, as a result of which patterns may themselves be very large subgraphs. An algorithm called *TSMiner* was proposed in [110] to determine frequent structures in very large scale graphs.

Graph pattern mining has numerous applications for a variety of applications. For example, in the case of labeled data, such pattern mining techniques can be used in order to determine *structural classification rules*. For example,

the technique in [205] uses this approach for the purpose of XML data classification. In this case, we have a data set consisting of multiple (XML) graphs,
each of which is associated with a class label. The method in [205] determines
the rules in which the left hand side is a structure and the right hand side is a
class label. This is used for the purposes of classification. Another application
of frequent pattern mining is studied in [121], in which these patterns are used
in order to create *gBoost*, which is a classifier designed as an application of
boosting. Frequent pattern mining has been found to be particularly useful in
the chemical and biological domain [28, 65, 101, 120]. Frequent pattern mining techniques have been used to perform important functions in this domain
such as classification or determination of metabolic pathways.

Frequent graph pattern mining is also useful for the purpose of creating
graph indexes. In [201], the frequent structures in a graph collection are mined,
so that they can be used as features for an indexing process. The similarity of
frequent pattern membership behavior across graphs is used to define a rough
similarity function for the purpose of filtering. An inverted representation is
constructed on this feature based representation in order to filter out irrelevant graphs for the similarity search process. The technique of [201] is much
more efficient than other competitive techniques because of its feature based
approach. In general, frequent pattern mining algorithms are useful for any
application which can be defined effectively on the basis of aggregate characteristics. In general graph pattern mining techniques have the same range of
applicability as they do for the case of vanilla frequent pattern mining.

3.2   Clustering Algorithms for Graph Data

In this section, we will discuss a variety of algorithms for clustering graph
data. This includes both classical graph clustering algorithms as well as algorithms for clustering XML data. Clustering algorithms have significant applications in a variety of graph scenarios such as congestion detection, facility
location, and XML data integration [126]. Within the context of graph algorithms, the clustering can be of two types:

- **Node Clustering Algorithms:** In this case, we have one large graph,
  and we attempt to cluster the underlying nodes with the use of a distance
  (or similarity) value on the edges. In this case, the edges of the graph are
  labeled with numerical distance values. These numerical distance values
  are used in order to create clusters of nodes. A particular case is one in
  which the presence of an edge refers to a similarity value of 1, whereas
  the absence of an edge refers to a similarity value of 0. We note that the
  problem of minimizing the inter-cluster similarity for a fixed number of
  clusters essentially reduces to the problem of *graph partitioning* or the
  minimum multi-way cut problem. This is also referred to as the prob-

lem of mining dense graphs and pseudo-cliques. Recently, the problem has also been studied in the database literature as that of *quasi-clique determination*. In this problem, we determine groups of nodes which are "almost cliques". In other words, an edge exists between any pair of nodes in the set with high probability. We will study the different classes of node clustering algorithms in a different section.

- **Graph Clustering Algorithms:** In this case, we have a (possibly large) number of graphs which need to be clustered based on their underlying structural behavior. This problem is challenging because of the need to match the structures of the underlying graphs, and use these structures for clustering purposes. Such algorithms are discussed both in the context of classical graph data sets as well as semi-structured data. Therefore, we will discuss both of these variations.

In the following subsections, we will discuss each of the above kinds of graph clustering algorithms.

Node Clustering Algorithms.   A number of algorithms for graph node clustering are discussed in [78]. In [78], the graph clustering problem is related to the minimum cut and graph partitioning problems. In this case, it is assumed that the underlying graphs have weights on the edges. It is desired to partition the graph in such a way so as to minimize the weights of the edges across the partitions. The simplest case is the 2-way minimum cut problem, in which we wish to partition the graph into two clusters, so as to minimize the weight of the edges across the partitions. This version of the problem is efficiently solvable, and can be resolved by repeated applications of the *maximum flow problem* [13]. This is because the maximum flow between source s and sink t determines the minimum s-t cut. By using different source and sink combinations, it is also possible to find the global minimum cut. A second way of determining a minimum cut is by using a contraction-based edge-sampling approach. This is a probabilistic technique in which we successively sample edges in order to collapse nodes into larger sets of nodes. By successively sampling different sequences of edges and picking the optimum value [177], it is possible to determine a global minimum cut. Both of the above techniques are quite efficient and the time-complexity is polynomial in terms of the number of nodes and edges. An interesting discussion of this problem may be found in [78].

The *multi-way graph partitioning problem* is significantly more difficult, and is NP-hard [80]. In this case, we wish to partition a graph into $k > 2$ components, so that the total weight of the edges whose ends lie in different partitions is minimized. A well known technique for graph partitioning is the Kerninghan-Lin algorithm [116]. This classical algorithm is based on a hill-climbing (or more generally neighborhood-search technique) for determining the optimal graph partitioning. Initially, we start off with a random cut of the graph. In each iteration, we exchange a pair of vertices in two partitions, to see if the overall cut value is reduced. In the event that the cut value is reduced, then the interchange is performed. Otherwise, we pick another pair of vertices in order to perform the interchange. This process is repeated until we converge to a optimal solution. We note that this optimum may not be a global optimum, but may only be a local optimum of the underlying data. The main variation in different versions of the Kerninghan-Lin algorithm is the policy which is used for performing the interchanges on the vertices. We note that the use of more sophisticated strategies allows a better improvement in the objective function for each interchange, but also requires more time for each interchange. This is a natural tradeoff which may work out differently depending upon the nature of the application at hand. We note that the problem of graph partitioning is studied widely in the literature. A detailed survey may be found in [77].

A closely related problem is that of dense subgraph determination in massive graphs. This problem is frequently encountered in large graph data sets. For example, the problem of determining large subgraphs of web graphs was studied in [82]. In this paper, a min-hash approach was used to determine the *shingles* which represent dense subgraphs. The broad idea is to represent the outlinks of a particular node as sets. Two nodes are considered similar, if they share many outlinks. Thus, consider a node A with an outlink set S_A and a node B with outlink set S_B. Then the similarity between the two nodes is defined by the *Jaccard coefficient*, which is defined as $\frac{S_A \cap S_B}{S_A \cup S_B}$. We note that explicit enumeration of all the edges in order to compute this can be computationally inefficient. Rather, a *min-hash approach* is used in order to perform the estimation. This *min-hash approach* is as follows. We sort the universe of nodes in a random order. For any set of nodes in random sorted order, we determine the first node $First(A)$ for which an outlink exists from A to $First(A)$. We also determine the first node $First(B)$ for which an outlink exists from B to $First(B)$. It can be shown that the Jaccard coefficient is an unbiased estimate of the probability that $First(A)$ and $First(B)$ are the same node. By repeating this process over different permutations over the universe of nodes, it is possible to accurately estimate the Jaccard Coefficient. This is done by using a constant number of permutations c of the node order. Thus, for each node, a fingerprint of size c can be constructed. By comparing the fingerprints of two nodes, the Jaccard coefficient can be estimated. This approach can be further generalized with the use of every s element set contained entirely with S_A and S_B. By using different values of s and c, it is possible to design an algorithm which distinguishes between two sets that are above or below a certain threshold of similarity.

The overall technique in [82] first generates a set of c shingles of size s for each node. The process of generating the c shingles is extremely straightforward. Each node is processed independently. We use the min-wise hash function approach in order to generate subsets of size s from the outlinks at each node. This results in c subsets for each node. Thus, for each node, we have a set of c shingles. Thus, if the graph contains a total of n nodes, the total size of this shingle fingerprint is $n \times c \times sp$, where sp is the space required for each shingle. Typically sp will be $O(s)$, since each shingle contains s nodes. For each distinct shingle thus created, we can create a list of nodes which contain it. In general, we would like to determine groups of shingles which contain a large number of common nodes. In order to do so, the method in [82] performs a second-order shingling in which the meta-shingles are created from the shingles. Thus, this further compresses the graph in a data structure of size $c \times c$. This is essentially a constant size data structure. We note that this group of meta-shingles has the the property that they contain a large number of common nodes. The dense subgraphs can then be extracted from these meta-shingles. More details on this approach may be found in [82].

A related problem is that of determining quasi-cliques in the underlying data. Quasi-cliques are essentially relaxations on the concept of cliques. In the case of a clique, the subgraph induced on a set of nodes is *complete*. On the other hand, in the case of a γ-quasi-clique, each vertex in that subset of nodes has a degree of at least $\gamma \cdot k$, where γ is a fraction, and k is the number of nodes in that set. The first work on determining γ-quasi-cliques was discussed in [5], in which a randomized algorithm is used in order to determine a quasi-clique with the largest size. A closely related problem is that of finding *frequently occurring cliques* in *multiple data sets*. In other words, when multiple graphs are obtained from different data sets, some dense subgraphs occur frequently together in the different data sets. Such graphs help in determining *important dense patterns of behavior in different data sources*. Such techniques find applicability in mining important patterns in graphical representations of customers. The techniques are also helpful in mining cross-graph quasi-cliques in gene expression data. A description of the application of the technique to the problem of gene-expression data may be found in [153]. An efficient algorithm for determining cross graph quasi-cliques was proposed in [148].

Classical Algorithms for Clustering XML and Graph Data. In this section, we will discuss a variety of algorithms for clustering XML and graph data. We note that XML data is quite similar to graph data in terms of how the data is organized structurally. In has been shown in [8, 63, 126, 133] that the use of this structural behavior is more critical for effective processing. There are two main techniques used for clustering of XML documents. These techniques are as follows:

- **Structural Distance-based Approach:** This approach computes structural distances between documents and uses them in order to compute clusters of documents. Such distance-based approaches are quite general and effective techniques over a wide variety of non-numerical domains such as categorical and string data. It is therefore natural to explore this technique in the context of graph data. One of the earliest work on clustering tree structured data is the *XClust algorithm* [126], which was designed to cluster XML schemas for efficient integration of large numbers of Document Type Definitions (DTDs) of XML sources. It adopts the agglomerative hierarchical clustering method which starts with clusters of single DTDs and gradually merges the two most similar clusters into one larger cluster. The similarity between two DTDs is based on their element similarity, which can be computed according to the semantics, structure, and context information of the elements in the corresponding DTDs. One of the shortcomings of the XClust algorithm is that it does not make full use of the structure information of the DTDs, which is quite important in the context of clustering tree-like structures. The method in [45] computes similarity measures based on the structural edit-distance between documents. This edit-distance is used in order to compute the distances between clusters of documents.

  Another clustering technique which falls in this general class of methods is the *S-GRACE* algorithm. The main idea is to use the element-subelement relationships in the distance function rather than the simple use of the tree-edit distance as in [45]. S-GRACE is a hierarchical clustering algorithm [133]. In [133], an XML document is converted to a structure graph (or s-graph), and the distance between two XML documents is defined according to the number of the common element-subelement relationships, which can capture better structural similarity relationships than the tree edit distance in some cases [133].

- **Structural Summary Based Approach:** In many cases, it is possible to create summaries from the underlying documents. These summaries are used for creating groups of documents which are similar to these summaries. The first summary-based approach for clustering XML documents was presented in [63]. In [63], the XML documents are modeled as rooted ordered labeled trees. A framework for clustering XML documents by using structural summaries of trees is presented. The aim is to improve algorithmic efficiency without compromising cluster quality.

  A second approach for clustering XML documents is presented in [8], and is referred to as *XProj*. This technique is a partition-based algorithm. The primary idea in this approach is to use frequent-pattern mining algorithms in order to determine the summaries of frequent structures in the data. The technique uses a k-means type approach in which each cluster center comprises a set of frequent patterns which are local to the partition for that cluster. The frequent patterns are mined using the documents assigned to a cluster center in the last iteration. The documents are then further re-assigned to a cluster center based on the average similarity between the document and the newly created cluster centers from the local frequent patterns. In each iteration the document-assignment and the mined frequent patterns are iteratively re-assigned, until the cluster centers and document partitions converge to a final state. It has been shown in [8] that such a structural summary based approach is significantly superior to a similarity function based approach as presented in [45]. The method is also superior to the structural approach in [63] because of its use of more robust representations of the underlying structural summaries.

3.3   Classification Algorithms for Graph Data

Classification is a central task in data mining and machine learning. As graphs are used to represent entities and their relationships in an increasing variety of applications, the topic of graph classification has attracted much attention in both academia and industry. For example, in pharmaceutics and drug design, we are interested to know the relationship between the activity of a chemical compound and the structure of the compound, which is represented by a graph. In social network analysis, we study the relationship between the health of a community (e.g., whether it is expanding or shrinking) and its structure, which again is represented by graphs.

Graph classification is concerned with two different but related learning tasks.

- **Label Propagation.** A subset of nodes in a graph are labeled. The task is to learn a model from the labeled nodes and use the model to classify the unlabeled nodes.

- **Graph classification.** A subset of graphs in a graph dataset are labeled. The task is to learn a model from the labeled graphs and use the model to classify the unlabeled graphs.

Label Propagation.   The concept of *label or belief propagation* [174, 209, 210] is a fundamental technique which is used in order to leverage graph structure in the context of classification in a number of relational domains. The scenario of label propagation [44] occurs in many applications. As an example, social network analysis is being used as a mean for targeted marketing. Retailers track customers who have received promotions from them. Those customers who respond to the promotion (by making a purchase) are labeled as positive nodes in the graph representing the social network, and those who do not respond are labeled as negative. The goal of target marketing is to send promotions to customers who are most likely to respond to promotions. It boils down to learning a model from customers who have received promotions and predicting the responses of other potential customers in the social network. Intuitively, we want to find out how existing positive and negative labels propagate in the graph to unlabeled nodes.

Based on the assumption that "similar" nodes should have similar labels, the core challenge for label propagation lies in devising a distance function that measures the similarity between two nodes in the graph. One common approach of defining the distance between two nodes is to count the average number of steps it takes to reach one node from the other using a random walk [119, 178]. However, it has a significant drawback: it takes $O(n^3)$ time to derive the distances and $O(n^2)$ space to store the distances between all pairs. However, many graphs in real life applications are sparse, which reduces the complexity of computing the distance [211, 210]. For example, Zhou et al [210] introduces a method whose complexity is nearly linear to the number of non-zero entries of the sparse coefficient matrix. A survey of label propagation methods can be found in [179].

Kernel-based Graph Classification Methods.   Kernel-based graph classification employs a graph kernel to measure the similarity between two labeled graphs. The method is based on random walks. For each graph, we enumerate its paths, and we derive probabilities for such paths. The graph kernel compares the set of paths and their probabilities between the two graphs. A random path (represented as a sequence of node and edge labels) is generated via a random walk: First, we randomly select a node from the graph. During the next and each of the subsequent steps, we either stop (the path ends) or randomly select an adjacent node to continue the random walk. The choices we make are subject to a given stopping probability and a node transition probability. By repeating the random walks, we derive a table of paths, each of which is associated with a probability.

In order to measure the similarity between two graphs, we need to measure the similarity between nodes, edges, and paths.

- **Node/Edge kernel.** An example of a node/edge kernel is the identity kernel. If two nodes/edges have the same label, then the kernel returns 1 otherwise 0. If the node/edge labels take real values, then a Gaussian kernel can be used instead.

- **Path kernel.** A path is a sequence of node and edge labels. If two paths are of the same length, the path kernel can be constructed as the product of node and edge kernels. If two paths are of different lengths, the path kernel simply returns 0.

- **Graph kernel.** As each path is associated with a probability, we can define the graph kernel as the expectation of the path kernel over all possible paths in the two graphs.

The above definition of a graph kernel is straightforward. However, it is computationally infeasible to enumerate all the paths. In particular, in cyclic graphs, the length of a path is unbounded, which makes enumeration impossible. Thus, more efficient approaches are needed to compute the kernel. It turns out that the definition of the kernel can be reformulated to show a nested structure. In the case of directed acyclic graphs the nodes can be topologically ordered such that there is no path from node j to i if $i < j$, the kernel can be redefined as a recursive function, and dynamic programming can handle this problem in $O(|\mathcal{X}| \cdot |\mathcal{X}'|)$, where $\mathcal{X}$ and $\mathcal{X}'$ are the set of nodes in the two graphs. In the case of cyclic graphs, the kernel's feature space (label sequences) is possibly infinite because of loops. The computation of cyclic graph kernel can still be done with linear system theory and convergence properties of the kernel.

Boosting-based Graph Classification Methods. While the kernel-based method provides an elegant solution to graph classification, it does not explicitly reveal what graph features (substructures) are relevant for classification. To address this issue, a new approach of graph classification based on pattern mining is introduced. The idea is to perform graph classification based on a graph's important substructures. We can create a binary feature vector based on the presence or absence of a certain substructure (subgraph) and apply an off-the-shelf classifier.

Since the entire set of subgraphs is often very large, we must focus on a small subset of features that are relevant. The most straightforward approach for finding interesting features is through frequent pattern mining. However, frequent patterns are not necessarily relevant patterns. For instance, in chemical graphs, ubiquitous patterns such as C-C or C-C-C are frequent, but have almost no significance in predicting important characteristics of chemical compounds such as activity, toxicity, etc.

Boosting is used to automatically select a relevant set of subgraphs as features for classification. LPBoost (Linear Program Boost) learns a linear discriminant function for feature selection. To obtain an interpretable rule, we need to obtain a sparse weight vector, where only a few weights are nonzero. It was shown [162] that graph boosting can achieve better accuracy than graph kernels, and it has the advantage of discovering key substructures explicitly at the same time.

The problem of graph classification is closely related to that of XML classification. This is because XML data can be considered an instance of *rich graphs*, in which nodes and edges have features associated with them. Consequently, many of the methods for XML classification can also be used for structural graph classification. In [205], a rule-based classifier (called *XRules*) was proposed in which we associate structural features on the left-hand side with class labels on the right-hand side. The structural features on the left-hand side are determined by computing the structural features in the graph which are both *frequent* and *discriminative* for classification purposes. These structural features are used in order to construct a prioritized list of rules which are used for classification purposes. The top-k rules are determined based on the discriminative behavior and the majority class label on the right hand side of these k rules is reported as the final result.

Other Related Work.   The problem of node classification arises in a number of different application contexts such as relational data classification, social network classification, and blog classification. A technique has been proposed in [138], which uses link-based similarity for node-classification in the context of relational data. This approach constructs *link features* from the underlying structure and uses them in order to create an effective model for classification. Recently, this technique has also been used in the context of link-based classification of blogs [23]. However, all of these techniques use link-based methods only. Since many of these techniques arise in the context of text data, it is natural to examine whether such content can be used in order to improve classification accuracy. A method to perform *collective classification* of email speech acts has been proposed in [39]. It has been shown that the analysis of relational aspects of emails (such as emails in a particular thread) significantly improves the classification accuracy. It has also been shown in [206] that the use of graph structures during categorization improves the classification accuracy of web pages. Another work [25] discusses the problem of label acquisition in the context of collective classification.

3.4   The Dynamics of Time-Evolving Graphs

Many networks in real applications arise in the context of networked entities such as the web, mobile networks, military networks, and social networks. In such cases, it is useful to examine various aspects of the *evolution dynamics* of *typical networks*, such as the web or social networks. Thus, this line of research focusses on modeling the general evolution properties of very large graphs which are *typically* encountered. Considerable study has been devoted to that of examining generic evolution properties which *hold across massive networks such as web networks, citation networks and social networks*. Some examples of such properties are as follows:

Densification: Most real networks such as the web and social networks continue to become more dense over time [129]. This essentially means that these networks continue to add more links over time (than are deleted). This is a natural consequence of the fact that much of the web and social media is a relatively recent phenomenon for which new applications continue to be found over time. In fact most real graphs are known to exhibit a *densification power law*, which characterizes the variation in densification behavior over time. This law states that the number of nodes in the network increases superlinearly with the number of nodes over time, whereas the number of edges increases superlinearly over time. In other words, if $n(t)$ and $e(t)$ represent the number of edges and nodes in the network at time t, then we have:

$$e(t) \propto n(t)^{\alpha} \tag{2.1}$$

The value of the exponent α lies between 1 and 2.

Shrinking Diameters: The *small world* phenomenon of graphs is well known. For example, it was shown in [130] that the average path length between two MSN messenger users is 6.6. This can be considered a verification of the (internet version of the) widely known rule of "six degrees of separation" in (generic) social networks. It was further shown in [129], that the diameters of massive networks such as the web continue to shrink over time. This may seem surprising, because one would expect that the diameter of the network should grow as more nodes are added. However, it is important to remember that edges are added more rapidly to the network than nodes (as suggested by Equation 2.1 above). As more edges are added to the graph it becomes possible to traverse from one node to another with the use of a fewer number of edges.

While the above observations provide an understanding of some key aspects of specific aspects of long-term evolution of massive graphs, they do not provide an idea of how the evolution in social networks can be *modeled* in a comprehensive way. A method which was proposed in [131] uses the *maximum likelihood principle* in order to characterize the evolution behavior of massive social networks. This work uses data-driven strategies in order to model the online behavior of networks. The work studies the behavior of four different networks, and uses the observations from these networks in order to create a model of the underlying evolution. It also shows that edge locality plays an important role in the evolution of social networks. A complete model of a node's behavior during its lifetime in the network is studied in this work.

Another possible line of work in this domain is to study methods for characterizing the evolution of specific graphs. For example, in a social network, it may be useful to determine the newly forming or decaying communities in the underlying network [9, 16, 50, 69, 74, 117, 131, 135, 171, 173]. It was shown in [9] how expanding or contracting communities in a social network may be characterized by examining the relative behavior of edges, as they are received in a dynamic graph stream. The techniques in this paper characterize the structural behavior of the incremental graph within a given time window, and uses it in order to determine the birth and death of communities in the graph stream. This is the first piece of work which studies the problem of evolution in *fast streams of graphs*. It is particularly challenging to study the stream case, because of the inherent combinatorial complexity of graph structural analysis, which does not lend itself well to the stream scenario.

The work in [69] uses statistical analysis and visualization in order to provide a better idea of the changing community structure in an evolving social network. A method in [171] performs parameter-free mining of large time-evolving graphs. This technique can determine the evolving communities in the network, as well as the critical change-points in time. A key property of this method is that it is *parameter-free*, and this increases the usability of the method in many scenarios. This is achieved with the use of the MDL principle in the mining process. A related technique can also perform parameter-free analysis of evolution in massive networks [74] with the use of the MDL principle. The method can determine which communities have shrunk, split, or emerged over time.

The problem of evolution in graphs is usually studied in the context of clustering, because clusters provide a natural summary for understanding both the underlying graph and the changes inherent during the evolution process. The need for such characterization arises in the context of massive networks, such as interaction graphs [16], community detection in social networks [9, 50, 135, 173], and generic clustering changes in linked information networks [117]. The work by [16] provides an *event based framework*, which provides an understanding of the typical events which occur in real networks, when new communities may form, evolve, or dissolve. Thus, this method can provide an easy way of making a quick determination of whether specific kinds of changes may be occurring in a particular network. A key technique used by many methods is to analyze the communities in the data over specific time slices and then determine the change between the slices to diagnose the nature of the underlying evolution. The method in [135] deviates from this two-step approach and constructs a unified framework for the determination of communities with the use of a best fit to a temporal-smoothness model. The work in [50] presents a spectral method for evolutionary clustering, which is also based on the temporal-smoothness concept. The method in [173] studies techniques for evolutionary characterization of networks in multi-modal graphs. Finally, a recent method proposed in [117] combines the problem of clustering and evolutionary analysis into one framework, and shows how to determine evolving clusters in a dynamic environment. The method in [117] uses a density-based characterization in order to construct *nano-clusters* which are further leveraged for evolution analysis.

A different approach is to use association rule-based mining techniques [22]. The algorithm takes a sequence of snapshots of an evolving graph, and then attempts to determine rules which define the changes in the underlying graph. Frequently occurring sequences of changes in the underlying graph are considered important indicators for rule determination. Furthermore, the frequent patterns are decomposed in order to study the confidence that a particular sequence of steps in the past will lead to a particular transition. The probability of such a transition is referred to as *confidence*. The rules in the underlying graph are then used in order to characterize the overall network evolution.

Another form of evolution in the networks is in terms of the underlying *flow of communication (or information)*. Since the flow of communication and information implicitly defines a graph (stream), the dynamics of this behavior can be very interesting to study for a number of different applications. Such behaviors arise often in a variety of information networks such as social networks, blogs, or author citation graphs. In many cases, the evolution may take the form of cascading information through the underlying graphs. The idea is that information propagates through the social network through contact between the different entities in the network. The evolution of this information flow shares a number of similarities with the spread of diseases in networks. We will discuss more on this issue in a later section of this paper. Such evolution has been studied in [128], which studies how to characterize the evolution behavior in blog graphs.

4.   Graph Applications

In this section, we will study the application of many of the aforementioned mining algorithms to a variety of graph applications. Many data domains such as chemical data, biological data, and the web are naturally structured as graphs. Therefore, it is natural that many of the mining applications discussed earlier can be leveraged for these applications. In this section, we will study the diverse applications that graph mining techniques can support. We will also see that even though these applications are drawn from different domains, there are some common threads which can be leveraged in order to improve the quality of the underlying results.

4.1   Chemical and Biological Applications

Drug discovery is a time consuming and extremely expensive undertaking. Graphs are natural representations for chemical compounds. In chemical graphs, nodes represent atoms and edges represent bonds between atoms. Biology graphs are usually on a higher level where nodes represent amino acids and edges represent connections or contacts among amino acids. An important assumption, which is known as the structure activity relationship (SAR) principle, is that the properties and biological activities of a chemical compound are related to its structure. Thus, graph mining may help reveal chemical and biology characteristics such as activity, toxicity, absorption, metabolism, etc. [30], and facilitate the process of drug design. For this reason, academia and pharmaceutical industry have stepped up efforts in chemical and biology graph mining, in the hope that it will dramatically reduce the time and cost in drug discovery.

Although graphs are natural representations for chemical and biology structures, we still need a computationally efficient representation, known as descriptors, that is conducive to operations ranging from similarity search to various structure driven predictions. Quite a few descriptors have been proposed. For example, hash fingerprints [2, 1] are a vectorized representation. Given a chemical graph, we create a a hash fingerprint by enumerating certain types of basic structures (e.g., cycles and paths) in the graph, and hashing them into a bit-string. In another line of work, researchers use data mining methods to find frequent subgraphs [150] in a chemical graph database, and represent each chemical graph as a vector in the feature space created by the set of frequent subgraphs. A detailed description and comparison of various descriptors can be found in [190].

One of the most fundamental operations on chemical compounds is similarity search. Various graph matching algorithms have been employed for i) *rank-retrieval*, that is, searching a large database to find chemical compounds that share the same bioactivity as a query compound; and ii) *scaffold-hopping*, that is, finding compounds that have similar bioactivity but different structure from the query compound. Scaffold-hopping is used to identify compounds that are good "replacement" for the query compound, which either has some undesirable properties (e.g., toxicity), or is from the existing patented chemical space. Since chemical structure determines bioactivity (the SAR principle), scaffold-hopping is challenging, as the identified compounds must be structurally similar enough to demonstrate similar bioactivity, but different enough to be a novel chemotype. Current approaches for similarity matching can be classified into two categories. One category of approaches perform similarity matching directly on the descriptor space [192, 170, 207]. The other category of approaches also consider indirect matching: if a chemical compound c is structurally similar to the query compound q, and another chemical compound c' is structurally similar to c, then c' and q are indirect matches. Clearly, indirect macthing has the potential to indentify compounds that are functionally similar but structurally different, which is important to scaffold-hopping [189, 191].

Another important application area for chemical and biology graph mining is structure-driven prediction. The goal is to predict whether a chemical structure is active or inactive, or whether it has certain properties, for example, toxic or nontoxic, etc. SVM (Support Vector Machines) based methods have proved effective for this task. Various vector space based kernel functions, including the widely used radial basis function and the Min-Max kernel [172, 192], are used to measure the similarity between chemical compounds that are represented by vectors. Instead of working on the vector space, another category of SVM methods use graph kernels to compare two chemical structures. For instance, in [160], the size of the maximum common subgraph of two graphs is used as a similarity measure.

In late 1980's, the pharmaceutical industry embraced a new drug discovery paradigm called target-based drug discovery. Its goal is to develop a drug that selectively modulates the effects of the disease-associated gene or gene product without affecting other genes or molecular mechanisms in the organism. This is made possible by the High Throughput Screening (HTS) technique, which is able to rapidly testing a large number of compounds based on their binding activity against a given target. However, instead of increasing the productivity of drug design, HTS slowed it down. One reason is that a large number of screened candidates may have unsatisfactory phenotypic effects such as toxity and promiscuity, which may dramatically increase the validation cost in later stage drug discovery [163]. Target Fishing [109] tackles the above issues by employing computational techniques to directly screen molecules for desirable phenotype effects. In [190], we offer a detailed description of various such methods, including multi-category Bayesian models [149], SVM rank [188], Cascade SVM [188, 84], and Ranking Perceptron [62, 188].

4.2   Web Applications

The world wide web is naturally structured in the form of a graph in which the web pages are the nodes and the links are the edges. The linkage structure of the web holds a wealth of information which can be exploited for a variety of data mining purposes. The most famous application which exploits the linkage structure of the web is the *PageRank* algorithm [29, 151]. This algorithm has been one of the key secrets to the success of the well known *Google* search engine. The basic idea behind the page rank algorithm is that the importance of a page on the web can be gauged from the number and importance of the hyperlinks pointing to it. The intuitive idea is to model a random surfer who follows the links on the pages with equal likelihood. Then, it is evident that the surfer will arrive more frequently at web pages which have a large number of paths leading to them. The intuitive interpretation of page rank is the probability that a random surfer arrives at a given web page during a random walk. Thus, the page rank essentially forms a probability distribution over web pages, so that the sum of the page rank over all the web pages sums to 1. In addition, we sometimes add teleportation, in which we can transition *any* web page in the collection uniformly at random.

Let A be the set of edges in the graph. Let π_i denote the steady state probability of node i in a random walk, and let $P = [p_{ij}]$ denote the transition matrix for the random-walk process. Let α denote the *teleportation probability* at a given step, and let q_i be the ith value of a probability vector defined over all the nodes which defines the probability that the teleportation takes place to node i at any given step (conditional on the fact that teleportation does take place). For the time-being, we assume that each value of q_i is the same, and is equal to $1/n$, where n is the total number of nodes. Then, for a given node i, we can derive the following steady-state relationship:

$$\pi_i = \sum_{j:(j,i)\in A} \pi_j \cdot p_{ji} \cdot (1 - \alpha) + \alpha \cdot q_i \qquad (2.2)$$

Note that we can derive such an equation for each node; this will result in a linear system of equations on the transition probabilities. The solutions to this system provides the page rank vector $\overline{\pi}$. This linear system has n variables, and n different constraints, and can therefore be expressed in n^2 space in the worst-case. The solution to such a linear systems requires matrix operations which are at least quadratic (and at most cubic) in the total number of nodes. This can be quite expensive in practice. Of course, since the page rank needs to be computed only once in a while in batch phase, it is possible to implement it reasonably well with the use of a few carefully designed matrix techniques. The *PageRank* algorithm [29, 151] uses an iterative approach which computes the principal eigenvectors of the normalized link matrix of the web. A description of the page rank algorithm may be found in [151].

We note that the page-rank algorithm only looks at the link structure during the ranking process, and does not include any information about the content of the underlying web pages. A closely related concept is that of *topic-sensitive page rank* [95], in which we use the topics of the web pages during the ranking process. The key idea in such methods is to allow for *personalized teleportation* (or jumps) during the random-walk process. At each step of the random walk, we allow a transition (with probability α) to a sample set S of pages which are related to the topic of the search. Otherwise, the random walk continues in its standard way with probability $(1 - \alpha)$. This can be easily achieved by modifying the vector $\overline{q} = (q_1 \ldots q_n)$, so that we set the appropriate components in this vector to 1, and others to 0. The final steady-state probabilities with this modified random-walk defines the topic-sensitive page rank. The greater the probability α, the more the process biases the final ranking towards the sample set S. Since each topic-sensitive personalization vector requires the storage of a very large page rank vector, it is possible to pre-compute it in advance only in a limited way, with the use of some representative or authoritative pages. The idea is that we use a limited number of such personalization vectors $\overline{q}$ and determine the corresponding *personalized* page rank vectors $\overline{\pi}$

for these authoritative pages. A judicious combination of these different personalized page rank vectors (for the authoritative pages) is used in order to define the response for a given query set. Some examples of such approaches are discussed in [95, 108]. Of course, such an approach has limitations in terms of the level of granularity in which it can perform personalization. It has been shown in [79] that fully personalized page rank, in which we can precisely bias the random walk towards an *arbitrary* set of web pages will always require at least quadratic space in the worst-case. Therefore, the approach in [79] observes that the use of Monte-Carlo sampling can greatly reduce the space requirements without sufficiently affecting quality. The work in [79] pre-stores Monte-Carlo samples of node-specific random walks, which are also referred to as *fingerprints*. It has been shown in [79] that a very high level of accuracy can be achieved in limited space with the use of such fingerprints. Subsequent recent work [42, 87, 175, 21] has built on this idea in a variety of scenarios, and shown how such dynamic personalized page rank techniques can be made even more efficient and effective. Detailed surveys on different techniques for page rank computation may be found in [20].

Other relevant approaches include the use of measures such as the *hitting time* in order to determine and rank the context sensitive proximity of nodes. The hitting time between node i to j is defined as the expected number of hops that a random surfer would require to reach node j from node i. Clearly, the hitting time is a function of not just the length of the shortest paths, but also the number of possible paths which exist from node i to node j. Therefore, in order to determine similarity among linked objects, the hitting time is a much better measurement of proximity as compared to the use of shortest-path distances. A truncated version of the hitting time defines the objective function by restricting only to the instances in which the hitting time is below a given threshold. When the hitting time is larger than a given threshold, the contribution is simply set at the threshold value. Fast algorithms for computing a truncated variant of the hitting time are discussed in [164]. The issue of scalability in random-walk algorithms is critical because such graphs are large and dynamic, and we would like to have the ability to rank quickly for particular kinds of queries. A method in [165] proposes a fast dynamic re-ranking method, when user feedback is incorporated. A related problem is that of investigating the behavior of random walks of fixed length. The work in [203] investigates the problem of neighborhood aggregation queries. The aggregation query can be considered an "inverse version" of the hitting time, where we are fixing the number of hops and attempting to determine the *number* of hits, rather than the number of hops to hit. One advantage of this definition is that it automatically considers only truncated random walks in which the length of the walk is below a given threshold h; it is also a cleaner definition than the truncated hitting time by treating different walks in a uniform way. The work in [203] determines nodes that have the top-k highest aggregate values over their h-hop neighbors with the use of a Local Neighborhood Aggregation framework called LONA. The framework exploits locality properties in network space to create an efficient index for this query.

Another related idea on determining authoritative ranking is that of the *hub-authority model* [118]. The page-rank technique determines authority by using linkage behavior as indicative of authority. The work in [118] proposes that web pages are one of two kinds:

- **Hubs** are pages which link to authoritative pages.

- **Authorities** are pages which are linked to by good hubs.

A score is associated with both hubs and authorities corresponding to their goodness for being hubs and authorities respectively. The hubs scores affect the authority scores and vice-versa. An iterative approach is used in order to compute both the hub and authority scores. The HITS algorithm proposed in [118] uses these two scores in order to compute the hubs and authorities in the web graph.

Many of these applications arise in the context of dynamic graphs in which the nodes and edges of the graph are received over time. For example, in the context of a social network in which new links are being continuously created, the estimation of page rank is inherently a dynamic problem. Since the page rank algorithm is critically dependent upon the behavior of random walks, the streaming page rank algorithm [166] samples nodes independently in order to create short random walks from each node. This walks can then be merged to create longer random walks. By running several such random walks, the page rank can be effectively estimated. This is because the page rank is simply the probability of visiting a node in a random walk, and the sampling algorithm simulates this process well. The key challenge for the algorithm is that it is possible to get stuck during the process of random walks. This is because the sampling process picks both nodes and edges in the sample, and it is possible to traverse an edge such that the end point of that edge is not present in the node sample. Furthermore, we do not allow repeated traversal of nodes in order to preserve randomness. Such stuck nodes can be handled by keeping track of the set S of sampled nodes whose walks have already been used for extending the random walk. New edges are sampled out of both the stuck node and the nodes in S. These are used in order to extend the walk further as much as possible. If the new end-point is a sampled node whose walk is not in S, then we continue the merging process. Otherwise, we repeat the process of sampling edges out of S and all the stuck nodes visited since the last walk was used.

Another application commonly encountered in the context of graph mining is the analysis of query flow logs. We note that a common way for many users to navigate on the web is to use search engines to discover web pages and then click some of the hyperlinks in the search results. The behavior of the resulting graphs can be used to determine the topic distributions of interest, and semantic relationships between different topics.

In many web applications, it is useful to determine clusters of web pages or blogs. For this purpose, it is helpful to leverage the linkage structure of the web. A common technique which is often used for web document clustering is that of *shingling* [32, 82]. In this case, the min-hash approach is used in order to determine densely connected regions of the web. In addition, any of a number of quasi-clique generation techniques [5, 148, 153] can be used for the purpose of determination of dense regions of the graph.

Social Networking. Social networks are very large graphs which are defined by people who appear as nodes, and links which correspond to communications or relationships between these different people. The links in the social network can be used to determine relevant communities, members with particular expertise sets, and the flow of information in the social network. We will discuss these applications one by one.

The problem of community detection in social networks is related to the problem of *node clustering* of very large graphs. In this case, we wish to determine dense clusters of nodes based on the underlying linkage structure [158]. Social networks are a specially challenging case for the clustering problem because of the typically massive size of the underlying graph. As in the case of web graphs, any of the well known shingling or quasi-clique generation methods [5, 32, 82, 148, 153] can be used in order to determine relevant communities in the network. A technique has been proposed in [167] to use stochastic flow simulations for determining the clusters in the underlying graphs. A method for determining the clustering structure with the use of the eigen-structure of the linkage matrix in order to determine the community structure is proposed in [146]. An important characteristic of large networks is that they can often be characterized by the nature of the underlying subgraphs. In [27], a technique has been proposed for counting the number of subgraphs of a particular type in a large network. It has been shown that this characterization is very useful for clustering large networks. Such precision cannot be achieved with the use of other topological properties. Therefore, this approach can also be used for community detection in massive networks. The problem of community detection is particularly interesting in the context of *dynamic analysis* of evolving networks in which we try to determine how the communities in the graph may change over time. For example, we may wish to determine *newly forming communities*, *decaying communities*, or *evolving communities*. Some recent methods for such problems may be found in [9, 16, 50, 69, 74, 117, 131, 135, 171, 173]. The work in [9] also examines this problem in the context of evolving graph streams. Many of these techniques examine the problem of community detection and change detection in a single framework. This provides the ability to present the changes in the underlying network in a summarized way.

Node clustering algorithms are closely related to the concept of *centrality analysis* in networks. For example, the technique discussed in [158] uses a k-medoids approach which yields k central points of the network. This kind of approach is very useful in different kinds of networks, though in different contexts. In the case of social networks, these central points are typically key members in the network which are well connected to other members of the community. Centrality analysis can also be used in order to determine the central points in information flows. Thus, it is clear that the same kind of structural analysis algorithm can lead to different kinds of insights in different networks.

Centrality detection is closely related to the problem of information flow spread in social networks. It was observed that many recently developed viral flow analysis techniques [40, 127, 147] can be used in the context of a variety of other social networking information flow related applications. This is because information flow applications can be understood with similar behavior models as viral spread. These applications are: (1) We would like to determine the most influential members of the social network; i.e. members who cause the most flow of information outwards. (2) Information in the social behavior often cascades through it in the same way as an epidemic. We would like to measure the information cascade rate through the social network, and determine the effect of different sources of information. The idea is that monitoring promotes the early detection of information flows, and is beneficial to the person who can detect it. The cascading behavior is particularly visible in the case of blog graphs, in which the cascading behavior is reflected in the form of added links over time. Since it is not possible to monitor all blogs simultaneously, it is desirable to minimize the monitoring cost over the different blogs, by assuming a fixed monitoring cost per node. This problem is NP-hard [127], since the vertex-cover problem can be reduced to it. The main idea in [128] is to use an approximation heuristic in order to minimize the monitoring cost. Such an approach is not restricted to the blog scenario, but it is also applicable to other scenarios such as monitoring information exchange in social networks, and monitoring outages in communication networks. (3) We would like to determine the conditions which lead to the critical mass necessary for uncontrolled information transmission. Some techniques for characterizing these conditions are discussed in [40, 187]. The work in [187] relates the structure of the adjacency matrix to the transmissibility rate in order to measure the threshold for an epidemic. Thus, the connectivity structure of the underlying graph is critical in measuring the rate of information dissemination in the underlying network. It has been shown in [187] that the eigenstructure of the adjacency matrix can be directly related to the threshold for an epidemic.

Other Computer Network Applications.   Many of these techniques can also be used for other kinds of networks such as communication networks. Structural analysis and robustness of communication networks is highly dependent upon the design of the underlying network graph. Careful design of the underlying graph can help avoid network failures, congestions, or other weaknesses in the overall network. For example, centrality analysis [158] can be used in the context of a communication network in order to determine critical points of failure. Similarly, the techniques for flow dissemination in social networks can be used to model viral transmission in communication networks as well. The main difference is that we model viral infection probability along an edge in a communication network instead of the information flow probability along an edge in a social network.

Many reachability techniques [10, 48, 49, 53, 54, 184] can be used to determine optimal routing decisions in computer networks. This is also related to the problem of determining pairwise node-connectivity [7] in computer networks. The technique in [7] uses a compression-based synopsis to create an effective connectivity index for massive disk-resident graphs. This is useful in communication networks in which we need to determine the minimum number of edges to be deleted in order to disconnect a particular pair of nodes from one another.

4.3   Software Bug Localization

A natural application of graph mining algorithms is that of software bug localization. Software bug localization is an important application from the perspective of software reliability and testing. The control flow of programs can be modeled in the form of call-graphs. The goal of software bug localization techniques is to mine such call graphs in order to determine the bugs in the underlying programs. Call graphs are of two types:

- **Static call graphs** can be inferred from the source code of a given program. All the methods, procedures and functions in the program are nodes, and the relationships between the different methods are defined as edges. It is also possible to define nodes for data elements and model relationships between different data elements and edges. In the case of static call graphs, it is often possible to use *typical examples* of the structure of the program in order to determine portions of the software where atypical anamolies may occur.

- **Dynamic call graphs** are created during program execution, and they represent the invocation structure. For example, a call from one procedure to another creates an edge which represents the invocation relationship between the two procedures. Such call graphs can be extremely large in massive software programs, since such programs may contain thousands of invocations between the different procedures. In such cases, the difference in structural, frequency or sequence behavior of successful and failing invocations can be used to localize software bugs. Such call graphs can be particularly useful in localizing bugs which are occasional in nature and may occur in some invocations and not others.

We further note that bug localization is not exhaustive in terms of the kinds of errors it can catch. For example, logical errors in a program which are not a result of the program structure, and which do not affect the sequence or structure of execution of the different methods cannot be localized with such techniques. Furthermore software bug localization is not an exact science. Rather, it can be used in order to provide software testing experts with possible bugs, and they can use this in order to make relevant corrections.

An interesting case is one in which different program executions lead to different structure, sequence and frequency of executions which are specific to failures and successes of the final program execution. These failures and successes may be a result of logical errors, which lead to changes in structure and frequency of method calls. In such cases, the software bug-localization can be modeled as a classification problem. The first step is to create call graphs from the executions. This is achieved by tracing the program executions during the testing process. We note that such call graphs may be huge and unwieldy for use with graph mining algorithms. The large sizes of call-graphs creates a challenge for graph mining procedures. This is because graph mining algorithms are often designed for relatively small graphs, whereas such call graphs may be huge. Therefore, a natural solution is to reduce the size of the call graph with the use of a compression based approach. This naturally results in loss of information, and in some cases, it also results in an inability to use the localization approach effectively when the loss of information is extensive.

The next step is to use frequent subgraph mining techniques on the training data in order to determine those patterns which occur more frequently in faulty executions. We note that this is somewhat similar to the technique often utilized in rule-based classifiers which attempt to link particular patterns and conditions to specific class labels. Such patterns are then associated with the different methods and are used in order to provide a ranking of the methods and functions in the program which may possibly contain bugs. This also provides a causality and understanding of the bugs in the underlying programs.

We note that the compression process is critical in providing the ability to efficiently process the underlying graphs. One natural method for reducing the size of the corresponding graphs is to map multiple nodes in the call graph into a single node. For example, in *total reduction*, we map every node in the call node which corresponds to the same method onto one node in the compressed graph. Thus, the total number of nodes in the graph is at most equal to the number of methods. Such a technique has been used in [136] in order to reduce the size of the call graph. A second method which may be used is to compress the iteratively executed structures such as loops into a single node. This is a natural approach, since an iteratively executed structure is one of the most commonly occurring blocks in call graphs. Another technique is to reduce subtrees into single nodes. A variety of localization strategies with the use of such reduction techniques are discussed in [67, 68, 72].

Finally, the reduced graphs are mined in order to determine discriminative structures for bug localization. The method in [72] is based on determining discriminative subtrees from the data. Specifically, the method finds all subtrees which are frequent to failing executions, but are not frequent in correct executions. These are then used in order to construct rules which may be used for specific instances of classification of program runs. More importantly, such rules provide an understanding of the causality of the bugs, and this understanding can be used in order to support the correction of the underlying errors.

The above technique is designed for finding structural characteristics of the execution which can be used for isolating software bugs. However, in many cases the structural characteristics may not be the only features which may be relevant to localization of bugs. For example, an important feature which may be used in order to determine the presence of bugs is the *relative frequency* of the invocation of different methods. For example, invocations which have bugs may call a particular method more frequently than others. A natural way to learn this is to associate edge weights with the call graph. These edge weights correspond to the frequency of invocation. Then, we use these edge weights in order to analyze the calls which are most relevant to discriminating between correct and failing executions. A number of methods for this class of techniques is discussed in [67, 68].

We note that both structure and frequency are different aspects of the data which can be leveraged in order to perform the localization. Therefore, it makes sense to combine these approaches in order to improve the localization process. The techniques in [67, 68] create a score for both the structure-based and frequency-based features. A combination of these scores is then used for the bug localization process. It has been shown [67, 68] that such an approach is more effective than the use of either of the two features.

Another important characteristic which can be explored in future work is to analyze the *sequence of program calls*, rather than simply analyzing the dynamic call structure or the frequency of calls of the different methods. Some initial work [64] in this direction shows that sequence mining encodes excellent information for bug localization even with the use of simple methods.

However, this technique does not use sophisticated graph mining techniques in order to further leverage this sequence information. Therefore, it can be a fruitful avenue for future research to incorporate sequential information into the graph mining techniques which are currently available.

Another line of analysis is the analysis of static source code rather than the dynamic call graphs. In such cases, it makes more sense to look particular classes of bugs, rather than try to isolate the source of the execution error. For example, neglected conditions in software programs [43] can create failing conditions. For example, a *case* statement in a software program with a missing condition is a commonly occurring bug. In such cases, it makes sense to design domain-specific techniques for localizing the bug. For this purpose, techniques based on *static* program-dependence graphs are used. These are distinguished from the dynamic call graphs discussed above, in the sense that the latter requires execution of the program to create the graphs, whereas in this case the graphs are constructed in a static fashion. Program dependence graphs essentially create a graphical representation of the relationships between the different methods and data elements of a program. Different kinds of edges are used to denote control and data dependencies. The first step is to determine conditional rules [43] in a program which illustrates the program dependencies which are frequently occurring in a project. Then we search for (static) instantiations within the project which violate these rules. In many cases, such instantiations could correspond to neglected conditions in the software program.

The field of software bug localization faces a number of key challenges. One of the main challenges is that the work in the field has mostly focussed on smaller software projects. Larger programs are a challenge, because the corresponding call graphs may be huge and the process of graph compression may lose too much information. While some of these challenges may be alleviated with the development of more efficient mining techniques for larger graphs, some advantages may also be obtained with the use of better representations at the *modeling level*. For example, the nodes in the graph can be represented at a coarser level of granularity at the modeling phase. Since the modeling process is done with a better level of understanding of the possibilities for the bugs (as compared to an automated compression process), it is assumed that such an approach would lose much less information for bug localization purposes. A second direction is to combine the graph-based techniques with other effective statistical techniques [137] in order to create more robust classifiers. In future research, it should be reasonable to expect that larger software projects can be analyzed only with the use of such combined techniques which can make use of different characteristics of the underlying data.

5. Conclusions and Future Research

In this chapter, we presented a survey of graph mining and management applications. We also provide a survey of the common applications which arise in the context of graph mining applications. Much of the work in recent years has focussed on small and memory-resident graphs. Much of the future challenges arise in the context of *very large disk-resident graphs*. Other important applications are designed in the context of *massive graphs streams*. Graph streams arise in the context of a number of applications such as social networking, in which the communications between large groups of users are captured in the form of a graph. Such applications are very challenging, since the entire data cannot be localized on disk for the purpose of structural analysis. Therefore, new techniques are required to summarize the structural behavior of graph streams, and use them for a variety of analytical scenarios. We expect that future research will focus on the large-scale and stream-based scenarios for graph mining.

Notes

1. FLWOR is an acronym for FOR-LET-WHERE-ORDER BY-RETURN.

References

[1] Chemaxon. *Screen, Chemaxon Inc.*, 2005.

[2] Daylight. *Daylight Toolkit, Daylight Inc, Mission Viejo, CA, USA*, 2008.

[3] Oracle Spatial Topology and Network Data Models 10g Release 1 (10.1) **URL:** *http://www.oracle.com/technology/products/spatial /pdf/10g_network_model_twp.pdf*

[4] Semantic Web Challenge. **URL:** *http://challenge.semanticweb.org/*

[5] J. Abello, M. G. Resende, S. Sudarsky, Massive quasi-clique detection. *Proceedings of the 5th Latin American Symposium on Theoretical Informatics (LATIN) (Cancun, Mexico).* 598-612, 2002.

[6] S. Abiteboul, P. Buneman, D. Suciu. *Data on the web: from relations to semistructured data and XML.* Morgan Kaufmann Publishers, Los Altos, CA 94022, USA, 1999.

[7] C. Aggarwal, Y. Xie, P. Yu. GConnect: A Connectivity Index for Massive Disk-Resident Graphs, *VLDB Conference*, 2009.

[8] C. Aggarwal, N. Ta, J. Feng, J. Wang, M. J. Zaki. XProj: A Framework for Projected Structural Clustering of XML Documents, *KDD Conference*, 2007.

[9] C. Aggarwal, P. Yu. Online Analysis of Community Evolution in Data Streams. *SIAM Conference on Data Mining*, 2005.

[10] R. Agrawal, A. Borgida, H.V. Jagadish. Efficient Maintenance of Transitive Relationships in Large Data and Knowledge Bases, *ACM SIGMOD Conference*, 1989.

[11] R. Agrawal, R. Srikant. Fast algorithms for mining association rules in large databases, *VLDB Conference*, 1994.

[12] S. Agrawal, S. Chaudhuri, G. Das. DBXplorer: A system for keyword-based search over relational databases. *ICDE Conference*, 2002.

[13] R. Ahuja, J. Orlin, T. Magnanti. Network Flows: Theory, Algorithms, and Applications, *Prentice Hall*, Englewood Cliffs, NJ, 1992.

[14] S. Alexaki, V. Christophides, G. Karvounarakis, D. Plexousakis. On Storing Voluminous RDF Description Bases. In *WebDB*, 2001.

[15] S. Alexaki, V. Christophides, G. Karvounarakis, D. Plexousakis. The ICS-FORTH RDFSuite: Managing Voluminous RDF Description Bases. In *SemWeb*, 2001.

[16] S. Asur, S. Parthasarathy, and D. Ucar. An event-based framework for characterizing the evolutionary behavior of interaction graphs. *ACM KDD Conference*, 2007.

[17] R. Baeza-Yates, A Tiberi. Extracting semantic relations from query logs. *ACM KDD Conference*, 2007.

[18] Z. Bar-Yossef, R. Kumar, D. Sivakumar. Reductions in streaming algorithms, with an application to counting triangles in graphs. *ACM SODA Conference*, 2002.

[19] D. Beckett. The Design and Implementation of the Redland RDF Application Framework. *WWW Conference*, 2001.

[20] P. Berkhin. A survey on pagerank computing. *Internet Mathematics*, 2(1), 2005.

[21] P. Berkhin. Bookmark-coloring approach to personalized pagerank computing. *Internet Mathematics*, 3(1), 2006.

[22] M. Berlingerio, F. Bonchi, B. Bringmann, A. Gionis. Mining Graph-Evolution Rules, *PKDD Conference*, 2009.

[23] S. Bhagat, G. Cormode, I. Rozenbaum. Applying link-based classification to label blogs. *WebKDD/SNA-KDD*, pages 97–117, 2007.

[24] G. Bhalotia, C. Nakhe, A. Hulgeri, S. Chakrabarti, S. Sudarshan. Keyword searching and browsing in databases using BANKS. *ICDE Conference*, 2002.

[25] M. Bilgic, L. Getoor. Effective label acquisition for collective classification. *ACM KDD Conference*, pages 43–51, 2008.

[26] S. Boag, D. Chamberlin, M. F. Fernandez, D. Florescu, J. Robie, J. Simeon. XQuery 1.0: An XML query language. **URL:** W3C, http://www.w3.org/TR/xquery/, 2007.

[27] I. Bordino, D. Donato, A. Gionis, S. Leonardi. Mining Large Networks with Subgraph Counting. *IEEE ICDM Conference*, 2008.

[28] C. Borgelt, M. R. Berthold. Mining molecular fragments: Find- ing Relevant Substructures of Molecules. *ICDM Conference*, 2002.

[29] S. Brin, L. Page. The Anatomy of a Large Scale Hypertextual Search Engine, *WWW Conference*, 1998.

[30] H.J. Bohm, G. Schneider. *Virtual Screening for Bioactive Molecules.* Wiley-VCH, 2000.

[31] B. Bringmann, S. Nijssen. What is frequent in a single graph? *PAKDD Conference*, 2008.

[32] A. Z. Broder, M. Charikar, A. Frieze, M. Mitzenmacher. Syntactic clustering of the web, *WWW Conference, Computer Networks*, 29(8–13):1157–1166, 1997.

[33] J. Broekstra, A. Kampman, F. V. Harmelen. Sesame: A Generic Architecture for Storing and Querying RDF and RDF Schema. In *ISWC Conference*, 2002.

[34] H. Bunke. On a relation between graph edit distance and maximum common subgraph. *Pattern Recognition Letters*, 18: pp. 689–694, 1997.

[35] H. Bunke, G. Allermann. Inexact graph matching for structural pattern recognition. *Pattern Recognition Letters*, 1: pp. 245–253, 1983.

[36] H. Bunke, X. Jiang, A. Kandel. On the minimum common supergraph of two graphs. *Computing*, 65(1): pp. 13–25, 2000.

[37] H. Bunke, K. Shearer. A graph distance metric based on the maximal common subgraph. *Pattern Recognition Letters*, 19(3): pp. 255–259, 1998.

[38] J. J. Carroll, I. Dickinson, C. Dollin, D. Reynolds, A. Seaborne, K. Wilkinson. Jena: implementing the Semantic Web recommendations. In *WWW Conference*, 2004.

[39] V. R. de Carvalho, W. W. Cohen. On the collective classification of email "speech acts". *ACM SIGIR Conference*, pages 345–352, 2005.

[40] D. Chakrabarti, Y. Wang, C. Wang, J. Leskovec, C. Faloutsos. Epidemic thresholds in real networks. *ACM Transactions on Information Systems and Security*, 10(4), 2008.

[41] D. Chakrabarti, Y. Zhan, C. Faloutsos R-MAT: A Recursive Model for Graph Mining. *SDM Conference*, 2004.

[42] S. Chakrabarti. Dynamic Personalized Pagerank in Entity-Relation Graphs, *WWW Conference*, 2007.

[43] R.-Y. Chang, A. Podgurski, J. Yang. Discovering Neglected Conditions in Software by Mining Dependence Graphs. *IEEE Transactions on Software Engineering*, 34(5):579–596, 2008.

[44] O. Chapelle, A. Zien, B. Schelkopf, editors. *Semi-Supervised Learning.* MIT Press, Cambridge, MA, 2006.

[45] S. S. Chawathe. Comparing Hierachical data in external memory. *Very Large Data Bases Conference*, 1999.

[46] C. Chen, C. Lin, M. Fredrikson, M. Christodorescu, X. Yan, J. Han, Mining Graph Patterns Efficiently via Randomized Summaries, *VLDB Conference*, 2009.

[47] L. Chen, A. Gupta, M. E. Kurul. Stack-based algorithms for pattern matching on dags. *VLDB Conference*, 2005.

[48] J. Cheng, J. Xu Yu, X. Lin, H. Wang, P. S. Yu. Fast Computing of Reachability Labelings for Large Graphs with High Compression Rate, *EDBT Conference*, 2008.

[49] J. Cheng, J. Xu Yu, X. Lin, H. Wang, P. S. Yu. Fast Computation of Reachability Labelings in Large Graphs, *EDBT Conference*, 2006.

[50] Y. Chi, X. Song, D. Zhou, K. Hino, B. L. Tseng. Evolutionary spectral clustering by incorporating temporal smoothness. *KDD Conference*, 2007.

[51] C. Chung, J. Min, K. Shim. APEX: An adaptive path index for XML data. In *SIGMOD Conference*, 2002.

[52] J. Clark, S. DeRose. XML Path Language (XPath). **URL:** W3C, http://www.w3.org/TR/xpath/, 1999.

[53] E. Cohen. Size-estimation Framework with Applications to Transitive Closure and Reachability, *Journal of Computer and System Sciences*, v.55 n.3, p.441-453, Dec. 1997.

[54] E. Cohen, E. Halperin, H. Kaplan, U. Zwick. Reachability and Distance Queries via 2-hop Labels, *ACM Symposium on Discrete Algorithms*, 2002.

[55] S. Cohen, J. Mamou, Y. Kanza, Y. Sagiv. XSEarch: A semantic search engine for XML. *VLDB Conference*, 2003.

[56] M. P. Consens, A. O. Mendelzon. GraphLog: a visual formalism for real life recursion. In *PODS Conference*, 1990.

[57] D. Conte, P. Foggia, C. Sansone, M. Vento. Thirty Years of Graph Matching in Pattern Recognition. *International Journal of Pattern Recognition and Artificial Intelligence*, 18(3): pp. 265–298, 2004.

[58] D. Cook, L. Holder. Mining Graph Data, *John Wiley & Sons Inc*, 2007.

[59] B. F. Cooper, N. Sample, M. Franklin, G. Hjaltason, M. Shadmon. A fast index for semistructured data. In *VLDB Conference*, pages 341–350, 2001.

[60] L.P. Cordella, P. Foggia, C. Sansone, M. Vento. A (Sub)graph Isomorphism Algorithm for Matching Large Graphs. *IEEE Transactions on Pattern Analysis and Machine Intelligence*, 26(20): pp. 1367–1372, 2004.

[61] G. Cormode, S. Muthukrishnan. Space efficient mining of multigraph streams. *ACM PODS Conference*, 2005.

[62] K. Crammer Y. Singer. A new family of online algorithms for category ranking. *Journal of Machine Learning Research.*, 3:1025–1058, 2003.

[63] T. Dalamagas, T. Cheng, K. Winkel, T. Sellis. Clustering XML Documents Using Structural Summaries. *Information Systems*, Elsevier, January 2005.

[64] V. Dallmeier, C. Lindig, A. Zeller. Lightweight Defect Localization for Java. In *Proc. of the 19th European Conf. on Object-Oriented Programming (ECOOP)*, 2005.

[65] M. Deshpande, M. Kuramochi, N. Wale, G. Karypis. Frequent Substructure-based Approaches for Classifying Chemical Compounds. *IEEE Transactions on Knowledge and Data Engineering*, 17: pp. 1036–1050, 2005.

[66] E. W. Dijkstra. A note on two problems in connection with graphs. *Numerische Mathematik*, 1 (1959), S. 269-271.

[67] F. Eichinger, K. Bøhm, M. Huber. Improved Software Fault Detection with Graph Mining. *Workshop on Mining and Learning with Graphs*, 2008.

[68] F. Eichinger, K. Bøhm, M. Huber. Mining Edge-Weighted Call Graphs to Localize Software Bugs. *PKDD Conference*, 2008.

[69] T. Falkowski, J. Bartelheimer, M. Spilopoulou. Mining and Visualizing the Evolution of Subgroups in Social Networks, *ACM International Conference on Web Intelligence*, 2006.

[70] M. Faloutsos, P. Faloutsos, C. Faloutsos. On Power Law Relationships of the Internet Topology. *SIGCOMM Conference*, 1999.

[71] W. Fan, K. Zhang, H. Cheng, J. Gao. X. Yan, J. Han, P. S. Yu O. Verscheure. Direct Mining of Discriminative and Essential Frequent Patterns via Model-based Search Tree. *ACM KDD Conference*, 2008.

[72] G. Di Fatta, S. Leue, E. Stegantova. Discriminative Pattern Mining in Software Fault Detection. *Workshop on Software Quality Assurance*, 2006.

[73] J. Feigenbaum, S. Kannan, A. McGregor, S. Suri, J. Zhang. Graph Distances in the Data-Stream Model. *SIAM Journal on Computing*, 38(5): pp. 1709–1727, 2008.

[74] J. Ferlez, C. Faloutsos, J. Leskovec, D. Mladenic, M. Grobelnik. Monitoring Network Evolution using MDL. *IEEE ICDE Conference*, 2008.

[75] M. Fiedler, C. Borgelt. Support computation for mining frequent subgraphs in a single graph. *Workshop on Mining and Learning with Graphs (MLG'07)*, 2007.

[76] M.A. Fischler, R.A. Elschlager. The representation and matching of pictorial structures. *IEEE Transactions on Computers*, 22(1): pp 67–92, 1973.

[77] P.-O. Fjallstrom. Algorithms for Graph Partitioning: A Survey, *Linkoping Electronic Articles in Computer and Information Science*, Vol 3, no 10, 1998.

[78] G. Flake, R. Tarjan, M. Tsioutsiouliklis. Graph Clustering and Minimum Cut Trees, *Internet Mathematics*, 1(4), 385–408, 2003.

[79] D. Fogaras, B. Racz, K. Csalogany, T. Sarlos. Towards scaling fully personalized pagerank: Algorithms, lower bounds, and experiments. *Internet Mathematics*, 2(3), 2005.

[80] M. S. Garey, D. S. Johnson. Computers and Intractability: A Guide to the Theory of NP-completeness,*W. H. Freeman*, 1979.

[81] T. Gartner, P. Flach, S. Wrobel. On graph kernels: Hardness results and efficient alternatives. *16th Annual Conf. on Learning Theory*, pp. 129–143, 2003.

[82] D. Gibson, R. Kumar, A. Tomkins, Discovering Large Dense Subgraphs in Massive Graphs, *VLDB Conference*, 2005.

[83] R. Giugno, D. Shasha, GraphGrep: A Fast and Universal Method for Querying Graphs. *International Conference in Pattern recognition (ICPR)*, 2002.

[84] S. Godbole, S. Sarawagi. Discriminative methods for multi-labeled classification. *PAKDD Conference*, pages 22–30, 2004.

[85] R. Goldman, J. Widom. DataGuides: Enable query formulation and optimization in semistructured databases. *VLDB Conference*, pages 436–445, 1997.

[86] L. Guo, F. Shao, C. Botev, J. Shanmugasundaram. XRANK: ranked keyword search over XML documents. *ACM SIGMOD Conference*, pages 16–27, 2003.

[87] M. S. Gupta, A. Pathak, S. Chakrabarti. Fast algorithms for top-k personalized pagerank queries. *WWW Conference*, 2008.

[88] R. H. Guting. GraphDB: Modeling and querying graphs in databases. In *VLDB Conference*, pages 297–308, 1994.

[89] M. Gyssens, J. Paredaens, D. van Gucht. A graph-oriented object database model. In *PODS Conference*, pages 417–424, 1990.

[90] J. Han, J. Pei, Y. Yin. Mining Frequent Patterns without Candidate Generation. *SIGMOD Conference*, 2000.

[91] S. Harris, N. Gibbins. 3store: Efficient bulk RDF storage. In *PSSS Conference*, 2003.

[92] S. Harris, N. Shadbolt. SPARQL query processing with conventional relational database systems. In *SSWS Conference*, 2005.

[93] M. Al Hasan, V. Chaoji, S. Salem, J. Besson, M. J. Zaki. ORIGAMI: Mining Representative Orthogonal Graph Patterns. *ICDM Conference*, 2007.

[94] D. Haussler. Convolution kernels on discrete structures. *Technical Report UCSC-CRL-99-10*, University of California, Santa Cruz, 1999.

[95] T. Haveliwala. Topic-Sensitive Page Rank, *World Wide Web Conference*, 2002.

[96] H. He, A. K. Singh. Query Language and Access Methods for Graph Databases, appears as a chapter in *Managing and Mining Graph Data, ed. Charu Aggarwal, Springer*, 2010.

[97] H. He, Querying and mining graph databases. *Ph.D. Thesis, UCSB*, 2007.

[98] H. He, A. K. Singh. Efficient Algorithms for Mining Significant Substructures from Graphs with Quality Guarantees. *ICDM Conference*, 2007.

[99] H. He, H. Wang, J. Yang, P. S. Yu. BLINKS: Ranked keyword searches on graphs. *SIGMOD Conference*, 2007.

[100] J. Huan, W. Wang, J. Prins, J. Yang. Spin: Mining Maximal Frequent Subgraphs from Graph Databases. *KDD Conference*, 2004.

[101] J. Huan, W. Wang, D. Bandyopadhyay, J. Snoeyink, J. Prins, A. Tropsha. Mining Spatial Motifs from Protein Structure Graphs. *Research in Computational Molecular Biology (RECOMB)*, pp. 308–315, 2004.

[102] V. Hristidis, N. Koudas, Y. Papakonstantinou, D. Srivastava. Keyword proximity search in XML trees. *IEEE Transactions on Knowledge and Data Engineering*, 18(4):525–539, 2006.

[103] V. Hristidis, Y. Papakonstantinou. Discover: Keyword search in relational databases. *VLDB Conference*, 2002.

[104] A. Inokuchi, T. Washio, H. Motoda. An Apriori-based Algorithm for Mining Frequent Substructures from Graph Data. *PKDD Conference*, pages 13–23, 2000.

[105] H. V. Jagadish. A compression technique to materialize transitive closure. *ACM Trans. Database Syst.*, 15(4):558–598, 1990.

[106] H. V. Jagadish, S. Al-Khalifa, A. Chapman, L. V. S. Lakshmanan, A. Nierman, S. Paparizos, J. M. Patel, D. Srivastava, N. Wiwatwattana, Y. Wu, C. Yu. TIMBER: A native XML database. In *VLDB Journal*, 11(4):274–291, 2002.

[107] H. V. Jagadish, L. V. S. Lakshmanan, D. Srivastava, K. Thompson. TAX: A tree algebra for XML. *DBPL Conference*, 2001.

[108] G. Jeh, J. Widom. Scaling personalized web search. In *WWW*, pages 271–279, 2003.

[109] J. L. Jenkins, A. Bender, J. W. Davies. In silico target fishing: Predicting biological targets from chemical structure. *Drug Discovery Today*, 3(4):413–421, 2006.

[110] R. Jin, C. Wang, D. Polshakov, S. Parthasarathy, G. Agrawal. Discovering Frequent Topological Structures from Graph Datasets. *ACM KDD Conference*, 2005.

[111] R. Jin, H. Hong, H. Wang, Y. Xiang, N. Ruan. Computing Label-Constraint Reachability in Graph Databases. *Under submission*, 2009.

[112] R. Jin, Y. Xiang, N. Ruan, D. Fuhry. 3-HOP: A high-compression indexing scheme for reachability query. *SIGMOD Conference*, 2009.

[113] V. Kacholia, S. Pandit, S. Chakrabarti, S. Sudarshan, R. Desai, H. Karambelkar. Bidirectional expansion for keyword search on graph databases. *VLDB Conference*, 2005.

[114] H. Kashima, K. Tsuda, A. Inokuchi. Marginalized Kernels between Labeled Graphs, *ICML*, 2003.

[115] R. Kaushik, P. Bohannon, J. Naughton, H. Korth. Covering indexes for branching path queries. In *SIGMOD Conference*, June 2002.

[116] B.W. Kernighan, S. Lin. An efficient heuristic procedure for partitioning graphs, *Bell System Tech. Journal*, vol. 49, Feb. 1970, pp. 291-307.

[117] M.-S. Kim, J. Han. A Particle-and-Density Based Evolutionary Clustering Method for Dynamic Networks, *VLDB Conference*, 2009.

[118] J. M. Kleinberg. Authoritative Sources in a Hyperlinked Environment. *Journal of the ACM*, 46(5):pp. 604–632, 1999.

[119] R.I. Kondor, J. Lafferty. Diffusion kernels on graphs and other discrete input spaces. *ICML Conference*, pp. 315–322, 2002.

[120] M. Koyuturk, A. Grama, W. Szpankowski. An Efficient Algorithm for Detecting Frequent Subgraphs in Biological Networks. *Bioinformatics*, 20:I200–207, 2004.

[121] T. Kudo, E. Maeda, Y. Matsumoto. An Application of Boosting to Graph Classification, *NIPS Conf.* 2004.

[122] R. Kumar, P Raghavan, S. Rajagopalan, D. Sivakumar, A. Tomkins, E. Upfal. The Web as a Graph. *ACM PODS Conference*, 2000.

[123] M. Kuramochi, G. Karypis. Frequent subgraph discovery. *ICDM Conference*, pp. 313–320, Nov. 2001.

[124] M. Kuramochi, G. Karypis. Finding frequent patterns in a large sparse graph. *Data Mining and Knowledge Discovery*, 11(3): pp. 243–271, 2005.

[125] J. Larrosa, G. Valiente. Constraint satisfaction algorithms for graph pattern matching. *Mathematical Structures in Computer Science*, 12(4): pp. 403–422, 2002.

[126] M. Lee, W. Hsu, L. Yang, X. Yang. XClust: Clustering XML Schemas for Effective Integration. *CIKM Conference*, 2002.

[127] J. Leskovec, A. Krause, C. Guestrin, C. Faloutsos, J. VanBriesen, N. S. Glance. Cost-effective outbreak detection in networks. *KDD Conference*, pp. 420–429, 2007.

[128] J. Leskovec, M. McGlohon, C. Faloutsos, N. Glance, M. Hurst. Cascading Behavior in Large Blog Graphs, *SDM Conference*, 2007.

[129] J. Leskovec, J. Kleinberg, C. Faloutsos. Graphs over time: Densification laws, shrinking diameters and possible explanations. *ACM KDD Conference*, 2005.

[130] J. Leskovec, E. Horvitz. Planetary-Scale Views on a Large Instant-Messaging Network, *WWW Conference*, 2008.

[131] J. Leskovec, L. Backstrom, R. Kumar, A. Tomkins. Microscopic Evolution of Social Networks, *ACM KDD Conference*, 2008.

[132] Q. Li, B. Moon. Indexing and querying XML data for regular path expressions. In *VLDB Conference*, pages 361–370, September 2001.

[133] W. Lian, D.W. Cheung, N. Mamoulis, S. Yiu. An Efficient and Scalable Algorithm for Clustering XML Documents by Structure, *IEEE Transactions on Knowledge and Data Engineering*, Vol 16, No. 1, 2004.

[134] L. Lim, H. Wang, M. Wang. Semantic Queries in Databases: Problems and Challenges. *CIKM Conference*, 2009.

[135] Y.-R. Lin, Y. Chi, S. Zhu, H. Sundaram, B. L. Tseng. FacetNet: A framework for analyzing communities and their evolutions in dynamic networks. *WWW Conference*, 2008.

[136] C. Liu, X. Yan, H. Yu, J. Han, P. S. Yu. Mining Behavior Graphs for "Backtrace" of Noncrashing Bugs. *SDM Conference*, 2005.

[137] C. Liu, X. Yan, L. Fei, J. Han, S. P. Midkiff. SOBER: Statistical Model-Based Bug Localization. *SIGSOFT Software Engineering Notes*, 30(5):286–295, 2005.

[138] Q. Lu, L. Getoor. Link-based classification. *ICML Conference*, pages 496–503, 2003.

[139] F. Manola, E. Miller. RDF Primer. W3C, http://www.w3.org/TR/rdf-primer/, 2004.

[140] A. McGregor. Finding Graph Matchings in Data Streams. *APPROX-RANDOM*, pp. 170–181, 2005.

[141] T. Milo and D. Suciu. Index structures for path expression. In *ICDT Conference*, pages 277–295, 1999.

[142] S. Navlakha, R. Rastogi, N. Shrivastava. Graph Summarization with Bounded Error. *ACMSIGMOD Conference*, pp. 419–432, 2008.

[143] M. Neuhaus, H. Bunke. Self-organizing maps for learning the edit costs in graph matching. *IEEE Transactions on Systems, Man, and Cybernetics*, 35(3) pp. 503–514, 2005.

[144] M. Neuhaus, H. Bunke. Automatic learning of cost functions for graph edit distance. *Information Sciences*, 177(1), pp 239–247, 2007.

[145] M. Neuhaus, H. Bunke. Bridging the Gap Between Graph Edit Distance and Kernel Machines. *World Scientific*, 2007.

[146] M. Newman. Finding community structure in networks using the eigenvectors of matrices. *Physical Review E*, 2006.

[147] M. E. J. Newman. The spread of epidemic disease on networks, *Phys. Rev. E 66, 016128, 2002.*

[148] J. Pei, D. Jiang, A. Zhang. On Mining Cross-Graph Quasi-Cliques, *ACM KDD Conference*, 2005.

[149] Nidhi, M. Glick, J. Davies, J. Jenkins. Prediction of biological targets for compounds using multiple-category bayesian models trained on chemogenomics databases. *J Chem Inf Model*, 46:1124–1133, 2006.

[150] S. Nijssen, J. Kok. A quickstart in frequent structure mining can make a difference. *Proceedings of SIGKDD*, pages 647–652, 2004.

[151] L. Page, S. Brin, R. Motwani, T. Winograd. The PageRank Citation Ranking: Bringing Order to the Web. *Technical report, Stanford Digital Library Technologies Project*, 1998.

[152] Z. Pan, J. Heflin. DLDB: Extending relational databases to support Semantic Web queries. In *PSSS Conference*, 2003.

[153] J. Pei, D. Jiang, A. Zhang. Mining Cross-Graph Quasi-Cliques in Gene Expression and Protein Interaction Data, *ICDE Conference*, 2005.

[154] E. Prud'hommeaux and A. Seaborne. SPARQL query language for RDF. W3C, **URL:** http://www.w3.org/TR/rdf-sparql-query/, 2007.

[155] L. Qin, J.-X. Yu, L. Chang. Keyword search in databases: The power of RDBMS. *SIGMOD Conference*, 2009.

[156] S. Raghavan, H. Garcia-Molina. Representing web graphs. *ICDE Conference*, pages 405-416, 2003.

[157] S. Ranu, A. K. Singh. GraphSig: A scalable approach to mining significant subgraphs in large graph databases. *ICDE Conference*, 2009.

[158] M. Rattigan, M. Maier, D. Jensen. Graph Clustering with Network Sructure Indices. *ICML*, 2007.

[159] P. R. Raw, B. Moon. PRIX: Indexing and querying XML using prüfer sequences. *ICDE Conference*, 2004.

[160] J. W. Raymond, P. Willett. Maximum common subgraph isomorphism algorithms for the matching of chemical structures. *J. Comp. Aided Mol. Des.*, 16(7):521–533, 2002.

[161] K. Riesen, X. Jiang, H. Bunke. Exact and Inexact Graph Matching: Methodology and Applications, appears as a chapter in *Managing and Mining Graph Data, ed. Charu Aggarwal, Springer*, 2010.

[162] H. Saigo, S. Nowozin, T. Kadowaki, T. Kudo, and K. Tsuda. GBoost: A mathematical programming approach to graph classification and regression. *Machine Learning*, 2008.

[163] F. Sams-Dodd. Target-based drug discovery: is something wrong? *Drug Discov Today*, 10(2):139–147, Jan 2005.

[164] P. Sarkar, A. Moore, A. Prakash. Fast Incremental Proximity Search in Large Graphs, *ICML Conference*, 2008.

[165] P. Sarkar, A. Moore. Fast Dynamic Re-ranking of Large Graphs, *WWW Conference*, 2009.

[166] A. D. Sarma, S. Gollapudi, R. Panigrahy. Estimating PageRank in Graph Streams, *ACM PODS Conference*, 2008.

[167] V. Satuluri, S. Parthasarathy. Scalable Graph Clustering Using Stochastic Flows: Applications to Community Discovery, *ACM KDD Conference*, 2009.

[168] R. Schenkel, A. Theobald, G. Weikum. Hopi: An efficient connection index for complex XML document collections. *EDBT Conference*, 2004.

[169] J. Shanmugasundaram, K. Tufte, C. Zhang, G. He, D. J. DeWitt, J. F. Naughton. Relational databases for querying XML documents: Limitations and opportunities. *VLDB Conference*, 1999.

[170] N. Stiefl, I. A. Watson, K. Baumann, A. Zaliani. Erg: 2d pharmacophore descriptor for scaffold hopping. *J. Chem. Info. Model.*, 46:208–220, 2006.

[171] J. Sun, S. Papadimitriou, C. Faloutsos, P. Yu. GraphScope: Parameter Free Mining of Large Time-Evolving Graphs, *ACM KDD Conference*, 2007.

[172] S. J. Swamidass, J. Chen, J. Bruand, P. Phung, L. Ralaivola, P. Baldi. Kernels for small molecules and the prediction of mutagenicity, toxicity and anti-cancer activity. *Bioinformatics*, 21(1):359–368, 2005.

[173] L. Tang, H. Liu, J. Zhang, Z. Nazeri. Community evolution in dynamic multi-mode networks. *ACM KDD Conference*, 2008.

[174] B. Taskar, P. Abbeel, D. Koller. Discriminative probabilistic models for relational data. In *UAI*, pages 485–492, 2002.

[175] H. Tong, C. Faloutsos, J.-Y. Pan. Fast random walk with restart and its applications. In *ICDM*, pages 613–622, 2006.

[176] S. Trißl, U. Leser. Fast and practical indexing and querying of very large graphs. *SIGMOD Conference*, 2007.

[177] A. A. Tsay, W. S. Lovejoy, D. R. Karger. Random Sampling in Cut, Flow, and Network Design Problems, *Mathematics of Operations Research*, 24(2):383-413, 1999.

[178] K. Tsuda, W. S. Noble. Learning kernels from biological networks by maximizing entropy. *Bioinformatics*, 20(Suppl. 1):i326–i333, 2004.

[179] K. Tsuda, H. Saigo. Graph Classification, appears as a chapter in *Managing and Mining Graph Data, Springer*, 2010.

[180] J.R. Ullmann. An Algorithm for Subgraph Isomorphism. *Journal of the Association for Computing Machinery*, 23(1): pp. 31–42, 1976.

[181] N. Vanetik, E. Gudes, S. E. Shimony. Computing Frequent Graph Patterns from Semi-structured Data. *IEEE ICDM Conference*, 2002.

[182] R. Volz, D. Oberle, S. Staab, and B. Motik. KAON SERVER : A Semantic Web Management System. In *WWW Conference*, 2003.

[183] H. Wang, C. Aggarwal. A Survey of Algorithms for Keyword Search on Graph Data. appears as a chapter in *Managing and Mining Graph Data, Springer*, 2010.

[184] H. Wang, H. He, J. Yang, J. Xu-Yu, P. Yu. Dual Labeling: Answering Graph Reachability Queries in Constant Time. *ICDE Conference*, 2006.

[185] H. Wang, S. Park, W. Fan, P. S. Yu. ViST: A Dynamic Index Method for Querying XML Data by Tree Structures. *In SIGMOD Conference*, 2003.

[186] H. Wang, X. Meng. On the Sequencing of Tree Structures for XML Indexing. *In ICDE Conference*, 2005.

[187] Y. Wang, D. Chakrabarti, C. Wang, C. Faloutsos. Epidemic Spreading in Real Networks: An Eigenvalue Viewpoint, SRDS, pp. 25-34, 2003.

[188] N. Wale, G. Karypis. Target identification for chemical compounds using target-ligand activity data and ranking based methods. Technical Report TR-08-035, University of Minnesota, 2008.

[189] N. Wale, G. Karypis, I. A. Watson. Method for effective virtual screening and scaffold-hopping in chemical compounds. *Comput Syst Bioinformatics Conf*, 6:403–414, 2007.

[190] N. Wale, X. Ning, G. Karypis. Trends in Chemical Graph Data Mining, appears as a chapter in *Managing and Mining Graph Data, Springer*, 2010.

[191] N. Wale, I. A. Watson, G. Karypis. Indirect similarity based methods for effective scaffold-hopping in chemical compounds. *J. Chem. Info. Model.*, 48(4):730–741, 2008.

[192] N. Wale, I. A. Watson, G. Karypis. Comparison of descriptor spaces for chemical compound retrieval and classification. *Knowledge and Information Systems*, 14:347–375, 2008.

[193] C. Weiss, P. Karras, A. Bernstein. Hexastore: Sextuple Indexing for Semantic Web Data Management. In *VLDB Conference*, 2008.

[194] K. Wilkinson. Jena property table implementation. In *SSWS Conference*, 2006.

[195] K. Wilkinson, C. Sayers, H. A. Kuno, and D. Reynolds. Efficient RDF storage and retrieval in Jena2. In SWDB Conference, 2003.

[196] Y. Xu, Y. Papakonstantinou. Efficient LCA based keyword search in XML data. *EDBT Conference*, 2008.

[197] Y. Xu, Y.Papakonstantinou. Efficient keyword search for smallest LCAs in XML databases. *ACM SIGMOD Conference*, 2005.

[198] X. Yan, J. Han. CloseGraph: Mining Closed Frequent Graph Patterns, *ACM KDD Conference*, 2003.

[199] X. Yan, H. Cheng, J. Han, P. S. Yu. Mining Significant Graph Patterns by Scalable Leap Search, *SIGMOD Conference*, 2008.

[200] X. Yan, J. Han. Gspan: Graph-based Substructure Pattern Mining. *ICDM Conference*, 2002.

[201] X. Yan, P. S. Yu, J. Han. Graph indexing: A frequent structure-based approach. *SIGMOD Conference*, 2004.

[202] X. Yan, P. S. Yu, J. Han. Substructure similarity search in graph databases. *SIGMOD Conference*, 2005.

[203] X. Yan, B. He, F. Zhu, J. Han. Top-K Aggregation Queries Over Large Networks, *IEEE ICDE Conference*, 2010.

[204] J. X. Yu, J. Cheng. Graph Reachability Queries: A Survey, appears as a chapter in *Managing and Mining Graph Data, Springer*, 2010.

[205] M. J. Zaki, C. C. Aggarwal. XRules: An Effective Structural Classifier for XML Data, *KDD Conference*, 2003.

[206] T. Zhang, A. Popescul, B. Dom. Linear prediction models with graph regularization for web-page categorization. *ACM KDD Conference*, pages 821–826, 2006.

[207] Q. Zhang, I. Muegge. Scaffold hopping through virtual screening using 2d and 3d similarity descriptors: Ranking, voting and consensus scoring. *J. Chem. Info. Model.*, 49:1536–1548, 2006.

[208] P. Zhao, J. Yu, P. Yu. Graph indexing: tree + delta >= graph. *VLDB Conference*, 2007.

[209] D. Zhou, J. Huang, B. Schelkopf. Learning from labeled and unlabeled data on a directed graph. *ICML Conference*, pages 1036–1043, 2005.

[210] D. Zhou, O. Bousquet, J. Weston, B. Schelkopf. Learning with local and global consistency. *Advances in Neural Information Processing Systems (NIPS) 16*, pages 321–328. MIT Press, 2004.

[211] X. Zhu, Z. Ghahramani, J. Lafferty. Semi-supervised learning using gaussian fields and harmonic functions. *ICML Conference*, pages 912–919, 2003.

Chapter 3

GRAPH MINING: LAWS AND GENERATORS

Deepayan Chakrabarti
Yahoo! Research
deepay@yahoo-inc.com

Christos Faloutsos
School of Computer Science
Carnegie Mellon University
christos@cs.cmu.edu

Mary McGlohon
School of Computer Science
Carnegie Mellon University
mmcgloho@cs.cmu.edu

Abstract   *How does the Web look? How could we tell an "abnormal" social network from a "normal" one?* These and similar questions are important in many fields where the data can intuitively be cast as a graph; examples range from computer networks, to sociology, to biology, and many more. Indeed, any $M : N$ relation in database terminology can be represented as a graph. Many of these questions boil down to the following: "How can we generate synthetic but *realistic* graphs?" To answer this, we must first understand what *patterns* are common in real-world graphs, and can thus be considered a mark of normality/realism. This survey gives an overview of the incredible variety of work that has been done on these problems. One of our main contributions is the integration of points of view from physics, mathematics, sociology and computer science.

Keywords: Power laws, structure, generators

C.C. Aggarwal and H. Wang (eds.), *Managing and Mining Graph Data,*
Advances in Database Systems 40, DOI 10.1007/978-1-4419-6045-0_3,
© Springer Science+Business Media, LLC 2010

1. Introduction

Informally, a graph is set of nodes, pairs of which might be connected by edges. In a wide array of disciplines, data can be intuitively cast into this format. For example, computer networks consist of routers/computers (nodes) and the links (edges) between them. Social networks consist of individuals and their interconnections (business relationships, kinship, trust, etc.) Protein interaction networks link proteins which must work together to perform some particular biological function. Ecological food webs link species with predator-prey relationships. In these and many other fields, graphs are seemingly ubiquitous.

The problems of detecting abnormalities ("outliers") in a given graph, and of *generating* synthetic but realistic graphs, have received considerable attention recently. Both are tightly coupled to the problem of finding the distinguishing characteristics of real-world graphs, that is, the "patterns" that show up frequently in such graphs and can thus be considered as marks of "realism." A good generator will create graphs which match these patterns. Patterns and generators are important for many applications:

- *Detection of abnormal subgraphs/edges/nodes:* Abnormalities should deviate from the "normal" patterns, so understanding the patterns of naturally occurring graphs is a prerequisite for detection of such outliers.

- *Simulation studies:* Algorithms meant for large real-world graphs can be tested on synthetic graphs which "look like" the original graphs. For example, in order to test the next-generation Internet protocol, we would like to simulate it on a graph that is "similar" to what the Internet will look like a few years into the future.

- *Realism of samples:* We might want to build a small sample graph that is similar to a given large graph. This smaller graph needs to match the "patterns" of the large graph to be realistic.

- *Graph compression:* Graph patterns represent regularities in the data. Such regularities can be used to better compress the data.

Thus, we need to detect patterns in graphs, and then generate synthetic graphs matching such patterns automatically.

This is a hard problem. What patterns should we look for? What do such patterns mean? How can we generate them? Due to the ubiquity and wide applicability of graphs, a lot of research ink has been spent on this problem, not only by computer scientists but also physicists, mathematicians, sociologists and others. However, there is little interaction among these fields, with the result that they often use different terminology and do not benefit from each other's advances. In this survey, we attempt to give an overview of the main

Symbol	Description
N	Number of nodes in the graph
E	Number of edges in the graph
k	Degree for some node
$<k>$	Average degree of nodes in the graph
CC	Clustering coefficient of the graph
$CC(k)$	Clustering coefficient of degree-k nodes
γ	Power law exponent: $y(x) \propto x^{-\gamma}$
t	Time/iterations since the start of an algorithm

Table 3.1. *Table of symbols*

ideas. Our focus is on combining sources from all the different fields, to gain a coherent picture of the current state-of-the-art. The interested reader is also referred to some excellent and entertaining books on the topic [12, 81, 35].

The organization of this chapter is as follows. In section 2, we discuss graph patterns that appear to be common in real-world graphs. Then, in section 3, we describe some graph generators which try to match one or more of these patterns. Typically, we only provide the main ideas and approaches; the interested reader can read the relevant references for details. In all of these, we attempt to collate information from several fields of research. Table 3.1 lists the symbols we will use.

2. Graph Patterns

What are the distinguishing characteristics of graphs? What "rules" and "patterns" hold for them? When can we say that two different graphs are *similar* to each other? In order to come up with models to generate graphs, we need some way of comparing a natural graph to a synthetically generated one; the better the match, the better the model. However, to answer these questions, we need to have some basic set of graph attributes; these would be our vocabulary in which we can discuss different graph types. Finding such attributes will be the focus of this section.

What is a "good" pattern? One that can help distinguish between an actual real-world graph and any fake one. However, we immediately run into several problems. First, given the plethora of different natural and man-made phenomena which give rise to graphs, can we expect all such graphs to follow any particular patterns? Second, is there any *single* pattern which can help differentiate between all real and fake graphs? A third problem (more of a constraint than a problem) is that we want to find patterns which can be computed efficiently; the graphs we are looking at typically have at least around 10^5 nodes and 10^6 edges. A pattern which takes $O(N^3)$ or $O(N^2)$ time in the number of nodes N might easily become impractical for such graphs.

The best answer we can give today is that while there are many differences between graphs, some patterns show up regularly. Work has focused on finding several such patterns, which *together* characterize naturally occurring graphs. A large portion of the literature focuses on two major properties: power laws and small diameters. Our discussion will address both of these properties. For each pattern, we also give the computational requirements for finding/computing the pattern, and some real-world examples of that pattern. Definitions are provided for key ideas which are used repeatedly. Next, we will discuss other patterns of interest, both in static snapshots of graphs and in evolving graphs. Finally, we discuss patterns specific to some well-known graphs, like the Internet and the WWW.

2.1   Power Laws and Heavy-Tailed Distributions

While the Gaussian distribution is common in nature, there are many cases where the probability of events far to the right of the mean is significantly higher than in Gaussians. In the Internet, for example, most routers have a very low degree (perhaps "home" routers), while a few routers have extremely high degree (perhaps the "core" routers of the Internet backbone) [43]. Power-law distributions attempt to model this.

We will divide the following discussion into two parts. First, we will discuss "traditional" power laws: their definition, how to compute them, and real-world examples of their presence. Then, we will discuss deviations from pure power laws, and some common methods to model these.

"Traditional" Power Laws.

Definition 3.1 (Power Law). *Two variables x and y are related by a power law when:*

$$y(x) = Ax^{-\gamma} \tag{3.1}$$

where A and γ are positive constants. The constant γ is often called the power law exponent.

Definition 3.2 (Power Law Distribution). *A random variable is distributed according to a power law when the probability density function (pdf) is given by:*

$$p(x) = Ax^{-\gamma}, \quad \gamma > 1, x \geq x_{min} \tag{3.2}$$

The extra $\gamma > 1$ requirement ensures that $p(x)$ can be normalized. Power laws with $\gamma < 1$ rarely occur in nature, if ever [66].

Skewed distributions, such as power laws, occur very often. In the Internet graph, the degree distribution follows such a power law [43]; that is, the count

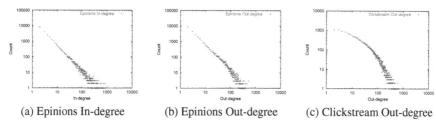

(a) Epinions In-degree   (b) Epinions Out-degree   (c) Clickstream Out-degree

Figure 3.1. *Power laws and deviations:* Plots (a) and (b) show the in-degree and out-degree distributions on a log-log scale for the *Epinions* graph (an online social network of $75, 888$ people and $508, 960$ edges [34]). Both follow power-laws. In contrast, plot (c) shows the out-degree distribution of a *Clickstream* graph (a bipartite graph of users and the websites they surf [63]), which deviates from the power-law pattern.

c_k of nodes with degree k, versus the degree k, is a line on a log-log scale. The eigenvalues of the adjacency matrix of the Internet graph also show a similar behavior: when eigenvalues are plotted versus their rank on a log-log scale (called the scree plot), the result is a straight line. A possible explanation of this is provided by Mihail and Papadimitriou [61]. The World Wide Web graph also obeys power laws [51]: the in-degree and out-degree distributions both follow power-laws, as well as the number of the so-called "bipartite cores" ($\approx$ communities, which we will see later) and the distribution of PageRank values [23, 73]. Redner [76] shows that the citation graph of scientific literature follows a power law with exponent 3. Figures 3.1(a) and 3.1(b) show two examples of power laws.

The significance of a power law distribution $p(x)$ lies in the fact that it decay only polynomially quickly as $x \to \infty$, instead of exponential decay for the Gaussian distribution. Thus, a power law degree distribution would be much more likely to have nodes with a very high degree (much larger than the mean) than the Gaussian distribution. Graphs exhibiting such degree distributions are called *scale-free* graphs, because the form of $y(x)$ in Equation 3.1 remains unchanged to within a multiplicative factor when the variable x is multiplied by a scaling factor (in other words, $y(ax) = by(x)$). Thus, there is no special "characteristic scale" for the variables; the functional form of the relationship remains the same for all scales.

Computation issues:. The process of finding a power law pattern can be divided into three parts: creating the scatter plot, computing the power law exponent, and checking for goodness of fit. We discuss these issues below, using the detection of power laws in degree distributions as an example.

Creating the scatter plot (for the degree distribution): The algorithm for calculating the degree distributions (irrespective of whether they are power laws or not) can be expressed concisely in SQL. Assuming that the graph is represented as a table with the schema `Graph(fromnode, tonode)`, the code for calculating in-degree and out-degree is given below. The case for weighted graphs, with the schema `Graph(fromnode, tonode, weight)`, is a simple extension of this.

```
SELECT outdegree, count(*)              SELECT indegree, count(*)
FROM                                    FROM
    (SELECT count(*) AS outdegree           (SELECT count(*) AS indegree
    FROM Graph                              FROM Graph
    GROUP BY fromnode)                      GROUP BY tonode)
GROUP BY outdegree                      GROUP BY indegree
```

Computing the power law exponent   This is no simple task: the power law could be only in the tail of the distribution and not over the entire distribution, estimators of the power law exponent could be biased, some required assumptions may not hold, and so on. Several methods are currently employed, though there is no clear "winner" at present.

1 *Linear regression on the log-log scale:* We could plot the data on a log-log scale, then optionally "bin" them into equal-sized buckets, and finally find the slope of the linear fit. However, there are at least three problems: (i) this can lead to biased estimates [45], (ii) sometimes the power law is only in the *tail* of the distribution, and the point where the tail begins needs to be hand-picked, and (iii) the right end of the distribution is very noisy [66]. However, this is the simplest technique, and seems to be the most popular one.

2 *Linear regression after logarithmic binning:* This is the same as above, but the bin widths increase exponentially as we go towards the tail. In other words, the number of data points in each bin is counted, and then the height of each bin is then divided by its width to normalize. Plotting the histogram on a log-log scale would make the bin sizes equal, and the power-law can be fitted to the heights of the bins. This reduces the noise in the tail buckets, fixing problem (iii). However, binning leads to loss of information; all that we retain in a bin is its average. In addition, issues (i) and (ii) still exist.

3 *Regression on the cumulative distribution:* We convert the pdf $p(x)$ (that is, the scatter plot) into a *cumulative distribution* $F(x)$:

$$F(x) = P(X \geq x) = \sum_{z=x}^{\infty} p(z) = \sum_{z=x}^{\infty} A z^{-\gamma} \qquad (3.3)$$

The approach avoids the loss of data due to averaging inside a histogram bin. To see how the plot of $F(x)$ versus x will look like, we can bound $F(x)$:

$$\int_x^\infty Az^{-\gamma}dz < F(x) < Ax^{-\gamma} + \int_x^\infty Az^{-\gamma}dz$$

$$\Rightarrow \quad \frac{A}{\gamma - 1}x^{-(\gamma-1)} < F(x) < Ax^{-\gamma} + \frac{A}{\gamma - 1}x^{-(\gamma-1)}$$

$$\Rightarrow \quad F(x)\,\text{sim}\,x^{-(\gamma-1)} \tag{3.4}$$

Thus, the cumulative distribution follows a power law with exponent $(\gamma - 1)$. However, successive points on the cumulative distribution plot are not mutually independent, and this can cause problems in fitting the data.

4 *Maximum-Likelihood Estimator (MLE):* This chooses a value of the power law exponent γ such that the likelihood that the data came from the corresponding power law distribution is maximized. Goldstein et al [45] find that it gives good unbiased estimates of γ.

5 *The Hill statistic:* Hill [48] gives an easily computable estimator, that seems to give reliable results [66]. However, it also needs to be told where the tail of the distribution begins.

6 *Fitting only to extreme-value data:* Feuerverger and Hall [44] propose another estimator which is claimed to reduce bias compared to the Hill statistic without significantly increasing variance. Again, the user must provide an estimate of where the tail begins, but the authors claim that their method is robust against different choices for this value.

7 *Non-parametric estimators:* Crovella and Taqqu [31] propose a non-parametric method for estimating the power law exponent without requiring an estimate of the beginning of the power law tail. While there are no theoretical results on the variance or bias of this estimator, the authors empirically find that accuracy increases with increasing dataset size, and that it is comparable to the Hill statistic.

Checking for goodness of fit The correlation coefficient has typically been used as an informal measure of the goodness of fit of the degree distribution to a power law. Recently, there has been some work on developing statistical "hypothesis testing" methods to do this more formally. Beirlant et al. [15] derive a bias-corrected Jackson statistic for measuring goodness of fit of the data to a generalized Pareto distribution. Goldstein et al. [45] propose a Kolmogorov-Smirnov test to determine the fit. Such measures need to be used more often in the empirical studies of graph datasets.

Examples of power laws in the real world.   Examples of power law degree distributions include the Internet AS[1] graph with exponent $2.1 - 2.2$ [43], the Internet router graph with exponent $sim\,2.48$ [43, 46], the in-degree and out-degree distributions of subsets of the WWW with exponents 2.1 and $2.38-2.72$ respectively [13, 54, 24], the in-degree distribution of the African web graph with exponent 1.92 [19], a citation graph with exponent 3 [76], distributions of website sizes and traffic [2], and many others. Newman [66] provides a comprehensive list of such work.

Deviations from Power Laws.

Informal description.   While power laws appear in a large number of graphs, deviations from a pure power law are sometimes observed. We discuss these below.

Detailed description.   Pennock et al. [75] and others have observed deviations from a pure power law distribution in several datasets. Two of the more common deviations are exponential cutoffs and lognormals.

Exponential cutoffs Sometimes, the distribution looks like a power law over the lower range of values along the x-axis, but decays very fast for higher values. Often, this decay is exponential, and this is usually called an exponential cutoff:

$$y(x = k) \propto e^{-k/\kappa} k^{-\gamma} \tag{3.5}$$

where $e^{-k/\kappa}$ is the exponential cutoff term and $k^{-\gamma}$ is the power law term. Amaral et al. [10] find such behaviors in the electric power-grid graph of Southern California and the network of airports, the vertices being airports and the links being non-stop connections between them. They offer two possible explanations for the existence of such cutoffs. One, high-degree nodes might have taken a long time to acquire all their edges and now might be "aged", and this might lead them to attract fewer new edges (for example, older actors might act in fewer movies). Two, high-degree nodes might end up reaching their "capacity" to handle new edges; this might be the case for airports where airlines prefer a small number of high-degree hubs for economic reasons, but are constrained by limited airport capacity.

Lognormals or the "DGX" distribution Pennock et al. [75] recently found while the whole WWW does exhibit power law degree distributions, subsets of the WWW (such as university homepages and newspaper homepages) deviate significantly. They observed unimodal distributions on the log-log scale. Similar distributions were studied by Bi et al. [17], who found that a discrete truncated lognormal (called the Discrete Gaussian Exponential or "DGX" by the authors) gives a very good fit. A lognormal is a distribution whose logarithm is a Gaussian; it looks like a truncated parabola in log-log scales. The DGX distribution extends the lognormal to discrete distributions (which is what we get in degree distributions), and can be expressed by the formula:

$$y(x = k) = \frac{A(\mu, \sigma)}{k} \exp\left[-\frac{(\ln k - \mu)^2}{2\sigma^2}\right] \quad k = 1, 2, \ldots \quad (3.6)$$

where μ and σ are parameters and $A(\mu, \sigma)$ is a constant (used for normalization if $y(x)$ is a probability distribution). The DGX distribution has been used to fit the degree distribution of a bipartite "clickstream" graph linking websites and users (Figure 3.1(c)), telecommunications and other data.

Examples of deviations from power laws in the real world   Several data sets have shown deviations from a pure power law [10, 75, 17, 62]: examples include the electric power-grid of Southern California, the network of airports, several topic-based subsets of the WWW, Web "clickstream" data, sales data in retail chains, file size distributions, and phone usage data.

2.2   Small Diameters

Informal description:.   Travers and Milgram [80] conducted a famous experiment where participants were asked to reach a randomly assigned target individual by sending a chain letter. They found that for all the chains that completed, the average length of such chains was six, which is a very small number considering the large population the participants and targets were chosen from. This leads us to believe in the concept of "six degrees of separation": the diameter of a graph is an attempt to capture exactly this.

Detailed description.   Several (often related) terms have been used to describe the idea of the "diameter" of a graph:

- *Expansion and the "hop-plot"*: Tangmunarunkit et al. [78] use a well-known metric from theoretical computer science called "expansion," which measures the rate of increase of neighborhood with increasing h. This has been called the "hop-plot" elsewhere [43].

  Definition 3.3 (Hop-plot). *Starting from a node u in the graph, we find the number of nodes $N_h(u)$ in a neighborhood of h hops. We repeat this starting from each node in the graph, and sum the results to find the total*

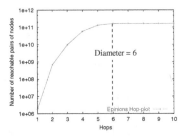

Figure 3.2. *Hop-plot and effective diameter* This is the hop-plot of the *Epinions* graph [34, 28]. We see that the number of reachable pairs of nodes flattens out at around 6 hops; thus the effective diameter of this graph is 6.

neighborhood size N_h for h hops ($N_h = \sum_u N_h(u)$). The hop-plot is just the plot of N_h versus h.

- *Effective diameter or Eccentricity:* The hop-plot can be used to calculate the *effective diameter* (also called the *eccentricity*) of the graph.

  Definition 3.4 (Effective diameter). *This is the minimum number of hops in which some fraction (say, 90%) of all connected pairs of nodes can reach each other [79].*

  Figure 3.2 shows the hop-plot and effective diameter of an example graph.

- *Characteristic path length:* For each node in the graph, consider the shortest paths from it to every other node in the graph. Take the average length of all these paths. Now, consider the average path lengths for *all* possible starting nodes, and take their median. This is the characteristic path length [25].

- *Average diameter:* This is calculated in the same way as the characteristic path length, except that we take the mean of the average shortest path lengths over all nodes, instead of the median.

While the use of "expansion" as a metric is somewhat vague[2], most of the other metrics are quite similar. The advantage of eccentricity is that its definition works, as is, even for disconnected graphs, whereas we must consider only the largest component for the characteristic and average diameters. Characteristic path length and eccentricity are less vulnerable to outliers than average diameter, but average diameter might be the better if we want worst case analysis.

A concept related to the hop-plot is that of the *hop-exponent*: Faloutsos et al. [43] conjecture that for many graphs, the neighborhood size N_h

grows exponentially with the number of hops h. In other words, $N_h = ch^{\mathcal{H}}$ for h much less than the diameter of the graph. They call the constant $\mathcal{H}$ the hop-exponent. However, the diameter is so small for many graphs that there are too few points in the hop-plot for this premise to be verified and to calculate the hop-exponent with any accuracy.

Computational issues. One major problem with finding the diameter is the computational cost: all the definitions essentially require computing the "neighborhood size" of each node in the graph. One approach is to use repeated matrix multiplications on the adjacency matrix of the graph; however, this takes asymptotically $O(N^{2.88})$ time and $O(N^2)$ memory space. Another technique is to do breadth-first searches from each node of the graph. This takes $O(N + E)$ space but requires $O(NE)$ time. Another issue with breadth-first search is that edges are not accessed sequentially, which can lead to terrible performance on disk-resident graphs. Palmer et al. [71] find that randomized breadth-first search algorithms are also ill-suited for large graphs, and they provide a randomized algorithm for finding the hop-plot which takes $O((N+E)d)$ time and $O(N)$ space (apart from the storage for the graph itself), where N is the number of nodes, E the number of edges and d the diameter of the graph (typically very small). Their algorithm offers provable bounds on the quality of the approximated result, and requires only sequential scans over the data. They find the technique to be far faster than exact computation, and providing much better estimates than other schemes like sampling.

Examples in the real world. The diameters of several naturally occurring graphs have been calculated, and in almost all cases they are very small compared to the graph size. Faloutsos et al. [43] find an effective diameter of around 4 for the Internet AS level graph and around 12 for the Router level graph. Govindan and Tangmunarunkit [46] find a 97%-effective diameter of around 15 for the Internet Router graph. Broder et al. [24] find that the average path length in the WWW (when a path exists at all) is about 16 if we consider the directions of links, and around 7 if all edges are considered to be undirected. Albert et al. [8] find the average diameter of the webpages in the nd.edu domain to be 11.2. Watts and Strogatz [83] find the average diameters of the power grid and the network of actors to be 18.7 and 3.65 respectively. Many other such examples can be found in the literature; Tables 1 and 2 of [7] and table 3.1 of [65] list some such work.

2.3 Other Static Graph Patterns

Apart from power laws and small diameters, some other patterns have been observed in large real-world graphs. These include the resilience of such graphs to random failures, and correlations found in the *joint* degree distributions of the graphs. Additionally, we observe structural patterns in the *edge weights* in static snapshots of graphs. We will explore these topics below.

Resilience.

Informal description.   The resilience of a graph is a measure of its robustness to node or edge failures. Many real-world graphs are resilient against random failures but vulnerable to *targeted* attacks.

Detailed description.   There are at least two definitions of resilience:

- Tangmunarunkit et al. [78] define resilience as a function of the number of nodes n: the resilience $R(n)$ is the "minimum cut-set" size within an n-node ball around any node in the graph (a ball around a node X refers to a group of nodes within some fixed number of hops from node X). The "minimum cut-set" is the minimum number of edges that need to be cut to get two disconnected components of roughly equal size; intuitively, if this value is large, then it is hard to disconnect the graph and disrupt communications between its nodes, implying higher resilience. For example, a 2D grid graph has $R(n) \propto \sqrt{n}$ while a tree has $R(n) = 1$; thus, a tree is less resilient than a grid.

- Resilience can be related to the graph diameter: a graph whose diameter does not increase much on node or edge removal has higher resilience [71, 9].

Computation issues.   Calculating the "minimum cut-set" size is NP-hard, but approximate algorithms exist [49]. Computing the graph diameter is also costly, but fast randomized algorithms exist [71].

Examples in the real world.   In general, most real-world networks appear to be resilient against random node/edge removals, but are susceptible to targeted attacks: examples include the Internet Router-level and AS-level graphs, as well as the WWW [71, 9, 78].

Patterns in weighted graphs.

Informal description.   Edges in a graph often have *edge weights*. For instance, the size of packets transferred in a computer network, or length of phone calls (in seconds) in a phone-call network. These edge weights often follow patterns, as described in [59] and [5].

Detailed description. The first pattern we observe is the *Weight Power Law* (WPL). Let $E(t)$, $W(t)$ be the number of edges and total weight of a graph, at time t. They, they follow a power law

$$W(t) = E(t)^w$$

where w is the *weight* exponent.

The weight exponent w ranges from 1.01 to 1.5 for the real graphs studied in [59], which included blog graphs, computer network graphs, and political campaign donation graphs, suggesting that this pattern is universal to real social network-like graphs.

In other words, the more edges that are added to the graph, *superlinearly* more weight is added to the graph. This is counterintuitive, as one would expect the average weight-per-edge to remain constant or to increase linearly.

We find the same pattern for each node. If a node i has out-degree out_i, its out-weight $outw_i$ exhibits a "fortification effect"– there will be a power-law relationship between its degree and weight. We call this the *Snapshot Power Law* (SPL), and it applies to both in- and out- degrees.

Specifically, at a given point in time, we plot the scatterplot of the in/out weight versus the in/out degree, for all the nodes in the graph, at a given time snapshot. Here, every point represents a node and the x and y coordinates are its degree and total weight, respectively. To achieve a good fit, we bucketize the x axis with logarithmic binning [64], and, for each bin, we compute the median y.

Examples in the real world. We find these patterns apply in several real graphs, including network traffic, blogs, and even political campaign donations. A plot of WPL and SPL may be found in Figure 3.3.

Several other weighted power laws, such as the relationship between the eigenvalues of the graph and the weights of the edges, may be found in [5].

Other metrics of measurement. We have discussed a number of patterns found in graphs, many more can be found in the literature. While most of the focus regarding node degrees has fallen on the in-degree and the out-degree distributions, there are "higher-order" statistics that could also be considered. We combine all these statistics under the term *joint distributions*, differentiating them from the degree-distributions which are the *marginal distributions*. Some of these statistics include:

- *In and out degree correlation* The in and out degrees might be independent, or they could be (anti)correlated. Newman et al. [67] find a positive correlation in email networks, that is, the email addresses of individuals with large address books appear in the address books of many others.

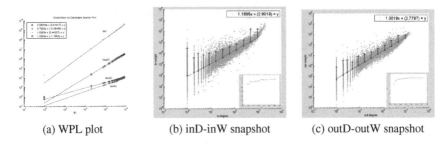

(a) WPL plot   (b) inD-inW snapshot   (c) outD-outW snapshot

Figure 3.3. Weight properties of the campaign donations graph: (a) shows all weight properties, including the densification power law and WPL. (b) and (c) show the Snapshot Power Law for in- and out-degrees. Both have slopes > 1 ("fortification effect"), that is, that the more campaigns an organization supports, the superlinearly-more money it donates, and similarly, the more donations a candidate gets, the more average amount-per-donation is received. Inset plots on (c) and (d) show iw and ow versus time. Note they are very stable over time.

However, it is hard to measure this with good accuracy. Calculating this well would require a lot of data, and it might be still be inaccurate for high-degree nodes (which, due to power law degree distributions, are quite rare).

- *Average neighbor degree* We can measure the average degree $d_{av}(i)$ of the neighbors of node i, and plot it against its degree $k(i)$. Pastor-Satorras et al. [74] find that for the Internet AS level graph, this gives a power law with exponent 0.5 (that is, $d_{av}(i) \propto k(i)^{-0.5}$).

- *Neighbor degree correlation* We could calculate the joint degree distributions of adjacent nodes; however this is again hard to measure accurately.

2.4   Patterns in Evolving Graphs

The search for graph patterns has focused primarily on static patterns, which can be extracted from one snapshot of the graph at some time instant. Many graphs, however, evolve over time (such as the Internet and the WWW) and only recently have researchers started looking for the patterns of graph evolution. Some key patterns have emerged:

- *Densification Power Law:* Leskovec et al. [58] found that several real graphs grow over time according to a power law: the number of nodes $N(t)$ at time t is related to the number of edges $E(t)$ by the equation:

$$E(t) \propto N(t)^{\alpha} \quad 1 \leq \alpha \leq 2 \tag{3.7}$$

where the parameter α is called the Densification Power Law exponent, and remains stable over time. They also find that this "law" exists for

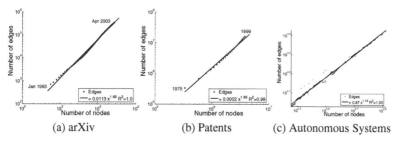

Figure 3.4. *The Densification Power Law* The number of edges $E(t)$ is plotted against the number of nodes $N(t)$ on log-log scales for (a) the arXiv citation graph, (b) the patents citation graph, and (c) the Internet Autonomous Systems graph. All of these grow over time, and the growth follows a power law in all three cases [58].

several different graphs, such as paper citations, patent citations, and the Internet AS graph. This quantifies earlier empirical observations that the average degree of a graph increases over time [14]. It also agrees with theoretical results showing that only a law like Equation 3.7 can maintain the power-law degree distribution of a graph as more nodes and edges get added over time [37]. Figure 3.4 demonstrates the densification law for several real-world networks.

- *Shrinking Diameters:* Leskovec et al. [58] also find that the effective diameters (definition 3.4) of graphs are actually *shrinking* over time, even though the graphs themselves are growing. This can be observed after the *gelling point*– before a certain point a graph is still building to normal properties. This is illustrated in Figure 3.5(a)– for the first few time steps the diameter grows, but it quickly peaks and begins shrinking.

- *Component Size Laws* As a graph evolves, a giant connected component forms: that is, most nodes are reachable to each other through some path. This phenomenon is present both in random and real graphs. What is also found, however, is that once the largest component gels and edges continue to be added, the sizes of the *next-largest connected components* remain constant or oscillating. This phenomenon is shown in Figure 3.5, and discussed in [59].

- *Patterns in Timings:* There are also several interesting patterns regarding the timestamps of edge additions. We find that edge *weight* additions to a graph are bursty: over time, edges are not added to the overall graph uniformly over time, but are uneven yet self-similar [59]. We illustrate this in Figure 3.6. However, in the case of many graphs, timeliness of a particular *node* is important in its edge additions. As shown in [56], incoming edges to a blog post decay with a surprising power-law expo-

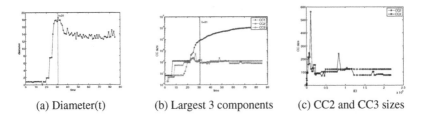

| (a) Diameter(t) | (b) Largest 3 components | (c) CC2 and CC3 sizes |

Figure 3.5. Connected component properties of Postnet network, a network of blog posts. Notice that we experience an early gelling point at (a), where the diameter peaks. Note in (b), a log-linear plot of component size vs. time, that at this same point in time the giant connected component takes off, while the sizes of the second and third-largest connected components (CC2 and CC3) stabilize. We focus on these next-largest connected components in (c).

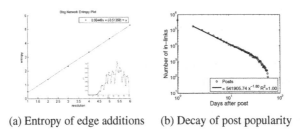

(a) Entropy of edge additions   (b) Decay of post popularity

Figure 3.6. Timing patterns for a network of blog posts. (a) shows the entropy plot of edge additions, showing burstiness. The inset shows the addition of edges over time. (b) describes the decay of post popularity. The horizontal axis indicates time since a post's appearance (aggregated over all posts), while the vertical axis shows the number of links acquired on that day.

nent of -1.5, rather than exponentially or linearly as one might expect. This is shown in Figure 3.6.

These surprising patterns are probably just the tip of the iceberg, and there may be many other patterns hidden in the dynamics of graph growth.

2.5   The Structure of Specific Graphs

While most graphs found naturally share many features (such as the small-world phenomenon), there are some specifics associated with each. These might reflect properties or constraints of the domain to which the graph belongs. We will discuss some well-known graphs and their specific features below.

The Internet.   The networking community has studied the structure of the Internet for a long time. In general, it can be viewed as a collection of interconnected routing domains; each domain is a group of nodes (such routers, switches etc.) under a single technical administration [26]. These domains can be considered as either a *stub* domain (which only carries traffic originating or

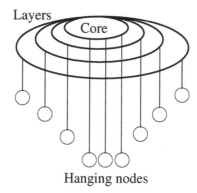

Layers

Core

Hanging nodes

Figure 3.7. *The Internet as a "Jellyfish"* The Internet AS-level graph can be thought of as a core, surrounded by concentric layers around the core. There are many one-degree nodes that hang off the core and each of the layers.

terminating in one of its members) or a *transit* domain (which can carry any traffic). Example stubs include campus networks, or small interconnections of Local Area Networks (LANs). An example transit domain would be a set of backbone nodes over a large area, such as a wide-area network (WAN).

The basic idea is that stubs connect nodes locally, while transit domains interconnect the *stubs*, thus allowing the flow of traffic between nodes from different stubs (usually distant nodes). This imposes a *hierarchy* in the Internet structure, with transit domains at the top, each connecting several stub domains, each of which connects several LANs.

Apart from hierarchy, another feature of the Internet topology is its apparent *Jellyfish* structure at the AS level (Figure 3.7), found by Tauro et al. [79]. This consists of:

- *A core*, consisting of the highest-degree node and the clique it belongs to; this usually has 8–13 nodes.

- *Layers around the core*. These are organized as concentric circles around the core; layers further from the core have lower importance.

- *Hanging nodes*, representing one-degree nodes linked to nodes in the core or the outer layers. The authors find such nodes to be a large percentage (about 40–45%) of the graph.

The World Wide Web (WWW). Broder et al. [24] find that the Web graph is described well by a "bowtie" structure (Figure 3.8(a)). They find that the Web can be broken in 4 approximately equal-sized pieces. The core of the bowtie is the *Strongly Connected Component* (SCC) of the graph: each node in the SCC has a directed path to any other node in the SCC. Then, there is the IN component: each node in the IN component has a directed path to all the nodes in the SCC. Similarly, there is an OUT component, where each node can be reached by directed paths from the SCC. Apart from these, there are webpages which can reach some pages in OUT and can be reached from pages in IN without going through the SCC; these are the TENDRILS. Occasionally, a tendril can connect nodes in IN and OUT; the tendril is called a TUBE in this case. The remainder of the webpages fall in *disconnected components*. A similar study focused on only the Chilean part of the Web graph found that the disconnected component is actually very large (nearly 50% of the graph size) [11].

Dill et al. [33] extend this view of the Web by considering subgraphs of the WWW at different scales (Figure 3.8(b)). These subgraphs are groups of webpages sharing some common trait, such as content or geographical location. They have several remarkable findings:

1 *Recursive bowtie structure*: Each of these subgraphs forms a bowtie of its own. Thus, the Web graph can be thought of as a hierarchy of bowties, each representing a specific subgraph.

2 *Ease of navigation*: The SCC components of all these bowties are tightly connected together via the SCC of the whole Web graph. This provides a navigational backbone for the Web: starting from a webpage in one bowtie, we can click to its SCC, then go via the SCC of the entire Web to the destination bowtie.

3 *Resilience*: The union of a random collection of subgraphs of the Web has a large SCC component, meaning that the SCCs of the individual subgraphs have strong connections to other SCCs. Thus, the Web graph is very resilient to node deletions and does not depend on the existence of large taxonomies such as yahoo.com; there are several alternate paths between nodes in the SCC.

We have discussed several patterns occurring in real graphs, and given some examples. Next, we would like to know, how can we re-create these patterns? What sort of mechanisms can help explain real-world behaviors? To answer these questions we turn to *graph generators*.

3.   Graph Generators

Graph generators allow us to create synthetic graphs, which can then be used for, say, simulation studies. But when is such a generated graph "realistic?" This happens when the synthetic graph matches all (or at least several) of the patterns mentioned in the previous section. Graph generators can provide insight into graph creation, by telling us which processes can (or cannot) lead to the development of certain patterns.

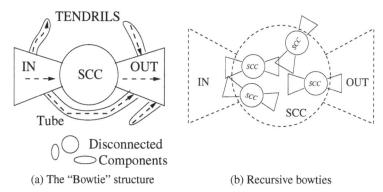

(a) The "Bowtie" structure   (b) Recursive bowties

Figure 3.8. *The "Bowtie" structure of the Web*: Plot (a) shows the 4 parts: IN, OUT, SCC and TENDRILS [24]. Plot (b) shows *Recursive Bowties*: subgraphs of the WWW can each be considered a bowtie. All these smaller bowties are connected by the navigational backbone of the main SCC of the Web [33].

Graph models and generators can be broadly classified into five categories:

1 *Random graph models:* The graphs are generated by a random process. The basic random graph model has attracted a lot of research interest due to its phase transition properties.

2 *Preferential attachment models:* In these models, the "rich" get "richer" as the network grows, leading to power law effects. Some of today's most popular models belong to this class.

3 *Optimization-based models:* Here, power laws are shown to evolve when risks are minimized using limited resources. This may be particularly relevant in the case of real-world networks that are constrained by geography. Together with the preferential attachment models, optimization-based models try to provide mechanisms that automatically lead to power laws.

4 *Tensor-based models:* Because many patterns in real graphs are self-similar, one can generate realistic graphs by using self-similar mechanisms through tensor multiplication.

5 *Internet-specific models* As the Internet is one of the most important graphs in computer science, special-purpose generators have been developed to model its special features. These are often hybrids, using ideas from the other categories and melding them with Internet-specific requirements.

We will discuss graph generators from each of these categories in this section. This is not a complete list, but we believe it includes most of the key ideas

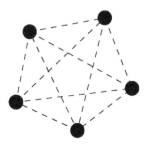

Figure 3.9. *The Erdos-Renyi model* The black circles represent the nodes of the graph. Every possible edge occurs with equal probability.

from the current literature. For each group of generators, we will try to provide the specific problem they aim to solve, followed by a brief description of the generator itself and its properties, and any open questions. We will also note variants on each major generator and briefly address their properties. While we will not discuss in detail all generators, we provide citations and a summary.

3.1   Random Graph Models

Random graphs are generated by picking nodes under some random probability distribution and then connecting them by edges. We first look at the basic Erdos-Renyi model, which was the first to be studied thoroughly [40], and then we discuss modern variants of the model.

The Erdos-Renyi Random Graph Model.

Problem being solved.   Graph theory owes much of its origins to the pioneering work of Erdos and Renyi in the 1960s [40, 41]. Their random graph model was the first and the simplest model for generating a graph.

Description and Properties.   We start with N nodes, and for every pair of nodes, an edge is added between them with probability p (as in Figure 3.9). This defines a *set* of graphs $G_{N,p}$, all of which have the same parameters (N, p).

Degree Distribution The probability of a vertex having degree k is

$$p_k = \binom{N}{k} p^k (1-p)^{N-k} \approx \frac{z^k e^{-z}}{k!} \quad \text{with } z = p(N-1) \qquad (3.8)$$

For this reason, this model is often called the "Poisson" model.

Size of the largest component Many properties of this model can be solved exactly in the limit of large N. A property is defined to hold for parameters (N, p) if the probability that the property holds on every graph in $G_{N,p}$ approaches 1 as $N \to \infty$. One of the most noted properties concerns the size of the largest component (subgraph) of the graph. For a low value of p, the graphs in $G_{N,p}$ have low density with few edges and all the components are small, having an exponential size distribution and finite mean size. However, with a high value of p, the graphs have a *giant component* with $O(N)$ of the nodes in the graph belonging to this component. The rest of the components again have an exponential size distribution with finite mean size. The changeover (called the *phase transition*) between these two regimes occurs at $p = \frac{1}{N}$. A heuristic argument for this is given below, and can be skipped by the reader.

Finding the phase transition point Let the fraction of nodes not belonging to the giant component be u. Thus, the probability of random node not belonging to the giant component is also u. But the neighbors of this node also do not belong to the giant component. If there are k neighbors, then the probability of this happening is u^k. Considering all degrees k, we get

$$
\begin{aligned}
u &= \sum_{k=0}^{\infty} p_k u^k \\
&= e^{-z} \sum_{k=0}^{\infty} \frac{(uz)^k}{k!} \quad \text{(using Eq 3.8)} \\
&= e^{-z} e^{uz} = e^{z(u-1)}
\end{aligned}
\tag{3.9}
$$

Thus, the fraction of nodes in the giant component is

$$
S = 1 - u = 1 - e^{-zS}
\tag{3.10}
$$

Equation 3.10 has no closed-form solutions, but we can see that when $z < 1$, the only solution is $S = 0$ (because $e^{-x} > 1 - x$ for $x \in (0,1)$). When $z > 1$, we can have a solution for S, and this is the size of the giant component. The phase transition occurs at $z = p(N-1) = 1$. Thus, a giant component appears only when p scales faster than N^{-1} as N increases.

---

[1] $P(k) \propto k^{-2.255} / \ln k$; [18] study a special case, but other values of the exponent γ may be possible with similar models.

[2] Inet-3.0 matches the Internet AS graph very well, but formal results on the degree-distribution are not available.

[3] $\gamma = 1 + \frac{1}{\alpha}$ as $k \to \infty$ (Eq. 3.16)

Tree-shaped subgraphs Similar results hold for the appearance of trees of different sizes in the graph. The critical probability at which almost every graph contains a subgraph of k nodes and l edges is achieved when p scales as N^z where $z = -\frac{k}{l}$ [20]. Thus, for $z < -\frac{3}{2}$, almost all graphs consist of isolated nodes and edges; when z passes through $-\frac{3}{2}$, trees of order 3 suddenly appear, and so on.

Diameter Random graphs have a diameter concentrated around $\log N / \log z$, where z is the average degree of the nodes in the graph. Thus, the diameter grows slowly as the number of nodes increases.

Clustering coefficient The probability that any two neighbors of a node are themselves connected is the connection probability $p = \frac{<k>}{N}$, where $<k>$ is the average node degree. Therefore, the clustering coefficient is:

$$CC_{random} = p = \frac{<k>}{N} \qquad (3.11)$$

Open questions and discussion. It is hard to exaggerate the importance of the Erdos-Renyi model in the development of modern graph theory. Even a simple graph generation method has been shown to exhibit phase transitions and criticality. Many mathematical techniques for the analysis of graph properties were first developed for the random graph model.

However, even though random graphs exhibit such interesting phenomena, they do not match real-world graphs particularly well. Their degree distribution is Poisson (as shown by Equation 3.8), which has a very different shape from power-laws or lognormals. There are no correlations between the degrees of adjacent nodes, nor does it show any form of "community" structure (which often shows up in real graphs like the WWW). Also, according to Equation 3.11, $\frac{CC_{random}}{<k>} = \frac{1}{N}$; but for many real-world graphs, $\frac{CC}{<k>}$ is independent of N (See figure 9 from [7]).

Thus, even though the Erdos-Renyi random graph model has proven to be very useful in the early development of this field, it is not used in most of the recent work on modeling real graphs. To address some of these issues, researchers have extended the model to the so-called Generalized Random Graph Models, where the degree distribution can be set by the user (typically, set to be a power law).

Analytic techniques for studying random graphs involve generating functions. A good reference is by Wilf [85].

Generalized Random Graph Models. Erdos-Renyi graphs result in a Poisson degree distribution, which often conflicts with the degree distributions of many real-world graphs. Generalized random graph models extend the basic random graph model to allow arbitrary degree distributions.

Given a degree distribution, we can randomly assign a degree to each node of the graph so as to match the given distribution. Edges are formed by randomly linking two nodes till no node has extra degrees left. We describe two different models below: the PLRG model and the Exponential Cutoffs model. These differ only in the degree distributions used; the rest of the graph-generation process remains the same. The graphs thus created can, in general, include self-graphs and multigraphs (having multiple edges between two nodes).

The PLRG model   One of the obvious modifications to the Erdos-Renyi model is to change the degree distribution from Poisson to power-law. One such model is the Power-Law Random Graph (PLRG) model of Aiello et al. [3] (a similar model is the *Power Law Out Degree* (PLOD) model of Palmer and Steffan [72]). There are two parameters: α and β. The number of nodes of degree k is given by e^{α}/k^{β}.

By construction, the degree distribution is specifically a power law:

$$p_k \propto k^{-\beta} \qquad (3.12)$$

where β is the power-law exponent.

The authors show that graphs generated by this model can have several possible properties, based only on the value of β. When $\beta < 1$, the graph is almost surely connected. For $1 < \beta < 2$, a giant component exists, and smaller components are of size $O(1)$. For $2 < \beta < \beta_0 \ \text{sim}\ 3.48$, the giant component exists and the smaller components are of size $O(\log N)$. At $\beta = \beta_0$, the smaller components are of size $O(\log N/\log\log N)$. For $\beta > \beta_0$, no giant component exists. Thus, for the giant component, we have a *phase transition* at $\beta = \beta_0 = 3.48$; there is also a change in the size of the smaller components at $\beta = 2$.

The Exponential cutoffs model   Another generalized random graph model is due to Newman et al. [69]. Here, the probability that a node has k edges is given by

$$p_k = Ck^{-\gamma}e^{-k/\kappa} \qquad (3.13)$$

where C, γ and κ are constants.

This model has a power law (the $k^{-\gamma}$ term) augmented by an exponential cutoff (the $e^{-k/\kappa}$ term). The exponential cutoff, which is believed to be present in some social and biological networks, reduces the heavy-tail behavior of a pure power-law degree distribution. The results of this model agree with those of [3] when $\kappa \to \infty$.

Analytic expressions are known for the average path length of this model, but this typically tends to be somewhat less than that in real-world graphs [7].

Apart from PLRG and the exponential cutoffs model, some other related models have also been proposed, a notable model generalization being dot-product models [70]. Another important model is that of Aiello et al. [4], who assign weights to nodes and then form edges probabilistically based on the product of the weights of their end-points. The exact mechanics are, however, close to preferential attachment, and we will discuss later.

Similar models have also been proposed for generating directed and bipartite random graphs. Recent work has provided analytical results for the sizes of the strongly connected components and cycles in such graphs [30, 37]. We do not discuss these any further; the interested reader is referred to [69].

Open questions and discussion. Generalized random graph models retain the simplicity and ease of analysis of the Erdos-Renyi model, while removing one of its weaknesses: the unrealistic Poisson degree distribution. However, most such models only attempt to match the degree distribution of real graphs, and no other patterns. For example, in most random graph models, the probability that two neighbors of a node are themselves connected goes as $O(N^{-1})$. This is exactly the clustering coefficient of the graph, and goes to zero for large N; but for many real-world graphs, $\frac{CC}{<k>}$ is independent of N (See figure 9 from [7]). Also, many real world graphs (such as the WWW) exhibit the existence of communities of nodes, with stronger ties within the community than outside; random graphs do not appear to show any such behavior. Further work is needed to accommodate these patterns into the random graph generation process.

3.2   Preferential Attachment and Variants

Problem being solved. Generalized random graph models try to model the power law or other degree distribution of real graphs. However, they do not make any statement about the *processes* generating the network. The search for a mechanism for network generation was a major factor in fueling the growth of the preferential attachment models, which we discuss below.

Basic Preferential Attachment. In the mid-1950s, Herbert Simon [77] showed that power law tails arise when "the rich get richer." Derek Price applied this idea (which he called *cumulative advantage*) to the case of networks [32], as follows. We grow a network by adding vertices over time. Each vertex gets a certain out-degree, which may be different for different vertices but whose mean remains at a constant value m over time. Each outgoing edge from the new vertex connects to an old vertex with a probability proportional to the in-degree of the old vertex. This, however, leads to a problem since all

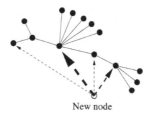

New node

Figure 3.10. *The Barabasi-Albert model* New nodes are added; each new node prefers to connect to existing nodes of high degree. The dashed lines show some possible edges for the new node, with thicker lines implying higher probability.

nodes initially start off with in-degree zero. Price corrected this by adding a constant to the current in-degree of a node in the probability term, to get

$$P(\text{edge to existing vertex } v) = \frac{k(v) + k_0}{\sum_i (k(i) + k_0)}$$

where $k(i)$ represents the current in-degree of an existing node i, and k_0 is a constant.

A similar model was proposed by Barabasi and Albert [13]. It has been a very influential model, and formed the basis for a large body of further work. Hence, we will look at the Barabasi-Albert model (henceforth called the BA model) in detail.

Description of the BA model.   The BA model proposes that structure emerges in network topologies as the result of two processes:

1 *Growth*: Contrary to several other existing models (such as random graph models) which keep a fixed number of nodes during the process of network formation, the BA model starts off with a small set of nodes and *grows* the network as nodes and edges are added over time.

2 *Preferential Attachment*: This is the same as the "rich get richer" idea. The probability of connecting to a node is proportional to the current degree of that node.

Using these principles, the BA model generates an *undirected* network as follows. The network starts with m_0 nodes, and grows in stages. In each stage, one node is added along with m edges which link the new node to m existing nodes (Figure 3.10). The probability of choosing an existing node as an endpoint for these edges is given by

$$P(\text{edge to existing vertex } v) = \frac{k(v)}{\sum_i k(i)} \qquad (3.14)$$

where $k(i)$ is the degree of node i. Note that since the generated network is undirected, we do not need to distinguish between out-degrees and in-degrees. The effect of this equation is that nodes which already have more edges connecting to them, get even more edges. This represents the "rich get richer" scenario.

There are a few differences from Price's model. One is that the number of edges per new node is fixed at m (a positive integer); in Price's model only the mean number of added edges needed to be m. However, the major difference is that while Price's model generates a directed network, the BA model is undirected. This avoids the problem of the initial in-degree of nodes being zero; however, many real graphs are directed, and the BA model fails to model this important feature.

Properties of the BA model.   We will now discuss some of the known properties of the BA model. These include the degree distribution, diameter, and correlations hidden in the model.

Degree distribution The degree distribution of the BA model [36] is given by:

$$p_k \approx k^{-3}$$

for large k. In other words, the degree distribution has a power law "tail" with exponent 3, independent of the value of m.

Diameter Bollobás and Riordan [22] show that for large N, the diameter grows as $O(\log N)$ for $m = 1$, and as $O(\log N / \log \log N)$ for $m \geq 2$. Thus, this model displays the *small-world* effect: the distance between two nodes is, on average, far less than the total number of nodes in the graph.

Correlations between variables Krapivsky and Redner [52] find two correlations in the BA model. First, they find that degree and age are positively correlated: older nodes have higher mean degree. The second correlation is in the degrees of neighboring nodes, so that nodes with similar degree are more likely to be connected. However, this asymptotically goes to 0 as $N \rightarrow \infty$.

Open questions and discussion.   The twin ideas of *growth* and *preferential attachment* are definitely an immense contribution to the understanding of network generation processes. However, the BA model attempts to explain graph structure using *only* these two factors; most real-world graphs are probably generated by a slew of different factors. The price for this is some inflexibility in graph properties of the BA model.

- The power-law exponent of the degree distribution is fixed at $\gamma = 3$, and many real-world graphs deviate from this value.

- The BA model generates undirected graphs only; this prevents the model from being used for the many naturally occurring directed graphs.

- While Krapivsky and Redner show that the BA model should have correlations between node degree and node age (discussed above), Adamic and Huberman [1] apparently find no such correlations in the WWW.

- The generated graphs have exactly one connected component. However, many real graphs have several isolated components. For example, websites for companies often have private set of webpages for employees/projects only. These are a part of the WWW, but there are no paths to those webpages from outside the set. Military routers in the Internet router topology are another example.

- The BA model has a constant average degree of m; however, the average degree of some graphs (such as citation networks) actually increases over time according to a Densification Power Law [14, 58, 37]

- The diameter of the BA model increases as N increases; however, many graphs exhibit shrinking diameters.

Also, further work is needed to confirm the existence or absence of a community structure in the generated graphs.

While the basic BA model does have these limitations, its simplicity and power make it an excellent base on which to build extended models. In fact, the bulk of graph generators in use today can probably trace their lineage back to this model. In the next few sections, we will look at some of these extensions and variations; as we will see, most of these are aimed at removing one or the other of the aforementioned limitations.

Variants on Preferential Attachment.

Initial attractiveness. While the BA model generates graphs with a power law degree distribution, the power law exponent is stuck at $\gamma = 3$. Dorogovtsev et al. [36, 35] propose a simple one-parameter extension of the basic model which allows $\gamma \in [2, \infty)$. Other methods, such as the AB model described later, also do this, but they require more parameters. In initial attractiveness, an extra "initial attractiveness" parameter is added which governs the probability of "young" sites gaining new edges. Adjusting this parameter will vary the degree distribution, adding significant flexibility to the BA model.

Internal edges and Rewiring. Albert and Barabasi [6] proposed another method to add flexibility in the power law exponent. In the original BA model, one node and m edges are added to the graph every iteration. Albert and

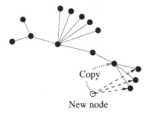

Figure 3.11. *The edge copying model* New nodes can choose to copy the edges of an existing node. This models the copying of links from other peoples' websites to create a new website.

Barabasi decouple this addition of nodes and edges, and also extend the model by introducing the concept of edge rewiring. Starting with a small set of m_0 nodes, the resulting model (henceforth called the AB model) combines 3 processes: adding internal edges, removing/reconnecting ("rewiring") edges, and adding new nodes with some edges. This model exhibits either a power-law or exponential degree distribution, depending on the parameters used.

Edge Copying Models.   Several graphs show community behavior, such as topic-based communities of websites on the WWW. Kleinberg et al. [51] and Kumar et al. [54] try to model this by using the intuition that most webpage creators will be familiar with webpages on topics of interest to them, and so when they create new webpages, they will link to some of these existing topical webpages. Thus, most new webpages will enhance the "topical community" effect of the WWW.

The Kleinberg [51] generator creates a directed graph. In this generator, nodes are independently created and deleted in each distribution, and edges incident on deleted nodes are also removed. Also, edges may be added to or deleted from existing nodes. Then, there is the key edge copying mechanism, where a node may copy edges from another node. An illustration is shown in Figure 3.11. This is similar to preferential attachment because the pages with high-degree will be linked to by many other pages, and so have a greater chance of getting copied.

Kumar et al. [54] propose a very similar model. However, there are some important differences. Whenever a new node is added, only *one* new edge is added. The copying process takes place when head or tail of some existing edge gets chosen as the endpoint of the new edge. This model may serve to create "communities" as there may be important nodes on each "topic".

This and similar models by analyzed by Kumar et al. [53]. In-degree distribution of Kleinberg's model follows a power law, and both in-and out-degree of Kumar et al.'s model follow power laws.

The Kleinberg model [51] generates a tree; no "back-edges" are formed from the old nodes to the new nodes. Also, in the model of Kumar et al. [54], a fixed fraction of the nodes have zero in-degree or zero out-degree; this might not be the case for all real-world graphs (see Aiello et al. [4] for related issues). However, the simple idea of copying edges can clearly lead to both power laws as well as community effects. "Edge copying" models are, thus, a very promising direction for future research.

Modifying the preferential attachment equation. Chen et al. [29] had found the AB model somewhat lacking in modeling the Web. Specifically, they found that the preference for connecting to high-degree nodes is stronger than that predicted by linear preferential attachment. Bu and Towsley [25] attempt to address this issue.

The AB model [6] is changed by removing the edge rewiring process, and modifying the linear preferential attachment equation of the AB model to show higher preference for nodes with high degrees (as in [29]). This is called the GLP (Generalized Linear Preference) model. The degree distribution follows a power law. Also, they also find empirically that the clustering coefficient for a GLP graph is much closer to that of the Internet than the BA, AB and Power-Law Random Graph (PLRG [3]) models.

Others such as Krapivsky and Redner [52] have studied *non-linear* preferential attachment, finding this tended to produce degree decay faster than a power law.

Modeling increasing average degree. The average degree of several real-world graphs (such as citation graphs) increases over time [37, 14, 58], according to a Densification Power Law. Barabasi et al. [14] attempt to modify the basic BA model to accommodate this effect. In the model, a new edge chooses *both* its endpoints by preferential attachment. The number of internal nodes added per iteration is proportional to the the current number of nodes in the graph. Thus, it leads to the phenomenon of *accelerated growth*: the average degree of the graph increases linearly over time.

However, the analysis of this model shows that it has two power-law regimes. The power law exponent is $\gamma = 2$ for low degrees, and $\gamma = 3$ for high degrees. In fact, over a long period of time, the exponent converges to $\gamma = 2$.

Node fitness measures. The preferential attachment models noted above tend to have a correlation between the age of a node and its degree: higher the age, more the degree [52]. However, Adamic and Huberman find that this does not hold for the WWW [1]. There are websites which were created late but still have far higher in-degree than many older websites. Bianconi and Barabasi [18] try to model this. Their model attaches a *fitness parameter* to each node, which does not change over time. The idea is that even a node which is added late could overtake older nodes in terms of degree, if the newer node has a much higher fitness value.

The authors analyze the case when the fitness parameters are drawn randomly from a uniform $[0, 1]$ distribution. The resulting degree distribution is a power law with an extra inverse logarithmic factor. For the case where all fitness values are the same, this model becomes the simple BA model.

Having a node's popularity depend on its "fitness" intuitively makes a lot of sense. Further research is needed to determine the distribution of node fitness values in real-world graphs.

Generalizing preferential attachment.   The BA model is undirected. A simple adaptation to the directed case is: new edges are created to point from the new nodes to existing nodes chosen preferentially according to their *in-degree*. However, the out-degree distribution of this model would not be a power law. Aiello et al. [4] propose a very general model for generating directed graphs which give power laws for both in-degree and out-degree distributions. A similar model was also proposed by Bollobas et al. [21]. The work shows that even a very general version of preferential attachment can lead to power law degree distributions. Further research is needed to test for all the other graph patterns, such as diameter, community effects and so on.

PageRank-based preferential attachment.   Pandurangan et al. [73] found that the *PageRank* [23] values for a snapshot of the Web graph follow a power law. They propose a model that tries to match this *PageRank* distribution of real-world graphs, *in addition to* the degree distributions. They modify the basic preferential attachment mechanism by adding a *PageRank*-based preferential attachment component– not only do edges preferentially connect to high degree nodes, but also high PageRank nodes. They empirically show that this model can match both the degree distributions as well as the *PageRank* distribution of the Web graph. However, closed-form formulas for the degree distributions are not provided for this model. The authors also found that the plain edge-copying model of Kumar et al. [54] could *also* match the *PageRank* distribution (in addition to the degree distributions) without specifically attempting to do so. Thus, this work might be taken to be another alternative model of the Web.

The Forest Fire model.   Leskovec et al. [58] develop a preferential-attachment based model which matches the Densification Power Law and the shrinking diameter patterns of graph evolution, in addition to the power law degree distribution. A node chooses an *ambassador* node uniformly at random, and then links recursively to the ambassador node's neighbors.

This creates preferential linking without explicitly assigning such probability. This method is similar to the edge copying model discussed earlier because existing links are "copied" to the new node v as the fire spreads. This leads to a community of nodes, which share similar edges.

The Butterfly model.   Most preferential-attachment based models will form a single connected component, when, in real graphs, there are many smaller components that evolve and occasionally join with each other. Mc-Glohon et al. [59] develop a model that addresses this. Like in the Forest Fire model, there is an ambassador mechanism. However, there is no guarantee of linkage, so a node may become isolated and form its own new component for other nodes to join to. Additionally, instead of a single ambassador, a node may choose multiple ambassadors. This will allow components to join together.

The Butterfly model empirically produces power laws for both in- and out-degree, as well as reproducing the Densification Power Law and shrinking diameter. Furthermore, it reproduces oscillating patterns of the next-largest connected components mentioned earlier.

Deviations from power laws.

Problem being solved.   Pennock et al. [75] find that while the WWW as a whole might exhibit power-law degree distributions, subgraphs of web-pages belonging to specific categories or topics often show significant deviations from a power law. They attempt to model this deviation from power-law behavior.

Description and properties.   Their model is similar to the BA model, except for two differences:

- *Internal edges* The m new edges added in each iteration need not be incident on the new node being added that iteration. Thus, the new edges could be *internal* edges.

- *Combining random and preferential attachment* Instead of pure preferential attachment, the endpoints of new edges are chosen according to a linear combination of preferential attachment and uniform random attachment. The probability of a node v being chosen as one endpoint of an edge is given by:

$$p(v) = \alpha \frac{k(v)}{2mt} + (1 - \alpha) \frac{1}{m_0 + t} \tag{3.15}$$

Here, $k(v)$ represents the current degree of node v, $2mt$ is the total number of edges at time t, $(m_0 + t)$ is the current number of nodes at time

t, and $\alpha \in [0, 1]$ is a free parameter. To rephrase the equation, in order to choose a node as an endpoint for a new edge, we either do preferential attachment with probability α, or we pick a node at random with probability $(1 - \alpha)$.

One point of interest is that even if a node is added with degree 0, there is always a chance for it to gain new edges via the uniform random attachment process. The preferential attachment and uniform attachment parts of Equation 3.15 represent two different behaviors of webpage creators (according to the authors):

- The preferential attachment term represents adding links which the creator became aware of because they were popular.

- The uniform attachment term represents the case when the author adds a link because it is relevant to him, and this is irrespective of the popularity of the linked page. This allows even the poorer sites to gain some edges.

Degree distribution The authors derive a degree distribution function for this model:

$$P(k) \propto (k + c)^{-1-\frac{1}{\alpha}} \qquad (3.16)$$

where c is a function of m and α. This gives a power-law of exponent $(1+1/\alpha)$ in the tail. However, for low degrees, it deviates from the power-law, as the authors wanted.

Power-law degree distributions have shown up in many real-world graphs. However, it is clear that deviations in this do show up in practice. This is one of the few models we are aware of that specifically attempt to model such deviations, and as such, is a step in the right direction.

Open questions and discussion.   This model can match deviations from power laws in degree distributions. However, further work is needed to test for other graph patterns, like diameter, community structure and such.

Implementation issues.   Here, we will briefly discuss certain implementation aspects. Consider the BA model. In each iteration, we must choose edge endpoints according to the linear preferential attachment equation. Naively, each time we need to add a new edge, we could go over all the existing nodes and find the probability of choosing each node as an endpoint, based on its current degree. However, this would take $O(N)$ time each iteration, and $O(N^2)$ time to generate the entire graph. A better approach [65] is to keep an array: whenever a new edge is added, its endpoints are appended to the array. Thus, each node appears in the array as many times as its degree. Whenever we must choose a node according to preferential attachment, we can choose any cell of the array uniformly at random, and the node stored in that cell can be considered to have been chosen under preferential attachment. This requires $O(1)$ time for each iteration, and $O(N)$ time to generate the entire graph; however, it needs extra space to store the edge list.

This technique can be easily extended to the case when the preferential attachment equation involves a constant β, such as $P(v) \propto (k(v) - \beta)$ for the GLP model. If the constant β is a negative integer (say, $\beta = -1$ as in the AB model), we can handle this easily by adding $|\beta|$ entries for every existing node into the array. However, if this is not the case, the method needs to be modified slightly: with some probability α, the node is chosen according to the simple preferential attachment equation (like in the BA model). With probability $(1 - \alpha)$, it is chosen uniformly at random from the set of existing nodes. For each iteration, the value of α can be chosen so that the final effect is that of choosing nodes according to the modified preferential attachment equation.

Summary of Preferential Attachment Models. All preferential attachment models use the idea that the "rich get richer": high-degree nodes attract more edges, or high-PageRank nodes attract more edges, and so on. This simple process, along with the idea of network growth over time, *automatically* leads to the power-law degree distributions seen in many real-world graphs. As such, these models made a very important contribution to the field of graph mining. Still, most of these models appear to suffer from some limitations: for example, they do not seem to generate any "community" structure in the graphs they generate. Also, apart from the work of Pennock et al. [75], little effort has gone into finding reasons for deviations from power-law behaviors for some graphs. It appears that we need to consider additional processes to understand and model such characteristics.

3.3 Optimization-based generators

Most of the methods described above have approached power-law degree distributions from the preferential-attachment viewpoint: if the "rich get richer", power-laws might result. However, another point of view is that power laws can result from *resource optimizations*. There may be a number of constraints applied to the models– cost of connections, geographical distance, etc. We will discuss some models based on optimization of resources next.

The Highly Optimized Tolerance model.

Problem being solved:. Carlson and Doyle [27, 38] have proposed an optimization-based reason for the existence of power laws in graphs. They say that power laws may arise in systems due to *tradeoffs* between yield (or profit), resources (to prevent a risk from causing damage) and tolerance to risks.

Description and properties:.   As an example, suppose we have a forest which is prone to forest fires. Each portion of the forest has a different
chance of starting the fire (say, the dryer parts of the forest are more likely to
catch fire). We wish to minimize the damage by assigning resources such as
firebreaks at different positions in the forest. However, the total available resources are limited. The problem is to place the firebreaks so that the expected
cost of forest fires is minimized.

In this model, called the *Highly Optimized Tolerance* (HOT) model, we have
n possible events (starting position of a forest fire), each with an associated
probability $p_i (1 \leq i \leq n)$ (dryer areas have higher probability). Each event
can lead to some *loss* l_i, which is a function of the resources r_i allocated for
that event: $l_i = f(r_i)$. Also, the total resources are limited: $\sum_i r_i \leq R$ for
some given R. The aim is to minimize the expected cost

$$ J = \left\{ \sum_i p_i l_i \mid l_i = f(r_i), \sum_i r_i \leq R \right\} \qquad (3.17) $$

Degree distribution: The authors show that if we assume that cost and resource
usage are related by a power law $l_i \propto r_i^{\beta}$, then, under certain assumptions
on the probability distribution p_i, resources are spent on places having higher
probability of costly events. In fact, resource placement is related to the probability distribution p_i by a power law. Also, the probability of events which
cause a loss greater than some value k is related to k by a power law.

The salient points of this model are:

- high efficiency, performance and robustness to designed-for uncertain-
  ties

- hypersensitivity to design flaws and unanticipated perturbations

- nongeneric, specialized, structured configurations, and

- power laws.

Resilience under attack: This concurs with other research regarding the vulnerability of the Internet to attacks. Several researchers have found that while
a large number of randomly chosen nodes and edges can be removed from the
Internet graph without appreciable disruption in service, attacks *targeting* important nodes can disrupt the network very quickly and dramatically [71, 9].
The HOT model also predicts a similar behavior: since routers and links are
expected to be down occasionally, it is a "designed-for" uncertainty and the
Internet is impervious to it. However, a *targeted* attack is not designed for, and
can be devastating.

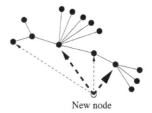

New node

Figure 3.12. *The Heuristically Optimized Tradeoffs model* A new node prefers to link to existing nodes which are both close in distance and occupy a "central" position in the network.

Newman et al. [68] modify HOT using a utility function which can be used to incorporate "risk aversion." Their model (called *Constrained Optimization with Limited Deviations* or COLD) truncates the tails of the power laws, lowering the probability of disastrous events.

HOT has been used to model the sizes of files found on the WWW. The idea is that dividing a single file into several smaller files leads to faster load times, but increases the cost of navigating through the links. They show good matches with this dataset.

Open questions and discussion. The HOT model offers a completely new recipe for generating power laws; power laws can result as a by-product of resource optimizations. However, this model requires that the resources be spread in an *globally-optimal* fashion, which does not appear to be true for several large graphs (such as the WWW). This led to an alternative model by Fabrikant et al. [42], which we discuss next.

Modification: The Heuristically Optimized Tradeoffs model. Fabrikant et al. [42] propose an alternative model in which the graph grows as a result of trade-offs made *heuristically* and locally (as opposed to optimally, for the HOT model).

The model assumes that nodes are spread out over a geographical area. One new node is added in every iteration, and is connected to the rest of the network with *one* link. The other endpoint of this link is chosen to optimize between two conflicting goals: (1) minimizing the "last-mile" distance, that is, the *geographical* length of wire needed to connect a new node to a pre-existing graph (like the Internet), and, (2) minimizing the transmission delays based on number of hops, or, the distance along the network to reach other nodes. The authors try to optimize a linear combination of the two (Figure 3.12). Thus, a new node i should be connected to an existing node j chosen to minimize

$$\alpha.d_{ij} + h_j \ (j < i) \tag{3.18}$$

where d_{ij} is the distance between nodes i and j, h_j is some measure of the "centrality" of node j, and α is a constant that controls the relative importance of the two.

The authors find that the characteristics of the network depend greatly on the value of α, and may be a single hub or have an exponential degree distribution, but for a range of values power-law degree distribution results.

As in the *Highly Optimized Tolerance* model described before (Subsection 3.3.0), power laws are seen to fall off as a by-product of resource optimizations. However, only local optimizations are now needed, instead of global optimizations. This makes the *Heuristically Optimized Tradeoffs* model very appealing.

Other research in this direction is the recent work of Berger et al. [16], who generalize the *Heuristically Optimized Tradeoffs* model, and show that it is equivalent to a form of preferential attachment; thus, competition between opposing forces can give rise to preferential attachment, and we already know that preferential attachment can, in turn, lead to power laws and exponential cutoffs.

Incorporating Geographical Information.   Both the random graph and preferential attachment models have neglected one attribute of many real graphs: the constraints of geography. For example, it is easier (cheaper) to link two routers which are physically close to each other; most of our social contacts are people we meet often, and who consequently probably live close to us (say, in the same town or city), and so on. In the following paragraphs, we discuss some important models which try to incorporate this information.

The Small-World Model.

Problem being solved.   The small-world model is motivated by the observation that most real-world graphs seem to have low average distance between nodes (a global property), but have high clustering coefficients (a local property). Two experiments from the field of sociology shed light on this phenomenon.

Travers and Milgram [80] conducted an experiment where participants had to reach randomly chosen individuals in the U.S.A. using a chain letter between close acquaintances. Their surprising find was that, for the chains that completed, the average length of the chain was only six, in spite of the large population of individuals in the "social network." While only around 29% of the chains were completed, the idea of small paths in large graphs was still a landmark find.

The reason behind the short paths was discovered by Mark Granovetter [47], who tried to find out how people found jobs. The expectation was that the job

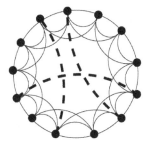

Figure 3.13. *The small-world model* Nodes are arranged in a ring lattice; each node has links to its immediate neighbors (solid lines) and some long-range connections (dashed lines).

seeker and his eventual employer would be linked by long paths; however, the actual paths were empirically found to be very short, usually of length one or two. This corresponds to the low average path length mentioned above. Also, when asked whether a friend had told them about their current job, a frequent answer of the respondents was *"Not a friend, an acquaintance"*. Thus, this low average path length was being caused by acquaintances, with whom the subjects only shared *weak ties*. Each acquaintance belonged to a different social circle and had access to different information. Thus, while the social graph has high clustering coefficient (i.e., is "clique-ish"), the low diameter is caused by weak ties joining faraway cliques.

Description and properties.   Watts and Strogatz [83] independently came up with a model with these characteristics: it has *high clustering coefficient* but *low diameter* . Their model (Figure 3.13), which has only one parameter p, consists of the following: begin with a ring lattice where each node has a set of "close friendships". Then rewire: for each node, each edge is rewired with probability p to a new random destination– these are the "weak ties".

Distance between nodes, and Clustering coefficient For $p = 0$ the graph remains a ring lattice, where both clustering coefficient and average distance between nodes are high. For $p = 1$, both values are very low. For a range of values in between, the average distance is low while clustering coefficient is high– as one would expect in real graphs. The reason for this is that the introduction of a few long-range edges (which are exactly the weak ties of Granovetter) leads to a highly nonlinear effect on the average distance L. Distance is contracted not only between the endpoints of the edge, but also their immediate neighborhoods (circles of friends). However, these few edges lead to a very small change in the clustering coefficient. Thus, we get a broad range of p for which the small-world phenomenon coexists with a high clustering coefficient.

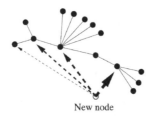

New node

Figure 3.14. *The Waxman model* New nodes prefer to connect to existing nodes which are closer in distance.

Degree distribution All nodes start off with degree k, and the only changes to their degrees are due to rewiring. The shape of the degree distribution is similar to that of a random graph, with a strong peak at k, and it decays exponentially for large k.

Open questions and discussion.   The small-world model is very successful in combining two important graph patterns: small diameters and high clustering coefficients. However, the degree distribution decays exponentially, and does not match the power-law distributions of many real-world graphs. Extension of the basic model to power law distributions is a promising research direction.

Other geographical models.

The Waxman Model.   While the Small World model begins by constraining nodes to a local neighborhood, the Waxman model [84] explicitly builds the graph based on optimizing geographical constraints, to model the Internet graph.

   The model is illustrated in Figure 3.14. Nodes (representing routers) are placed randomly in Cartesian 2-D space. An edge (u, v) is placed between two points u and v with probability

$$P(u, v) = \beta \exp \frac{-d(u, v)}{L\alpha} \qquad (3.19)$$

Here, α and β are parameters in the range $(0, 1)$, $d(u, v)$ is the Euclidean distance between points u and v, and L is the maximum Euclidean distance between points. The parameters α and β control the geographical constraints. The value of β affects the *edge density*: larger values of β result in graphs with higher edge densities. The value of α relates the short edges to longer ones: a small value of α increases the density of short edges relative to longer edges. While it does not yield a power-law degree distribution, it has been popular in the networking community.

The BRITE generator. Medina et al. [60] try to combine the geographical properties of the Waxman generator with the incremental growth and preferential attachment techniques of the BA model. Their graph generator, called BRITE, has been extensively used in the networking community for simulating the structure of the Internet.

Nodes are placed on a square grid, with some m links per node. Growth occurs either all at once (as in Waxman) or incrementally (as in BA). Edges are wired randomly, preferentially, or combined preferential and geographical constraints as follows: Suppose that we want to add an edge to node u. The probability of the other endpoint of the edge being node v is a *weighted* preferential attachment equation, with the weights being the the probability of that edge existing in the pure Waxman model (Equation 3.19)

$$P(u, v) = \frac{w(u, v)k(v)}{\sum_i w(u, i)k(i)} \qquad (3.20)$$

$$\text{where } w(u, v) = \beta \exp \frac{-d(u, v)}{L\alpha} \text{ as in Eq. 3.19}$$

The emphasis of BRITE is on creating a system that can be used to generate different kinds of topologies. This allows the user a lot of flexibility, and is one reason behind the widespread use of BRITE in the networking community. However, one limitation is that there has been little discussion of parameter fitting, an area for future research.

Yook et al. Model. Yook et al. [87] find two interesting linkages between geography and networks (specifically the Internet): First, the geographical distribution of Internet routers and Autonomous Systems (AS) is a fractal, and is strongly correlated with population density. Second, the probability of an edge occurring is *inversely proportional* to the Euclidean distance between the endpoints of the edge, likely due to cost of physical wire (which dominates over administrative cost for long links). However, in the Waxman and BRITE models, this probability decays exponentially with length (Equation 3.19).

To remedy the first problem, they suggest using a self-similar geographical distribution of nodes. For the second problem, they propose a modified version of the BA model. Each new node u is placed on the map using the self-similar distribution, and adds edges to m existing nodes. For each of these edges, the probability of choosing node v as the endpoint is given by a modified preferential attachment equation:

$$P(\text{node } u \text{ links to existing node } v) \propto \frac{k(v)^\alpha}{d(u, v)^\sigma} \qquad (3.21)$$

where $k(v)$ is the current degree of node v and $d(u, v)$ is the Euclidean distance between the two nodes. The values α and σ are parameters, with $\alpha = \sigma = 1$

giving the best fits to the Internet. They show that varying the values of α and σ can lead to significant differences in the topology of the generated graph.

Similar geographical constraints may hold for social networks as well: individuals are more likely to have friends in the same city as compared to other cities, in the same state as compared to other states, and so on recursively. Watts et al. [82] and (independently) Kleinberg [50] propose a hierarchical model to explain this phenomenon.

PaC - utility based.   Du et al. proposed an agent-based model "Pay and Call" or *PaC*, where agents make decisions about forming edges based on a perceived "profit" of an interaction. Each agent has a "friendliness" parameter. Calls are made with some "emotional dollars" cost, and agents may derive some benefit from each call. If two "friendly" agents interact, there is a higher benefit than if one or both agents are "unfriendly". The specific procedures are detailed in [39]. *PaC* generates degree, weight, and clique distributions as found in most real graphs.

3.4   Tensor-based

The R-MAT (Recursive MATrix) graph generator.   We have seen that most of the current graph generators focus on only one graph pattern – typically the degree distribution – and give low importance to all the others. There is also the question of how to fit model parameters to match a given graph. What we would like is a tradeoff between parsimony (few model parameters), realism (matching most graph patterns, if not all), and efficiency (in parameter fitting and graph generation speed). In this section, we present the R-MAT generator, which attempts to address all of these concerns.

Problem being solved.   The R-MAT [28] generator tries to meet several desiderata:

- The generated graph should match several graph patterns, including *but not limited to* power-law degree distributions (such as hop-plots and eigenvalue plots).

- It should be able to generate graphs exhibiting deviations from power-laws, as observed in some real-world graphs [75].

- It should exhibit a strong "community" effect.

- It should be able to generate directed, undirected, bipartite or weighted graphs with the same methodology.

- It should use as few parameters as possible.

- There should be a fast parameter-fitting algorithm.

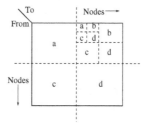

Figure 3.15. *The R-MAT model* The adjacency matrix is broken into four equal-sized partitions, and one of those four is chosen according to a (possibly non-uniform) probability distribution. This partition is then split recursively till we reach a single cell, where an edge is placed. Multiple such edge placements are used to generate the full synthetic graph.

- The generation algorithm should be efficient and scalable.

Description and properties. The R-MAT generator creates directed graphs with 2^n nodes and E edges, where both values are provided by the user. We start with an empty adjacency matrix, and divide it into four equal-sized partitions. One of the four partitions is chosen with probabilities a, b, c, d respectively ($a + b + c + d = 1$), as in Figure 3.15. The chosen partition is again subdivided into four smaller partitions, and the procedure is repeated until we reach a simple cell (=1×1 partition). The nodes (that is, row and column) corresponding to this cell are linked by an edge in the graph. This process is repeated E times to generate the full graph. There is a subtle point here: we may have *duplicate* edges (i.e., edges which fall into the same cell in the adjacency matrix), but we only keep one of them when generating an un-weighted graph. To smooth out fluctuations in the degree distributions, some noise is added to the (a, b, c, d) values at each stage of the recursion, followed by renormalization (so that $a + b + c + d = 1$). Typically, $a \geq b, a \geq c, a \geq d$.

Degree distribution There are only 3 parameters (the partition probabilities a, b, and c; $d = 1 - a - b - c$). The skew in these parameters ($a \geq d$) leads to lognormals and the DGX [17] distribution, which can successfully model both power-law and "unimodal" distributions [75] under different parameter settings.

Communities Intuitively, this technique is generating "communities" in the graph:

- The partitions a and d represent separate groups of nodes which correspond to communities (say, "Linux" and "Windows" users).

- The partitions b and c are the *cross-links* between these two groups; edges there would denote friends with separate preferences.

- The recursive nature of the partitions means that we automatically get sub-communities within existing communities (say, "RedHat" and "Mandrake" enthusiasts within the "Linux" group).

Diameter, singular values and other properties   We show experimentally that graphs generated by R-MAT have small diameter and match several other criteria as well.

Extensions to undirected, bipartite and weighted graphs   The basic model generates directed graphs; all the other types of graphs can be easily generated by minor modifications of the model. For undirected graphs, a directed graph is generated and then made symmetric. For bipartite graphs, the same approach is used; the only difference is that the adjacency matrix is now rectangular instead of square. For weighted graphs, the number of *duplicate* edges in each cell of the adjacency matrix is taken to be the weight of that edge. More details may be found in [28].

Parameter fitting algorithm   Given some input graph, it is necessary to fit the R-MAT model parameters so that the generated graph matches the input graph in terms of graph patterns.

We can calculate the expected degree distribution: the probability p_k of a node having outdegree k is given by

$$p_k = \frac{1}{2^n} \binom{E}{k} \sum_{i=0}^{n} \binom{n}{i} \left[\alpha^{n-i}(1-\alpha)^i\right]^k \left[1 - \alpha^{n-i}(1-\alpha)^i\right]^{E-k}$$

where 2^n is the number of nodes in the R-MAT graph, E is the number of edges, and $\alpha = a + b$. Fitting this to the outdegree distribution of the input graph provides an estimate for $\alpha = a + b$. Similarly, the indegree distribution of the input graph gives us the value of $b + c$. Conjecturing that the $a : b$ and $a : c$ ratios are approximately $75 : 25$ (as seen in many real world scenarios), we can calculate the parameters (a, b, c, d).

Chakrabarti et al. showed experimentally that R-MAT can match both power-law distributions as well as deviations from power-laws [28], using a number of real graphs. The patterns matched by R-MAT include both in- and out-degree distributions, "hop-plot" and "effective diameter", singular value vs. rank plots, "Network value" vs. rank plots, and "stress" distribution. Authors also compared R-MAT fits to those achieved by *AB*, *GLP*, and *PG* models.

Open questions and discussion.   While the R-MAT model shows promise, there has not been any thorough analytical study of this model. Also, it seems that only 3 parameters might not provide enough "degrees of freedom" to match all varieties of graphs; extensions of this model should be investigated. A step in this direction is the *Kronecker graph generator* [57], which generalizes the R-MAT model and can match several interesting patterns such as the Densification Power Law and the shrinking diameters effect in addition to all the patterns that R-MAT matches.

Graph Generation by Kronecker Multiplication. The R-MAT generator described in the previous paragraphs achieves its power mainly via a form of recursion: the adjacency matrix is recursively split into equal-sized quadrants over which edges are distributed unequally. One way to generalize this idea is via Kronecker matrix multiplication, wherein one small initial matrix is recursively "multiplied" with itself to yield large graph topologies. Unlike R-MAT, this generator has simple closed-form expressions for several measures of interest, such as degree distributions and diameters, thus enabling ease of analysis and parameter-fitting.

Description and properties. We first recall the definition of the Kronecker product.

Definition 3.5 (Kronecker product of matrices). *Given two matrices $\mathcal{A} = [a_{i,j}]$ and $\mathcal{B}$ of sizes $n \times m$ and $n' \times m'$ respectively, the Kronecker product matrix $\mathcal{C}$ of dimensions $(n * n') \times (m * m')$ is given by*

$$\mathcal{C} = \mathcal{A} \otimes \mathcal{B} \doteq \begin{pmatrix} a_{1,1}\mathcal{B} & a_{1,2}\mathcal{B} & \dots & a_{1,m}\mathcal{B} \\ a_{2,1}\mathcal{B} & a_{2,2}\mathcal{B} & \dots & a_{2,m}\mathcal{B} \\ \vdots & \vdots & \ddots & \vdots \\ a_{n,1}\mathcal{B} & a_{n,2}\mathcal{B} & \dots & a_{n,m}\mathcal{B} \end{pmatrix} \quad (3.22)$$

In other words, for any nodes X_i and X_j in $\mathcal{A}$ and X_k and X_ℓ in $\mathcal{B}$, we have nodes $X_{i,k}$ and $X_{j,\ell}$ in the Kronecker product $\mathcal{C}$, and an edge connects them iff the edges (X_i, X_j) and (X_k, X_ℓ) exist in $\mathcal{A}$ and $\mathcal{B}$. The Kronecker product of two graphs is the Kronecker product of their adjacency matrices.

Let us consider an example. Figure 3.16(a–c) shows the recursive construction of $G \otimes H$, when $G = H$ is a 3-node path. Consider node $X_{1,2}$ in Figure 3.16(c): It belongs to the H graph that replaced node X_1 (see Figure 3.16(b)), and in fact is the X_2 node (i.e., the center) within this small H-graph. Thus, the graph H is recursively embedded "inside" graph G.

The Kronecker graph generator simply applies the Kronecker product multiple times over. Starting with a binary *initiator* graph, successively larger graphs are produced by repeated Kronecker multiplication. The properties of the generated graph thereby depend on those of the initiator graph.

There are several interesting properties of the Kronecker generator which are discussed in detail in [55]. Kronecker graphs have multinomial degree dis-

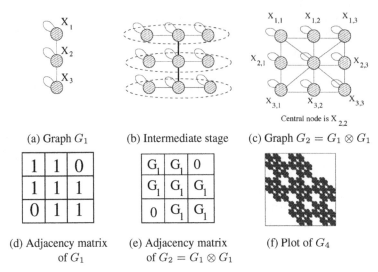

(a) Graph G_1 (b) Intermediate stage   (c) Graph $G_2 = G_1 \otimes G_1$

1	1	0
1	1	1
0	1	1

G_1	G_1	0
G_1	G_1	G_1
0	G_1	G_1

(d) Adjacency matrix   (e) Adjacency matrix   (f) Plot of G_4
of G_1 of $G_2 = G_1 \otimes G_1$

Figure 3.16. *Example of Kronecker multiplication* Top: a "3-chain" and its Kronecker product with itself; each of the X_i nodes gets expanded into 3 nodes, which are then linked together. Bottom row: the corresponding adjacency matrices, along with matrix for the fourth Kronecker power G_4.

tributions, static diameter/effective diameter (if nodes have self-loops), multinomial distributions of eigenvalues, and community structure. Additionally, it provably follows the Densification Power Law.

Thanks to its simple mathematical structure, Kronecker graph generation allows the derivation of closed-form formulas for several important patterns. Of particular importance are the "temporal" patterns regarding changes in properties as the graph grows over time: both the constant diameter and the densification power law patterns are similar to those observed in real-world graphs [58], and are not matched by most graph generators.

While Kronecker multiplication allows several patterns to be computed analytically, its discrete nature leads to "staircase effects" in the degree and spectral distributions. A modification of the aforementioned generator avoids these effects: instead of a 0/1 matrix, the initiator graph adjacency matrix is chosen to have *probabilities* associated with edges. The edges are then chosen based on these probabilities.

RTM: Recursive generator for weighted, evolving graphs. Akoglu et al. [5] extend the Kronecker model to allow for multi-edges, or weighted edges. To the initial adjacency matrix, another dimension, or mode, is added to represent time. Then, in each iteration the *Kronecker tensor product* of the graph is taken. This will produce a growing graph that is self-similar in structure.

Since it shares many properties of the Kronecker generator, all static properties as well as densification are followed. Additionally, the weight additions over time will also be self-similar, as shown in real graphs in [59]. It was also shown to mimic other patterns for weighted graphs, such as the Weight Power Law and Snapshot Power Laws, as discussed in the previous section.

3.5 Generators for specific graphs

Generators for the Internet Topology. While the generators described above are applicable to any graphs, some special-purpose generators have been proposed to specifically model the Internet topology. Structural generators exploit the hierarchical structure of the Internet, while the Inet generator modifies the basic preferential attachment model to better fit the Internet topology. We look at both of these below.

Structural Generators.

Problem being solved. Work done in the networking community on the structure of the Internet has led to the discovery of *hierarchies* in the topology. At the lowest level are the Local Area Networks (LANs); a group of LANs are connected by *stub domains*, and a set of *transit domains* connect the stubs and allow the flow of traffic between nodes from different stubs. However, the previous models do not explicitly enforce such hierarchies on the generated graphs.

Description and properties. Calvert et al. [26] propose a graph generation algorithm which specifically models this hierarchical structure. The general topology of a graph is specified by six parameters, which are the numbers of transit domains, stub domains and LANs, and the number of nodes in each. More parameters are needed to model the connectivities within and across these hierarchies. To generate a graph, points in a plane are used to represent the locations of the centers of the transit domains. The nodes for each of these domains are spread out around these centers, and are connected by edges. Now, the stub domains are placed on the plane and are connected to the corresponding transit node. The process is repeated with nodes representing LANs.

The authors provide two implementations of this idea. The first, called *Transit-Stub*, does not model LANs. Also, the method of generating connected subgraphs is to keep generating graphs till we get one that is connected. The second, called *Tiers*, allows multiple stubs and LANs, but allows only one transit domain. The graph is made connected by connecting nodes using a minimum spanning tree algorithm.

Open questions and discussion. These models can specifically match the hierarchical nature of the Internet, but they make no attempt to match any other graph pattern. For example, the degree distributions of the generated graphs need not be power laws. Also, the models use many parameters but provide only limited flexibility: what if we want a hierarchy with more than 3 levels? Hence, while these models have been widely used in the networking community, the need modifications to be as useful in other settings.

Tangmunarunkit et al. [78] compare such structural generators against generators which focus only on power-law distributions. They find that even though power-law generators do not explicitly model hierarchies, the graphs generated by them have a substantial level of hierarchy, though not as strict as with the generators described above. Thus, the hierarchical nature of the structural generators can also be mimicked by other generators.

The Inet topology generator.

Problem being solved.   Winick and Jamin [86] developed the Inet generator to model only the Internet Autonomous System (AS) topology, and to match features specific to it.

Description and properties.   Inet-2.2 generates the graph by the following steps:

- Each node is assigned a degree from a power-law distribution with an exponential cutoff (as in Equation 3.13).

- A spanning tree is formed from all nodes with degree greater than 1.

- All nodes with degree one are attached to his spanning tree using linear preferential attachment.

- All nodes in the spanning tree get extra edges using linear preferential attachment till they reach their assigned degree.

The main advantage of this technique is in ensuring that the final graph remains connected.

However, they find that under this scheme, too many of the low degree nodes get attached to other low-degree nodes. For example, in the Inet-2.2 topology, 35% of degree 2 nodes have adjacent nodes with degree 3 or less; for the Internet, this happens only for 5% of the degree-2 nodes. Also, the highest degree nodes in Inet-2.2 do not connect to as many low-degree nodes as the Internet. To correct this, Winick and Jamin come up with the Inet-3 generator, with a modified preferential attachment system.

The preferential attachment equation now has a weighting factor which uses the degrees of the nodes on both ends of some edge. The probability of a degree

i node connecting to a degree j node is

$$P(\text{degree } i \text{ node connects to degree } j \text{ node}) \propto w_i^j.j \qquad (3.23)$$

$$\text{where } w_i^j = MAX \left(1, \sqrt{\left(\log \frac{i}{j} \right)^2 + \left(\log \frac{f(i)}{f(j)} \right)^2} \right) \qquad (3.24)$$

Here, $f(i)$ and $f(j)$ are the number of nodes with degrees i and j respectively, and can be easily obtained from the degree distribution equation. Intuitively, what this weighting scheme is doing is the following: when the degrees i and j are close, the preferential attachment equation remains linear. However, when there is a large difference in degrees, the weight is the Euclidean distance between the points on the log-log plot of the degree distribution corresponding to degrees i and j, and this distance increases with increasing difference in degrees. Thus, edges connecting nodes with a big difference in degrees are preferred.

Open questions and discussion.   Inet has been extensively used in the networking literature. However, the fact that it is so specific to the Internet AS topology makes it somewhat unsuitable for any other topologies.

3.6   Graph Generators: A summary

We have seen many graph generators in the preceding pages. Is any generator the "best?" Which one should we use? The answer seems to depend on the application area: the *Inet* generator is specific to the Internet and can match its properties very well, the *BRITE* generator allows geographical considerations to be taken into account, "edge copying" models provide a good intuitive mechanism for modeling the growth of the Web along with matching degree distributions and community effects, and so on. However, the final word has not yet been spoken on this topic. Almost all graph generators focus on only one or two patterns, typically the degree distribution; there is a need for generators which can combine many of the ideas presented in this subsection, so that they can match most, if not all, of the graph patterns. R-MAT is a step in this direction.

4.   Conclusions

Naturally occurring graphs, perhaps collected from a variety of different sources, still tend to possess several common patterns. The most common of these are:

- Power laws, in degree distributions, in PageRank distributions, in eigenvalue-versus-rank plots and many others,

- Small diameters, such as the "six degrees of separation" for the US social network, 4 for the Internet AS level graph, and 12 for the Router level graph, and

- "Community" structure, as shown by high clustering coefficients, large numbers of bipartite cores, etc.

Graph generators attempt to create synthetic but "realistic" graphs, which can mimic these patterns found in real-world graphs. Recent research has shown that generators based on some very simple ideas can match some of the patterns:

- *Preferential attachment* Existing nodes with high degree tend to attract more edges to themselves. This basic idea can lead to power-law degree distributions and small diameter.

- *"Copying" models* Popular nodes get "copied" by new nodes, and this leads to power law degree distributions as well as a community structure.

- *Constrained optimization* Power laws can also result from optimizations of resource allocation under constraints.

- *Small-world models* Each node connects to all of its "close" neighbors and a few "far-off" acquaintances. This can yield low diameters and high clustering coefficients.

These are only some of the models; there are many other models which add new ideas, or combine existing models in novel ways. We have looked at many of these, and discussed their strengths and weaknesses. In addition, we discussed the recently proposed R-MAT model, which can match most of the graph patterns for several real-world graphs.

While a lot of progress has been made on answering these questions, a lot still needs to be done. More patterns need to be found; though there is probably a point of "diminishing returns" where extra patterns do not add much information, we do not think that point has yet been reached. Also, typical generators try to match only one or two patterns; more emphasis needs to be placed on matching the entire gamut of patterns. This cycle between finding more patterns and better generators which match these new patterns should eventually help us gain a deep insight into the formation and properties of real-world graphs.

Notes

1. Autonomous System, typically consisting of many routers administered by the same entity.
2. Tangmunarunkit et al. [78] use it only to differentiate between exponential and sub-exponential growth

References

[1] Lada A. Adamic and Bernardo A. Huberman. Power-law distribution of the World Wide Web. *Science*, 287:2115, 2000.

[2] Lada A. Adamic and Bernardo A. Huberman. The Web's hidden order. *Communications of the ACM*, 44(9):55–60, 2001.

[3] William Aiello, Fan Chung, and Linyuan Lu. A random graph model for massive graphs. In *ACM Symposium on Theory of Computing*, pages 171–180, New York, NY, 2000. ACM Press.

[4] William Aiello, Fan Chung, and Linyuan Lu. Random evolution in massive graphs. In *IEEE Symposium on Foundations of Computer Science*, Los Alamitos, CA, 2001. IEEE Computer Society Press.

[5] Leman Akoglu, Mary Mcglohon, and Christos Faloutsos. Rtm: Laws and a recursive generator for weighted time-evolving graphs. In *International Conference on Data Mining*, December 2008.

[6] Reka Albert and Albert-Laszlo Barabasi. Topology of evolving networks: local events and universality. *Physical Review Letters*, 85(24):5234–5237, 2000.

[7] Reka Albert and Albert-Laszlo Barabasi. Statistical mechanics of complex networks. *Reviews of Modern Physics*, 74(1):47–97, 2002.

[8] Reka Albert, Hawoong Jeong, and Albert-Laszlo Barabasi. Diameter of the World-Wide Web. *Nature*, 401:130–131, September 1999.

[9] Reka Albert, Hawoong Jeong, and Albert-Laszlo Barabasi. Error and attack tolerance of complex networks. *Nature*, 406:378–381, 2000.

[10] Lu"s A. Nunes Amaral, Antonio Scala, Marc Barthelemy, and H. Eugene Stanley. Classes of small-world networks. *Proceedings of the National Academy of Sciences*, 97(21):11149–11152, 2000.

[11] Ricardo Baeza-Yates and Barbara Poblete. Evolution of the Chilean Web structure composition. In *Latin American Web Congress*, Los Alamitos, CA, 2003. IEEE Computer Society Press.

[12] Albert-Laszlo Barabasi. *Linked: The New Science of Networks*. Perseus Books Group, New York, NY, first edition, May 2002.

[13] Albert-Laszlo Barabasi and Reka Albert. Emergence of scaling in random networks. *Science*, 286:509–512, 1999.

[14] Albert-Laszlo Barabasi, Hawoong Jeong, Z. Neda, Erzsebet Ravasz, A. Schubert, and Tamas Vicsek. Evolution of the social network of scientific collaborations. *Physica A*, 311:590–614, 2002.

[15] Jan Beirlant, Tertius de Wet, and Yuri Goegebeur. A goodness-of-fit statistic for Pareto-type behaviour. *Journal of Computational and Applied Mathematics*, 186(1):99–116, 2005.

[16] Noam Berger, Christian Borgs, Jennifer T. Chayes, Raissa M. D'Souza, and Bobby D. Kleinberg. Competition-induced preferential attachment. *Combinatorics, Probability and Computing*, 14:697–721, 2005.

[17] Zhiqiang Bi, Christos Faloutsos, and Flip Korn. The DGX distribution for mining massive, skewed data. In *Conference of the ACM Special Interest Group on Knowledge Discovery and Data Mining*, pages 17–26, New York, NY, 2001. ACM Press.

[18] Ginestra Bianconi and Albert-Laszle Barabasi. Competition and multi-scaling in evolving networks. *Europhysics Letters*, 54(4):436–442, 2001.

[19] Paolo Boldi, Bruno Codenotti, Massimo Santini, and Sebastiano Vigna. Structural properties of the African Web. In *International World Wide Web Conference*, New York, NY, 2002. ACM Press.

[20] Bela Bollobas. *Random Graphs*. Academic Press, London, 1985.

[21] Bela Bollobas, Christian Borgs, Jennifer T. Chayes, and Oliver Riordan. Directed scale-free graphs. In *ACM-SIAM Symposium on Discrete Algorithms*, Philadelphia, PA, 2003. SIAM.

[22] Bela Bollobas and Oliver Riordan. The diameter of a scale-free random graph. Combinatorica, 2002.

[23] Sergey Brin and Lawrence Page. The anatomy of a large-scale hyper-textual Web search engine. *Computer Networks and ISDN Systems*, 30(1–7):107–117, 1998.

[24] Andrei Z. Broder, Ravi Kumar, Farzin Maghoul, Prabhakar Raghavan, Sridhar Rajagopalan, Raymie Stata, Andrew Tomkins, and Janet Wiener. Graph structure in the web: experiments and models. In *International World Wide Web Conference*, New York, NY, 2000. ACM Press.

[25] Tian Bu and Don Towsley. On distinguishing between Internet power law topology generators. In *IEEE INFOCOM*, Los Alamitos, CA, 2002. IEEE Computer Society Press.

[26] Kenneth L. Calvert, Matthew B. Doar, and Ellen W. Zegura. Modeling Internet topology. *IEEE Communications Magazine*, 35(6):160–163, 1997.

[27] Jean M. Carlson and John Doyle. Highly optimized tolerance: A mechanism for power laws in designed systems. *Physical Review E*, 60(2):1412–1427, 1999.

[28] Deepayan Chakrabarti, Yiping Zhan, and Christos Faloutsos. R-MAT: A recursive model for graph mining. In *SIAM Data Mining Conference*, Philadelphia, PA, 2004. SIAM.

[29] Q. Chen, H. Chang, Ramesh Govindan, Sugih Jamin, Scott Shenker, and Walter Willinger. The origin of power laws in Internet topologies revisited.

In *IEEE INFOCOM*, Los Alamitos, CA, 2001. IEEE Computer Society Press.

[30] Colin Cooper and Alan Frieze. The size of the largest strongly connected component of a random digraph with a given degree sequence. *Combinatorics, Probability and Computing*, 13(3):319–337, 2004.

[31] Mark Crovella and Murad S. Taqqu. Estimating the heavy tail index from scaling properties. *Methodology and Computing in Applied Probability*, 1(1):55–79, 1999.

[32] Derek John de Solla Price. A general theory of bibliometric and other cumulative advantage processes. *Journal of the American Society for Information Science*, 27:292–306, 1976.

[33] Stephen Dill, Ravi Kumar, Kevin S. McCurley, Sridhar Rajagopalan, D. Sivakumar, and Andrew Tomkins. Self-similarity in the Web. In *International Conference on Very Large Data Bases*, San Francisco, CA, 2001. Morgan Kaufmann.

[34] Pedro Domingos and Matthew Richardson. Mining the network value of customers. In *Conference of the ACM Special Interest Group on Knowledge Discovery and Data Mining*, New York, NY, 2001. ACM Press.

[35] Sergey N. Dorogovtsev and José Fernando Mendes. *Evolution of Networks: From Biological Nets to the Internet and WWW*. Oxford University Press, Oxford, UK, 2003.

[36] Sergey N. Dorogovtsev, José Fernando Mendes, and Alexander N. Samukhin. Structure of growing networks with preferential linking. *Physical Review Letters*, 85(21):4633–4636, 2000.

[37] Sergey N. Dorogovtsev, José Fernando Mendes, and Alexander N. Samukhin. Giant strongly connected component of directed networks. *Physical Review E*, 64:025101 1–4, 2001.

[38] John Doyle and Jean M. Carlson. Power laws, Highly Optimized Tolerance, and Generalized Source Coding. *Physical Review Letters*, 84(24):5656–5659, June 2000.

[39] Nan Du, Christos Faloutsos, Bai Wang, and Leman Akoglu. Large human communication networks: patterns and a utility-driven generator. In *KDD '09: Proceedings of the 15th ACM SIGKDD international conference on Knowledge discovery and data mining*, pages 269–278, New York, NY, USA, 2009. ACM.

[40] Paul Erdős and Alfréd Rényi. On the evolution of random graphs. *Publication of the Mathematical Institute of the Hungarian Academy of Science*, 5:17–61, 1960.

[41] Paul Erdős and Alfréd Rényi. On the strength of connectedness of random graphs. *Acta Mathematica Scientia Hungary*, 12:261–267, 1961.

[42] Alex Fabrikant, Elias Koutsoupias, and Christos H. Papadimitriou. Heuristically Optimized Trade-offs: A new paradigm for power laws in the Internet. In *International Colloquium on Automata, Languages and Programming*, pages 110–122, Berlin, Germany, 2002. Springer Verlag.

[43] Michalis Faloutsos, Petros Faloutsos, and Christos Faloutsos. On power-law relationships of the Internet topology. In *Conference of the ACM Special Interest Group on Data Communications (SIGCOMM)*, pages 251–262, New York, NY, 1999. ACM Press.

[44] Andrey Feuerverger and Peter Hall. Estimating a tail exponent by modelling departure from a Pareto distribution. *The Annals of Statistics*, 27(2):760–781, 1999.

[45] Michael L. Goldstein, Steven A. Morris, and Gary G. Yen. Problems with fitting to the power-law distribution. *The European Physics Journal B*, 41:255–258, 2004.

[46] Ramesh Govindan and Hongsuda Tangmunarunkit. Heuristics for Internet map discovery. In *IEEE INFOCOM*, pages 1371–1380, Los Alamitos, CA, March 2000. IEEE Computer Society Press.

[47] Mark S. Granovetter. The strength of weak ties. *The American Journal of Sociology*, 78(6):1360–1380, May 1973.

[48] Bruce M. Hill. A simple approach to inference about the tail of a distribution. *The Annals of Statistics*, 3(5):1163–1174, 1975.

[49] George Karypis and Vipin Kumar. Multilevel algorithms for multi-constraint graph partitioning. Technical Report 98-019, University of Minnesota, 1998.

[50] Jon Kleinberg. Small world phenomena and the dynamics of information. In *Neural Information Processing Systems Conference*, Cambridge, MA, 2001. MIT Press.

[51] Jon Kleinberg, Ravi Kumar, Prabhakar Raghavan, Sridhar Rajagopalan, and Andrew Tomkins. The web as a graph: Measurements, models and methods. In *International Computing and Combinatorics Conference*, Berlin, Germany, 1999. Springer.

[52] Paul L. Krapivsky and Sidney Redner. Organization of growing random networks. *Physical Review E*, 63(6):066123 1–14, 2001.

[53] Ravi Kumar, Prabhakar Raghavan, Sridhar Rajagopalan, D. Sivakumar, Andrew Tomkins, and Eli Upfal. Stochastic models for the Web graph. In *IEEE Symposium on Foundations of Computer Science*, Los Alamitos, CA, 2000. IEEE Computer Society Press.

[54] Ravi Kumar, Prabhakar Raghavan, Sridhar Rajagopalan, and Andrew Tomkins. Extracting large-scale knowledge bases from the web. In *Inter-*

national Conference on Very Large Data Bases, San Francisco, CA, 1999. Morgan Kaufmann.

[55] Jure Leskovec, Deepayan Chakrabarti, Jon Kleinberg, Christos Faloutsos, and Zoubin Gharamani. Kronecker graphs: an approach to modeling networks, 2008.

[56] Jure Leskovec, Mary Mcglohon, Christos Faloutsos, Natalie Glance, and Matthew Hurst. Cascading behavior in large blog graphs. *SIAM International Conference on Data Mining (SDM)*, 2007.

[57] Jure Leskovec, Deepayan Chakrabarti, Jon Kleinberg, and Christos Faloutsos. Realistic, mathematically tractable graph generation and evolution, using Kronecker Multiplication. In *Conference on Principles and Practice of Knowledge Discovery in Databases*, Berlin, Germany, 2005. Springer.

[58] Jure Leskovec, Jon Kleinberg, and Christos Faloutsos. Graphs over time: Densification laws, shrinking diameters and possible explanations. In *Conference of the ACM Special Interest Group on Knowledge Discovery and Data Mining*, New York, NY, 2005. ACM Press.

[59] Mary Mcglohon, Leman Akoglu, and Christos Faloutsos. Weighted graphs and disconnected components: Patterns and a generator. In *ACM Special Interest Group on Knowledge Discovery and Data Mining (SIG-KDD)*, August 2008.

[60] Alberto Medina, Ibrahim Matta, and John Byers. On the origin of power laws in Internet topologies. In *Conference of the ACM Special Interest Group on Data Communications (SIGCOMM)*, pages 18–34, New York, NY, 2000. ACM Press.

[61] Milena Mihail and Christos H. Papadimitriou. On the eigenvalue power law. In *International Workshop on Randomization and Approximation Techniques in Computer Science*, Berlin, Germany, 2002. Springer Verlag.

[62] Michael Mitzenmacher. A brief history of generative models for power law and lognormal distributions. In *Proc. 39th Annual Allerton Conference on Communication, Control, and Computing*, Urbana-Champaign, IL, 2001. UIUC Press.

[63] Alan L. Montgomery and Christos Faloutsos. Identifying Web browsing trends and patterns. *IEEE Computer*, 34(7):94–95, 2001.

[64] M. E. J. Newman. Power laws, pareto distributions and zipf's law, December 2004.

[65] Mark E. J. Newman. The structure and function of complex networks. *SIAM Review*, 45:167–256, 2003.

[66] Mark E. J. Newman. Power laws, pareto distributions and Zipf's law. *Contemporary Physics*, 46:323–351, 2005.

[67] Mark E. J. Newman, Stephanie Forrest, and Justin Balthrop. Email networks and the spread of computer viruses. *Physical Review E*, 66(3):035101 1–4, 2002.

[68] Mark E. J. Newman, Michelle Girvan, and J. Doyne Farmer. Optimal design, robustness and risk aversion. *Physical Review Letters*, 89(2):028301 1–4, 2002.

[69] Mark E. J. Newman, Steven H. Strogatz, and Duncan J. Watts. Random graphs with arbitrary degree distributions and their applications. *Physical Review E*, 64(2):026118 1–17, 2001.

[70] Christine Nickel. *Random Dot Product Graphs: A Model for Social Networks*. PhD thesis, The Johns Hopkins University, 2007.

[71] Christopher Palmer, Phil B. Gibbons, and Christos Faloutsos. ANF: A fast and scalable tool for data mining in massive graphs. In *Conference of the ACM Special Interest Group on Knowledge Discovery and Data Mining*, New York, NY, 2002. ACM Press.

[72] Christopher Palmer and J. Gregory Steffan. Generating network topologies that obey power laws. In *IEEE Global Telecommunications Conference*, Los Alamitos, CA, November 2000. IEEE Computer Society Press.

[73] Gopal Pandurangan, Prabhakar Raghavan, and Eli Upfal. Using PageRank to characterize Web structure. In *International Computing and Combinatorics Conference*, Berlin, Germany, 2002. Springer.

[74] Romualdo Pastor-Satorras, Alexei Vasquez, and Alessandro Vespignani. Dynamical and correlation properties of the Internet. *Physical Review Letters*, 87(25):258701 1–4, 2001.

[75] David M. Pennock, Gary W. Flake, Steve Lawrence, Eric J. Glover, and C. Lee Giles. Winners don't take all: Characterizing the competition for links on the Web. *Proceedings of the National Academy of Sciences*, 99(8):5207–5211, 2002.

[76] Sidney Redner. How popular is your paper? an empirical study of the citation distribution. *The European Physics Journal B*, 4:131–134, 1998.

[77] Herbert Simon. On a class of skew distribution functions. *Biometrika*, 42(3/4):425–440, 1955.

[78] Hongsuda Tangmunarunkit, Ramesh Govindan, Sugih Jamin, Scott Shenker, and Walter Willinger. Network topologies, power laws, and hierarchy. Technical Report 01-746, University of Southern California, 2001.

[79] Sudhir L. Tauro, Christopher Palmer, Georgos Siganos, and Michalis Faloutsos. A simple conceptual model for the Internet topology. In *Global Internet*, Los Alamitos, CA, 2001. IEEE Computer Society Press.

[80] Jeffrey Travers and Stanley Milgram. An experimental study of the Small World problem. *Sociometry*, 32(4):425–443, 1969.

[81] Duncan J. Watts. *Six Degrees: The Science of a Connected Age.* W. W. Norton and Company, New York, NY, 1st edition, 2003.

[82] Duncan J. Watts, Peter Sheridan Dodds, and Mark E. J. Newman. Identity and search in social networks. *Science*, 296:1302–1305, 2002.

[83] Duncan J. Watts and Steven H. Strogatz. Collective dynamics of 'small-world' networks. *Nature*, 393:440–442, 1998.

[84] Bernard M. Waxman. Routing of multipoint connections. *IEEE Journal on Selected Areas in Communications*, 6(9):1617–1622, December 1988.

[85] H. S. Wilf. *Generating Functionology.* Academic Press, 1990.

[86] Jared Winick and Sugih Jamin. Inet-3.0: Internet Topology Generator. Technical Report CSE-TR-456-02, University of Michigan, Ann Arbor, 2002.

[87] Soon-Hyung Yook, Hawoong Jeong, and Albert-Laszlo Barabasi. Modeling the Internet's large-scale topology. *Proceedings of the National Academy of Sciences*, 99(21):13382–13386, 2002.

Chapter 4

QUERY LANGUAGE AND ACCESS METHODS FOR GRAPH DATABASES*

Huahai He*

Google Inc.
Mountain View, CA 94043, USA
huahai@google.com

Ambuj K. Singh

Department of Computer Science
University of California, Santa Barbara
Santa Barbara, CA 93106, USA
ambuj@cs.ucsb.edu

Abstract   With the prevalence of graph data in a variety of domains, there is an increasing need for a language to query and manipulate graphs with heterogeneous attributes and structures. We present a graph query language (GraphQL) that supports bulk operations on graphs with arbitrary structures and annotated attributes. In this language, graphs are the basic unit of information and each query manipulates one or more collections of graphs at a time. The core of GraphQL is a graph algebra extended from the relational algebra in which the selection operator is generalized to graph pattern matching and a composition operator is introduced for rewriting matched graphs. Then, we investigate access methods of the selection operator. Pattern matching over large graphs is challenging due to the NP-completeness of subgraph isomorphism. We address this by a combination of techniques: use of neighborhood subgraphs and profiles, joint reduction of the search space, and optimization of the search order. Experimental results on real and synthetic large graphs demonstrate that graph specific optimizations outperform an SQL-based implementation by orders of magnitude.

---

*This is a revised and extended version of the article "Graphs-at-a-time: Query Language and Access Methods for Graph Databases", Huahai He and Ambuj K. Singh, In Proceedings of the 2008 ACM SIGMOD Conference, http://doi.acm.org/10.1145/1376616.1376660. Reprinted with permission of ACM.
*Work done while at the University of California, Santa Barbara.

C.C. Aggarwal and H. Wang (eds.), *Managing and Mining Graph Data*,
Advances in Database Systems 40, DOI 10.1007/978-1-4419-6045-0_4,
© Springer Science+Business Media, LLC 2010

Keywords: Graph query language, Graph algebra, Graph pattern matching

1.   Introduction

Data in multiple domains can be naturally modeled as graphs. Examples include the Semantic Web [32], GIS, images [3], videos [24], social networks, Bioinformatics and Cheminformatics. Semantic Web standardizes information on the web as a graph with a set of entities and explicit relationships. In Bioinformatics, graphs represent several kinds of information: a protein structure can be modeled as a set of residues (nodes) and their spatial proximity (edges); a protein interaction network can be similarly modeled by a set of genes/proteins (nodes) and physical interactions (edges). In Cheminformatics, graphs are used to represent atoms and bonds in chemical compounds.

The growing heterogeneity and size of the above data has spurred interest in diverse applications that are centered on graph data. Existing data models, query languages, and database systems do not offer adequate support for the modeling, management, and querying of this data. There are a number of reasons for developing native graph-based data management systems. Considering expressiveness of queries: we need query languages that manipulate graphs in their full generality. This means the ability to define constraints (graph-structural and value) on nodes and edges *not* in an iterative one-node-at-a-time manner but simultaneously on the entire object of interest. This also means the ability to return a graph (or a set of graphs) as the result and not just a set of nodes. Another need for native graph databases is prompted by efficiency considerations. There are heuristics and indexing techniques that can be applied only if we operate in the domain of graphs.

1.1   Graphs-at-a-time Queries

Generally, a graph query takes a graph pattern as input, retrieves graphs from the database which contain (or are similar to) the query pattern, and returns the retrieved graphs or new graphs composed from the retrieved graphs. Examples of graph queries can be found in various domains:

- Find all heterocyclic chemical compounds that contain a given aromatic ring and a side chain. Both the ring and the side chain are specified as graphs with atoms as nodes and bonds as edges.

- Find all protein structures that contain the α-β-barrel motif [5]. This motif is specified as a cycle of β strands embraced by another cycle of α helices.

- Given a query protein complex from one species, is it functionally conserved in another species? The protein complex may be specified as a graph with nodes (proteins) labeled by Gene Ontology [14] terms.

- Find all instances from an RDF (Resource Description Framework [26]) graph where two departments of a company share the same shipping company. The query graph (of three nodes and two edges) has the constraints that nodes share the same company attribute and the edges are labeled by a "shipping" attribute. Report the result as a single graph with departments as nodes and edges between nodes that share a shipper.

- Find all co-authors from the DBLP dataset (a collection of papers represented as small graphs) in a specified set of conference proceedings. Report the results as a co-authorship graph.

As illustrated above, there is an increasing need for a language to query and manipulate graphs with heterogeneous attributes and structures. The language should be native to graphs, general enough to meet the heterogeneous nature of real world data, declarative, and yet implementable. Most importantly, a graph query language needs to support the following feature.

- Graphs should be the basic unit of information. The language should explicitly address graphs and queries should be graphs-at-a-time, taking one or more collections of graphs as input and producing a collection of graphs as output.

1.2   Graph Specific Optimizations

A graph query language is useful only if it can be efficiently implemented. This is especially important since one encounters the usual bottlenecks of subgraph isomorphism. As graphs are special cases of relations, graph queries can still be reduced to the relational model. However, the general-purpose relational model allows little opportunity for graph specific optimizations since it breaks down the graph structures into individual relations. Let us consider a simple example as follows. Figure 4.1 shows a graph query and a graph where each node has a single label as its attribute (nodes with the same label are distinguished by subscripts).

Consider an SQL-based approach to the sample graph query. The graph in the database can be modeled in two tables. Table V(vid, label) stores the set of nodes[1] where vid is the node identifier. Table E(vid1, vid2) stores the set of edges where vid1 and vid2 are end points of each edge. The graph query can then be expressed as an SQL query with multiple joins:

---

[1]For convenience, the terms "vertex" and "node" are used interchangeably in this chapter.

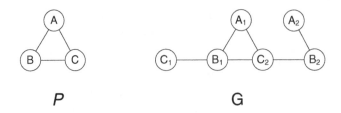

Figure 4.1. A sample graph query and a graph in the database

```
SELECT V1.vid, V2.vid, V3.vid
FROM   V AS V1, V AS V2, V AS V3,
       E AS E1, E AS E2, E AS E3
WHERE V1.label = 'A' AND V2.label = 'B' AND V3.label = 'C'
  AND V1.vid = E1.vid1 AND V1.vid = E3.vid1
  AND V2.vid = E1.vid2 AND V2.vid = E2.vid1
  AND V3.vid = E2.vid2 AND V3.vid = E3.vid2
  AND V1.vid <> V2.vid AND V1.vid <> V3.vid
  AND V2.vid <> V3.vid;
```

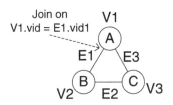

Figure 4.2. SQL-based implementation

As can be seen in the above example, although the graph query can be expressed by an SQL query, the global view of graph structures is lost. This prevents pruning of the search space that utilizes local or global graph structural information. For instance, nodes A_2 and C_1 in G can be safely pruned since they have only one neighbor. Node B_2 can also be pruned after A_2 is pruned. Furthermore, the SQL query involves many join operations. Traditional query optimization techniques such as dynamic programming do not scale well with the number of joins. This makes SQL-based implementations inefficient.

1.3   GraphQL

This chapter presents *GraphQL*, a graph query language in which graphs are the basic unit of information from the ground up. GraphQL uses a graph pattern as the main building block of a query. A graph pattern consists of a graph structure and a predicate on attributes of the graph. Graph pattern matching is defined by combining subgraph isomorphism and predicate evaluation. The core of GraphQL is a bulk *graph algebra* extended from the relational algebra in which the selection operator is generalized to graph pattern matching and a composition operator is introduced for rewriting matched graphs. In terms of expressive power, GraphQL is relationally complete and is contained in Datalog [28]. The nonrecursive version of GraphQL is equivalent to the relational algebra.

The chapter then describes efficient processing of the selection operator over large graph databases (either a single large graph or a large collection of graphs). We first present a basic graph pattern matching algorithm, and then apply three graph specific optimization techniques to the basic algorithm. The first technique prunes the search space locally using neighborhood subgraphs or their profiles. The second technique performs global pruning using an approximation algorithm called pseudo subgraph isomorphism [17]. The third technique optimizes the search order based on a cost model for graphs. Experimental study shows that the combination of these three techniques allows us to scale to both large queries and large graphs.

GraphQL has a number of distinct features:

1 Graph structures and structural operations are described by the notion of formal languages for graphs. This notion is useful for manipulating graphs and is the basis of the query language (Section 2).

2 A graph algebra is defined along the line of the relational algebra. Each graph algebraic operator manipulates graphs or sets of graphs. The graph algebra generalizes the selection operator to graph pattern matching and introduces a composition operator for rewriting matched graphs. In terms of expressive power, the graph algebra is relationally complete and is contained in Datalog (Section 3.3).

3 An efficient implementation of the selection operator over large graphs is presented. Experimental results on large real and synthetic graphs show that graph specific optimizations outperform an SQL-based implementation by orders of magnitude (Sections 4 and 5).

2.   Operations on Graph Structures

In order to define graph patterns and operations on graph structures, we need a formal way to describe graph structures and how they can be combined into new graph structures. As such we extend the notion of formal languages [20] from the string domain to the graph domain. The notion deals with graph structures only. Description of attributes on graphs will be discussed in the next section.

In existing formal languages (e.g., regular expressions, context-free languages), a formal grammar consists of a finite set of terminals and nonterminals, and a finite set of production rules. A production rule consists of a nonterminal on the left hand side and a sequence of terminals and nonterminals on the right hand side. The production rules are used to derive strings of characters. Strings are the basic units of information.

In a *formal language for graphs*, the basic units are graph structures instead of strings. The nonterminals, called *graph motifs*, are either simple graphs or composed of other graph motifs by means of *concatenation, disjunction,* or *repetition*. A *graph grammar* is a finite set of graph motifs. The *language* of a graph grammar is the set of all graphs derivable from graph motifs of that grammar.

A simple graph motif represents a graph with constant structure. It consists of a set of nodes and a set of edges. Each node, edge, or graph is identified by a *variable* if it needs to be referenced elsewhere. Figure 4.3 shows a simple graph motif and its graphical representation.

Figure 4.3. A simple graph motif

A complex graph motif consists of one or more graph motifs by concatenation, disjunction, or repetition. In the string domain, a string connects to other strings implicitly through its head and tail. In the graph domain, a graph may connect to other graphs in a structural way. These interconnections need to be explicitly specified.

2.1 Concatenation

A graph motif can be composed of two or more graph motifs. The constituent motifs are either left unconnected or concatenated in one of two ways. One way is to connect nodes in each motif by new edges. Figure 4.4(a) shows an example of concatenation by edges. Graph motif G_2 is composed of two motifs G_1 of Figure 4.3. The two motifs are connected by two edges. To avoid name conflicts, alias names of G_1 are used.

The other way of concatenation is to *unify* nodes in each motif. Two edges are unified automatically if their respective end nodes are unified. Figure 4.4(b) shows an example of concatenation by unification.

Concatenation is useful for defining Cartesian product and join operations on graphs.

2.2   Disjunction

A graph motif can be defined as a disjunction of two or more graph motifs. Figure 4.5 shows an example of disjunction. In graph motif G_4, two anonymous graph motifs are declared (comprising of node v_3 or nodes v_3 and v_4). Only one of them is selected and connected to the rest of G_4. In disjunction, all the constituent graph motifs should have the same "interface" to the outside.

2.3   Repetition

A graph motif may be defined by itself to derive recursive graph structures. Figure 4.6(a) shows the construction of a path and a cycle. In the base case, the path has two nodes and one edge. In the recurrence step, the path contains itself as a member, adds a new node v_1 which connects to v_1 of the nested path, and exports the nested v_2 so that the new path has the same "interface." The keyword **"export"** is equivalent to declaring a new node and unifying it with the nested node. Graph motif $Cycle$ is composed of motif $Path$ with an additional edge that connects the end nodes of the $Path$.

Recursions in the graph domain are not limited to paths and cycles. Figure 4.6(b) illustrates an example where the repetition unit is a graph motif. Motif G_5 contains an arbitrary number of motif G_1 and a root node v_0. The

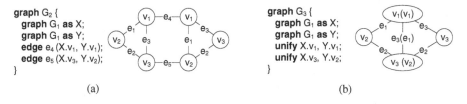

```
graph G2 {
    graph G1 as X;
    graph G1 as Y;
    edge e4 (X.v1, Y.v1);
    edge e5 (X.v3, Y.v2);
}
```

(a)

```
graph G3 {
    graph G1 as X;
    graph G1 as Y;
    unify X.v1, Y.v1;
    unify X.v3, Y.v2;
}
```

(b)

Figure 4.4. (a) Concatenation by edges, (b) Concatenation by unification

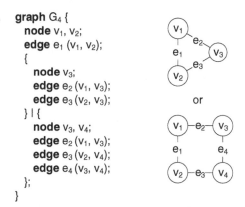

```
graph G4 {
    node v1, v2;
    edge e1 (v1, v2);
    {
        node v3;
        edge e2 (v1, v3);
        edge e3 (v2, v3);
    } | {
        node v3, v4;
        edge e2 (v1, v3);
        edge e3 (v2, v4);
        edge e4 (v3, v4);
    };
}
```

or

Figure 4.5. Disjunction

declaration recursively contains G_5 itself and a new G_1, with $G_1.v_1$ connected to v_0, where v_0 is exported from the nested G_5. The first resulting graph consists of node v_0 alone, the second consists of node v_0 connected to G_1 through edge e_1, the third consists of node v_0 connected to two instances of G_1 through edge e_1, and so on.

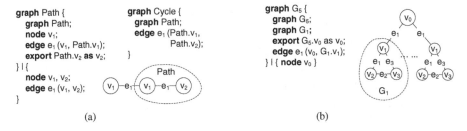

(a)   (b)

Figure 4.6. (a) Path and cycle, (b) Repetition of motif G_1

3.   Graph Query Language

This section presents the GraphQL query language. We first describe the data model. Next, we define graph patterns and graph pattern matching. We then present a graph algebra and its bulk operators which is the core of the graph query language. Finally, we illustrate the syntax of the graph query language through an example.

3.1   Data Model

Graphs in the real world contain not only graph structural information, but also attributes on nodes and edges. In GraphQL, we use a *tuple*, a list of name and value pairs, to represent the attributes of each node, edge, or graph. A tuple may have an optional *tag* that denotes the tuple type. Tuples are *annotated* to the graph structures so that the representations of attributes and structures are clearly separate. Figure 4.7 shows a sample graph that represents a paper (the graph has no edges). Node v_1 has two attributes "title" and "year". Nodes v_2 and v_3 have a tag "author" and an attribute "name".

```
graph G <inproceedings> {
    node v₁ <title="Title1", year=2006>;
    node v₂ <author name="A">;
    node v₃ <author name="B">;
};
```

Figure 4.7. A sample graph with attributes

In the relational model, tuples are the basic unit of information. Each algebraic operator manipulates collections of tuples. A relational query is always equivalent to an algebraic expression which is a combination of the operators. A relational database consists of one or more tables (relations) of tuples.

In GraphQL, graphs are the basic unit of information. Each operator takes one or more collections of graphs as input and generates a collection of graphs as output. A graph database consists of one or more collections of graphs. Unlike the relational model, graphs in a collection do not necessarily have identical structures and attributes. However, they can still be processed in a uniform way by binding to a graph pattern.

The GraphQL data model is similar to the TAX model [22] as for XML. In TAX, trees are the basic unit and the operators work on collections of trees. Trees in a collection have similar but not identical structures and attributes. This is captured by a pattern tree.

3.2 Graph Patterns

A graph pattern is the main building block of a graph query. Essentially, it consists of a graph motif and a predicate on attributes of the motif. The graph motif specifies constraints on graph structures and the predicate specifies constraints on attributes. A graph pattern is used to select graphs of interest.

Definition 4.1. *(Graph Pattern) A graph pattern is a pair* $\mathcal{P} = (\mathcal{M}, \mathcal{F})$, *where* $\mathcal{M}$ *is a graph motif and* $\mathcal{F}$ *is a predicate on the attributes of the motif.*

The predicate $\mathcal{F}$ is a combination of boolean or arithmetic comparison expressions. Figure 4.8 shows a sample graph pattern. The predicate can be broken down to predicates on individual nodes or edges, as shown on the right side of the figure.

```
graph P {                              graph P {
    node v₁;                 or            node v₁ where name="A";
    node v₂;                               node v₂ where year>2000;
} where v₁.name="A"                    };
  and v₂.year>2000;
```

Figure 4.8. A sample graph pattern

Next, we define the notion of graph pattern matching which generalizes subgraph isomorphism with evaluation of the predicate.

Definition 4.2. *(Graph Pattern Matching) A graph pattern* $\mathcal{P}(\mathcal{M}, \mathcal{F})$ *is matched with a graph G if there exists an injective mapping* ϕ: $V(\mathcal{M}) \rightarrow V(G)$ *such that i) For* $\forall\, e(u, v) \in E(\mathcal{M})$, $(\phi(u), \phi(v))$ *is an edge in G, and ii) predicate* $\mathcal{F}_\phi(G)$ *holds.*

A graph pattern is recursive if its motif is recursive (see Section 2.3). A recursive graph pattern is matched with a graph if one of its derived motifs is matched with the graph.

$$\text{Mapping } \Phi{:}$$
$$\Phi(P.\mathsf{v}_1) \to \mathsf{G}.\mathsf{v}_2$$
$$\Phi(P.\mathsf{v}_2) \to \mathsf{G}.\mathsf{v}_1$$

Figure 4.9. A mapping between the graph pattern in Figure 4.8 and the graph in Figure 4.7

Figure 4.9 shows an example of graph pattern matching between the pattern in Figure 4.8 and the graph in Figure 4.7.

If a graph pattern is matched to a graph, the binding between them can be used to access the graph (either graph structural information or attributes on the graph). As a graph pattern can match many graphs, this allows us to access a collection of graphs uniformly even though the graphs may have heterogenous structures and attributes. We use a *matched graph* to denote the binding between a graph pattern and a graph.

Definition 4.3. *(Matched Graph) Given an injective mapping ϕ between a pattern P and a graph G, a matched graph is a triple $\langle \phi, P, G \rangle$ and is denoted by $\phi_P(G)$.*

Although a matched graph is formally defined by a triple, it has all characteristics of a graph. Thus, all terms and conditions that apply to a graph also apply to a matched graph. For example, a collection of matched graphs is also a collection of graphs. As such it can match another graph pattern, resulting in another collection of matched graphs (two levels of bindings).

A graph pattern can match a graph in multiple places, resulting in multiple bindings (matched graphs). This is considered further when we discuss the selection operator in Section 3.3.0.

3.3   Graph Algebra

We define a graph algebra along the lines of the relational algebra. This allows us to inherit the solid foundation and experience of the relational model. All relational operators have their counterparts or alternatives in the graph algebra. These operators are defined directly on graphs since graphs are now the basic units of information. In particular, the selection operator is generalized to graph pattern matching; a composition operator is introduced to generate new graphs from matched graphs.

Selection (σ).   A selection operator σ takes a graph pattern P and a collection of graphs C as arguments, and produces a collection of matched graphs as output. The result is denoted by $\sigma_P(C)$:

$$\sigma_P(C) = \{\phi_P(G) \mid G \in C\}$$

A graph database may consist of a single large graph, e.g., a social network. A single large graph and a collection of graphs are treated in the same way. A collection of graphs is a special case of a single large graph, whereas a single large graph is considered as many inter-connected or overlapping small graphs. These small graphs are captured by the graph pattern of the selection operator.

A graph pattern can match a graph many times. Thus, a selection could return many instances for each graph in the input collection. We use an option "exhaustive" to specify whether it should return one or all possible mappings between the graph pattern and the graph. Whether one or all mappings are required depends on the application.

Cartesian Product ($\times$) and Join ($\bowtie$). A Cartesian product operator takes two collections of graphs C and D as input, and produces a collection of graphs as output. Each graph in the output collection is composed of a graph from C and another from D. The constituent graphs are unconnected:

$$C \times D = \{ \text{ graph } \{ \text{ graph } G_1, G_2; \} \mid G_1 \in C, G_2 \in D \}$$

As in the relational algebra, the join operator in the graph algebra can be defined by a Cartesian product followed by a selection:

$$C \bowtie_P D = \sigma_P(C \times D)$$

In a *valued join*, the join condition is a predicate on attributes of the constituent graphs. The constituent graphs are unconnected in the resultant graph. No new graph structures are generated. Figure 4.10 shows an example of valued join.

> **graph** {
>   **graph** G₁, G₂;
> } **where** G₁.id = G₂.id;

Figure 4.10. An example of valued join

In a *structural join*, the constituent graphs can be concatenated by edges or unification. New graph structures are generated in the resultant graph. This is specified through a composition operator which is described next.

Composition (ω). Composition operators are used to generate new graphs from existing (matched) graphs. In order to specify the composition operators, we introduce the concept of graph templates.

Definition 4.4. *(Graph Template) A graph template T consists of a list of formal parameters which are graph patterns, and a template body which is defined by referring to the graph patterns.*

Once actual parameters (matched graphs) are given, a graph template is *instantiated* to a real graph. This is similar to invoking a function: the template body is the function body; the graph patterns are the formal parameters; the matched graphs are the actual parameters. The resulting graph can be denoted by $\mathcal{T}_{\mathcal{P}_1..\mathcal{P}_k}(G_1, ..., G_k)$.

T_P = graph {
    node v_1 <label=P.v_1.name>;
    node v_2 <label=P.v_2.title>;
    edge e_1 (v_1, v_2);
}
	T_P (G) = graph {
	    node v_1 <label="A">;
	    node v_2 <label="Title1">;
	    edge e_1 (v_1, v_2);
	}

(a)	(b)

Figure 4.11. (a) A graph template with a single parameter $\mathcal{P}$, (b) A graph instantiated from the graph template. $\mathcal{P}$ and G are shown in Figure 4.8 and Figure 4.7.

Figure 4.11 shows a sample graph template and a graph instantiated from the graph template. $\mathcal{P}$ is the formal parameter of the template. The template body consists of two nodes constructed from $\mathcal{P}$ and an edge between them. Given the actual parameter G, the template is instantiated to a graph.

Now we can define the composition operator. A *primitive composition* operator ω takes a graph template $\mathcal{T}_{\mathcal{P}}$ with a single parameter, and a collection of matched graphs $\mathcal{C}$ as input. It produces a collection of instantiated graphs as output:

$$\omega_{\mathcal{T}_{\mathcal{P}}}(\mathcal{C}) = \{\mathcal{T}_{\mathcal{P}}(G) \mid G \in \mathcal{C}\}$$

Generally, a composition operator allows two or more collections of graphs as input. This can be expressed by a primitive composition operator and a Cartesian product operator, the latter of which combines multiple collections of graphs into one:

$$\omega_{\mathcal{T}_{\mathcal{P}_1, \mathcal{P}_2}}(\mathcal{C}_1, \mathcal{C}_2) = \omega_{\mathcal{T}_{\mathcal{P}}}(\mathcal{C}_1 \times \mathcal{C}_2),$$

where $\mathcal{P} = $ **graph** { **graph** $\mathcal{P}_1$, $\mathcal{P}_2$; }.

Other operators. Projection and Renaming, two other operators of the relational algebra, can be expressed using the composition operator. The set operators (union, difference, intersection) can also be defined easily. In terms of expressive power, the five basic operators (selection, Cartesian product, primitive composition, union, and difference) are complete. Other operators and any algebraic expressions can be expressed as combinations of these five operators.

Algebraic laws are important for query optimization as they provide equivalent transformations of query plans. Since the graph algebra is defined along the lines of the relational algebra, laws of relational algebra carry over.

3.4   FLWR Expressions

We adopt the FLWR (For, Let, Where, and Return) expressions in XQuery [4] as the syntax of our graph query language. The query syntax is shown in Appendix 4.A. We illustrate the syntax through an example.

```
graph P {
    node v₁ <author>;
    node v₂ <author>;
} where P.booktitle="SIGMOD";
C:= graph {};
for P exhaustive in doc("DBLP")
let C:= graph {
    graph C;
    node P.v₁, P.v₂;
    edge e₁ (P.v₁, P.v₂);
    unify P.v₁, C.v₁ where P.v₁.name=C.v₁.name;
    unify P.v₂, C.v₂ where P.v₂.name=C.v₂.name;
}
```

Figure 4.12. A graph query that generates a co-authorship graph from the DBLP dataset

Figure 4.12 shows an example that generates a co-authorship graph C from a collection of papers. The query states that any pair of authors in a paper should appear in the co-authorship graph with an edge between them. The graph pattern P matches a pair of authors in a paper. The for clause selects all such pairs from the data source. The let clause places each pair in the co-authorship graph and adds an edge between them. The unifications ensure that each author appears only once. Again, two edges are unified automatically if their end nodes are unified.

Figure 4.13 shows a running example of the query. The DBLP collection consists of two graphs G_1 and G_2. The pair of author nodes (A, B) is first chosen and an edge is inserted between them. The pair (C, D) is chosen next and the (C, D) subgraph is inserted. When the third pair (A, C) is chosen, unification ensures that the old nodes are reused and an edge is added between existing A and C. The processing of the fourth pair adds one more edge and completes the execution.

The query can be translated into a recursive algebraic expression:

$$C = \sigma_J(\omega_{\tau_{P,C}}(\sigma_P(\text{``DBLP''}), \{C\}))$$

where $\sigma_P(\text{``DBLP''})$ corresponds to the for clause, $\tau_{P,C}$ is the graph template in the let clause, and J is a graph pattern for the join condition: $P.v_1.name = C.v_1.name$ & $P.v_2.name = C.v_2.name$. The algebraic expression turns out to be a structural join that consists of three primitive operators: Cartesian product, primitive composition, and selection.

```
DBLP: graph G₁ {
         node v₁ <author name="A">;
         node v₂ <author name="B">;
      };
      graph G₂ {
         node v₁ <author name="C">;
         node v₂ <author name="D">;
         node v₃ <author name="A">;
      };
```

Iteration	Mapping	Co-authorship graph C
1	$\Phi(P.v_1) \rightarrow G_1.v_1$ $\Phi(P.v_2) \rightarrow G_1.v_2$	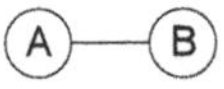 (A)——(B)
2	$\Phi(P.v_1) \rightarrow G_2.v_1$ $\Phi(P.v_2) \rightarrow G_2.v_2$	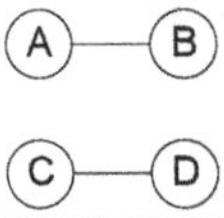 (A)——(B) (C)——(D)
3	$\Phi(P.v_1) \rightarrow G_2.v_1$ $\Phi(P.v_2) \rightarrow G_2.v_3$	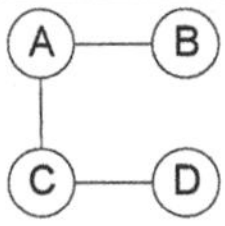 (A)——(B) | (C)——(D)
4	$\Phi(P.v_1) \rightarrow G_2.v_2$ $\Phi(P.v_2) \rightarrow G_2.v_3$	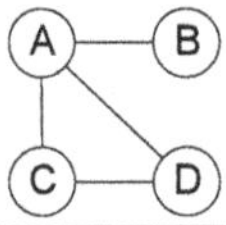 (A)——(B) (C)——(D)

Figure 4.13. A possible execution of the Figure 4.12 query

3.5   Expressive Power

We now discuss the expressive power of GraphQL. We first show that the relational algebra (RA) is contained in GraphQL.

Theorem 4.5. *(RA ⊆ GraphQL) For any RA expression, there exists an equivalent GraphQL algebra expression.*

Proof: We can represent a relation (tuple) in GraphQL using a graph that has a single node with attributes as the tuple. The primitive operations of RA (selection, projection, Cartesian product, union, difference) can then be expressed in GraphQL. The selection operator can be simulated using a graph pattern with the given predicate as the selection condition. For projection, one rewrites the projected attributes to a new node using the composition operator. Other operations (product, union, difference) are straightforward as well.   □

Next, we show that GraphQL is contained in Datalog. This is proved by translating graphs, graph patterns, and graph templates into facts and rules of Datalog.

Theorem 4.6. *(GraphQL $\subseteq$ Datalog) For any GraphQL algebra expression, there exists an equivalent Datalog program.*

Proof: We first translate all graphs of the database into facts of Datalog. Figure 4.14 shows an example of the translation. Essentially, we rewrite each variable of the graph as a unique constant string, and then establish a connection between the graph and each node and edge. Note that for undirected graphs, we need to write an edge twice to permute its end nodes.

```
graph G <attr1=value1> {            graph('G').
    node v₁, v₂, v₃;                node('G', 'G.v₁').
    edge e₁(v₁, v₂);                node('G', 'G.v₂').
};                                  node('G', 'G.v₃').
                         ⟹         edge('G', 'G.e₁', 'G.v₁', 'G.v₂').
                                    edge('G', 'G.e₁', 'G.v₂', 'G.v₁').
                                    attribute('G', 'attr1', value1).
```

Figure 4.14. The translation of a graph into facts of Datalog

For each graph pattern, we translate it into a rule of Datalog. Figure 4.15 gives an example of such translation. The body of the rule is a conjunction of the constituent elements of the graph pattern. The predicate of the graph pattern is written naturally. It can then be shown that a graph pattern matches a graph if and only if the corresponding rule matches the facts that represent the graph.

Subsequently, one can translate the graph algebraic operations into Datalog in a way similar to translating RA into Datalog. Thus, we can translate any GraphQL algebra expression into an equivalent Datalog program. □

```
graph P {                           Pattern(P, V₂, V₃, E₁):-
    node v₂, v₃;                        graph(P),
    edge e₁(v₃, v₂);                    node(P, V₂),
} where P.attr1 > value1;   ⟹         node(P, V₃),
                                        edge(P, E₁, V₃, V₂),
                                        attribute(P, 'attr1', Temp),
                                        Temp > value1.
```

Figure 4.15. The translation of a graph pattern into a rule of Datalog

It is well known that nonrecursive Datalog (nr-Datalog) is equivalent to RA. Consequently, the nonrecursive version of GraphQL (nr-GraphQL) is also equivalent to RA.

Corollary 4.7. *nr-GraphQL $\equiv$ RA.*

4.   Implementation of the Selection Operator

We now discuss efficient implementation of the selection operator. Other graph algebraic operators can find their counterpart implementations in relational databases, and future research opportunities are open for graph specific optimizations.

Generally, graph databases can be classified into two categories. One category is a large collection of small graphs, e.g., chemical compounds. The selection operator returns a subset of the collection as answers. The main challenge in this category is to reduce the number of pairwise graph pattern matchings. A number of graph indexing techniques have been proposed to address this challenge [17, 34, 40]. Graph indexing plays a similar role for graph databases as B-trees for relational databases: only a small number of graphs need to be accessed. Scanning of the whole collection of graphs is not necessary.

In the second category, the graph database consists of one or a few very large graphs, e.g., protein interaction networks, Web information, social networks. Graphs in the answer set are not readily present in the database and need to be constructed from the single large graph. The challenge here is to accelerate the graph pattern matching itself. In this chapter, we focus on the second category.

We first describe the basic graph pattern matching algorithm in Section 4.1, and then discuss accelerations to the basic algorithm in Sections 4.2, 4.3, and 4.4. We restrict our attention to nonrecursive graph patterns and in-memory processing. Recursive graph pattern matching and disk-based access methods remain as future research directions.

4.1   Graph Pattern Matching

Graph pattern matching is essentially an extension of subgraph isomorphism with predication evaluation (Definition 4.2). Algorithm 4.1 outlines the basic graph pattern matching algorithm.

The predicate of graph pattern $\mathcal{P}$ is rewritten as predicates on individual nodes $\mathcal{F}_u$'s and edges $\mathcal{F}_e$'s. Predicates that cannot be pushed down, e.g., "$u_1.label = u_2.label$", remain in the graph-wide predicate $\mathcal{F}$. For each node u in pattern $\mathcal{P}$, there is a set of candidate matched nodes in G with respect to $\mathcal{F}_u$. These nodes are called *feasible mates* of node u and is denoted by $\Phi(u)$:

Definition 4.8. *(Feasible Mates) The feasible mates $\Phi(u)$ of node u is the set of nodes in graph G that satisfies predicate F_u:*

$$\Phi(u) = \{v | v \in V(G), \mathcal{F}_u(v) = \textbf{\textit{true}}\}.$$

The feasible mates of all nodes in the pattern define the search space of graph pattern matching:

Definition 4.9. *(Search Space) The search space of a graph pattern matching is defined as the product of feasible mates for each node of the graph pattern:*

$$\Phi(u_1) \times .. \times \Phi(u_k),$$

where k is the number of nodes in the graph pattern.

Algorithm 4.1: Graph Pattern Matching

Input: Graph Pattern $\mathcal{P}$, Graph G
Output: One or all feasible mappings $\phi_{\mathcal{P}}(G)$

1 **foreach** *node $u \in V(\mathcal{P})$* **do**
2     $\Phi(u) \leftarrow \{v | v \in V(G), \mathcal{F}_u(v) = \textbf{true}\}$
3   // Local pruning and retrieval of $\Phi(u)$ (Section 4.2)
4 **end**
5 // Reduce $\Phi(u_1) \times .. \times \Phi(u_k)$ globally (Section 4.3)
6 // Optimize search order of $u_1, .., u_k$ (Section 4.4)
7 Search(1);

8 **void** Search(i)
9 **begin**
10   **foreach** $v \in \Phi(u_i)$, *v is free* **do**
11   **if not** *Check(u_i, v)* **then continue**;
12         $\phi(u_i) \leftarrow v$;
13   **if** $i < |V(\mathcal{P})|$ **then** Search($i + 1$);
14   **else if** $\mathcal{F}_\phi(G)$ **then**
15   Report ϕ ;
16   **if not** *exhaustive* **then stop**;
17   **end**
18 **end**

19 **boolean** Check(u_i, v)
20 **begin**
21   **foreach** *edge $e(u_i, u_j) \in E(\mathcal{P}), j < i$* **do**
22   **if** *edge $e'(v, \phi(u_j)) \notin E(G)$* **or not** $\mathcal{F}_e(e')$ **then**
23   **return false**;
24   **end**
25   **return true**;
26 **end**

Algorithm 4.1 consists of two phases. The first phase (lines 1–4) retrieves the feasible mates for each node u in the pattern. The second phase (Lines 7–26) searches over the product $\Phi(u_1) \times .. \times \Phi(u_k)$ in a depth-first manner for subgraph isomorphism. Procedure Search(i) iterates on the ith node to find feasible mappings for that node. Procedure Check(u_i, v) examines if u_i can be mapped to v by considering their edges. Line 12 maps u_i to v. Lines 13–16 continue to search for the next node or if it is the last node, evaluate the graph-wide predicate. If it is true, then a feasible mapping $\phi : V(\mathcal{P}) \to V(G)$ has been found and is reported (line 15). Line 16 stops searching immediately if only one mapping is required.

The graph pattern and the graph are represented as a vertex set and an edge set, respectively. In addition, adjacency lists of the graph pattern are used to support line 21. For line 22, edges of graph G can be represented in a hashtable where keys are pairs of the end points. To avoid repeated evaluation of edge predicates (line 22), another hashtable can be used to store evaluated pairs of edges.

The worst-case time complexity of Algorithm 4.1 is $O(n^k)$ where n and k are the sizes of graph G and graph pattern $\mathcal{P}$, respectively. This complexity is a consequence of subgraph isomorphism that is known to be NP-hard. In practice, the running time depends on the size of the search space.

We now consider possible ways to accelerate Algorithm 4.1:

1 How to reduce the size of $\Phi(u_i)$ for each node u_i? How to efficiently retrieve $\Phi(u_i)$?

2 How to reduce the overall search space $\Phi(u_1) \times .. \times \Phi(u_k)$?

3 How to optimize the search order?

We present three techniques that respectively address the above questions. The first technique prunes each $\Phi(u_i)$ individually and retrieves it efficiently through indexing. The second technique prunes the overall search space by considering all nodes in the pattern simultaneously. The third technique applies ideas from traditional query optimization to find the right search order.

4.2   Local Pruning and Retrieval of Feasible Mates

Node attributes can be indexed directly using traditional index structures such as B-trees. This allows for fast retrieval of feasible mates and avoids a full scan of all nodes. To reduce the size of feasible mates $\Phi(u_i)$'s even further, we can go beyond nodes and consider neighborhood subgraphs of the nodes. The neighborhood information can be exploited to prune infeasible mates at an early stage.

Definition 4.10. *(Neighborhood Subgraph) Given graph G, node v and radius r, the neighborhood subgraph of node v consists of all nodes within distance r (number of hops) from v and all edges between the nodes.*

Node v is a feasible mate of node u_i only if the neighborhood subgraph of u_i is sub-isomorphic to that of v (with u_i mapped to v). Note that if the radius is 0, then the neighborhood subgraphs degenerate to nodes.

Although neighborhood subgraphs have high pruning power, they incur a large computation overhead. This overhead can be reduced by representing neighborhood subgraphs by their light-weight *profiles*. For instance, one can define the profile as a sequence of the node labels in lexicographic order. The pruning condition then becomes whether a profile is a subsequence of the other.

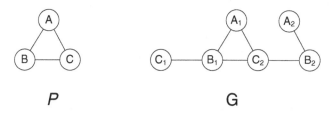

P   *G*

Figure 4.16. A sample graph pattern and graph

Nodes of G	Neighborhood subgraphs of radius 1	Profiles	Search space
A_1		ABC	Retrieve by nodes: $\{A_1, A_2\} \times \{B_1, B_2\} \times \{C_1, C_2\}$
A_2		AB	Retrieve by neighborhood subgraphs: $\{A_1\} \times \{B_1\} \times \{C_2\}$
B_1		ABCC	Retrieve by profiles of neighborhood subgraphs: $\{A_1\} \times \{B_1, B_2\} \times \{C_2\}$
B_2		ABC	
C_1		BC	
C_2		ABBC	

Figure 4.17. Feasible mates using neighborhood subgraphs and profiles. The resulting search spaces are also shown for different pruning techniques.

Figure 4.16 shows the sample graph pattern $\mathcal{P}$ and the database graph G again for convenience. Figure 4.17 shows the neighborhood subgraphs of radius 1 and their profiles for nodes of G. If the feasible mates are retrieved using node attributes, then the search space is $\{A_1, A_2\} \times \{B_1, B_2\} \times \{C_1, C_2\}$. If the feasible mates are retrieved using neighborhood subgraphs, then the search space is $\{A_1\} \times \{B_1\} \times \{C_2\}$. Finally, if the feasible mates are retrieved using profiles, then the search space is $\{A_1\} \times \{B_1, B_2\} \times \{C_2\}$. These are shown in the right side of Figure 4.17.

If the node attributes are selective, e.g., many unique attribute values, then one can index the node attributes using a B-tree or hashtable, and store the neighborhood subgraphs or profiles as well. Retrieval is done by indexed access to the node attributes, followed by pruning using neighborhood subgraphs or profiles. Otherwise, if the node attributes are not selective, one may have to index the neighborhood subgraphs or profiles. Recent graph indexing techniques [9, 17, 23, 34, 36, 39–42] or multi-dimensional indexing methods such as R-trees can be used for this purpose.

4.3   Joint Reduction of Search Space

We reduce the overall search space iteratively by an approximation algorithm called Pseudo Subgraph Isomorphism [17]. This prunes the search space by considering the whole pattern and the space $\Phi(u_1) \times .. \times \Phi(u_k)$ simultaneously. Essentially, this technique checks for each node u in pattern $\mathcal{P}$ and its feasible mate v in graph G whether the adjacent subtree of u is sub-isomorphic to that of v. The check can be defined recursively on the depth of the adjacent subtrees: the level l subtree of u is sub-isomorphic to that of v only if the level $l - 1$ subtrees of u's neighbors can all be matched to those of v's neighbors. To avoid subtree isomorphism tests, a bipartite graph $\mathcal{B}_{u,v}$ is defined between neighbors of u and v. If the bipartite graph has a semi-perfect matching, i.e., all neighbors of u are matched, then u is level l sub-isomorphic to v. In the bipartite graph, an edge is present between two nodes u' and v' only if the level $l - 1$ subtree of u' is sub-isomorphic to that of v', or equivalently the bipartite graph $\mathcal{B}_{u',v'}$ at level $l - 1$ has a semi-perfect matching. A more detailed description can be found in [17].

Algorithm 4.2 outlines the refinement procedure. At each iteration (lines 3–20), a bipartite graph $\mathcal{B}_{u,v}$ is constructed for each u and its feasible mate v (lines 5–9). If $\mathcal{B}_{u,v}$ has no semi-perfect matching, then v is removed from $\Phi(u)$, thus reducing the search space (line 13).

The algorithm has two implementation improvements on the refinement procedure discussed in [17]. First, it avoids unnecessary bipartite matchings. A pair $\langle u, v \rangle$ is marked if it needs to be checked for semi-perfect matching (lines 2, 4). If the semi-perfect matching exists, then the pair is unmarked (lines 10–11). Otherwise, the removal of v from $\Phi(u)$ (line 13) may affect the existence of semi-perfect matchings of the neighboring $\langle u', v' \rangle$ pairs. As a result,

---

Algorithm 4.2: Refine Search Space

Input: Graph Pattern $\mathcal{P}$, Graph G, Search space $\Phi(u_1) \times .. \times \Phi(u_k)$,
level l
Output: Reduced search space $\Phi'(u_1) \times .. \times \Phi'(u_k)$

1 **begin**
2   **foreach** $u \in \mathcal{P}, v \in \Phi(u)$ **do** Mark $\langle u, v \rangle$;
3   **for** $i \leftarrow 1$ *to* l **do**
4   **foreach** $u \in \mathcal{P}, v \in \Phi(u)$, $\langle u, v \rangle$ *is marked* **do**
5   //Construct bipartite graph $\mathcal{B}_{u,v}$
6        $N_{\mathcal{P}}(u), N_G(v)$: neighbors of u, v;
7   **foreach** $u' \in N_{\mathcal{P}}(u), v' \in N_G(v)$ **do**
8          $\mathcal{B}_{u,v}(u', v') \leftarrow \begin{cases} 1 & \text{if } v' \in \Phi(u'); \\ 0 & \text{otherwise.} \end{cases}$
9   **end**
10   **if** $\mathcal{B}_{u,v}$ *has a semi-perfect matching* **then**
11   Unmark $\langle u, v \rangle$;
12   **else**
13   Remove v from $\Phi(u)$;
14   **foreach** $u' \in N_{\mathcal{P}}(u), v' \in N_G(v), v' \in \Phi(u')$ **do**
15   Mark $\langle u', v' \rangle$;
16   **end**
17   **end**
18   **end**
19   **if** *there is no marked* $\langle u, v \rangle$ **then break**;
20 **end**
21 **end**

---

these pairs are marked and checked again (line 14). Second, the $\langle u, v \rangle$ pairs are stored and manipulated using a hashtable instead of a matrix. This reduces the space and time complexity from $O(k \cdot n)$ to $O(\sum_{i=1}^{k} |\Phi(u_i)|)$. The overall time complexity is $O(l \cdot \sum_{i=1}^{k} |\Phi(u_i)| \cdot (d_1 d_2 + M(d_1, d_2)))$ where l is the refinement level, d_1 and d_2 are maximum degrees of $\mathcal{P}$ and G respectively, and $M()$ is the time complexity of maximum bipartite matching ($O(n^{2.5})$ for Hopcroft and Karp's algorithm [19]).

Figure 4.18 shows an execution of Algorithm 4.2 on the example in Figure 4.16. At level 1, A_2 and C_1 are removed from $\Phi(A)$ and $\Phi(C)$, respectively. At level 2, B_2 is removed from $\Phi(B)$ since the bipartite graph $\mathcal{B}_{B,B_2}$ has no semi-perfect matching (note that A_2 was already removed from $\Phi(A)$).

Whereas the neighborhood subgraphs discussed in Section 4.2 prune infeasible mates by using local information, the refinement procedure in Algo-

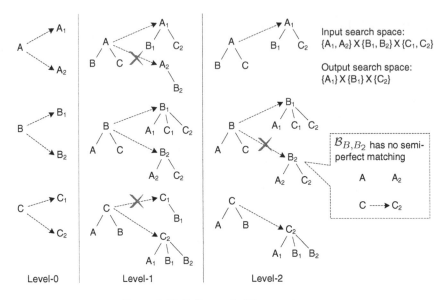

Figure 4.18. Refinement of the search space

rithm 4.2 prunes the search space globally. The global pruning has a larger
overhead and is dependent on the output of the local pruning. Therefore, both
pruning methods are indispensable and should be used together.

4.4   Optimization of Search Order

Next, we consider the search order of Algorithm 4.1. The goal here is to find
a good search order for the nodes. Since the search procedure is equivalent to
multiple joins, it is similar to a typical query optimization problem [7]. Two
principal issues need to be considered. One is the cost model for a given search
order. The other is the algorithm for finding a good search order. The cost
model is used as the objective function of the search algorithm. Since the
search algorithm is relatively standard (e.g., dynamic programming, greedy
algorithm), we focus on the cost model and illustrate that it can be customized
in the domain of graphs.

Cost Model.   A search order (a.k.a. a query plan) can be represented as a
rooted binary tree whose leaves are nodes of the graph pattern and each internal
node is a join operation. Figure 4.19 shows two examples of search orders.

We estimate the cost of a join (a node in the query plan tree) as the product
of cardinalities of the collections to be joined. The cardinality of a leaf node
is the number of feasible mates. The cardinality of an internal node can be
estimated as the product of cardinalities of collections reduced by a factor γ.

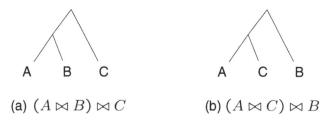

(a) $(A \bowtie B) \bowtie C$ (b) $(A \bowtie C) \bowtie B$

Figure 4.19. Two examples of search orders

Definition 4.11. *(Result size of a join) The result size of join i is estimated by*

$$Size(i) = Size(i.left) \times Size(i.right) \times \gamma(i)$$

where $i.left$ and $i.right$ are the left and right child nodes of i respectively, and $\gamma(i)$ is the reduction factor.

A simple way to estimate the reduction factor $\gamma(i)$ is to approximate it by a constant. A more elaborate way is to consider the probabilities of edges in the join: Let $\mathcal{E}(i)$ be the set of edges involved in join i, then

$$\gamma(i) = \prod_{e(u,v) \in \mathcal{E}(i)} P(e(u,v))$$

where $P(e(u,v))$ is the probability of edge $e(u,v)$ conditioned on u and v. This probability can be estimated as

$$P(e(u,v)) = \frac{freq(e(u,v))}{freq(u) \cdot freq(v)}$$

where $freq()$ denotes the frequency of the edge or node in the large graph.

Definition 4.12. *(Cost of a join) The cost of join i is estimated by*

$$Cost(i) = Size(i.left) \times Size(i.right)$$

Definition 4.13. *(Cost of a search order) The total cost of a search order Γ is estimated by*

$$Cost(\Gamma) = \sum_{i \in \Gamma} Cost(i)$$

For example, let the input search space be $\{A_1\} \times \{B_1, B_2\} \times \{C_2\}$. If we use a constant reduction factor γ, then $Cost(A \bowtie B) = 1 \times 2 = 2$, $Size(A \bowtie B) = 2\gamma$, $Cost((A \bowtie B) \bowtie C) = 2\gamma \times 1 = 2\gamma$. The total cost is $2 + 2\gamma$. Similarly, the total cost of $(A \bowtie C) \bowtie B$ is $1 + 2\gamma$. Thus, the search order $(A \bowtie C) \bowtie B$ is better than $(A \bowtie B) \bowtie C$.

Search Order. The number of all possible search orders is exponential in the number of nodes. It is expensive to enumerate all of them. As in many query optimization techniques, we consider only left-deep query plans, i.e., the outer node of each join is always a leaf node. The traditional dynamic programming would take an $O(2^k)$ time complexity for a graph pattern of size k. This is not scalable to large graph patterns. Therefore, we adopt a simple greedy approach in our implementation: at join i, choose a leaf node that minimizes the estimated cost of the join.

5. Experimental Study

In this section, we evaluate the performance of the presented graph pattern matching algorithms on large real and synthetic graphs. The graph specific optimizations are compared with an SQL-based implementation as described in Figure 4.2. MySQL server 5.0.45 is used and configured as: storage engine=MyISAM (non-transactional), key_buffer_size = 256M. Other parameters are set as default. For each large graph, two tables V(vid, label) and E(vid1, vid2) are created as in Figure 4.2. B-tree indices are built for each field of the tables.

The presented graph pattern matching algorithms were written in Java and compiled with Sun JDK 1.6. All the experiments were run on an AMD Athlon 64 X2 4200+ 2.2GHz machine with 2GB memory running MS Win XP Pro.

5.1 Biological Network

the real dataset is a yeast protein interaction network [2]. This graph consists of 3112 nodes and 12519 edges. Each node represents a unique protein and each edge represents an interaction between proteins.

To allow for meaningful queries, we add Gene Ontology (GO) [14] terms to the proteins. The Gene Ontology is a hierarchy of categories that describes cellular components, biological processes, and molecular functions of genes and their products (proteins). Each GO term is a node in the hierarchy and has one or more parent GO Terms. Each protein has one or more GO terms. We use high level GO terms as labels of the proteins (183 distinct labels in total). We index the node labels using a hashtable, and store the neighborhood subgraphs and profiles with radius 1 as well.

Clique Queries. The clique queries are generated with sizes (number of nodes) between 2 and 7 (sizes greater than 7 have no answers). For each size, a complete graph is generated with each node assigned a random label. The random label is selected from the top 40 most frequent labels. A total of 1000 clique queries are generated and the results are averaged. The queries are divided into two groups according to the number of answers returned: low hits (less than 100 answers) and high hits (more than 100 answers). Queries having no answers are not counted in the statistics. Queries having too many hits (more than 1000) are terminated immediately and counted in the group of high hits.

To evaluate the pruning power of the local pruning (Section 4.2) and the global pruning (Section 4.3), we define the *reduction ratio* of search space as

$$\gamma(\Phi, \Phi_0) = \frac{|\Phi(u_1)| \times .. \times |\Phi(u_k)|}{|\Phi_0(u_0)| \times .. \times |\Phi_0(u_k)|}$$

where Φ_0 refers to the baseline search space.

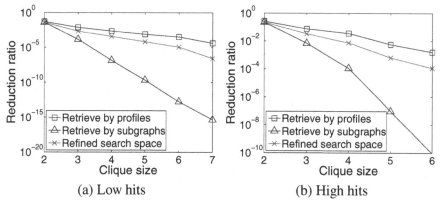

(a) Low hits   (b) High hits

Figure 4.20. Search space for clique queries

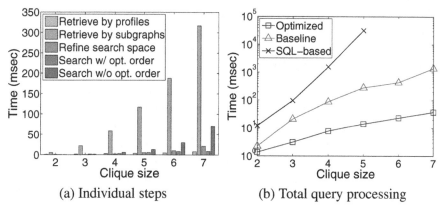

(a) Individual steps   (b) Total query processing

Figure 4.21. Running time for clique queries (low hits)

Figure 4.20 shows the reduction ratios of search space by different methods. "Retrieve by profiles" finds feasible mates by checking profiles and "Retrieve by subgraphs" finds feasible mates by checking neighborhood subgraphs (Section 4.2). "Refined search space" refers to the global pruning discussed in Section 4.3 where the input search space is generated by "Retrieve by profiles". The maximum refinement level ℓ is set as the size of the query. As can be seen from the figure, the refinement procedure always reduces the search space retrieved by profiles. Retrieval by subgraphs results in the smallest search space. This is due to the fact that neighborhood subgraphs for a clique query is actually the entire clique.

Figure 4.21(a) shows the average processing time for individual steps under varying clique sizes. The individual steps include retrieval by profiles, retrieval by subgraphs, refinement, search with the optimized order (Section 4.4), and search without the optimized order. The time for finding the optimized order is negligible since we take a greedy approach in our implementation. As shown in the figure, retrieval by subgraphs has a large overhead although it produces a smaller search space than retrieval by profiles. Another observation is that the optimized order improves upon the search time.

Figure 4.21(b) shows the average total query processing time in comparison to the SQL-based approach on low hits queries. The "Optimized" processing consists of retrieval by profiles, refinement, optimization of search order, and search with the optimized order. The "Baseline" processing consists of retrieval by node attributes and search without the optimized order on the baseline space. The query processing time in the "Optimized" case is improved greatly due to the reduced search space.

The SQL-based approach takes much longer time and does not scale to large clique queries. This is due to the unpruned search space and the large number of joins involved. Whereas our graph pattern matching algorithm (Section 4.1) is exponential in the number of nodes, the SQL-based approach is exponential in the number of edges. For instance, a clique of size 5 has 10 edges. This requires 20 joins between nodes and edges (as illustrated in Figure 4.2).

5.2   Synthetic Graphs

The synthetic graphs are generated using a simple Erdős-Rènyi [13] random graph model: generate n nodes, and then generate m edges by randomly choosing two end nodes. Each node is assigned a label (100 distinct labels in total). The distribution of the labels follows Zipf's law, i.e., probability of the xth label $p(x)$ is proportional to x^{-1}. The queries are generated by randomly extracting a connected subgraph from the synthetic graph.

We first fix the size of synthetic graphs n as $10K$, $m = 5n$, and vary the query size between 4 and 20. Figure 4.22 shows the search space and processing time for individual steps. Unlike clique queries, the global pruning produces the smallest search space, which outperforms the local pruning by full neighborhood subgraphs.

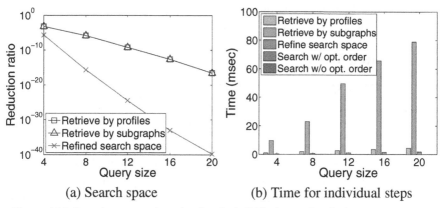

(a) Search space   (b) Time for individual steps

Figure 4.22. Search space and running time for individual steps (synthetic graphs, low hits)

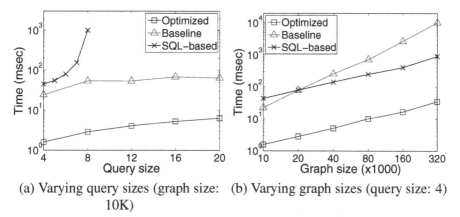

(a) Varying query sizes (graph size: (b) Varying graph sizes (query size: 4)
    10K)

Figure 4.23. Running time (synthetic graphs, low hits)

Figure 4.23 shows the total time with varying query sizes and graph sizes. As can be seen, The SQL-based approach is not scalable to large queries, though it scales to large graphs with small queries. In either case, the "Optimized" processing produces the smallest running time.

To summarize the experimental results, retrieval by profiles has much less overhead than that of retrieval by subgraphs. The refinement step (Section 4.3) greatly reduces the search space. The overhead of the search step is well compensated by the extensive reduction of search space. A practical combination would be retrieval by profiles, followed by refinement, and then search with an optimized order. This combination scales well with various query sizes and graph sizes. SQL-based processing is not scalable to large queries. Overall, the optimized processing performs orders of magnitude better than the SQL-based approach. While small improvements in SQL-based implementations can be achieved by careful tuning and other optimizations, the results show that query processing in the graph domain has clear advantages.

6.   Related Work

6.1   Graph Query Languages

A number of graph query languages have been historically available for representing and manipulating graphs. GraphLog [12] represents both data and queries graphically. Nodes and edges are labeled with one or more attributes. Edges in the queries are matched to either edges or paths in the data graphs. The paths can be regular expressions with possibly negation. A query graph is a graph with a distinguished edge. The distinguished edge introduces a new relation for nodes. The query graph can be naturally translated into a Datalog program where the distinguished edge corresponds to a new predicate (relation). A graphical query consists of one or more query graphs, each of which can use predicates defined in other query graphs. The predicates among them thus form a dependence graph of the graphical query. GraphLog queries are graphical queries in which the dependence graph must be acyclic. In terms of expressive power, GraphLog was shown to be equivalent to stratified linear Datalog [28]. GraphLog does not provide any algebraic operations on graphs, which is important for practical evaluation of queries.

In the category of object-oriented databases, GOOD [16] is a graph-oriented object data model. GOOD models an object database instance by a directed labeled graph, where objects in the database and attributes on the objects are both represented as nodes of the graph. GOOD does not distinguish between atomic, composed and set objects. There are only printable nodes and non-printable nodes. The printable nodes are used for graphical interfaces. As for edges, there are only functional edges and non-functional edges. The functional edges point to unique nodes in the graph. Both nodes and edges can have labels, which are defined by an object database scheme. GOOD defines a transformation language that contains five basic operations on graphs: node addition and deletion, edge addition and deletion, and abstraction that groups common nodes. These operations are defined using the notion of a pattern that describes subgraphs embedded in the object database instance. The transformation language is used for both querying and updates. In terms of expressive power, the transformation language can express operations on sets and recursive functions.

GraphDB [15] is another object-oriented data model and query language for graphs. In the GraphDB data model, the whole database is viewed as a single graph. Objects in the database are strong-typed and the object types support inheritance. Each object is associated with an object type and an object identity. The object can have data attributes or reference attributes to other objects. There are three kinds of object classes: simple classes, linked classes, and path classes. Objects of simple classes are nodes of the graph. Objects of link classes are edges and have two additional references to source and target simple objects. Objects of path classes have a list of references to node and edge objects in the graph. A query consists of several steps, each of which creates or manipulates a uniform sequence of objects, a heterogeneous sequence of objects, a single object, or a value of a data type. The uniform sequence of objects have a common tuple type, whereas the heterogenous sequence may belong to different object classes and tuple types. Queries are constructed in four fundamental ways: derive, rewrite, union, and custom graph operations. The derive statement is similar to the usual select...from...where statement, and can be used to specify a subgraph pattern, which is formulated as a list of node objects, edge objects, or either of them occurring in a path object. The rewrite operation transforms a heterogenous sequence of objects into a new sequence. The union operation transforms a heterogenous sequence into a uniform one by taking the least common tuple type. The graph operations are user-defined, e.g., shortest path search.

GOQL [35] also uses an object-oriented graph data model and is extended from OQL. Similar to GraphDB, GOQL defines object types for nodes, edges, paths, and graphs. As in OQL, GOQL uses the usual select...from...where statement to specify queries. In addition, it uses temporal operators next, until and connected to define path formulas. The path formulas can be used as predicates on sequences and paths in the queries. For query processing, GOQL translates queries into an object algebra (O-Algebra) with the extended temporal operators. PQL [25] is a pathway query language for biological networks. The language extends SQL with path expressions and is implemented on top of an RDBMS. In all these languages, the basic objects are nodes and edges as in the object-oriented data model, and paths as extended by the respective languages. Querying on graph structures are explicitly constructed from the basic objects.

More recently, XML databases have been studied intensively for tree-based data models and semistructured data. XML databases can be generally implemented in two approaches: mapping to relational database systems [33] or native XML implementations [21]. In the second approach, TAX [22] is a tree algebra for XML that operates natively on trees. TAX uses a pattern tree to match interesting nodes. The pattern tree consists of a tree structure and a predicate on nodes of the tree. Tree pattern matching thus plays an important role in XML query processing [1, 6]. GraphQL generalizes the idea of tree patterns to graph patterns. Graph patterns is the main building block of a graph query and graph pattern matching is an important part of graph query processing. Both GraphQL and TAX generalize the relational algebraic operators, including selection, product, set operations. TAX has additional operators such as copy-and-paste, value updates, node deletion and insertion. GraphQL can express these operations by the composition operator.

Some of the recent interest in Semantic Web has spurred Resource Description Framework (RDF) [26] and the accompanying SPARQL query language [27]. This model describes a graph by a set of triples, each of which describes an (attribute, value) pair or an interconnection between two nodes. The SPARQL query language works primarily through a pattern which is a constraint on a single node. All possible matchings of the pattern are returned from the graph database. A general graph query language could be more powerful by providing primitives for expressing constraints on the entire result graph simultaneously.

Table 4.1. Comparison of different query languages

Language	Basic unit	Query style	Semi-structured
GraphQL	graphs	set-oriented	yes
SQL	tuples	set-oriented	no
TAX	trees	set-oriented	yes
GraphLog	nodes/edges	logic pro.	-
OODB (GOOD, GraphDB, GOQL)	nodes/edges	navigational	no

Table 4.1 outlines the comparison between GraphQL and other query languages. GraphQL is different from other query languages in that graphs are chosen as the basic unit of information. This means graphs or sets of graphs are used as the operands and return types in all graph operations. Graph structures are thus preserved and carried over atomically. This is useful not only from a user's perspective but also for query optimizations that rely on graph structural information. In comparison to SQL, GraphQL has a similar algebraic system, but the algebraic operators are defined directly on graphs. In comparison to OODB, GraphQL queries are declarative and set-oriented, whereas OODB accesses single objects in a navigational manner (i.e., using references to access objects one after another in the object graph). With regard to data model and representation, GraphQL is semistructured and does not cast strict and predefined data types or schemas on nodes, edges, and graphs. In contrast, SQL presumes a strict schema in order to store data. OODB requires objects (nodes and edges) to be strong-typed. In comparison to XML databases, the main difference lies in the underlying data model. GraphQL deals with the graph (networked) data model, whereas XML databases deal with the hierarchical data model.

Graph grammars have been used previously for modeling visual languages and graph transformations in various domains [30, 29]. Our work is different in that our emphasis has been on a query language and database implementations.

6.2   Graph Indexing

Graph indexing is useful for graph pattern matching over a large collection of small graphs. GraphGrep [34] uses enumerated paths as index features to filter unmatched graphs. GIndex [40] uses discriminative frequent fragments as index features to improve filtering rates and reduce index sizes. Closure-tree [17] organizes graphs into a tree-based index structure using graph closures as the bounding boxes. GString [23] converts graph querying to subsequence matching. TreePi [41] uses frequent subtrees as index features. Williams et al. [39] decompose graphs and hash the canonical forms of the resulting subgraphs. SAGA [36] enumerates fragments of graphs and answers are generated by assembling hits of the query fragments. FG-index [9] uses frequent subgraphs as index features. Frequent graph queries are answered without verification and infrequent queries require only a small number of verifications. Zhao et al. [42] show that frequent tree-features plus a small number of discriminative graphs are better than frequent graph-features. While the above techniques can be used as access methods for the case of a large collection of small graphs, this chapter addresses graph pattern matching for the case of a single large graph.

Another line of graph indexing addresses reachability queries in large directed graphs [8, 10, 11, 31, 37, 38]. In a reachability query, two nodes are given and the answer is whether there exists a path between the two nodes. Reachability queries correspond to recursive graph patterns which are paths (Figure 4.6(a)). Indexing and processing of reachability queries are generally based on spanning trees with pre/post-order labeling [8, 37, 38] or 2-hop-cover [10, 11, 31]. These techniques can be incorporated into access methods for recursive graph pattern queries.

7.   Future Research Directions

Physical Storage of Graph Data.   Graphs in the real world are heterogeneous in both the structures and the underlying attributes. It is challenging to store graphs on disks for efficient storage and fast retrieval. What is the appropriate storage unit, nodes, edges, or graphs? In the category of a large collection of small graphs, how to store graphs with various sizes to fixed-length pages on disks? In the category of a single large graph, how to decompose the large graph into small chunks and preserve locality? Traditional storage techniques need to be re-considered, and new graph-specific heuristics might be devised to address these questions.

Implementation of Other Graph Operators. This chapter only addresses implementation of the selection operator. Other operators, such as joins on two collections of graphs, might be a challenge if the inter-graph join conditions are not trivial. In addition, operators such as ordering (ranking), aggregation (OLAP processing), are interesting research directions on their own.

Scalability to Very Large Graph Databases. The presented techniques consider graphs with millions of nodes and edges, or millions of small graphs. Graphs in some domains, such as Internet, social networks, are in the scale of tera-bytes or even larger. Graphs at this scale cannot be processed by single machines. Large-scale parallel and distributed schemes are needed for graph storage and query processing.

8.   Conclusion

We have presented GraphQL, a query language for graphs with arbitrary attributes and sizes. GraphQL has a number of appealing features. Graphs are the basic unit and graph structures are composable using the notion of formal languages for graphs. We developed efficient access methods for the selection operator using the idea of neighborhood subgraphs and profiles, refinement of the overall search space, and optimization of the search order. Experimental studies on real and synthetic graphs validated the access methods.

In summary, graphs are prevalent in multiple domains. This chapter has demonstrated the benefits of working with native graphs for queries and database implementations. Translations of graphs into relations are unnatural and cannot take advantage of graph-specific heuristics. The coupling of graph-based querying and native graph-based databases produces interesting possibilities from the point of view of expressiveness and implementation techniques. We have barely scratched the surface and much more needs to be done in matching characteristics of queries and databases to appropriate heuristics. The results of this chapter are an important first step in this regard.

Acknowledgments

This work was supported in part by NSF grants IIS-0612327.

Appendix: Query Syntax of GraphQL

```
Start ::= ( GraphPattern ";" | FLWRExpr ";" )* <EOF>

GraphPattern ::=   "graph" [<ID>] [Tuple] "{"
                       MemberDecl *
                   "}" ["where" Expr]

MemberDecl ::= "node" NodeDecl ("," NodeDecl)* ";"
```

```
            | "edge" EdgeDecl ("," EdgeDecl)* ";"
            | "graph" <ID>  ( "," <ID> )* ";"
            | "unify" Names "," Names ("," Names)* ";"


NodeDecl ::= [<ID>][Tuple] ["where" Expr]


EdgeDecl ::= [<ID>]"(" Names "," Names")" [Tuple] ["where" Expr]


Tuple ::= "<"[<ID>] (<ID>"="Literal)* ">"


FLWRExpr ::= "for" ( <ID> | GraphPattern )
             ["exhaustive"] "in" "doc" "(" string ")"
             ["where" Expr]
             ( "return" GraphTemplate |
               "let" <ID> "=" GraphTemplate )


GraphTemplate ::= "graph" [<ID>] [TupleTemplate] "{"
                      TMemberDecl *
                  "}" | <ID>


TMemberDecl ::= "node" TNodeDecl ("," TNodeDecl)* ";"
              | "edge" TEdgeDecl ("," TEdgeDecl)* ";"
              | "graph" <ID>  ( "," <ID> )* ";"
              | "unify" Names "," Names ("," Names)* ["where" Expr] ";"


TNodeDecl ::= [<ID>][TupleTemplate]


TEdgeDecl ::= [<ID>]"("Names "," Names")"[TupleTemplate]


TupleTemplate ::= "<"[<ID>] (<ID>"="Expr)* ">"


Expr ::= Term ( Op Expr )*


Op ::=    "|" | "&"  | "+" | "-"  | "*" | "/" |
          "==" | "!=" | ">" | ">=" | "<" |"<="


Term ::=  "(" Expr ")" | Literal | Names


Names ::= <ID> ("." <ID>)*


Literal ::= int | float | string
```

References

[1] S. Al-Khalifa, H. V. Jagadish, J. M. Patel, Y. Wu, N. Koudas, and D. Srivastava. Structural joins: A primitive for efficient xml query pattern matching. In *ICDE*, pages 141–, 2002.

[2] S. Asthana et al. Predicting protein complex membership using probabilistic network reliability. *Genome Research*, May 2004.

[3] S. Berretti, A. D. Bimbo, and E. Vicario. Efficient matching and index-ing of graph models in content-based retrieval. In *IEEE Trans. on Pattern Analysis and Machine Intelligence*, volume 23, 2001.

[4] S. Boag, D. Chamberlin, M. F. Fernandez, D. Florescu, J. Robie, and J. Simeon. XQuery 1.0: An XML query language. W3C, http://www.w3.org/TR/xquery/, 2007.

[5] C. Branden and J. Tooze. *Introduction to protein structure*. Garland, 2 edition, 1998.

[6] N. Bruno, N. Koudas, and D. Srivastava. Holistic twig joins: optimal XML pattern matching. In *SIGMOD Conference*, pages 310–321, 2002.

[7] S. Chaudhuri. An overview of query optimization in relational systems. In *PODS*, pages 34–43, 1998.

[8] L. Chen, A. Gupta, and M. E. Kurul. Stack-based algorithms for pattern matching on dags. In *Proc. of VLDB '05*, pages 493–504, 2005.

[9] J. Cheng, Y. Ke, W. Ng, and A. Lu. FG-Index: towards verification-free query processing on graph databases. In *Proc. of SIGMOD '07*, 2007.

[10] J. Cheng, J. X. Yu, X. Lin, H. Wang, and P. S. Yu. Fast computation of reachability labeling for large graphs. In *EDBT*, pages 961–979, 2006.

[11] E. Cohen, E. Halperin, H. Kaplan, and U. Zwick. Reachability and dis-tance queries via 2-hop labels. *SIAM J. Comput.*, 32(5):1338–1355, 2003.

[12] M. P. Consens and A. O. Mendelzon. GraphLog: a visual formalism for real life recursion. In *PODS*, 1990.

[13] P. Erdős and A. Renyi. On random graphs I. *Publ. Math. Debrecen*, (6):290–297, 1959.

[14] Gene Ontology. http://www.geneontology.org/.

[15] R. H. Guting. GraphDB: Modeling and querying graphs in databases. In *Proc. of VLDB'94*, pages 297–308, 1994.

[16] M. Gyssens, J. Paredaens, and D. van Gucht. A graph-oriented object database model. In *Proc. of PODS '90*, pages 417–424, 1990.

[17] H. He and A. K. Singh. Closure-Tree: An Index Structure for Graph Queries. In *Proc. of ICDE '06*, Atlanta, USA, 2006.

[18] H. He and A. K. Singh. Graphs-at-a-time: Query Language and Access Methods for Graph Databases. In *Proc. of SIGMOD '08*, pages 405–418, Vancouver, Canada, 2008.

[19] J. Hopcroft and R. Karp. An $n^{5/2}$ algorithm for maximum matchings in bipartite graphs. *SIAM J. Computing*, 1973.

[20] J. E. Hopcroft and J. D. Ullman. *Introduction to Automata Theory, Lan-guages, and Computation*. Addison Wesley, 1979.

[21] H. V. Jagadish, S. Al-Khalifa, A. Chapman, L. V. S. Lakshmanan, A. Nierman, S. Paparizos, J. M. Patel, D. Srivastava, N. Wiwatwattana, Y. Wu, and C. Yu. TIMBER: A native XML database. *VLDB J.*, 11(4):274–291, 2002.

[22] H. V. Jagadish, L. V. S. Lakshmanan, D. Srivastava, and K. Thompson. TAX: A tree algebra for XML. In *Proc. of DBPL'01*, 2001.

[23] H. Jiang, H. Wang, P. S. Yu, and S. Zhou. GString: A novel approach for efficient search in graph databases. In *ICDE*, 2007.

[24] J. Lee, J. Oh, and S. Hwang. STRG-Index: Spatio-temporal region graph indexing for large video databases. In *Proc. of SIGMOD*, 2005.

[25] U. Leser. A query language for biological networks. *Bioinformatics*, 21:ii33–ii39, 2005.

[26] F. Manola and E. Miller. RDF Primer. W3C, http://www.w3.org/TR/rdf-primer/, 2004.

[27] E. Prud'hommeaux and A. Seaborne. SPARQL query language for RDF. W3C, http://www.w3.org/TR/rdf-sparql-query/, 2007.

[28] R. Ramakrishnan and J. Gehrke. *Database Management Systems*, chapter 24 Deductive Databases. McGraw-Hill, third edition, 2003.

[29] J. Rekers and A. Schurr. A graph grammar approach to graphical parsing. In *11th International IEEE Symposium on Visual Languages*, 1995.

[30] G. Rozenberg (Ed.). *Handbook on Graph Grammars and Computing by Graph Transformation: Foundations*, volume 1. World Scientific, 1997.

[31] R. Schenkel, A. Theobald, and G. Weikum. Efficient creation and incremental maintenance of the HOPI index for complex XML document collections. In *Proc. of ICDE '05*, pages 360–371, 2005.

[32] N. Shadbolt, T. Berners-Lee, and W. Hall. The semantic web revisited. *IEEE Intelligent Systems*, 21(3):96–101, 2006.

[33] J. Shanmugasundaram, K. Tufte, C. Zhang, G. He, D. J. DeWitt, and J. F. Naughton. Relational databases for querying XML documents: Limitations and opportunities. In *VLDB*, pages 302–314, 1999.

[34] D. Shasha, J. T. L. Wang, and R. Giugno. Algorithmics and applications of tree and graph searching. In *Proc. of PODS*, 2002.

[35] L. Sheng, Z. M. Ozsoyoglu, and G. Ozsoyoglu. A graph query language and its query processing. In *ICDE*, 1999.

[36] Y. Tian, R. C. McEachin, C. Santos, D. J. States, and J. M. Patel. SAGA: a subgraph matching tool for biological graphs. *Bioinformatics*, 23(2), 2007.

[37] S. Trißl and U. Leser. Fast and practical indexing and querying of very large graphs. In *Proc. of SIGMOD '07*, pages 845–856, 2007.

[38] H. Wang, H. He, J. Yang, P. S. Yu, and J. X. Yu. Dual labeling: Answering graph reachability queries in constant time. In *Proc. of ICDE '06*, page 75, 2006.

[39] D. W. Williams, J. Huan, and W. Wang. Graph database indexing using structured graph decomposition. In *ICDE*, 2007.

[40] X. Yan, P. S. Yu, and J. Han. Graph Indexing: A frequent structure-based approach. In *Proc. of SIGMOD*, 2004.

[41] S. Zhang, M. Hu, and J. Yang. TreePi: A novel graph indexing method. In *ICDE*, 2007.

[42] P. Zhao, J. X. Yu, and P. S. Yu. Graph indexing: Tree + delta >= graph. In *Proc. of VLDB*, pages 938–949, 2007.

Chapter 5

GRAPH INDEXING

Xifeng Yan

Department of Computer Science
University of California at Santa Barbara
xyan@cs.ucsb.edu

Jiawei Han

Department of Computer Science
University of Illinois at Urbana-Champaign
hanj@cs.uiuc.edu

Abstract   Advanced database systems face a great challenge arising from the emergence
of massive, complex structural data in bioinformatics, chem-informatics, business processes, etc. One of the most important functions needed in these areas
is efficient search of complex graph data. Given a graph query, it is desirable
to retrieve relevant graphs quickly from a large database via efficient graph indices. This chapter gives an introduction to graph substructure search, approximate substructure search and their related graph indexing techniques, particularly feature-based graph indexing.

Keywords: Frequent pattern, graph index, graph query, similarity search

1.   Introduction

Development of scalable methods for analyzing large graph data sets, including graphs built from chemical structures and biological networks, poses
great challenges. At the core of many graph analysis applications, lies a common and critical problem: how to efficiently search graphs.

Given a graph database $D = \{G_1, G_2, \ldots, G_n\}$ and a graph query Q, *graph
search* returns a query answer set $D_Q = \{G | M(Q, G) = 1, G \in D\}$, where
M is a boolean function. M could be a function testing graph isomorphism
(full structure search), subgraph isomorphism (substructure search), approxi-

C.C. Aggarwal and H. Wang (eds.), *Managing and Mining Graph Data,*
Advances in Database Systems 40, DOI 10.1007/978-1-4419-6045-0_5,
© Springer Science+Business Media, LLC 2010

mate match (full structure similarity search), and subgraph approximate match (substructure similarity search). It is inefficient to perform a sequential scan on a graph database and check each graph to find answers to a query graph. Sequential scan is costly because one has to not only access the whole graph database but also check (sub)graph isomorphism. It is known that subgraph isomorphism is an NP-complete problem [8]. Therefore, high performance graph indexing is needed to quickly prune graphs that obviously violate the query requirement.

The problem of graph search has been addressed in different domains since it is a critical problem for many applications. In content-based image retrieval, Petrakis and Faloutsos [25] represented each graph as a vector of features and indexed graphs in a high dimensional space using R-trees. Shokoufandeh et al. [29] indexed graphs by a signature computed from the eigenvalues of adjacency matrices. Instead of casting a graph to a vector form, Berretti et al. [2] proposed a metric indexing scheme which organizes graphs hierarchically according to their mutual distances. The SUBDUE system developed by Holder et al. [17] uses minimum description length to discover substructures that compress graph data and represent structural concepts in the data. In 3D protein structure search, algorithms using hierarchical alignments on secondary structure elements [21], or geometric hashing [35], have already been developed. There are other literatures related to graph retrieval that we are not going to enumerate here.

In semistructured/XML databases, query languages built on path expressions become popular. Efficient indexing techniques for path expression were initially introduced in DataGuide [13] and 1-index [23]. A(k)-index [20] proposes k-bisimilarity to exploit local similarity existing in semistructured databases. APEX [7] and D(k)-index [5] consider the adaptivity of index structure to fit the query load. Index Fabric [9] represents every path in a tree as a string and stores it in a Patricia trie. For more complicated graph queries, Shasha et al. [28] extended the path-based technique to do full scale graph retrieval, which is also used in the Daylight system [18]. Srinivasa et al. [30] built indices based on multiple vector spaces with different abstract levels of graphs.

This chapter introduces feature-based graph indexing techniques that facilitate graph substructure search in graph databases with thousands of instances. Nevertheless, similar techniques can also be applied to indexing single massive graphs.

2.   Feature-Based Graph Index

Definition 5.1 (Substructure Search). *Given a graph database $D = \{G_1, G_2, \ldots, G_n\}$ and a query graph Q,* substructure search *is to find all the graphs that contain Q.*

Substructure search is one kind of basic graph queries, observed in many graph-related applications. Feature-based graph indexing is designed to answer substructure search queries, which consists of the following two major steps:

Index construction: It precomputes features from a graph database and builds indices based on these features. There are various kinds of features that could be used, including node/edge labels, paths, trees, and subgraphs. Let F be a feature set for a given graph database D. For any feature $f \in F$, D_f is the set of graphs containing f, $D_f = \{G | f \subseteq G, G \in D\}$. We define a null feature, $f_\varnothing$, which is contained by any graph. An inverted index is built between F and D: D_f could be the ids of graphs containing f, which is similar to inverted index in document retrieval [1].

Query processing: It has three substeps: (1) *Search*, which enumerates all the features in a query graph, Q, to compute the *candidate query answer set*, $C_Q = \bigcap_f D_f$ ($f \subseteq Q$ and $f \in F$); each graph in C_Q contains all of Q's features. Therefore, D_Q is a subset of C_Q. (2) *Fetching*, which retrieves the graphs in the candidate answer set from disks. (3) *Verification*, which checks the graphs in the candidate answer set to verify if they really satisfy the query. The candidate answer set is verified to prune false positives.

The *Query Response Time* of the above search framework is formulated as follows,

$$T_{search} + |C_Q| * (T_{io} + T_{iso\_test}), \tag{5.1}$$

where T_{search} is the time spent in the search step, T_{io} is the average I/O time of fetching a candidate graph from the disk, and T_{iso_test} is the average time of checking a subgraph isomorphism, which is conducted over query Q and graphs in the candidate answer set.

The candidate graphs are usually scattered around the entire disk. Thus, T_{io} is the I/O time of fetching a block on a disk (assume a graph can be accommodated in one disk block). The value of T_{iso_test} does not change much for a given query. Therefore, the key to improve the query response time is to minimize the size of the candidate answer set as much as possible. When a database is so large that the index cannot be held in main memory, T_{search} will affect the query response time.

Since all the features in the index contained by a query are enumerated, it is important to maintain a compact feature set in the memory. Otherwise, the cost of accessing the index may be even greater than that of accessing the database itself.

2.1 Paths

One solution to substructure search is to take paths as features to index graphs: Enumerate all the existing paths in a database up to a $maxL$ length and use them as features to index, where a path is a vertex sequence, $v_1, v_2, \ldots, v_k$, s.t., $\forall 1 \leq i \leq k - 1$, (v_i, v_{i+1}) is an edge. It uses the index to identify graphs that contain all the paths (up to the $maxL$ length) in the query graph.

This approach has been widely adopted in XML query processing. XML query is one kind of graph query, which is usually built around path expressions. Various indexing methods [13; 23; 9; 20; 7; 28; 5] have been developed to process XML queries. These methods are optimized for path expressions and tree-structured data. In order to answer arbitrary graph queries, Graph-Grep and Daylight systems were proposed in [28; 18]. All of these methods take *path* as the basic indexing unit; we categorize them as *path-based indexing*. The path-based approach has two advantages: (1) Paths are easier to manipulate than trees and graphs, and (2) The index space is predefined: All the paths up to the $maxL$ length are selected. In order to answer tree- or graph-structured queries, a path-based approach has to break query graphs into paths, search each path separately for the graphs containing the path, and join the results. Since the structural information could be lost when query graphs are decomposed to paths, likely many false positive candidates will be returned. In addition, a graph database may contain millions of different paths if it is large and diverse. These disadvantages motivate the search of new indexing features.

2.2   Frequent Structures

A straightforward approach of extending paths is to involve more complicated features, e.g., all of substructures extracted from a graph database. Unfortunately, the number of substructures could be even more than the number of paths, leaving an exponential index structure in practice. One solution is to set a threshold of substructures' frequency and only index those frequent ones.

Definition 5.2 (Frequent Structures). *Given a graph database $D = \{G_1, G_2, \ldots, G_n\}$ and a graph structure f, the* support *of f is defined as $sup(f) = |D_f|$, whereas D_f is referred as f's* supporting graphs. *With a predefined threshold min_sup, f is said to be* frequent *if $sup(f) \geq min\_sup$.*

Frequent structures could be used as features to index graphs. Given a query graph Q, if Q is frequent, the graphs containing Q can be retrieved directly since Q is indexed. Otherwise, we sort all Q's subgraphs in the support decreasing order: $f_1, f_2, \ldots, f_n$. There must exist a boundary between f_i and f_{i+1} where $|D_{f_i}| \geq min\_sup$ and $|D_{f_{i+1}}| < min\_sup$. Since all the frequent structures with minimum support min_sup are indexed, one can compute the candidate answer set C_Q by $\bigcap_{1 \leq j \leq i} D_{f_j}$, whose size is at most $|D_{f_i}|$. For many queries, $|D_{f_i}|$ is close to min_sup. Therefore, the cost of verifying C_Q is minimal when min_sup is low.

Unfortunately, for low support queries (i.e., queries whose answer set is small), the size of candidate answer set C_Q is related to the setting of *min_sup*. If *min_sup* is set too high, C_Q might be very large. If *min_sup* is set too low, it could be difficult to generate all the frequent structures due to the exponential pattern space.

Should a uniform *min_sup* be enforced for all the frequent structures? In order to reduce the overall index size, it is appropriate to have a *low* minimum support on *small* structures (for effectiveness) and a *high* minimum support on *large* structures (for compactness). This criterion of selecting frequent structures for effective indexing is called *size-increasing support constraint*.

Definition 5.3 (Size-increasing Support). *Given a monotonically nondecreasing function, $\psi(l)$, structure f is frequent under the* size-increasing support constraint *if and only if $|D_f| \geq \psi(size(f))$, and $\psi(l)$ is a* size-increasing support function.

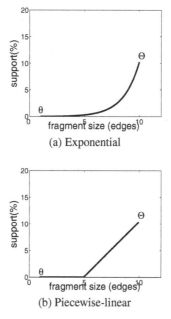

(a) Exponential

(b) Piecewise-linear

Figure 5.1. Size-increasing Support Functions

Figure 5.1 shows two size-increasing support functions: *exponential* and *piecewise-linear*. One could select size-1 structures with a minimum support θ and larger structures with a higher support until we exhaust structures up to the size of $maxL$ with a minimum support Θ.

The size-increasing support constraint will select and index small structures with low minimum supports and large structures with high minimum supports.

This method has two advantages: (1) the number of frequent structures so obtained is much smaller than that using a low uniform support, and (2) low-support large structures could be well indexed by their smaller subgraphs. The first advantage also shortens the mining process when graphs have big structures in common.

2.3   Discriminative Structures

Among similar structures with the same support, it is often sufficient to index only the *smallest common substructures* since more query graphs may contain these structures (higher coverage). That is to say, if f', a supergraph of f, has the same support as f, it will not be able to provide more information than f if both are selected as indexing features. That is, f' is not more *discriminative* than f. This concept can be extended to a collection of subgraphs.

Definition 5.4 (Redundant Structure). *Structure x is redundant with respect to a feature set F if D_x is close to $\bigcap_{f \in F \wedge f \subseteq x} D_f$.*

Each graph in $\bigcap_{f \in F \wedge f \subseteq x} D_f$ contains all x's subgraphs in the feature set F. If D_x is close to $\bigcap_{f \in F \wedge f \subseteq x} D_f$, it implies that the presence of structure x in a graph can be predicted well by the presence of its subgraphs. Thus, x should not be used as an indexing feature since it does not provide new benefits to pruning if its subgraphs are being indexed. In such case, x is a redundant structure. In contrast, there are structures that are not redundant, called *discriminative structures*.

Let $f_1, f_2, \ldots,$ and f_n be the indexing structures. Given a new structure x, the discriminative power of x can be measured by

$$Pr(x | f_{\varphi_1}, \ldots, f_{\varphi_m}), f_{\varphi_i} \subseteq x, 1 \leq \varphi_i \leq n. \tag{5.2}$$

Eq. (5.2) shows the probability of observing x in a graph given the presence of $f_{\varphi_1}, \ldots,$ and f_{φ_m}. *Discriminative ratio*, γ, is defined as $1/Pr(x | f_{\varphi_1}, \ldots, f_{\varphi_m})$, which could be calculated by the following formula:

$$\gamma = \frac{|\bigcap_i D_{f_{\varphi_i}}|}{|D_x|}, \tag{5.3}$$

where D_x is the set of graphs containing x and $\bigcap_i D_{f_{\varphi_i}}$ is the set of graphs containing the features belonging to x. In order to mine discriminative structures, a minimum discriminative ratio γ_{min} is selected; those structures whose discriminative ratio is at least γ_{min} are retained as indexing features. The structures are mined in a level-wise manner, from small size to large size. The concept of indexing discriminative frequent structures, called gIndex, was first introduced by Yan et al. [36]. gIndex is able to achieve better performance in comparison with path-based methods.

For a feature $x \subseteq Q$, the operation, $C_Q = C_Q \cap D_x$ could reduce the candidate answer set by intersecting the id lists of C_Q and D_x. One interesting question is how to reduce the number of intersection operations. Intuitively, if a query Q has two structures, $f_x \subset f_y$, then $C_Q \cap D_{f_x} \cap D_{f_y} = C_Q \cap D_{f_y}$. Thus, it is not necessary to intersect C_Q with D_{f_x}. Let $F(Q)$ be the set of discriminative structures contained in the query graph Q, i.e., $F(Q) = \{f_x | f_x \subseteq Q \wedge f_x \in F\}$. Let $F_m(Q)$ be the set of structures in $F(Q)$ that are not contained by other structures in $F(Q)$, i.e., $F_m(Q) = \{f_x | f_x \in F(Q), \nexists f_y, s.t., f_x \subset f_y \wedge f_y \in F(Q)\}$. The structures in $F_m(Q)$ are called *maximal discriminative structures*. In order to calculate C_Q, one only needs to perform intersection operations on the id lists of maximal discriminative structures.

2.4 Closed Frequent Structures

Graph query processing that applies feature-based graph indices often requires a post verification step that finds true answers from a candidate answer set. If the candidate answer set is large, the verification step might take a long time to finish. Fortunately, a query graph having a large answer set is likely a frequent graph, which can be very efficiently processed using the frequent structure based index without any post verification. If the query graph is not a frequent structure, the candidate answer set obtained from the frequent structure based index is likely small; hence the number of candidate verifications should be minimal. Based on this observation, Cheng et al. [6] investigated the issue arising from frequent structure based indexing. As discussed before, the number of frequent structures could be exponential, indicating a huge index, which might not fit into main memory. In this case, the query performance will be degraded, since graph query processing has to access disks frequently. Cheng et al. [6] proposed using δ-Tolerance Closed Frequent Subgraphs (δ-TCFGs) to compress the set of frequent structures. Each δ-TCFG can be regarded as a representative supergraph of a set of frequent structures. An outer inverted-index is built on the set of δ-TCFGs, which is resident in main memory. Then, an inner inverted-index is built on the cluster of frequent structures of each δ-TCFG, which is resident in disk. Using this two-level index structure, many graph queries could be processed directly without verification.

2.5 Trees

Zhao et al. [38] analyzed the effectiveness and efficiency of paths, trees, and graphs as indexing features from three aspects: feature size, feature selection cost, and pruning power. Like paths and graphs, tree features can be effectively and efficiently used as indexing features for graph databases. It was observed that the majority of frequent graph patterns discovered in many applications are tree structures. Furthermore, if the distribution of frequent trees and graphs is similar, likely they will share similar pruning power.

Since tree mining can be performed much more efficiently than graph mining, Zhao et al. [38] proposed a new graph indexing mechanism, called Tree+Δ, which first mines and indexes frequent trees, and then on-demand selects a small number of discriminative graph structures from a query, which might prune graphs more effectively than tree features. The selection of discriminative graph structures is done on-the-fly for a given query. In order to do so, the pruning power of a graph structure is estimated approximately by its subtree features with upper/lower bounds. Given a query, Tree+Δ enumerates all the frequent subtrees of Q up to the maximum size $maxL$. Based on the obtained frequent subtree feature set of Q, $T(Q)$, it computes the candidate answer set, C_Q, by intersecting the supporting graph set of t, for all $t \in T(Q)$. If Q is a non-tree cyclic graph, it obtains a set of discriminative non-tree features, F. These non-tree features, f, may be cached already in previous search. If not, Tree+Δ will scan the graph database and build an inverted index between f and graphs in D. Then it intersects C_Q with the supporting graph set D_f.

GCoding [39] is another tree-based graph indexing approach. For each node u, it extracts a level-n path tree, which consists of all n-step simple pathes from u in a graph. The node is then encoded with eigenvalues derived from this local tree structure. If a query graph Q is a subgraph of a graph G, for each vertex u in Q, there must exist a corresponding vertex u' in G such that the local structure around u in Q should be preserved around u' in G. There is a partial order relationship between the eigenvalues of these two local structures. Based on this property, GCoding could quickly prune graphs that violate the order.

GString [19] combines three basic structures together: path, star, and cycle for graph search. It first extracts all of cycles in a graph database and then finds the star and path structures in the remaining dataset. The indexing methodology of GString is different from the feature-based approach. It transforms graphs into string representations and treats the substructure search problem as a substring match problem. GString relies on suffix tree to perform indexing and search.

2.6   Hierarchical Indexing

Besides the feature-based indexing methodology, it is also possible to organize graphs in a hierarchical structure to facilitate graph search. Close-tree [15] and GDIndex [34] are two examples of hierarchical graph indexing.

Closure-tree organizes graphs hierarchically where each node in the hierarchical structure contains summary information about its descendants. Given two graphs and an isomorphism mapping between them, one can take an elementwise union of the two graphs and obtain a new graph where the attribute of vertices and edges is a union of their corresponding attribute values in the two graphs. This union graph summarizes the structural information of both graphs, and serves as their bounding box [15], akin to a Minimum Bounding Rectangle (MBR) in traditional index structures. There are two steps to process a graph query Q using the closure-tree index: (1) Traverse the closure tree and prune nodes (graphs) based on a pseudo subgraph isomorphism; (2) Verify the remaining graphs to find the real answers. The pseudo subgraph isomorphism performs approximate subgraph isomorphism testing with high accuracy and low cost.

GDIndex [34] proposes indexing the complete set of the induced subgraphs in a graph database. It organizes the induced subgraphs in a DAG structure and builds a hash table to cross-index the nodes in the DAG structure. Given a query graph, GDIndex first identifies the nodes in the DAG structure that share the same hash code with the query graph, and then their canonical codes are compared to find the right answers. Unfortunately, the index size of GDIndex could be exponential due to a large number of induced subgraphs. It was suggested to place a limit on the size of indexed subgraphs.

3.   Structure Similarity Search

A common problem in graph search is: what if there is no match or very few matches for a given query graph? In this situation, a subsequent query refinement process has to be taken in order to find the structures of interest. Unfortunately, it is often too time-consuming for a user to manually refine the query. One solution is to ask the system to find graphs that approximately contain the query graph. This structure similarity search problem has been studied in various fields. Willett et al. [33] summarized the techniques of fingerprint-based and graph-based similarity search in chemical compound databases. Raymond et al. [27] proposed a three tier algorithm for full structure similarity search. Nilsson[24] presented an algorithm for the pairwise approximate substructure matching. The matching is greedily performed to minimize a distance function for two graphs. Hagadone [14] recognized the importance of substructure similarity search in a large set of graphs. He used atom and edge labels to do screening. Messmer and Bunke [22] studied the reverse substructure similarity search problem in computer vision and pattern recognition. In [28], Shasha et al. also extended their substructure search algorithm to support queries with wildcards, i.e. don't care nodes and edges. In the following discussion, we will introduce feature-based graph indexing for substructure similarity search.

Definition 5.5 (Substructure Similarity Search). *Given a graph database* $D = \{G_1, G_2, \ldots, G_n\}$ *and a query graph Q, substructure similarity search is to discover all the graphs that approximately contain Q.*

Definition 5.6 (Substructure Similarity). *Given two graphs G and Q, if P is the maximum common subgraph of G and Q, then the substructure similarity between G and Q is defined by $\frac{|E(P)|}{|E(Q)|}$, and $\theta = 1 - \frac{|E(P)|}{|E(Q)|}$ is called relaxation ratio.*

Besides the common subgraph similarity measure, graph edit distance could also be used to measure the similarity between two graphs. It calculates the minimum number of edit operations (insertion, deletion, and substitution) needed to transform one graph into another [3].

3.1 Feature-Based Structural Filtering

Given a relaxed query graph, there is a connection between structure-based similarity and feature-based similarity, which could be used to leverage feature-based graph indexing techniques for similarity search.

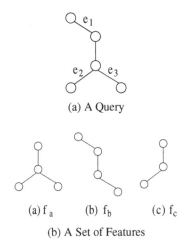

(a) A Query

(a) f $_a$ (b) f$_b$ (c) f$_c$

(b) A Set of Features

Figure 5.2. Query and Features

Figure 5.2(a) shows a query graph and Figure 5.2(b) depicts three structural fragments. Assume that these fragments are indexed as features in a graph database. Suppose there is no match for this query graph in a graph database. Then a user may relax one edge, e.g., e_1, e_2, or e_3, through a deletion operation. No matter which edge is relaxed, the relaxed query graph should have at least three embeddings of these features. That is, the relaxed query graph may *miss* at most four embeddings of these features in comparison with the seven embeddings in the original query graph: one f_a, two f_b's, and four f_c's. According to this constraint, graphs that do not contain at least three embeddings of these features could be safely pruned. This filtering concept is called *feature-based structural filtering*. In order to facilitate feature-based filtering, an index structure is developed, referred to *feature-graph matrix* [12; 28]. Each column of the feature-graph matrix corresponds to a target graph in the graph database, while each row corresponds to a feature being indexed. Each entry records the number of the embeddings of a specific feature in a target graph.

3.2　Feature Miss Estimation

	f_a	$f_{b(1)}$	$f_{b(2)}$	$f_{c(1)}$	$f_{c(2)}$	$f_{c(3)}$	$f_{c(4)}$
e_1	0	1	1	1	0	0	0
e_2	1	1	0	0	1	0	1
e_3	1	0	1	0	0	1	1

Figure 5.3. Edge-Feature Matrix

In order to calculate the maximum feature misses for a given relaxation ratio, we introduce *edge-feature matrix* that builds a map between edges and features for a query graph. In this matrix, each row represents an edge while each column represents an embedding of a feature. Figure 5.3 shows the matrix built for the query graph in Figure 5.2(a) and the features shown in Figure 5.2(b). All of the embeddings are recorded. For example, the second and the third columns are two embeddings of feature f_b in the query graph. The first embedding of f_b *covers* edges e_1 and e_2 while the second covers edges e_1 and e_3. The middle edge does not appear in the edge-feature matrix if a user prefers retaining it. We say that an edge e_i *hits* a feature f_j if f_j covers e_i.

The feature miss estimation problem is formulated as follows: *Given a query graph Q and a set of features contained in Q, if the relaxation ratio is θ, what is the maximum number of features that can be missed?* In fact, it is the maximum number of columns that can be hit by k rows in the edge-feature matrix, where $k = \lfloor \theta \cdot |G| \rfloor$. This is a classic maximum coverage (or set k-cover) problem, which has been proved NP-complete. The optimal solution that finds the maximal number of feature misses can be approximated by a greedy algorithm [16]. The greedy algorithm first selects a row that hits the largest number of columns and then removes this row and the columns covering it. This selection and deletion operation is repeated until k rows are removed. The number of columns removed by this greedy algorithm provides a way to estimate the upper bound of feature misses. Although the bound derived by the greedy algorithm cannot be improved asymptotically, it is possible to improve the greedy algorithm in practice by exhaustively searching the most selective features [37].

3.3   Frequency Difference

Once the upper bound of feature misses is obtained, it could be used to prune graphs. Let f_1, f_2, ..., f_n be the indexing features. Given a target graph G and a query graph Q, let $\mathbf{u} = [u_1, u_2, \ldots, u_n]^T$ and $\mathbf{v} = [v_1, v_2, \ldots, v_n]^T$ be their corresponding feature vectors, where u_i and v_i are the frequencies (i.e., the number of embeddings) of feature f_i in graphs G and Q. Figure 5.4 shows the two feature vectors $\mathbf{u}$ and $\mathbf{v}$. As mentioned before, for any feature set, the corresponding feature vector of a target graph can be obtained from the feature-graph matrix directly without scanning the graph database.

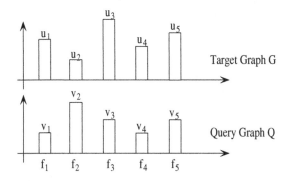

Figure 5.4. Frequency Difference

Eq. (5.4) calculates frequency difference of f_i between the query graph and the target graph,

$$r(u_i, v_i) = \begin{cases} 0, & if \ u_i \geq v_i, \\ v_i - u_i, & otherwise. \end{cases} \tag{5.4}$$

For the feature vectors shown in Figure 5.4, $r(u_1, v_1) = 0$; the extra embeddings from the target graph are not taken into account. The summed frequency difference of each feature in G and Q is written as $d(G, Q)$. Eq. (5.5) sums up all the frequency differences,

$$d(G, Q) = \sum_{i=1}^{n} r(u_i, v_i). \tag{5.5}$$

Suppose the query can be relaxed with k edges and the upper bound of allowed feature misses is then estimated using the greedy algorithm mentioned before. If $d(G, Q)$ is greater than that bound, it can be concluded that G does not contain Q within k edge relaxations. For this case, it is not necessary to perform any complicated structure comparison between G and Q. Since all the computations are done on the preprocessed information in the indices, the filtering process is fast.

3.4 Feature Set Selection

Though a bit counter-intuitive, using all the features together will not necessarily give the optimal solution; in some cases, it even deteriorates the performance rather than improving it. Given a query graph Q, let $F = \{f_1, f_2, \ldots, f_m\}$ be the set of features included in Q, and d_F^k the maximal number of features missed in F after Q is relaxed (either relabeled or deleted) with k edges. Relabeling and deleting an edge e in Q have the same effect: the features containing e are broken. Let $\mathbf{u} = [u_1, u_2, \ldots, u_m]^T$ and $\mathbf{v} = [v_1, v_2, \ldots, v_m]^T$ be the feature vectors built from a target graph G in the graph database and a query graph Q based on a chosen feature set F. Let $\Gamma_F = \{G | d(G, Q) > d_F^k\}$, which is the set of graphs pruned from the database by the feature set F. It is obvious that, for any feature set F, the greater the cardinality of Γ_F, the better.

In general, a candidate graph G passing a filter should satisfy the following inequality,

$$r(u_1, v_1) + r(u_2, v_2) + \ldots + r(u_n, v_n) \le d_F^k. \tag{5.6}$$

Let P be the maximum common subgraph of G and Q. Vector $\mathbf{u}' = [u_1', u_2', \ldots, u_n']^T$ is its feature vector. If G contains Q within the relaxation ratio, P should contain Q within the relaxation ratio as well, i.e.,

$$r(u_1', v_1) + r(u_2', v_2) + \ldots + r(u_n', v_n) \le d_F^k. \tag{5.7}$$

Since for any feature f_i, $u_i \ge u_i'$, we have

$$
\begin{aligned}
r(u_i, v_i) &\le r(u_i', v_i), \\
\sum_{i=1}^{n} r(u_i, v_i) &\le \sum_{i=1}^{n} r(u_i', v_i).
\end{aligned}
$$

Inequality (5.7) is stronger than Inequality (5.6). Assume that Inequality (5.7) does not hold for graph P, and there exists a feature f_i such that its frequency in P is too small to keep Inequality (5.7) true. However, Inequality (5.6) could still hold for graph G, if the misses of f_i is compensated by more occurrences of other features in G. This phenomenon is called *feature conjugation*. Feature conjugation likely takes place since the filtering does not distinguish the misses of individual features, but a collection of features. Due to feature conjugation, some graphs might not be pruned by the feature-based structural filtering method.

Definition 5.7 (Selectivity). *Given a graph database D, a query graph Q, and a feature f, the selectivity of f is defined by its average frequency difference within D and Q, written as $\delta_f(D, Q)$. $\delta_f(D, Q)$ is equal to the average of $r(u, v)$, where u is a variable denoting the frequency of f in a graph belonging to D, v is the frequency of f in Q, and r is defined in Eq. (5.4).*

There are three general feature set selection principles. The first principle is to select a large number of features. If only a small number of features are selected, the maximum allowed feature misses may become very close to $\sum_{i=1}^{n} v_i$. In that case, the filtering algorithm loses its pruning power. The second one is to make sure features cover the entire query graph. If most of the features cover several common edges, the relaxation of these edges will make the maximum allowed feature misses too big. The third one is to separate features with different selectivity. Low selective features deteriorate the potential filtering power from high selective ones due to frequency conjugation.

The above three criteria are not consistent with each other. For example, if all the features in a query graph are used, the second and the third principles will be violated since features often are concentrated in the center of a graph. On the other hand, one cannot use the most selective features alone because a query graph might not have enough highly selective features. The task of feature set selection is to make a trade-off among these principles. In practice, using a single filter with all the features included is not expected to perform well. Yan et al. [37] introduced a multi-filter strategy: Multiple filters are constructed and applied sequentially, where each filter uses a subset of features. This strategy was demonstrated to outperform a single filter based approach.

3.5   Structures with Gaps

The graph indexing methods introduced so far only consider connected subgraphs in a graph database. SAGA [31] proposes using fragments that do not always correspond to connected subgraphs and allows gaps in the indexing fragments.

The indexing unit in SAGA is a set of k nodes from the graphs in a database, where k is a user specified parameter, and is usually a small number. However, it could be expensive to enumerate all possible k-node sets in a large graph database. SAGA puts a limit on the diameter of each k-node set. If any pair of nodes in a k-node set are too far apart, this fragment does not correspond to a meaningful substructure, thus is not worth indexing. For a k-node set $\{v_1, v_2, \ldots, v_k\}$, if any two nodes v_i and v_j satisfy $d(v_i, v_j) \leq d_{max}$, where d_{max} is a diameter limit, SAGA connects the two nodes by a pseudo edge. Only those fragments that form a connected graph with the original edges or the newly introduced pseudo edges are indexed. Because of the pseudo edges, SAGA could index fragments with gaps.

The matching process of SAGA has three steps. The first step is to find small hits. In this step, the query graph is broken into small fragments and the graph index is probed to find database fragments that are similar to the query fragments. The second step is to assemble small hits retrieved in the first step to formulate larger matches. In this step, the small hits are first grouped by the database graph IDs and two neighbor hits are connected with each other to formulate a hit-compatible graph. This graph will tell which hits could be merged together to form a potential large match for the given query graph. The third step examines each candidate match and produces a set of real matches. SAGA allows users to specify a threshold to control the percentage of gap nodes in the subgraph match.

Different from Grafil [37] and SAGA [31], TALE [32] employs a new graph indexing method, called NH-Index (Neighborhood Index) for approximate subgraph matching of large query graphs efficiently. Instead of indexing various kinds of subgraphs in a graph database, NH-Index only considers the neighborhood structure of each node in a graph. Therefore, the number of indexing structures in NH-Index is equal to the number of nodes in the database, which is much smaller than the number of features used in many feature-based indexing methods. TALE also has an innovative matching paradigm for querying large graphs. Unlike the existing graph matching tools that treat every node in a graph equally, TALE distinguishes nodes by their importance in a graph structure. The algorithm first probes the NH-Index to match the important nodes in a query graph, and then progressively extends the matches by enclosing satisfiable nearby nodes of the matched nodes. TALE was applied to two real biological datasets and was able to produce meaningful results in both cases [32].

4. Reverse Substructure Search

In contrast to substructure search (Definition 5.1) which finds all graphs that contain a query graph, reverse substructure search finds all graphs that are contained by a query graph. Reverse substructure search finds applications in chem-informatics, pattern recognition [11] (visual surveillance, face recognition), cyber security (virus signature detection [10]), information management (user-interest mapping [26]), etc. For example, in chemistry, a descriptor is a set of atoms with designated bonds that has certain properties of chemical reactions. Given a new molecule, identifying "descriptor" structures can help researchers to understand its possible properties. In computer vision, attributed relational graphs (ARG) [11] are used to model images by transforming them into spatial entities such as points, lines, and shapes. ARG also connects these spatial entities (nodes) together with their mutual relationships (edges) such as distances, using a graph representation. The graph models of basic objects such as humans, animals, cars, airplanes, are built first. A recognition system could then query these models to identify objects, or perform large-scale video search for specific models if the key frames of videos are represented by ARGs. Such a system can also be used to automatically recognize and classify objects in technical drawings.

Definition 5.8 (Reverse Substructure Search). *Given a graph database* $\mathcal{D} = \{G_1, G_2, \ldots, G_n\}$ *and a graph query* Q, *find all graphs* G_i *in* $\mathcal{D}$, *s.t.,* $Q \supseteq G_i$.

Reverse substructure search has its unique characteristics. The pruning strategy employed in substructure search has *inclusion logic*: Given a query graph Q and a database graph $G \in \mathcal{D}$, if a feature $f \subseteq Q$ and $f \not\subseteq G$, then $Q \not\subseteq G$. That is, if feature f is in Q then the graphs not having f are pruned. The inclusion logic prunes graphs using features contained in the query graph. On the contrary, reverse substructure search has an exclusion logic: If a feature $f \not\subseteq Q$ and $f \subseteq G$, then $Q \not\supseteq G$. That is, if feature f is not in Q then the graphs having f are pruned.

According to the exclusion logic, given a graph database D, the best indexing features are those subgraphs contained by lots of graphs in D, but unlikely contained by a query graph. This kind of subgraph features are called *contrast features*. There is a connection between contrast subgraphs and their frequency: Both infrequent and very frequent subgraphs are likely not contrastive, and thus not useful for indexing. Therefore, one can apply frequent graph pattern mining and select those contrast subgraphs. The number of contrast subgraphs could be huge; most of them are very similar to each other. Since the index performance is determined by a set of indexing features, rather than individual ones, it is important to find a set of contrast subgraphs that collectively perform well. Chen et al. [4] developed a redundancy-aware selection mechanism, cIndex, to sort out a set of distinctive contrast subgraphs that can maximize the pruning performance for a set of query graphs. cIndex has a flat index structure, where each feature is tested sequentially against queries. Based on cIndex, cIndex-BottomUp and cIndex-TopDown were developed to support hierarchical indexing models that could further improve the pruning capability.

The bottom-up hierarchical index builds indices layer by layer starting from the bottom-level original graphs in a database. Figure 5.5(a) shows a bottom-up hierarchical index where the i_{th}-level index $\mathcal{I}_i$ is built by applying cIndex to features in the $(i-1)_{th}$-level index $\mathcal{I}_{i-1}$. For example, the first-level index $\mathcal{I}_1$ is built on the original graph database by cIndex. Once this is done, the features in $\mathcal{I}_1$ can be regarded as another graph database, where cIndex can be executed again to form a second-level index $\mathcal{I}_2$. Following this manner, one can continue building higher-level indices until the pruning gain becomes zero. This method is called cIndex-BottomUp. Note that in a bottom-up index, features on the i_{th}-level must be subgraphs of features on the $(i-1)_{th}$-level. In Figure 5.5(a), subgraph relationships are shown as edges. For example, f_1 is a subgraph of f_2, which is in turn a subgraph of f_3. Given a query graph Q, if $f1 \not\subseteq Q$, then the tree covered by f_1 need not be examined due to the exclusion logic. Since the index on each level will save some isomorphism tests for the

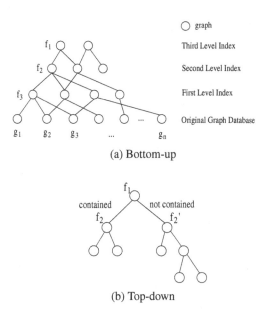

(a) Bottom-up

(b) Top-down

Figure 5.5. cIndex

graphs it indexes, it is obvious that cIndex-BottomUp should outperform the flat index of cIndex.

The top-down hierarchical index first puts f_1, the feature with the highest pruning power, at the top of the hierarchy (Figure 5.5(b)). Given a query graph Q, if f_1 is contained by Q, f_2 is further tested against Q; if f_1 is not contained by Q, all the graphs indexed by f_1 are pruned, and then the second feature f_2' is tested for the remaining graphs. In a flat index built by cIndex, f_2 and f_2' are forced to be the same: No matter whether f_1 is contained by Q or not, the same second feature will be examined next. However, in a top-down index, they can be different. As shown in [4], cIndex-TopDown achieved the best performance due to its differentiating index structure.

5. Conclusions

Graph indexing is one of the emerging important tasks in graph database management and graph data mining. It is fundamental to many graph related applications, especially when an application involves large scale graph databases. In this chapter, we introduced the concepts of substructure search, approximate substructure search, and feature-based graph indexing methods that mine and index a compact set of discriminative and selective structure features for fast graph retrieval. These methods are going to significantly improve the performance of advanced graph applications such as graph classification and clustering.

References

[1] R. Baeza-Yates and B. Ribeiro-Neto. *Modern Information Retrieval.* ACM Press/Addison-Wesley, 1999.

[2] S. Beretti, A. Bimbo, and E. Vicario. Efficient matching and indexing of graph models in content based retrieval. *IEEE Trans. on Pattern Analysis and Machine Intelligence*, 23:1089–1105, 2001.

[3] H. Bunke and G. Allermann. Inexact graph matching for structural pattern recognition. *Pattern Recognition Letters*, 1(4):245–253, 1983.

[4] C. Chen, X. Yan, P. S. Yu, J. Han, D.-Q. Zhang, and X. Gu. Towards graph containment search and indexing. In *Proc. of 2007 Int. Conf. on Very Large Data Bases (VLDB'07)*, pages 926 – 937, 2007.

[5] Q. Chen, A. Lim, and K. W. Ong. D(k)-Index: An adaptive structural summary for graph-structured data. In *Proc. of 2003 ACM-SIGMOD Int. Conf. Management of Data (SIGMOD'03)*, pages 134–144, 2003.

[6] J. Cheng, Y. Ke, W. Ng, and A. Lu. FG-Index: Towards verification-free query processing on graph databases. In *Proc. of 2007 ACM Int. Conf. on Management of Data (SIGMOD'07)*, pages 857 – 872, 2007.

[7] C. Chung, J. Min, and K. Shim. APEX: An adaptive path index for xml data. In *Proc. of 2002 ACM Int. Conf. on Management of Data (SIG-MOD'02)*, pages 121–132, 2002.

[8] S. Cook. The complexity of theorem-proving procedures. In *Proc. of the 3rd ACM Symp. on Theory of Computing (STOC'71)*, pages 151–158, 1971.

[9] B. Cooper, N. Sample, M. Franklin, G. Hjaltason, and M. Shadmon. A fast index for semistructured data. In *Proc. of 2001 Int. Conf. on Very Large Data Bases (VLDB'01)*, pages 341–350, 2001.

[10] Y. Fang, , R. Katz, and T. Lakshman. Gigabit rate packet pattern-matching using TCAM. In *Proc. of the 12th IEEE Int. Conf. on Network Protocols (ICNP'04)*, pages 174–183, 2004.

[11] K. Fu. A step towards unification of syntactic and statistical pattern recognition. *IEEE Trans. on Pattern Analysis and Machine Intelligence*, 8(3):398–404, 1986.

[12] R. Giugno and D. Shasha. GraphGrep: A fast and universal method for querying graphs. pages 112–115, 2002.

[13] R. Goldman and J. Widom. Dataguides: Enabling query formulation and optimization in semistructured databases. In *Proc. of 1997 Int. Conf. on Very Large Data Bases (VLDB'97)*, pages 436–445, 1997.

[14] T. Hagadone. Molecular substructure similarity searching: Efficient retrieval in two-dimensional structure databases. *J. Chem. Inf. Comput. Sci.*, 32:515–521, 1992.

[15] H. He and A. Singh. Closure-Tree: An index structure for graph queries. In *Proc. of 2006 Int. Conf. on Data Engineering (ICDE'06)*, 2006.

[16] D. Hochbaum. *Approximation Algorithms for NP-Hard Problems*. PWS Publishing, MA, 1997.

[17] L. Holder, D. Cook, and S. Djoko. Substructure discovery in the subdue system. In *Proc. of AAAI'94 Workshop on Knowledge Discovery in Databases (KDD'94)*, pages 169–180, 1994.

[18] C. James, D. Weininger, and J. Delany. *Daylight Theory Manual Version 4.82*. Daylight Chemical Information Systems, Inc, 2003.

[19] H. Jiang, H. Wang, P. Yu, and S. Zhou. GString: A novel approach for efficient search in graph databases. In *Proc. of 2007 Int. Conf. on Data Engineering (ICDE'07)*, pages 566–575, 2007.

[20] R. Kaushik, P. Shenoy, P. Bohannon, and E. Gudes. Exploiting local similarity for efficient indexing of paths in graph structured data. In *Proc. of 2002 Int. Conf. on Data Engineering (ICDE'02)*, pages 129–140, 2002.

[21] T. Madej, J. Gibrat, and S. Bryant. Threading a database of protein cores. *Proteins*, 3-2:289–306, 1995.

[22] B. Messmer and H. Bunke. A new algorithm for error-tolerant subgraph isomorphism detection. *IEEE Trans. on Pattern Analysis and Machine Intelligence*, 20:493–504, 1998.

[23] T. Milo and D. Suciu. Index structures for path expressions. *Lecture Notes in Computer Science*, 1540:277–295, 1999.

[24] N. Nilsson. *Principles of Artificial Intelligence*. Morgan Kaufmann, Palo Alto, CA, 1980.

[25] E. Petrakis and C. Faloutsos. Similarity searching in medical image databases. *Knowledge and Data Engineering*, 9(3):435–447, 1997.

[26] M. Petrovic, H. Liu, and H. Jacobsen. G-ToPSS: Fast filtering of graph-based metadata. In *Proc. of 2005 Int. Conf. on World Wide Web (WWW'05)*, pages 539–547, 2005.

[27] J. Raymond, E. Gardiner, and P. Willett. Rascal: Calculation of graph similarity using maximum common edge subgraphs. *The Computer Journal*, 45:631–644, 2002.

[28] D. Shasha, J. Wang, and R. Giugno. Algorithmics and applications of tree and graph searching. In *Proc. of the 21th ACM Symp. on Principles of Database Systems (PODS'02)*, pages 39–52, 2002.

[29] A. Shokoufandeh, S. Dickinson, K. Siddiqi, and S. Zucker. Indexing using a spectral encoding of topological structure. In *Proc. of IEEE Int. Conf. on Computer Vision and Pattern Recognition (CVPR'99)*, pages 2491–2497, 1999.

[30] S. Srinivasa and S. Kumar. A platform based on the multi-dimensional data model for analysis of bio-molecular structures. In *Proc. of 2003 Int. Conf. Very Large Data Bases (VLDB'03)*, pages 975–986, 2003.

[31] Y. Tian, R. McEachin, C. Santos, D. States, and J. Patel. SAGA: A subgraph matching tool for biological graphs. *Bioinformatics*, 23:232–239, 2007.

[32] Y. Tian and J. Patel. TALE: A tool for approximate large graph matching. *Proc. of 2008 Int. Conf. on Data Engineering (ICDE'08)*, pages 963–972, 2008.

[33] P. Willett, J. Barnard, and G. Downs. Chemical similarity searching. *J. Chem. Inf. Comput. Sci.*, 38:983–996, 1998.

[34] D. Williams, J. Huan, and W. Wang. Graph database indexing using structured graph decomposition. In *Proc. of 2007 Int. Conf. on Data Engineering (ICDE'07)*, pages 976–985, 2007.

[35] H. Wolfson and I. Rigoutsos. Geometric hashing: An introduction. *IEEE Computational Science and Engineering*, 4:10–21, 1997.

[36] X. Yan, P. S. Yu, and J. Han. Graph indexing: A frequent structure-based approach. In *Proc. of 2004 ACM-SIGMOD Int. Conf. on Management of Data (SIGMOD'04)*, pages 335–346, 2004.

[37] X. Yan, P. S. Yu, and J. Han. Substructure similarity search in graph databases. In *Proc. of 2005 ACM-SIGMOD Int. Conf. on Management of Data (SIGMOD'05)*, pages 766 – 777, 2005.

[38] P. Zhao, J. Yu, and P. Yu. Graph indexing: tree + delta $>=$ graph. In *Proc. of 2007 Int. Conf. on Very Large Data Bases (VLDB'07)*, pages 938–949, 2007.

[39] L. Zou, L. Chen, J. Yu, and Y. Lu. A novel spectral coding in a large graph database. In *Proc. of the 11th Int. Conf. on Extending Database Technology (EDBT'08)*, pages 181–192, 2008.

Chapter 6

GRAPH REACHABILITY QUERIES: A SURVEY

Jeffrey Xu Yu
The Chinese University of Hong Kong, China
yu@se.cuhk.edu.hk

Jiefeng Cheng
The Chinese University of Hong Kong, China
jfcheng@se.cuhk.edu.hk

Abstract   There are numerous applications that need to deal with a large graph, including bioinformatics, social science, link analysis, citation analysis, and collaborative networks. A fundamental query is to query whether a node is reachable from another node in a large graph, which is called a reachability query. In this survey, we discuss several existing approaches to process reachability queries. In addition, we will discuss how to answer reachability queries with the shortest distance, and graph pattern matching over a large graph.

Keywords: Graph, Reachability, Coding, Graph Pattern Matching.

1.   Introduction

Graph structured data is enjoying an increasing popularity as web technology and archiving techniques advance. Numerous emerging applications need to work with graph-like data due to its expressive power to handle complex relationships among objects. Instances include navigation behavior analysis for web usage mining [3], web site analysis [22], and biological network analysis for life science [33]. In addition, *RDF* allows users to explicitly describe semantic resources in graphs [6]. Querying and analyzing graph structured data becomes important. As a major standard for representing data on the World-Wide-Web, *XML* provides facilities for users to view data as graphs with two

C.C. Aggarwal and H. Wang (eds.), *Managing and Mining Graph Data*,
Advances in Database Systems 40, DOI 10.1007/978-1-4419-6045-0_6,
© Springer Science+Business Media, LLC 2010

different links, the parent-child links (document-internal links) and reference links (cross-document links), where the cross-document links are supported by value matching using ID/IDREF in *XML*. *XLink* (*XML* Linking Language) [19] and *XPointer* (*XML* Pointer Language) [20] provide more facilities for users to manage their complex data as graphs and integrate data effectively. The dominance of graphs in real-world applications demands new graph data management so that users can access graph data effectively and efficiently.

Graph reachability (or simply reachability) queries, to test whether there is a path from a node v to another node u in a large directed graph, have being studied [1, 24, 17, 28–30, 23, 13, 34, 32, 9, 14, 5, 26, 25, 10] and are deemed to be a very basic type of graph queries for many applications. Consider a semantic network that represents people as nodes in the graph and relationships among people as edges in the graph. There are needs to understand whether two people are related for security reasons [2]. On biological networks, where nodes are either molecules, or reactions, or physical interactions of living cells, and edges are interactions among them, there is an important question to "find all genes whose expressions are directly or indirectly influenced by a given molecule" [33]. All those questions can be mapped into reachability queries. The needs of such a reachability query can be also found in *XML* when two types of links (document-internal links and cross-document links) are treated the same. Recently, [8, 12, 35] studied graph matching problem on large graph data, where nodes in a match are connected by reachability relationships. Reachability queries are so common that fast processing is mandatory.

Reachability Queries: Let $G = (V, E)$ be a large directed graph that has n nodes and m edges. A *reachability queries* is denoted as $u \rightsquigarrow v$, where u and v are two nodes in G. Here, $u \rightsquigarrow v$ returns true if and only if there is a directed path in the directed graph G from u to v. In other words, let TC be the edge transitive closure of graph G, $u \rightsquigarrow v$ is true if and only if $(u, v) \in TC$. We call such a pair (u, v) a connection. Note: TC can be very large for a large and dense graph G. A reachability query over a directed graph G can be answered over a corresponding directed acyclic graph (DAG) of the graph G based on strongly connected components. Two nodes, u and v, are said to be in a strongly connected component, if and only if both $u \rightsquigarrow v$ and $v \rightsquigarrow u$ are true. And in a strongly connected component, for every two nodes, u and v, $u \rightsquigarrow v$ and $v \rightsquigarrow u$ are true. Given a directed graph $G(V, E)$, its strongly connected components, $C_1, C_2, \cdots$, can be efficiently identified in $O(n+m)$ time [18]. A DAG of the graph G, denoted G', can be constructed as follows. First, a strongly connected component C_i in G is replaced by a representative node v in G'. Second, all the edges between the nodes in the strongly connected component C_i are removed while all incoming edges and outgoing edges of C_i will be represented as incoming edges and outgoing edges of the representative node v in G'. A reachability query, $u \rightsquigarrow v$, over G can be processed over the

Table 6.1. The Time/Space Complexity of Different Approaches [25]

	Query Time	Index Construction Time	Index size
Transitive Closure [31]	$O(1)$	$O(nm)$	$O(n^2)$
Tree+SSPI [8]	$O(m-n)$	$O(n+m)$	$O(n+m)$
GRIPP [32]	$O(m-n)$	$O(n+m)$	$O(n+m)$
Dual-Labeling [34]	$O(1)$	$O(n+m+t^3)$	$O(n+t^2)$
Tree Cover [1]	$O(\log n)$	$O(nm)$	$O(n^2)$
Chain Cover [9]	$O(\log k)$	$O(n^2 + kn\sqrt{k})$	$O(nk)$
Path-Tree Cover [26]	$O(\log^2 k')$	$O(mk')$ or $O(nm)$	$O(nk')$
2-Hop Cover [17]	$O(m^{1/2})$	$O(n^3 \cdot |TC|)$	$O(nm^{1/2})$
3-Hop Cover [25]	$O(\log n + k)$	$O(kn^2 \cdot |Con(G)|)$	$O(nk)$

DAG G' by checking whether the corresponding strongly connected component, where v resides, is reachable from the corresponding strongly connected components, where u resides. In the following, without otherwise specified, we assume G is a DAG.

There are two possible approaches to process a reachability query, $u \rightsquigarrow v$, in a graph G. It can be processed as to traverse from u to v using breadth- or depth-first search over the graph G on demand, when a reachability query is issued. It incurs high cost as $O(n+m)$ time. On the other hand, it can be processed as to check whether (u,v) exists in the edge transitive closure of the graph G, TC, by precomputing and maintaining the edge transitive closure TC on disk. It results in high storage consumption in $O(n^2)$. The two approaches are infeasible. The former requires too much time in querying and the latter requires too much space.

In the literature, many approaches have been proposed to reduce the space consumption, and at the same time answer reachability queries efficiently. Recall that by precomputing and maintaining the edge transitive closure TC of G, it can answer a reachability query in $O(1)$ time at the expense of $O(n^2)$ space. Here, the edge transitive closure TC servers as an index to be used to answer reachability queries. The existing approaches attempt to increase the query processing time marginally in the range of $O(1)$ and $O(n+m)$, where $O(1)$ is the query time using the edge transitive closure TC and $O(n+m)$ is the query time using breadth- or depth-first search, by constructing an index that can significantly reduce the space consumption. For example, some approaches construct an index based on a spanning tree of the graph G plus some additional information to maintain reachability information over the graph G, and some construct an index that compresses the edge transitive closure TC. On this direction, the time of spending on constructing an index becomes an important issue too.

Table 6.1 shows a summary on the time/space complexity of different approaches [25]. Given a graph $G(V, E)$. Let $n = |V|$ and $m = |E|$. Simon proposes an algorithm to compute the edge transitive closure for a DAG, G, in $O(nm)$ time [31]. In other words, the time to construct an index based on the edge transitive closure of G is in $O(nm)$ time, and the index size is in $O(n^2)$ space, in the worst case. With the edge transitive closure constructed, the query time is constant $O(1)$.

In [8], Chen et al. propose an index by utilizing a spanning tree of the graph G. It takes $O(n + m)$ time to construct an index in $O(n + m)$ size. Given two nodes u and v in G, it can answer $u \rightsquigarrow v$ in $O(1)$ time if there is a path from u to v in the spanning tree, using a simple predicate, denoted $\mathcal{P}(,)$, between the codes (or labels) assigned to nodes over the spanning tree. We will discuss different encoding schema that assign codes (or labels) to nodes in G later in detail in this survey, and use codes and labels interchangeably. Let the codes for u and v be $\mathsf{code}(u)$ and $\mathsf{code}(v)$. If the predicate $\mathcal{P}(\mathsf{code}(u), \mathsf{code}(v))$ is true, then $u \rightsquigarrow v$ is true. However, because the codes are assigned based on the connections over the spanning tree of the graph G, it does not mean that $u \rightsquigarrow v$ is false if $\mathcal{P}(\mathsf{code}(u), \mathsf{code}(v))$ is false. There are edges in G that do not appear in the spanning tree. Chen et al. use an additional data structure called SSPI (Surrogate&Surplus Predecessor Index) to answer a reachability query in run time, which takes $O(m - n)$ time in the worst case. We call this approach Tree+SSPI. Like [8], a spanning tree of a graph G is also used in [32]. In [32], Trißl and Leser build an index, called GRIPP (GRaph Indexing based on Pre- and Postorder numbering), using a spanning tree of the graph G. Trißl and Leser discuss traversal strategies using the proposed GRIPP. The time and space complexities are the same to Tree+SSPI.

Wang et al. propose a dual-labeling approach in [34] for sparse graphs based on the observation that the majority of large graphs in real applications are sparse. It implies that the number of edges in the graph G that do not appear in a spanning tree of G is small. Let tree edges denote the edges that appear in the spanning tree, and non-tree edges denote the edges that do not appear in the spanning tree but appear in G. Let t be the number of such non-tree edges. Wang et al. consider to use a tree coding scheme (also called labeling) for tree edges and a graph coding (also called graph labeling) scheme for non-tree edges for sparse graphs where $t \ll n$. It handles the edge transitive closure over non-tree edges. The dual-labeling approach achieves $O(1)$ query time with an index of size $O(n + t^2)$ that is constructed in $O(n + m + t^3)$ time.

Agrawal et al. in [1] study a tree cover approach to assign labels to nodes in a DAG. In brief, if a node u can reach a node v, then u can reach any nodes in the subtree rooted at v. Agrawal et al. propose an optimal tree cover that maximally compresses the edge transitive closure. The index size is $O(n^2)$ in the worst case, but in practice, it can compress edge transitive closure which results in an even better compression rate than a chain cover [24, 9] which we will discuss next. The time complexity for index construction is $O(nm)$. It can construct an index for a large graph efficiently. The query time is $O(\log n)$.

Jagadish in [24] proposes a chain cover approach. The chain cover is to decompose a graph G into pairwise disjoint chains. A chain is more general than a path. Consider a path $a \rightarrow b \rightarrow c \rightarrow d$ in G, where $x \rightarrow y$ represents a directed edge in G. The path can be considered as a chain itself, $a \rightsquigarrow b \rightsquigarrow c \rightsquigarrow d$, where $x \rightsquigarrow y$ represents y is reachable from x. The path can be decomposed into two pairwise disjoint chains, $a \rightsquigarrow c$ and $b \rightsquigarrow d$. Both $a \rightsquigarrow c$ and $b \rightsquigarrow d$ are not paths. Like the tree cover, if a node u can reach a node v, then u can reach any nodes in the chain from the position of the node v. Jagadish proposes an algorithm in $O(n^3)$ to find the minimal number of chains, in G. The number of chains for G is called the width of G, denoted by k. Based on the chain cover, an index in $O(nk)$ size can be constructed. The query time is $O(\log k)$. In [9], Chen and Chen propose a new approach that can further reduce the time complexity of constructing the index based on the chain over to $O(n^2 + kn\sqrt{k})$.

Jin et al. propose path-tree cover in [26] along the line of tree cover [1]. Jin et al. decompose G into pairwise disjoint paths and build a tree over the paths by treading a decomposed path as a node in the tree. Let k' be the number of pairwise disjoint paths in G. Two algorithms are proposed, namely, PTree-1 and PTree-2. Both construct an index in $O(nk')$ space. PTree-1 constructs the index in $O(nm)$ time, whereas PTree-2 constructs it in $O(mk')$ time. The query time is in $O(\log^2 k')$.

Cohen et al. in [17] propose an index called 2-hop cover. A node, u, in a graph G is assigned two sets of nodes, as its label, called $L_{in}(u)$ and $L_{out}(u)$. $L_{in}(u)$ contains a set of nodes that can reach u and $L_{out}(u)$ contains a set of nodes that u can reach. The labels assigned to nodes are done in a way to ensure $u \rightsquigarrow v$ to be true if and only if $L_{out}(u) \cap L_{in}(v) \neq \emptyset$. It turns out to be a set cover problem. Cohen et al. propose an approximate algorithm to construct an index in $O(nm^{1/2})$ space. The time complexity for constructing such an index remains open. In [26], the conjecture is $O(n^3 \cdot |TC|)$ where $|TC|$ is the size of the edge transitive closure of G. Several efficient algorithms are proposed to compute 2-hop cover [29, 13, 14]. The 2-hop cover maintenance is studied in [30, 5]. Jin et al. in [25] further study a new approach, called 3-hop, that combines chain cover and 2-hop cover. The index construction time is $O(kn^2 \cdot |Con(G)|$. Here k is the number of pairwise disjoint paths in G, and $Con(G)$ is transitive closure contour of G defined in [25].

All the above are about how to answer reachability queries. Cohen et al. in [17] and Schenkel et al. in [30] address the distance-aware 2-hop cover which is to answer reachability queries with the shortest distance. Cheng and Yu in [10] propose efficient algorithms to fast compute distance-aware 2-hop cover.

The main difficult of computing distance-aware 2-hop cover is that it cannot condense a general directed graph into a DAG.

Before we discuss different graph coding schema, we explain a tree coding scheme for a tree. We call it single interval tree coding scheme in this survey. Many graph coding schema make use of the similar ideas used in the single interval tree coding scheme.

Single Interval Tree Coding Scheme: Let $G_S(V, E)$ be a tree. The single interval tree coding scheme (or simply SIT coding scheme) assigns a node $u \in G_S$ a code which is an interval, denoted $\mathsf{sitcode}(u) = [u_{start}, u_{end}]$, where u_{start} and u_{end} are two numbers such that $u_{start} < u_{end}$. The reachability, $u \rightsquigarrow v$, between two nodes, u and v, can be answered using the two corresponding codes, $\mathsf{sitcode}(u)$ and $\mathsf{sitcode}(v)$, in constant time $O(1)$. We denote it as a predicate $\mathcal{P}_{sit}(,)$

$$\mathcal{P}_{sit}(\mathsf{sitcode}(u), \mathsf{sitcode}(v)) = u_{start} < v_{start} \wedge v_{end} < u_{end}$$

Then, $u \rightsquigarrow v$ is true if and only if $\mathcal{P}_{sit}(\mathsf{sitcode}(u), \mathsf{sitcode}(v))$ is true. The codes can be assigned by traversing the tree G_S. Here, for a node, u, the u_{start} and u_{end} are the preorder and postorder values in a depth-first traversal of the tree. A counter is used with an initial value 0, and the counter value will increase by 1 before it visits another node in the traversal. In the tree traversal, a node will be visited twice. The u_{start} and u_{end} of a node u are assigned to be the counter values before and after all descendants of u have been traversed.

2.   Traversal Approaches

In this section, we introduce two approaches, namely, Tree+SSPI [8] and GRIPP [32]. Both approaches use the SIT coding scheme to assign codes to nodes in a spanning tree of a graph G, and attempt to reduce the query processing time in traversal using either additional data structures or processing strategies. It is worth noting that Tree+SSPI [8] is proposed for pattern matching in a general context, and can be used to answer reachability queries.

Let $T_S(V_S, E_S)$ be a spanning tree of a graph $G(V, E)$. Here V_S and E_S are sets of nodes and edges of the spanning tree T_S. Note that $V_S = V$ and $E_S \subseteq E$. We use E_S to denote the set of tree edges of the graph G, and $E_R = E - E_S$ to denote the set of non-tree edges of the graph G that do not appear in E_S. In addition, below in discussions of Tree+SSPI and GRIPP, we assume that every node in G is assigned a code based on the SIT coding scheme. Given a reachability query $u \rightsquigarrow v$, Tree+SSPI and GRIPP first check whether the predicate $\mathcal{P}_{sit}(\mathsf{sitcode}(u), \mathsf{sitcode}(v))$ is true or not. If it is true, then $u \rightsquigarrow v$ is true. Otherwise, Tree+SSPI and GRIPP need to take additional actions to further check the reachability $u \rightsquigarrow v$, because u can reach v through a combination of tree edges and non-tree edges. Below, we discuss the cases that $u \rightsquigarrow v$ cannot be answered simply using the SIT coding scheme.

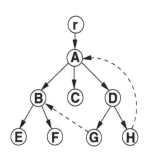

Node	Start	End	Type
r	0	21	tree
A	1	20	tree
B	2	7	tree
E	3	4	tree
F	5	6	tree
C	8	9	tree
D	10	19	tree
G	11	14	tree
B'	12	13	non-tree
H	15	18	tree
A'	16	17	non-tree

Figure 6.1. A Simple Graph G (left) and Its Index (right) (Figure 1 in [32])

2.1 Tree+SSPI

In [8], in addition to the SIT codes assigned to nodes, Chen et al. use another "space-economic" index, known as SSPI (Surrogate&Surplus Predecessor Index), to maintain information that needs to be used at run time to check reachability. The SSPI keeps a predecessor list for a node v in G, denoted as $PL(u)$. There are two types of predecessors. One is called *surrogate*, and the other is called *immediate surplus predecessor*. The two types of predecessors are explained in terms of the involvement of non-tree edges. Consider $u \rightsquigarrow v$ that must visit some non-tree edges on the path from u to v. Assume that (v_x, v_y) is the last non-tree edge on the path from u to v, then v_y is a surrogate predecessor of v if $v_y \neq v$ and v_x is an immediate surplus predecessor of v if $v_y = v$. SSPI can be constructed in a traversal of the spanning tree T_S of the graph G starting from the tree root. When a node v is visited, all its immediate surplus predecessors are added into $PL(v)$. Also, all nodes in $PL(u)$ are added into $PL(v)$, where u is the parent node of v in the spanning tree. It is sufficient to answer reachability queries using both SIT coding scheme and the SSPI.

To process a reachability query $u \rightsquigarrow v$, assuming that the SIT codes used return false when checking $u_{start} < v_{start} \wedge v_{end} < u_{end}$, Chen et al. design a *TwigStackD* algorithm. The *TwigStackD* algorithm checks the reachability via tree edges using run time stacks in traversing the spanning tree, and checks reachability via possible non-tree edges, using a partial solution pool that maintains some popped nodes from run time stacks temporally. The SSPI is used to answer which nodes can possibly reach a node v via non-tree edges.

2.2 GRIPP

Trißl and Leser in [32] use the SIT coding scheme in a different way. Instead of using SSPI and run time stacks, Trißl and Leser focus on how to traverse the graph using the SIT codes. The graph dealt in [32] is a directed graph. We explain it using the same example used in [32]. Figure 6.1 shows a simple directed graph G on the left side and the GRIPP index table on the right side. The solid arrows indicate tree edges in G, and dotted arrows indicate non-tree edges in G. As shown in the GRIPP index table, a node in G is assigned with one or more than one SIT codes depending on the number of incoming edges to the node. The type in the GRIPP index table indicates the type of the incoming edge based on which the node is assigned a SIT code. The nodes with a type of non-tree in GRIPP index table are also called hop-nodes. Consider the node A, its SIT code, sitcode(A) = $[A_{start}, A_{end}]$ = $[1, 20]$, is assigned when A is traversed from/to r via the tree edge (r, A), and the duplication of A, a hop-node, denoted A', has a different SIT code $[16, 17]$, which is assigned when A is traversed from/to H via the non-tree edge (H, A). It can be understood that a directed graph G is represented as a tree with node duplications. In other words, all the hop-nodes, such as A' and B' in the GRIPP index table, are node duplications and become the leaf nodes in such a tree.

Trißl and Leser in [32] study how to reduce the traversing time when processing a reachability query. Consider $D \rightsquigarrow r$. Based on SIT codes given in the GRIPP index table, D can reach the nodes, G, H, A', and B', where A' and B' are two hop-nodes, because, sitcode(D) = $[10, 19]$, sitcode(G) = $[11, 14]$, sitcode(H) = $[15, 18]$, sitcode(A') = $[16, 17]$, and sitcode(B') = $[12, 13]$. It implies that via the two hop-nodes, A' and B', there exists possibility that $D \rightsquigarrow r$ is true. Intuitively, it needs to hop to A and B to further traverse the graph G. Suppose it traverses A via the hop-node A' followed by traversing B via the hop-node B'. First, when it picks up A to traverse, it can traverse to A itself again, because A can reach H and then traverse to A via the hop-node A'. In this case, it does not need to traverse to A second time, because it cannot find any new possible reachability. Second, when it picks up B to traverse, it cannot find any new possible reachability, because A can reach B via tree edges and it has already explored all possible reachability via A that must include all the possible reachability via B. Based on the idea behind, Trißl and Leser study traversing order, pruning strategies, and and stop conditions. Because finding the optimal traversing order is NP-complete, Trißl and Leser propose some heuristics. For example, it attempts to traverse the giant strongly connected component first.

3.   Dual-Labeling

Wang et al. in [34] investigate a dual-labeling coding scheme for a graph G. They use a SIT coding scheme to encode nodes that can be reached via tree edges over a spanning tree of the graph G, and a new coding scheme to encode nodes that can be possibly reached via non-tree edges. The codes assigned to

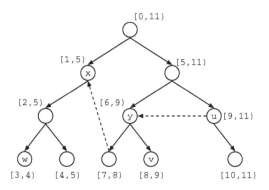

Figure 6.2. Tree Codes Used in Dual-Labeling (Figure 2 in [34])

nodes based on the tree edges over a spanning tree are slightly different from the SIT coding scheme used in GRIPP as seen in Figure 6.1. We also use the same example used in [34] to explain the main ideas.

Wang et al. assign modified SIT codes to nodes over a spanning tree of the graph G. We call it dual-tree code and denote it as dtcode(u) for $u \in G$, in the form of $[u_{start}, u_{end})$. An example is shown in Figure 6.2, where the solid arrows form a spanning tree and the dotted arrows are non-tree edges in G. The reachability $u \rightsquigarrow v$ over the spanning tree can be answered using dtcode(u) and dtcode(v) if $v_{start} \in$ dtcode(u) is true. We give a predicate $\mathcal{P}_{dt}(,)$ to test whether $u \rightsquigarrow v$ is true over the spanning tree.

$$\mathcal{P}_{dt}(\mathsf{dtcode}(u), \mathsf{dtcode}(v)) = v_{start} \in \mathsf{dtcode}(u)$$

Note: it does not mean that u cannot reach v if $\mathcal{P}_{dt}(\mathsf{dtcode}(u), \mathsf{dtcode}(v))$ is false, because there exist other non-tree edges via which u can possibly reach v. In [34], a non-tree edge (u', v') is represented as $u'_{star} \rightarrow [v'_{start}, v'_{end})$ in a link table. Consider Figure 6.2, there are two non-tree edges, such that $9 \rightarrow [6, 9)$ and $7 \rightarrow [1, 5)$. The link table maintains the edge transitive closure over the non-tree edges and therefore is also called a transitive link table. For example, the existence of the two non-tree edges, $9 \rightarrow [6, 9)$ and $7 \rightarrow [1, 5)$, in the transitive link table implies that $9 \rightarrow [1, 5)$ exists in the transitive link table. It is because the node with the dtcode $[7, 8)$ can be reached from the node with the dtcode $[6, 9)$ and therefore the node with dtcode $[9, 11)$ can reach the node with dtcode $[1, 5)$. Let t be the number of non-tree edges, the transitive link table is in $O(t^2)$ space. A reachability query, $u \rightsquigarrow v$, can be answered using the transitive link table. Let dtcode(u) $= [u_{start}, u_{end})$ and dtcode(v) $= [v_{start}, v_{end})$. Then, $u \rightsquigarrow v$ is true if it can find an entry, $i \rightarrow [j, k)$, in the transitive link table such as $i \in [u_{start}, u_{end})$ and $v_{start} \in [j, k)$. The former implies that u can reach the non-tree edge and the latter implies that from the non-tree edge v can be reached.

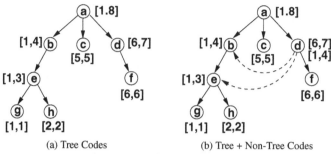

(a) Tree Codes   (b) Tree + Non-Tree Codes

Figure 6.3. Tree Cover (based on Figure 3.1 in [1])

In other to achieve $O(1)$ time, Wang et. al propose a transitive link count function (short for TLC function). As defined in Definition 1 in [34], *the proposed TLC function $N(x, y)$ computes the number of links $i \rightarrow [j, k)$ in the transitive link table that satisfy $i \geq x$ and $y \in [j, k)$.* Given two nodes, u and v, where $\text{dtcode}(u) = [u_{start}, u_{end})$ and $\text{dtcode}(u) = [u_{start}, u_{end})$. Assume that $\mathcal{P}_{dt}(\text{dtcode}(u), \text{dtcode}(t))$ is false. The following predicate $\mathcal{P}_{dg}(,)$ is defined over the graph via possible non-tree edges.

$$\mathcal{P}_{dg}(\text{dtcode}(u), \text{dtcode}(v)) = N(u_{start}, v_{start}) - N(u_{end}, v_{start}) > 0$$

$u \rightsquigarrow v$ is true over the possible non-tree edges if and only if the predicate $\mathcal{P}_{dg}(\text{dtcode}(u), \text{dtcode}(v))$ is true. Therefore, $u \rightsquigarrow v$ is true if and only if $\mathcal{P}_{dt}(\text{dtcode}(u), \text{dtcode}(v)) \lor \mathcal{P}_{dg}(\text{dtcode}(u), \text{dtcode}(v))$ is true.

Intuitively, it requires to maintain the TLC function $N(,)$ for every possible node pairs in G, which results in $O(n^2)$ space. In order to reduce it to $O(t^2)$ space, Wang et al. propose gridding and snapping techniques in [34]. Some techniques to trade off time for space are also discussed in [34].

4.   Tree Cover

As an early work, in 1989, Agrawal et al. proposed a tree cover code. It uses multiple intervals to encode every node in a graph G. Consider a tree shown in Figure 6.3(a). A node u is assigned an interval $[u_{start}, u_{end}]$, where u_{end} is the postorder in traversing the tree, and u_{start} is the smallest postorder in the descendants of the subtree rooted at the node u. Like the other tree coding, $u \rightsquigarrow v$ is true over the tree, if and only if $v_{end} \in [u_{start}, u_{end}]$ is true. Agrawal et al. consider how to assign codes to nodes in DAG by inheriting codes from a node v to another node u if there is a non-tree edge (u, v) in the graph G. Consider the DAG shown in Figure 6.3(b). There are two additional non-tree edges (d, b) and (d, e). The node d will inherit $[1, 4]$ and $[1, 3]$ from the nodes b and e respectively. Because $[1, 3] \subseteq [1, 4]$, d only needs to have an additional interval $[1, 4]$. Therefore, the code for a node u in G, denoted as $\text{tccode}(u) =$

Algorithm 1 Find-Tree-Cover(G)

1: let G' be a graph with an additional virtual root, γ, that links to all nodes in G that do not have any predecessors;
2: let L be the list of nodes in G' following a topological order;
3: $pred(\gamma) \leftarrow \emptyset$;
4: **for** each node v on L **do**
5:      **for** each pair of incoming edges (u, v) and (u', v) **do**
6:          **if** $|pred(u)| > |pred(u')|$ **then**
7:             delete the edge (u', v);
8:          **else**
9:             delete the edge (u, v);
10:          **end if**
11:      **end for**
12:      $pred(v) \leftarrow \{u\} \cup prev(u)$ for every incoming edge (u, v);
13: **end for**

$\{[u_{start_1}, u_{end_1}], [u_{start_2}, u_{end_2}], \cdots\}$, where u_{end_1} is the postorder when it traverses the spanning tree. In other words, $[u_{start_1}, u_{end_1}]$ is assigned to node u when traversing the spanning tree of the graph G, and the others are inherited from other nodes. Given the tree cover codes, $u \rightsquigarrow v$ is tree if and only if the postorder of v (v_{end_1}) is in an interval of the node u. The predicate $\mathcal{P}_{tc}(,)$ is given below.

$$\mathcal{P}_{tc}(\mathsf{tccode}(u), \mathsf{tccode}(v)) = \bigvee_i (v_{end_1} \in [u_{start_i}, u_{end_i}])$$

The total number of intervals for all codes in G becomes a factor to measure the quality of the tree cover. The total number varies depending on the selection of a spanning tree, known as tree cover, over the graph G. In [1], Agrawal et al. propose an algorithm to find the optimal tree cover. As shown in Algorithm 1, in order to achieve the optimal tree cover, for a node v, it retains the edge from the immediate predecessor of v with the maximum number of predecessors in the original DAG G, and delete the edges from the other immediate predecessors of v.

In [1], the storage issues and the tree-cover maintenance issue when a graph is updated are also discussed.

5. Chain Cover

Jagadish [24] proposes a chain cover coding scheme to answer a reachability query on a DAG G. A chain cover of G is a set of pairwise disjoint chains, $C_1, C_2, \cdots, C_k$. Here, a chain $C_i = v_{i_1} \rightsquigarrow v_{i_2} \rightsquigarrow \cdots \rightsquigarrow v_{i_k}$ where v_{i_j} is a node in G and $v_{i_{j+1}}$ is reachable from v_{i_j} in G. The union of the nodes in

---

Algorithm 2 Compute-Chain-Cover(G, $\{C_1, C_2, \cdots, C_k\}$)

Input: The DAG G, and a chain cover $\{C_1, \cdots, C_k\}$
Output: The chain cover code for every node in G

1: sort all nodes in G in topological order;
2: let every node v_i in G unmarked;
3: **while** there are unmarked node v_i in G that do not have unmarked immediate successors **do**
4:     chaincode(v_i) $\leftarrow \{(1, \infty), (2, \infty), \cdots, (k, \infty)\}$;
5:     let $L_{i,x}$ denote the x-th pair in chaincode(v_i);
6:     let $suc(v_i)$ denote the immediate successors of v_i in G;
7:     **for** every $v_j \in suc(v_i)$ **do**
8:         **for** $l = 1$ **to** k **do**
9:             $(l, p_{j,l}) \leftarrow L_{j,l}$;
10:             $(l, p_{i,l}) \leftarrow L_{i,l}$;
11:             **if** $p_{j,1} \leq p_{i,l}$ **then**
12:                 $L_{i,l} \leftarrow (l, p_{j,l})$;
13:             **end if**
14:         **end for**
15:     **end for**
16:     mark v_i;
17: **end while**
18: **return** the set of chaincode(v_i) for every $v_i \in G$;

---

all chains is the entire set of nodes in G, and the intersection of nodes in any two chains is empty. The optimal chain cover of G is a chain cover of G that contains the least number of chains among all possible chain covers of G.

Suppose the chain cover contains k chains, to answer the reachability queries, each node $v_i \in G$ is assigned a code, denote chaincode(v_i), which is a list of pairs, $\{(1, p_{i,1}), (2, p_{i,2}), \cdots, (k, p_{i,k})\}$. Each pair $(j, p_{i,j})$ means that the node v_i can reach any nodes from the position $p_{i,j}$ in the j-th chain. If v_i cannot reach any node in the j-th chain, then $p_{i,j} = +\infty$. The chain cover index contains chaincode(v_i) for every node v_i in G.

A reachability query $v_a \rightsquigarrow v_d$ can be answered using a predicate $\mathcal{P}_c(,)$ such that $v_a \rightsquigarrow v_d$ is true if and only if v_a appears at the $p_{a,j}$ position in a chain C_j and $p_{d,j} \leq p_{a,j}$. In other words, v_a can reach v_d in a chain C_j. All pairs in the chain cover index for G can be indexed and stored using a B+-tree. Answering a reachability query needs $O(\log(n))$ time with $O(n \cdot k)$ space.

Given a chain cover $C_1, C_2, \cdots, C_k$ of a DAG G, Algorithm 2 shows how to compute chaincode(v_i) for every $v_i \in G$. It visits every node in G in the reverse of topological order (line 3). For each node visited, its chaincode(v_i) is updated using its immediate successors if the corresponding position in the l-th chain, C_l, of an immediate successor is smaller than the current position v_i has in C_l. Let d_i be the out degree of node v_i (the number of immediate successors of v_i). The time complexity of Algorithm 2 is $O(\sum_{i=1}^{n}(d_i \cdot k)) = O(mk)$, where m is the number of edges in G. It becomes important to make k as small as possible. Below, we introduce two approaches that aim at computing the optimal chain cover with the minimal k.

5.1 Computing the Optimal Chain Cover

Jagadish in [24] proposes a min-flow approach to compute the optimal chain cover of a DAG G. The main idea is as follows. It constructs another graph H. For every node $v_i \in G$, it adds two nodes, x_i and y_i, in H and a directed edge (x_i, y_i) in H. In other words, a node in G is represented as an edge in H. For each edge (v_i, v_j) in G, it adds an edge (y_i, x_j) in H. A source node is added into H that links to every node with in-degree 0 in H, and a sink node is added that is linked by every node with out-degree 0 in H. Then, Jagadish proposes to find the min-flow from the source node to the sink node such that every edge (x_i, y_i) has a positive flow. It can be solved in time $O(n^3)$. Here, each flow corresponds to a chain in G. In such a way, it can get the chain cover of G. If a node may appear in several chains, it keeps one occurrence in any chain and removes the other occurrences.

Chen and Chen in [9] propose an approach using bipartite matching. All nodes in the DAG G are decomposed into several layers, $V_1, V_2, \cdots, V_h$, where h is the length of the longest path in G. The layers can be constructed as follows. V_1 is the set of nodes with out-degree 0 in G, and V_i is the set of nodes with out-degree 0 when the nodes in V_k, for $1 \leq k < i$ are removed from G. This can be done in $O(m)$ time.

Algorithm 3 shows how to find the optimal chain cover based on the layers. The main idea of Algorithm 3 is as follows. In each successive layers, it finds the maximum matching for the bipartite graph induced by the nodes in the two layers (line 1-4). For some unmatched node v, it adds a virtual node v' in the top of the two successive layer, in order to be further matched by nodes in the unseen upper layers (line 5-9). A potential edge (u, v') for some $u \in V_{i+2}$ is added, if and only if there is an edge from u to a node $x \in V_{i+1}$ and there is an alternating path from x to v'. A path is alternating with respect to M_i if and only if its edges alternately appear in $E_i \setminus M_i$ and M_i, where M_i is the maximum matching of the bipartite graph and E_i is the bipartite graph in the i-th iteration. Then, in line 10-13, each virtual node is resolved using the alternating paths by removing the virtual nodes, transferring the edges in the alternating paths, and adding the new edge from u to x as discussed above. An example for resolving a virtual node v' by an alternating path is illustrated in Figure 6.4. The optimal chain cover can be computed in time $O(n^2 + kn\sqrt{k})$

Algorithm 3 Optimal-Chain-Cover($G, \{V_1, V_2, \cdots, V_h\}$)

Input: a DAG G, and the layers $V_1, \cdots, V_h$
Output: The optimal chain cover $C_1, \cdots, C_k$

1: $V_1' \leftarrow V_1$;
2: **for** $i = 1$ **to** $h - 1$ **do**
3:     $V_{i+1}' \leftarrow V_{i+1}$;
4:     $M_i \leftarrow$ maximum matching of the bipartite graph induced by V_i' and V_{i+1}';
5:     **for all** unmatched node $v \in V_i'$ in M_i **do**
6:         create a virtual node v' in G;
7:         $V_{i+1}' \leftarrow V_{i+1}' \cup \{v'\}$;
8:         $M_i \leftarrow M_i \cup (v', v)$;
9:         create potential edges (u, v') for some $u \in V_{i+2}$;
10:     **end for**
11: **end for**
12: $CH \leftarrow M_1 \cup M_2 \cup \cdots \cup M_h$;
13: **for** $i = 1$ **to** $h - 1$ **do**
14:     **for all** virtual node $v' \in V_i'$ **do**
15:         resolve v' from CH using alternating paths in M_i;
16:     **end for**
17: **end for**
18: **return** CH;

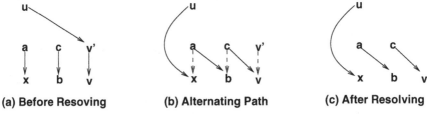

(a) Before Resoving   **(b) Alternating Path**   **(c) After Resolving**

Figure 6.4. Resolving a virtual node

where n is the number of nodes in G and k is the number of chains in the optimal chain cover (known as the width of G).

6.   Path-Tree Cover

Jin et al. in [26] propose a path-tree cover coding scheme to answer a reachability query on a DAG $G(V, E)$.

First, the graph $G(V, E)$ is decomposed into a set of pairwise disjoint paths, $P_1, P_2, \cdots, P_{k'}$. Here, a path $P_i = v_{i_1} \rightarrow v_{i_2} \rightarrow \cdots \rightarrow v_{i_k}$ where $v_{i_j} \rightarrow v_{i_{j+1}}$ is an edge in G. A path cover consists of k' paths such that (a) the union of the nodes in all the paths is the entire set of nodes in G and (b) the intersection of two paths is empty. The optimal path cover of G is a path cover of G that contains the least number of paths among all possible path covers of G. Such optimal path cover can be obtained using Simon's algorithm in [31].

Second, let P_i and P_j be two paths computed in the path cover. There may exist edges from some nodes in P_i to some nodes in P_j, denoted as $E_{P_i \to P_j}$, which is a subset of the edges in G. Some edges in $E_{P_i \to P_j}$ can be eliminated losslessly. For example, suppose $P_i = w$ and $P_j = u \to v$, and assume $E_{P_i \to P_j}$ consists of two edges from P_i to P_j, $\{w \to u, w \to v\}$. Then $w \to v$ can be eliminated, because there is a path $w \to u \to v$ that can answer the reachability query $w \rightsquigarrow v$. The similar can be done if there are edges from P_j to P_i in reverse order. The edge elimination in this way is lossless because it does not lose any reachability information. Let $E'_{P_i \to P_j}$ be a subset of $E_{P_i \to P_j}$ after edge elimination. Jin et al. show that all edges in $E'_{P_i \to P_j}$ are in parallel. Furthermore, Jin et al. use a single weighted edge from P_i to P_j, in order to represent how many nodes in P_i can reach a node in P_j. Based on the weighted edges from P_i to P_j, a weighted path-graph $G_P(V, E)$ is constructed. Here, V is a set of nodes representing paths, $P_1, P_2, \cdots, P_{k'}$, computed in the path cover, and E is a set of edges (P_i, P_j) with a weight, if $E'_{P_i \to P_j} \neq \emptyset$.

Third, based on the path-graph $G_P(V, E)$, Jin et al. construct a spanning tree $T_P(V, E)$, called path-tree, with two criteria: MaxEdgeCover and Min-PathIndex. The former means to cover as many edges in G as possible, and the latter means to reduce the size of a resulting path-tree cover as much as possible. The path tree is computed using the algorithm presented in [16, 21].

Finally, a path-tree cover code, ptcode(u), is assigned to node $u \in G$ based on the path-tree T_P. The ptcode(u) $= ((u_{start}, u_{end}), (u_x, u_y))$ consists of two pairs. The first pair is the interval $[u_{start}, u_{end}]$, like SIT code, assigned to the path P_i where u resides uniquely, because a node represents a path in T_P. The second pair (u_x, u_y) is used to record the position of the node u in the path P_i. A reachability query, $u \rightsquigarrow v$ is answered to be true, if the predicate $\mathcal{P}_{pt}(\text{ptcode}(u), \text{ptcode}(v))$ is true, such as $[v_{start}v_{end}] \subset [u_{start}, u_{end}] \wedge u_x < v_x \wedge u_y < u_y$. It is important to note that it does not mean $u \rightsquigarrow v$ is false if $\mathcal{P}_{pt}(\text{ptcode}(u), \text{ptcode}(v))$ is false, because the path-tree cover code and the predicate are both defined over the path-tree T_P. There may exist edges that cannot be fully covered by the path-tree.

The path-tree cover coding scheme is different from the tree cover [1] and the chain cover [24, 9]. Both tree cover and chain cover coding schema answer reachability queries only using the predicates, $\mathcal{P}_{tc}(,)$ and $\mathcal{P}_c(,)$, respectively. On the other hand, the path-tree cover coding scheme cannot answer reachability queries only using the predicate $\mathcal{P}_{pt}(,)$. The path-tree cover coding scheme shares similarity with the dual-labeling [34], and aims at covering as many non-tree edges as possible. Jin et al. in [26] show that the path-tree cover is superior over the optimal tree cover [1] and optimal chain cover [24] in terms of the compression ability.

7.   2-HOP Cover

Cohen et al. propose a 2-hop cover in [17] for a graph G. In a 2-hop cover, a node in G is assigned to a 2-hop code, $2hopcode(u) = (L_{in}(v), L_{out}(v))$, where $L_{in}(v)$ and $L_{out}(v)$ are subsets of the nodes in G. Based on the 2-hop cover, a reachability query $u \rightsquigarrow v$ is to be answered true if and only if $\mathcal{P}_{2hop}(2hopcode(u), 2hopcode(v))$ is true.

$$\mathcal{P}_{2hop}(2hopcode(u), 2hopcode(v)) = L_{out}(u) \cap L_{in}(v) \neq \emptyset$$

The main idea behind 2-hop cover coding scheme is to compress the edge transitive closure of G. Let $TC(G)$ be the edge transitive closure of G. A pair (u, v) in $TC(G)$ indicates that $u \rightsquigarrow v$ is true in G. Consider a node w in G as a center. All the ancestors of w, denoted as $ancs(w)$, can reach w, and w can reach any of its descendants, denoted as $desc(w)$. In other words, $ancs(w)$ is the set of nodes $\{u\}$ if $(u, w) \in TC(G)$ and $desc(w)$ is the set of nodes $\{v\}$ if $(w, v) \in TC(G)$. Let $A_w \subseteq ancs(w) \cup \{w\}$ and $D_w \subseteq desc(w) \cup \{w\}$. A complete bipartite graph, called a 2-hop cluster, is denoted $S(A_w, w, D_w)$, with the center w. A 2-hop cluster $S(A_w, w, D_w)$ indicates that every node, u in A_w can reach any node v in D_w, or $u \rightsquigarrow v$ is true for every $u \in A_w$ and $v \in D_w$. Given a cluster $S(A_w, w, D_w)$, it implies that if w is added into $L_{out}(u)$ for every $u \in A_w$ and is added into $L_{in}(v)$ for every $v \in D_w$, the reachability information presented by the complete bipartite graph $S(A_w, w, D_w)$ is completely preserved, because $u \rightsquigarrow v$ is true if and only if $L_{out}(u) \cap L_{in}(v) \neq \emptyset$. A $S(A_w, w, D_w)$ compactly represents $|A_w| \cdot |D_w| - 1$ pairs in $TC(G)$ in total with a space cost of $|A_w| + |D_w|$. A 2-hop cover is a set of 2-hop clusters that completely covers the edge transitive closure $TC(G)$.

The optimal 2-hop cover problem is to find the minimum size 2-hop cover, which is proved to be NP-hard [17]. Based on the greedy algorithm for minimum set cover problem [27], Cohen et al. give an approximation algorithm to get a nearly optimal 2-hop cover which is larger than the optimal one at most $O(\log n)$.

Algorithm 4 illustrates the ideas [17]. It computes the edge transitive closure $TC(G)$ (line 1). Let TC' be $TC(G)$ (line 2). In every iteration, it finds a 2-hop cluster $S(A_w, w, D_w)$ that has the maximum ratio, $(|S(A_w, w, D_w) \cap TC'|)/(|A_w| + |D_w|)$, among all possible 2-hop clusters. Here, TC' is used to indicate the set of pairs in $TC(G)$ that are not covered by any 2-hop clusters computed yet. After identifying the $S(A_w, w, D_w)$ with the maximum ratio in the current iteration, it removes all the pairs (u, v) from TC' if $u \in A_w$ and $v \in D_w$ (line 5). In line 6-7, it updates 2-hop cover codes.

Algorithm 4 2Hop-Cover(G)

1: compute the edge transitive closure $TC(G)$ of G;
2: $TC' \leftarrow TC(G)$;
3: **while** $TC' \neq \emptyset$ **do**
4:     find the max $S(A_w, w, D_w)$;
5:     remove all the pairs in TC' that are covered by $S(A_w, w, D_w)$;
6:     add w into $L_{out}(u)$ if $u \in A_w$;
7:     add w into $L_{in}(v)$ if $v \in D_w$;
8: **end while**

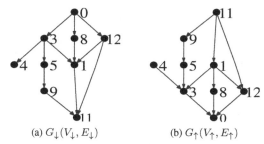

(a) $G_\downarrow(V_\downarrow, E_\downarrow)$ (b) $G_\uparrow(V_\uparrow, E_\uparrow)$

Figure 6.5. A Directed Graph, and its Two DAGs, $G_\downarrow$ and $G_\uparrow$ (Figure 2 in [13])

The computational cost is high as can be seen in Algorithm 4. First, it needs to compute the edge transitive closure. Second, it needs to rank all 2-hop clusters $S(A_w, w, D_w)$ based on $(|S(A_w, w, D_w) \cap TC'|)/(|A_w| + |D_w|)$ in every iteration. Third, it is difficult to compute 2-hop cover for a large graph.

7.1   A Heuristic Ranking

Schenkel et al. in [29] propose a heuristic ranking to avoid to recompute and rank all $(|S(A_w, w, D_w) \cap TC'|)/(|A_w| + |D_w|)$ for all possible centers $S(A_w, w, D_w)$ in every iteration. The idea is as follows. It computes all $|S(A_w, w, D_w) \cap TC'|/(|A_w| + |D_w|)$, for all nodes in G. Initially, $TC' = TC(G)$. Let d_w denote $|S(A_w, w, D_w) \cap TC'|/(|A_w| + |D_w|)$. It initially maintains all the pairs of (w, d_w) in a priority queue. The first is with the max ratio d_w value. In every iteration, it picks up the first (w, d_w) and recomputes $d'_w = |S(A_w, w, D_w) \cap TC'|/(|A_w| + |D_w|)$, if $d_w > d'_w$, the pair (w, d'_w) is enqueued into the priority queue. It repeats until it picks a node w such that $d_w = d'_w$. In practice, Schenkel et al. find that it only needs to repeat 2-3 times in every iteration on average.

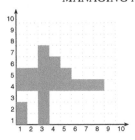

Figure 6.6. Reachability Map

w	tccode(w) for $w \in G_\downarrow$		tccode(w) for $w \in G_\uparrow$	
	$po_\downarrow(w)$	$I_\downarrow(w)$	$po_\uparrow(w)$	$I_\uparrow(w)$
0	9	[1,9]	4	[4,4]
1	1	[1,1],[3,3]	3	[1,5]
3	6	[1,6]	5	[4,5]
4	2	[2,2]	9	[4,5],[9,9]
5	5	[3,5]	6	[4,6]
8	7	[1,1],[3,3],[7,7]	1	[1,1],[4,4]
9	4	[3,4]	7	[4,7]
11	3	[3,3]	8	[1,8]
12	8	[1,1],[3,3],[8,8]	2	[2,2],[4,4]

Table 6.2. A Reachability Table for $G_\downarrow$ and $G_\uparrow$

7.2   A Geometrical-Based Approach

Cheng et al. in [13] propose a geometrical-based approach that does not need to compute the edge transitive closure of $TC(G)$ directly, and speeds up the computing of max ratio of the 2-hop clusters using an R-tree, in particular for a large dense graph G.

First, instead of computing the edge transitive closure $TC(G)$, Cheng et al. compute tree cover [1], because in practice the tree cover algorithm in [1] is very fast. The tree cover codes are used to compute 2-hop cover. Consider Figure 6.5(a) which shows a DAG $G_\downarrow(V_\downarrow, E_\downarrow)$. Suppose it needs to assign 2-hop codes to the graph shown in Figure 6.5(a). Cheng et al. compute the tree cover codes for $G_\downarrow(V_\downarrow, E_\downarrow)$, and compute the tree cover codes for another corresponding graph $G_\uparrow(V_\uparrow, E_\uparrow)$, which is a graph that by changing every edge $(u, v) \in G_\downarrow$ to (v, u). The Table 6.2 shows the tccode(w) for the node w in

$G_\downarrow$ and $G_\uparrow$. In particular, $po_\downarrow(w)$ and $po_\uparrow(w)$ indicate the postorder of w, and $I_\downarrow(w)$ and $I_\uparrow(w)$ indicate the intervals of w, in $G_\downarrow$ and $G_\uparrow$, respectively.

Second, based on the tree cover codes, Cheng et al. construct a 2-dimensional reachability map, a node w is mapped onto the (x_w, y_w) position in the reachability map as $(po_\downarrow(w), po_\uparrow(w))$. The reachability information $u \rightsquigarrow v$ is mapped onto 2-dimensional reachability map, (x_v, y_u). If $u \rightsquigarrow v$ is true, then $(x_v, y_u) = 1$, otherwise $(x_v, y_u) = 0$. Therefore, the same reachability information, that a 2-hop cluster $S(A_w, w, D_w)$ represents, is represented as a number of rectangles in the 2-dimensional reachability map.

With the assistance of the 2-dimensional reachability map, Cheng et al. find the max $S(A_w, w, D_w)$ in line 4 of Algorithm 4 as to find the max coverage of rectangles, which can be done using an R-tree. It is important to note that Cheng et al. in [13] try to maximize $|S(A_w, w, D_w) \cap TC'|$ instead of $|S(A_w, w, D_w) \cap TC'|/(|A_w| + |D_w|)$. Both are set cover problems.

7.3 Graph Partitioning Approaches

In this section, we discuss three graph partitioning approaches used in computing a 2-hop cover for a large graph G.

A Flat Partitioning Approach. Schenkel et al. propose a flat partitioning approach in [29] to compute 2-hop cover in three steps. First, it partitions the graph G into k subgraphs $G_1, G_2, \cdots, G_k$ depending on the available memory M. Second, it computes the edge transitive closure and the 2-hop cover for each subgraph G_i, for $1 \le i \le k$, using Algorithm 4 with the heuristic ranking discussed in the previous subsection. Third, it merges the k 2-hop covers computed for the k subgraphs, $G_1, G_2, \cdots, G_k$, by dealing with the edges that cross subgraphs. It is called a cover joining step, and the cover joining yields a 2-hop cover for the entire graph G. The cover joining is done as follows. Suppose the 2-hop covers for all k subgraphs are computed. Let (u, v) be a cross-partition edge where $u \in G_i$ and $v \in G_j$ and $G_i \ne G_j$. Schenkel et al. compute the 2-hop cover for G by encoding all reachability via (u, v) according to the following two operations.

- For all $a \in ancs(u)$, $L_{out}(a) \leftarrow L_{out}(a) \cup \{u\}$, and
- For all $d \in desc(v) \cup \{v\}$, $L_{in}(d) \leftarrow L_{in}(d) \cup \{u\}$.

It means that, 2-hop clusters, $(ancs(u), u, desc(u))$, for all cross-partition edges (u, v), are covered mandatorily to encode G. The compression rate of $TC(G)$ using the flat partitioning decreases. As reported in [29, 30], the cover joining becomes the bottleneck of the whole processing. Schenkel et al. in [30] propose an effective and efficient approach for the third step of cover joining, using a skeleton graph (SG).

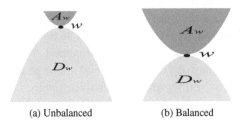

(a) Unbalanced   (b) Balanced

Figure 6.7. Balanced/Unbalanced $S(A_w, w, D_w)$

A skeleton graph is constructed at the partition-level. Suppose a graph $G(V, E)$ is partitioned into k subgraphs $G_1(V_1, E_1)$, $G_2(V_2, E_2)$, $\cdots$, $G_k(V_k, E_k)$. Here, $V = \cup_{i=1}^k V_i$ and $V_i \cap V_j = \emptyset$ if $i \neq j$. $E = E_C \cup (\cup_{i=1}^k E_i)$ where $E_i \cap E_j = \emptyset$ if $i \neq j$ and E_C is the set of cross-partition edges $E \setminus (\cup_{i=1}^k E_i)$. The skeleton graph $G_S(V_S, E_S)$ is constructed as follows. Here, V_S is a set of nodes u if u appears in a cross-partition edge in E_C. E_S contains all the cross-partition edges E_C, and in addition contains edges that explicitly indicate whether two cross-partition edges are connected via some paths in a subgraph. Consider a subgraph G_i, and let (v_i, v_j) and (v_k, v_l) be any two cross-partition edges such that v_j and v_k as nodes appear in G_i. There will be an edge (v_j, v_k) in E_S if $v_j \rightsquigarrow v_k$ is true in G_i. Schenkel et al. compute a 2-hop cover for G_S using Algorithm 4 with the heuristic ranking. At this stage, for a node $u \in G$ that does not appear in any cross-partition edges, u has a 2hopcode(u) which is computed in G_i where u resides. For a node $u \in G$ that appears in cross-partition edges, it has two 2-hop cover codes. One is computed because it appears in a subgraph G_i, 2hopcode(u). The other is the one computed in the skeleton graph G_S, denoted 2hopcode$'(u)$. Let 2hopcode$(u) = (L_{in}(u), L_{out}(u))$ and 2hopcode$'(u) = (L'_{in}(u), L'_{out}(u))$.

The final 2-hop cover code is computed by augmenting the 2-hop cover code computed for G_i using the 2-hop cover code computed over the skeleton graph. Let (u, v) be a cross-partition edge, where $u \in G_i$ and $v \in G_j$, and let $V(G_i)$ and $V(G_j)$ denote the sets of nodes in G_i and G_j. It is done using the following two operations.

- For all $a \in ancs(u) \cap V(G_i)$, $L_{out}(a) \leftarrow L_{out}(a) \cup L'_{out}(u)$, and
- For all $d \in desc(v) \cap V(G_j)$, $L_{in}(d) \leftarrow L_{in}(d) \cup L'_{in}(v)$.

The skeleton graph gives a global picture over the 2-hop cover and can compress the edge transitive closure effectively.

A Hierarchical Partitioning Approach. Cheng et al. in [14] consider the quality of the partitioning. The partitioning divides a large graph into smaller graphs and computes the 2-hop cover code for the large graph by augmenting

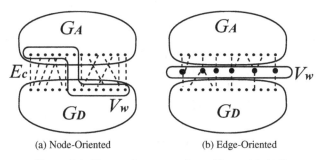

(a) Node-Oriented   (b) Edge-Oriented

Figure 6.8. Bisect G into G_A and G_D (Figure 6 in [14])

the 2-hop cover codes for smaller graphs. The main issue in the flat partition-ing [29, 30] is to find a way to compute 2-hop cover codes for a large graph with the limited memory. Because it is not easy to find an optimal partition-ing of graphs, Schenkel et al. take a simple approach. For a DAG graph G, it can start from the top or the bottom (refer to $G_\downarrow$ in Figure 6.5) to extract a subgraph that can be held in memory, and repeats it until the entire graph is decomposed into a set of smaller graphs. Consider a node w appearing in a cross-partition edge. The node w has potential power to compress the edge transitive closure effectively, because many nodes in one subgraph may con-nect to many nodes in another subgraph via the node w. However, there are two cases as illustrated in Figure 6.7. The flat partitioning may result a partitioning that result in many unbalanced 2-hop clusters $S(A_w, w, D_w)$ (Figure 6.7(a)). Cheng et al. attempt to partition a graph that results in balanced 2-hop clusters $S(A_w, w, D_w)$ (Figure 6.7(b)). Recall $S(A_w, w, D_w)$ uses $|A_w| + |D_w|$ space to compress $|A_w| \cdot |D_w| - 1$ entries in the edge transitive closure. Cheng et al. show that the compression rate $(|A_w| \cdot |D_w| - 1)/(|A_w| + |D_w|)$ is maximum when $|A_w| = |D_w|$.

Cheng et al. in [14] propose a hierarchical partitioning approach to partition a large graph G into two subgraphs, G_A and G_D, repeatedly in a top-down fashion. It repeats if a subgraph cannot be held in memory in such a manner.

The key idea presented in [14] is to select a set of centers, $V_w = \{w_1, w_2, \cdots\}$, as a cut to partition a graph G. Note that the set of centers implies a set of 2-hop clusters, $S(A_{w_1}, w_1, D_{w_1}), S(A_{w_2}, w_2, D_{w_2}), \cdots$. Sup-pose that G is partitioned into G_A and G_D. There exist a set of edges (u, v) where $u \in G_A$ and $v \in G_D$. Let E_C denote such a set of edges. Cheng et al. propose a node-oriented and an edge-oriented approach to identify V_w where $w_i \in V_w$ is selected from the set of nodes appearing in E_C. As illustrated in Figure 6.8(a), in the node-oriented approach, it selects a set of nodes in E_C as V_w. As illustrated in Figure 6.8(b), in the edge-oriented approach, it treats edges as virtual nodes and identify V_w. The set of V_w is computed as to find the minimum 2-hop cover to cover reachability cross G_A and G_D from the nodes appearing in E_C. It is important to note that reachability between the two subgraphs, G_A and G_D, are completely covered by the set of 2-hop clusters using the set of nodes V_w. Based on V_w, Cheng et al. extract an induced subgraph of G_A, denoted $G_\top$, which does not include any nodes in V_w, and extract an induced subgraph of G_D, denoted $G_\perp$, which does not include any nodes in V_w. Both $G_\top$ and $G_\perp$ are treated as G in the next steps to bisect.

7.4   2-Hop Cover Maintenance

A 2-hop cover is hard to compute. Schenkel et al. in [30] and Bramandia et al. in [5] study the 2-hop cover maintenance problem to minimize the effort of updating the 2-hop cover when updates occur, and avoid computing a 2-hop cover from the beginning. There are four operations, insertion/deletion of nodes/edges. It is straightforward to deal with insertions. Consider an insertion of a new edge between an existing node and a new node v to G. A simple solution is to insert $S(ancs(v), v, desc(v))$ into the 2-hop cover, i.e., inserting v to the L_{in} and L_{out} of all nodes in $desc(v)$ and $ancs(v)$, respectively. The deletion of nodes/edges becomes non-trivial, because a deletion of a node w may affect the reachability $u \rightsquigarrow v$ if $w \in L_{out}(u)$ and $w \in L_{in}(v)$. Removing w from $L_{out}(u)$ and $L_{in}(v)$ may make $u \rightsquigarrow v$ to be wrongly answered as false, because there may be other paths from u to v. The existing work focus on deletion operations. In this article, we mainly discuss their approaches to handle the deletion of an existing node. The similar idea can be applied to handling the deletion of an existing edge.

Re-labeling a subgraph.   When there is a deletion of an existing node, Schenkel et al. in [30] compute a 2-hop cover $\hat{L}$ of a subgraph G_{REL} of G, in order to reflect all the affected connections in G, due to the deletion of an existing node v. The existing 2-hop cover L for the graph G, before updating, will be updated to reflect all the affected connections by incorporating $\hat{L}$. The graph $G_{REL}(V_{REL}, E_{REL})$ is constructed as an induced graph of G, denoted as $G[V_{REL}]$. The set of nodes, V_{REL} is computed as follows. First, it includes all nodes in $ancs(v)$ in V_{REL}, which is shown as the striped region in Figure 6.9a. Second, it includes all nodes in $desc(u)$ into V_{REL} if $u \in ancs(v)$, which is shown as the gray region in Figure 6.9a. Note that G_{REL} represents all the affected connections.

The 2-hop cover $\hat{L}$ computed for G_{REL} is used to update the 2-hop cover L for the entire graph G as follows. It is obvious that all the connections (a, d), that exist in G, need to be updated if $a \in V_{REL}$. Note that $d \in V_{REL}$ in this case. All $L_{out}(a)$ for $a \in V_{REL}$ are updated as to be $\hat{L}_{out}(a)$. On the other hand, for a connection (a, d) that exists in G where $d \in V_{REL}$, the node a may or may not

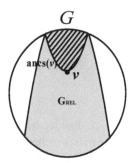

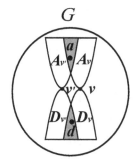

(a) Re-labeling a subgraph   (b) Reserving alternative paths

Figure 6.9. Two Maintenance Approaches

exist in V_{REL}. If $a \in V_{\mathrm{REL}}$, $\hat{L}_{in}(d)$ are used to reflect all (a, d), because a and d are both in G_{REL}. For the latter case, it keeps $L_{in}(d) \setminus V_{\mathrm{REL}}$, because such (a, d) are not affected by the deletion of v and are encoded by previous 2-hop clusters. Hence, $L_{in}(d)$ is updated as $(L_{in}(d) \setminus V_{\mathrm{REL}}) \cup \hat{L}_{in}(d)$.

A drawback of this approach is high maintenance cost, because G_{REL} can be as large as G itself. It means that the maintenance for the current 2-hop cover degrades into the re-computation of a new 2-hop cover for the entire graph. Bramandia et al. [4] show the 2-hop cover code maintenance using the geometrical-based approach [13].

Reserving all alternative paths. Bramandia et al. in [5] propose u2-hop that can work on a smaller set of affected connections online at the expense of a large space. It considers all connections (a, d), where $a \in ancs(v)$ and $d \in desc(v)$, and modifies $L_{out}(a)$ and $L_{in}(d)$ by removing (i) v, (ii) nodes that are on longer reachable from a or nodes that can not reach d any longer, due to the deletion of the node v. The operation (i) is to exclude $S(A_v, v, D_v)$ from the current 2-hop cover. The operation (ii) is to maintain $S(A_w, w, D_w)$, where $w \in ancs(v)$ or $w \in desc(v)$, by removing those nodes in A_w and D_w which no longer connect to w. In order to maintain the 2-hop cover, it is important to note that the succinct maintaining operations of [5] require redundancy in the 2-hop cover. Such redundancy comes from the requirement that for any connection (a, d) in G, it repeatedly encodes it with multiple 2-hop clusters for all different alternative paths from a to d, as illustrated by Figure 6.9b. The example shows that two alternative paths from a to d exist in G, and v and v' are contained in the two paths respectively. So both $S(A_v, v, D_v)$ and $S(A_{v'}, v', D_{v'})$ need to be maintain to cover (a, d).

In details, in encoding (a, d) for all alternative paths from a to d, a set of nodes W is used such that the removal of W disconnect all paths from a to d. It constructs 2-hop clusters based on $w \in W$ and any nodes that connect via

w are included in A_w and D_w. And all $w \in W$ are added into $L_{out}(a)$ and $L_{in}(d)$. Upon the deletion of a node w, it can safely remove w from all $L_{out}(a)$ and $L_{in}(d)$. It is because that if there is another path from a to d, there must be another $w' \in W$ such that $L_{out}(a)$ and $L_{in}(d)$ both contain w'. Note that the 2-hop cover compression ratio is in a relatively low priority in this regard.

8.   3-Hop Cover

Jin et al. in [25] propose a 3-Hop approach. Consider a transitive closure matrix for a DAG G (Figure 6.10). Suppose there exists a chain cover of G with k chains. Jin et al. show that the transitive closure matrix for G is a matrix of $k \times k$ blocks where each block is a Pseudo-upper triangular matrix. It can be done by ordering the nodes using their chain identifiers and then their positions in the chains. Jin et al. use $Con(G)$ to denote the set of pseudo-diagonal cells for all the blocks in the transitive closure matrix (the circled cells shown in Figure 6.10). It is easy to see that $Con(G)$ is enough to derive the transitive closure. $Con(G)$ can be easily calculated using Algorithm 2.

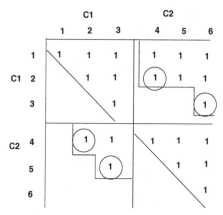

Figure 6.10. Transitive Closure Matrix

$Con(G)$ is already enough to answer a reachability query. But, the cost is high, because the number of nodes in $Con(G)$ can be large. Jin et al. encode $Con(G)$ using 3-hop cover codes. It is similar to the 2-hop cover codes. For every node u, there is a list of "entry points" $L_{in}(u)$ and a list of "exit points" $L_{out}(u)$. The difference between 2-hop and 3-hop is as follows. In a 2-hop cover code, u can reach v if any only if $L_{out}(u) \cap L_{in}(v) \neq \emptyset$. But in a 3-hop cover code, it allows a point in $L_{out}(u)$ reach another point in $L_{in}(v)$ via a chain. Suppose that there is a chain $\cdots \rightsquigarrow v_i \rightsquigarrow \cdots \rightsquigarrow v_j \rightsquigarrow \cdots$. Then, $u \rightsquigarrow v$ is true if u can reach v_i (1st hop), v_i can reach v_j (2nd hop), and v_j can reach v (3rd hop). The algorithm to compute the 3-hop cover codes is similar to the algorithm to compute the 2-hop cover codes. The only difference is that it needs to consider the set of pairs that can be encoded by each chain rather than each node. The time complexity for the 3-hop cover construction is $O(k \cdot n^2 \cdot |Con(G)|)$.

Given a 3-hop cover coding scheme encoding for $Con(G)$, it can answer a reachability query $u \rightsquigarrow v$ as follows: In the first step, it collects a set of entry points $L_{out}(u)$ can reach on the intermediate chains. In the second step, it collects a set of exit points which can reach v on the intermediate chains. Finally, it checks whether an entry point can reach an exit point using the chain ids and positions for nodes in the chain. The time complexity is $O(\log n + k)$ where n is the number of nodes in the graph G and k is the number of chains.

9. Distance-Aware 2-Hop Cover

The 2-hop cover coding schema discussed in the previous section can be used to answer reachability queries, $u \rightsquigarrow v$, but cannot be used to answer distance queries, $u \overset{\delta}{\rightsquigarrow} v$. A distance query $u \overset{\delta}{\rightsquigarrow} v$ is a reachability query $u \rightsquigarrow v$ with the shortest distance δ. In other words, it queries the shortest distance from u to v if it is reachable. Cohen et al. in [17] address this problem.

Consider an edge-weighted directed graph $G(E, V)$, where $\omega(u, v)$ represents the distance over the edge $(u, v) \in E$. Let $\delta(u, v)$ be the shortest distance from a node u to a node v. A 2-hop cover code of u is a pair of $L_{in}(u)$ and $L_{out}(u)$. Here, $L_{in}(u)$ is a set of pairs $\{(u_1, \delta(u_1, u)), (u_2, \delta(u_2, u)), \cdots\}$, and $L_{out}(u)$ is a set of pairs $\{(v_1, \delta(u, v_1)), (v_2, \delta(u, v_2)), \cdots\}$. A distance query $u \overset{\delta}{\rightsquigarrow} v$ is answered as

$$\min\{\delta(u, w) + \delta(w, v)|(w, \delta(u, w)) \in L_{out}(u) \wedge (w, \delta(w, v)) \in L_{in}(v)\}$$

It is worth nothing that the distance-aware 2-hop cover needs to maintain the additional shortest distance information.

Schenkel et al. in [30] discuss the distance-aware 2-hop cover. The algorithms in [30] can be used to compute the distance-aware 2-hop cover. However, in addition to the bottleneck in the third step, it needs high overhead to compute the shortest paths, and the resulting 2-hop cover can be unnecessarily large. Consider Figure 6.11. There is a subgraph G_i in which the node a is an ancestor of the nodes $x_1, x_2, \cdots, x_d$ in the subgraph G_i that appear in the cross-partition edges. As a result, all nodes, $x_1, x_2, \cdots, x_d$, appear in the skeleton graph. Assume that there is a 2-hop cluster, $S(A_w, w, D_w)$, in the skeleton graph, that contains all $x_1, x_2, \cdots, x_d$ in A_w. In computing the distance-aware 2-hop cover for G by augmenting the distance-aware 2-hop cover computed for the skeleton graph, it needs to identify the shortest path from a to w (Figure 6.11). There may exist many unnecessary pairs in the resulting distance-aware 2-hop cover such that $\delta(a, x) + \delta(x, w) > \delta(a, w)$.

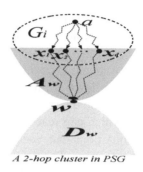

A 2-hop cluster in PSG

Figure 6.11. The 2-hop Distance Aware Cover (Figure 2 in [10])

Cheng and Yu in [10] discuss a new DAG-based approach and focus on two main issues.

- Issue-1: It cannot obtain a DAG G' for a directed graph G first, and compute the distance-aware 2-hop cover for G based on the distance-aware 2-hop cover computed for G'. In other words, it cannot represent a strongly connected component (SCC) in G as representative node in G'. It is because that a node w in a SCC on the shortest path from u to v does not necessarily mean that every node in the SCC is on the shortest path from u to v.

- Issue-2: The cost of dynamically selecting the best 2-hop cluster, in an iteration of the 2-hop cover program, cannot be reduced using the tree cover codes and R-tree as discussed in [13], because such techniques cannot handle distance information.

Cheng and Yu observe that if a 2-hop cluster, $S(A_w, w, D_w)$, is computed to cover all shortest paths containing the center node w, it can remove w from the underneath graph G, because there is no need to consider again any shortest paths via w any more.

Based on the observation, to deal with Issue-1, Cheng and Yu in [10] collapse every SCC into DAG by removing a small number of nodes from the SCC repeatedly until it obtains a DAG graph. To deal with Issue-2, when constructing 2-hop clusters, Cheng and Yu propose a new technique to reduce the 2-hop clusters by taking the already identified 2-hop clusters into consideration, to avoid storing unnecessary all-pairs of shortest paths.

Cheng and Yu propose a two-step solution. In the first phase, it attempts to obtain a DAG $G_\downarrow$ for a given graph G by removing a small number of nodes, $\hat{V}_{C_i}$, from every SCC, $C_i(V_{C_i}, E_{C_i})$. In computing a SCC $C_i(V_{C_i}, E_{C_i})$, every node, $w \in \hat{V}_{C_i}$ is taken as a center, and $S(A_w, w, D_w)$ is computed to cover shortest paths for the graph G. Then, all nodes in $\hat{V}_{C_i}$ will be removed, and

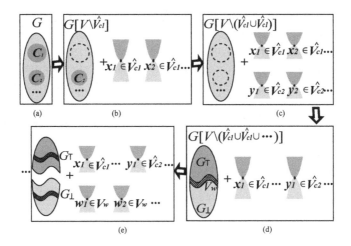

Figure 6.12. The Algorithm Steps (Figure 3 in [10])

a modified graph is constructed as an induced subgraph of $G(V, E)$, denoted as $G[V \setminus \hat{V}_{C_i}]$, with the set of nodes $V \setminus \hat{V}_{C_i}$. Figure 6.12(a) shows a graph G with several SCCs. Figure 6.12(b)-(d) illustrate the main idea of collapsing SCCs while computing 2-hop clusters. At the end, the original directed graph G is represented as a DAG G' plus a set of 2-hop clusters, $S(A_w, w, D_w)$, computed for every node, $w \in \hat{V}_{C_i}$. All shortest paths covered are the union of the shortest paths covered by all 2-hop clusters, $S(A_w, w, D_w)$, for every node, $w \in \hat{V}_{C_i}$, and the modified DAG G'. In the second phase, for the obtained DAG $G_\downarrow$, Cheng and Yu take the top-down partitioning approach to partition the DAG $G_\downarrow$, based on the early work in [14]. Figure 6.12(d)-(e) show that the graph can be partitioned hierarchically.

10. Graph Pattern Matching

In this section, we discuss several approaches to find graph patterns in a large data graph. A data graph is a directed node-labeled graph $G_D = (V, E, \Sigma, \phi)$. Here, V is a set of nodes, E is a set of edges (ordered pairs), Σ is a set of node labels, and ϕ is a mapping function which assigns each node, $v_i \in V$, a label $l_j \in \Sigma$. Below, we use $\mathsf{label}(v_i)$ to denote the label of node v_i. Given a label $l \in \Sigma$, the extent of l, denoted $\mathsf{ext}(l)$, is a set of nodes in G_D whose label is l. A graph pattern is a connected directed labeled graph $G_q = (V_q, E_q)$, where V_q is a subset of labels (Σ), and E_q is a set of edges (ordered pairs) between two nodes in V_q. There are two types of edges. Let $A, D \in V_q$. An edge $(A, D) \in E(G_q)$ represents a parent/child condition, denoted as $A \mapsto D$, which identifies all pairs of nodes, v_i and v_j, such that $(v_i, v_j) \in G_D$, $\mathsf{label}(v_i) = A$, and $\mathsf{label}(v_j) = D$. An edge $(A, D) \in E(G_q)$

represents a reachability condition, denoted as $A \hookrightarrow D$, that identifies all pairs of nodes, v_i and v_j, such that $v_i \rightsquigarrow v_j$ is true in G_D, for $\mathsf{label}(v_i) = A$, and $\mathsf{label}(v_j) = D$. A match in G_D matches the graph pattern G_q if it satisfies all the parent/child and reachability conditions conjunctively specified in G_q. A graph pattern matching query is to find all matches for a query graph. In this article, we focus on the reachability conditions, $A \hookrightarrow D$, and omit the discussions on parent/child conditions, $A \mapsto D$. We assume that a query graph G_p only consists of reachability conditions.

10.1   A Special Case: $A \hookrightarrow D$

In this section, we introduce three approaches to process $A \hookrightarrow D$ over a graph G_D.

Sort-Merge Join.   Wang et al. propose a sort-merge join algorithm in [36] to process $A \hookrightarrow D$ over a directed graph using the tree cover codes [1]. Recall that for a given node u, $\mathsf{tccode}(u) = \{[u_{start_1}, u_{end_1}], [u_{start_2}, u_{end_2}], \cdots \}$, where u_{end_1} is the postorder when it traverses the spanning tree. We use $post(u)$ to denote the postorder of node u.

Let $Alist$ and $Dlist$ be two lists of $\mathsf{ext}(A)$ and $\mathsf{ext}(D)$, respectively. In $Alist$, every node v_i keeps all its intervals in the $\mathsf{tccode}(v_i)$. In $Dlist$, every node v_j keeps its unique postorder $post(v)$. Also, $Alist$ is sorted on the intervals $[s, e]$ by the ascending order of s and then the descending order of e, and $Dlist$ is sorted by the postorder number in ascending order. The sort-merge join algorithm evaluates $A \hookrightarrow D$ over G_D by a single scan on $Alist$ and $Dlist$ using the predicate $\mathcal{P}_{tc}(,)$. Wang et al. [36] propose a naive GMJ algorithm and an IGMJ algorithm which uses a range search tree to improve the performance of the GMJ algorithm.

Hash Join.   Wang et al. also propose a hash join algorithm in [35] to process $A \hookrightarrow D$ over a directed graph using the tree cover codes. Unlike the sort-merge join algorithm, $Alist$ is a list of pairs $(val(u), post(u))$ for all $u \in ext(A)$. Here, $post(u)$ is the unique postorder of u, and $val(u)$ is either a start or an end of the intervals. Consider the node d in Figure 6.3(b), $post(d) = 7$, and there are two intervals, $[6, 7]$ and $[1, 4]$. In $Alist$, it keeps four pairs: $(6, 7)$, $(7, 7)$, $(1, 7)$, and $(4, 7)$. Like the sort-merge join algorithm, $Dlist$ keeps a list of postorders $post(v)$ for all $v \in \mathsf{ext}(D)$. $Alist$ is sorted in ascending order of $val(a)$ values, and $Dlist$ is sorted in ascending order of $post(d)$ values. The Hash Join algorithm, called HGJoin, is outline in Algorithm 5.

Join Index.   Cheng et al. in [15] study a join index approach to process $A \hookrightarrow D$ using a join index built on top of G_D. The join index is built based on the 2-hop cover codes. We explain it using the same example given in [15].

Algorithm 5 HGJoin($Alist$, $Dlist$)

1: $H \leftarrow \emptyset$;
2: $Output \leftarrow \emptyset$;
3: $a \leftarrow Alist.first$;
4: $d \leftarrow Dlist.first$;
5: **while** $a \neq Alist.last \wedge d \neq Dlist.last$ **do**
6:    **if** $val(a) \leq post(d)$ **then**
7:      **if** $post(a) \notin H$ **then**
8:        hash $post(a)$ into H;
9:        $a \leftarrow a.next$;
10:    **else if** $val(a) < post(d)$ **then**
11:      delete $post(a)$ from H;
12:      $a \leftarrow a.next$;
13:    **else**
14:      **for** all $post(a)$ in H **do**
15:        append $(post(a), post(d))$ to $Output$;
16:      **end for**
17:      $d \leftarrow d.next$;
18:    **end if**
19:   **else**
20:     **for** all $post(a)$ in H **do**
21:       append $(post(a), post(d))$ to $Output$;
22:     **end for**
23:     $d \leftarrow d.next$;
24:   **end if**
25: **end while**
26: **return** $Output$;

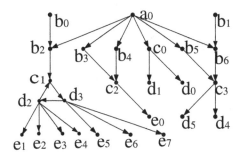

Figure 6.13. Data Graph (Figure 1(a) in [12])

A	A_{in}	A_{out}
a_0	$\emptyset$	$\{c_1, c_3\}$

B	B_{in}	B_{out}
b_0	$\emptyset$	$\{c_1\}$
b_1	$\emptyset$	$\{c_3, b_6\}$
b_2	$\{a_0, b_0\}$	$\{c_1\}$
b_3	$\{a_0\}$	$\{c_2\}$
b_4	$\{a_0\}$	$\{c_2\}$
b_5	$\{a_0\}$	$\{c_3\}$
b_6	$\{a_0\}$	$\{c_3\}$

C	C_{in}	C_{out}
c_0	$\{a_0\}$	$\emptyset$
c_1	$\emptyset$	$\emptyset$
c_2	$\{a_0\}$	$\emptyset$
c_3	$\emptyset$	$\emptyset$

D	D_{in}	D_{out}
d_0	$\{a_0, c_0\}$	$\emptyset$
d_1	$\{a_0, c_0\}$	$\emptyset$
d_2	$\{c_1\}$	$\{c_1\}$
d_3	$\{c_1\}$	$\{c_1\}$
d_4	$\{c_3\}$	$\emptyset$
d_5	$\{c_3\}$	$\emptyset$

E	E_{in}	E_{out}
e_0	$\{a_0, c_2\}$	$\emptyset$
e_1	$\{c_1\}$	$\emptyset$
$\vdots$	$\vdots$	$\vdots$
e_7	$\{c_1\}$	$\emptyset$

(a) Five Lists

(A,B)	$\{a_0\}$	(A,C)	$\{a_0, c_1, c_3\}$
(A,E)	$\{a_0, c_1\}$	(B,C)	$\{c_1, c_2, c_3\}$
(B,E)	$\{c_1, c_2\}$	(C,D)	$\{c_0, c_1, c_3\}$
(B,D)	$\{c_1, c_3\}$	(A,D)	$\{a_0, c_1, c_3\}$
(B,B)	$\{b_0, b_6\}$	(C,C)	$\{c_0, c_1, c_2, c_3\}$

(D,E)	$\{c_1\}$
(C,E)	$\{c_1, c_2\}$
(D,C)	$\{c_1\}$
(D,D)	$\{c_1\}$

(b) W-table

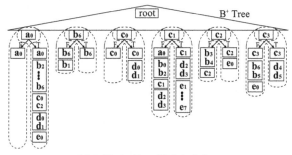

(c) A Cluster-Based R-Join-Index

Figure 6.14. A Graph Database for G_D (Figure 2 in [12])

Consider a graph G_D (Figure 6.13). The 2-hop cover codes for all nodes in G_D are shown in Figure 6.14(a). It is a compressed 2-hop cover code which removes $v \rightsquigarrow v$ from the 2-hop cover code computed. The predicate $\mathcal{P}_{2hop}(,)$ is slightly modified using the compressed 2-hop cover codes as follows.

$$\mathcal{P}_{2hop}(\text{2hopcode}(u), \text{2hopcode}(v)) = L_{out}(u) \cap L_{in}(v) \neq \emptyset \vee u \in L_{in}(v) \vee v \in L_{out}(u)$$

A cluster-based join index for a data graph G_D based on the 2-hop cover computed, $\mathcal{H} = \{S_{w_1}, S_{w_2}, \cdots\}$, where $S_{w_i} = S(A_{w_i}, w_i, D_{w_i})$ and all w_i are centers. It is a B^+-tree in which its non-leaf blocks are used for finding a given center w_i. In the leaf nodes, for each center w_i, its A_{w_i} and D_{w_i}, denoted *F-cluster* and *T-cluster*, are maintained. A w_i's *F-cluster* and *T-cluster* are further divided into labeled *F-subclusters/T-subclusters* where every node, a_i, in an *A*-labeled *F-subcluster* can reach every node d_j in a D-labeled *T-subcluster*, via w_i. Together with the cluster-based join index, it designs a W-table in which, an entry $W(X, Y)$ is a set of centers. A center w_i will be included in $W(A, B)$, if w_i has a non-empty A-labeled *F-subcluster* and a non-empty D-labeled *T-subcluster*. It helps to find the centers, w_i, in the cluster-based join index, that have an A-labeled *F-subcluster* and a D-labeled *T-subcluster*. For the cluster-based join index for G_D (Figure 6.13) is shown in Figure 6.14(c), and the W-table is shown in Figure 6.14(b). Consider $A \hookrightarrow B$. The entry $W(A, B)$ keeps $\{a_0\}$, which suggests that the answers can be only found in the clusters at the center a_0. As shown in Figure 6.14(c), the center a_0 has an A-labeled *F-subcluster* $\{a_0\}$, and a B-labeled *T-subcluster* $\{b_2, b_3, b_4, b_5, b_6\}$. The answer is the Cartesian product between these two labeled subclusters. It can process $A \hookrightarrow D$ queries efficiently.

Cheng et al in. [11] discuss performance issues between the sort-merge join approach and the index approach.

10.2 The General Cases

Chen et al. in [8] propose a holistic based approach for graph pattern matching. But, a query graph, G_q, is restricted to be a tree, which we introduce in brief in Section 2. Their *TwigStackD* algorithm process a tree-shaped G_q in two steps. In the first step, it uses *Twig-Join* algorithm in [7] to find all patterns in the spanning tree of G_D. In the second step, for each node popped out from the stacks used in *Twig-Join* algorithm, *TwigStackD* buffers all nodes which at least match a reachability condition in a bottom-up fashion, and maintains all the corresponding links among those nodes. When a top-most node that matches a reachability condition, *TwigStackD* enumerates the buffer pool and outputs all fully matched patterns. *TwigStackD* performs well for very sparse data graphs. But, its performance degrades noticeably when the G_D becomes dense, due to the high overhead of accessing edge transitive closures.

Cheng et al. in [11, 12] consider $A \hookrightarrow D$ as a R-join (like θ-join), and process a graph pattern matching as a sequence of R-joins. The issue is how to select join order. They propose a dynamic programming algorithm to determine the R-join order in [11]. They also propose an R-join/R-semijoin approach in [12]. The basic idea is to divide the join-index based approach into two steps namely filter and fetch. The filter steps shares the similarity with semijoin, and the fetch step is to join. Cheng et al. study how to select R-join/R-semijoin order by interleaving R-joins with R-semijoins, using dynamic programming in [12].

Wang et al. in [35] propose a query graph G_q based on the hash join approach, and consider how to share the processing cost when it needs to process several $Alist$ and $Dlist$ simultaneously. Wang et al. propose three basic join operators, namely, IT-HGJoin, T-HGJoin, and Bi-HGJoin. The IT-HGJoin processes a subgraph of a query with one descendant and multiple ancestors, for example, $A \hookrightarrow D \wedge B \hookrightarrow D$. The T-HGJoin process a subgraph of a query with one ancestor and multiple descendants, for example, $A \hookrightarrow C \wedge A \hookrightarrow D$. The Bi-HGJoin processes a complete bipartite subgraph of a query with multiple ancestors and multiple descendants, for example $A \hookrightarrow C \wedge A \hookrightarrow D \wedge B \hookrightarrow C \wedge B \hookrightarrow D$. A general query graph G_q will be processed by a set of subgraph queries using IT-HGJoin, T-HGJoin, and Bi-HGJoin.

11.   Conclusions and Summary

In this chapter, we presented a survey on reachability queries. We discussed several coding-based approaches using traversal, dual-labeling, tree cover, chain cover, path-tree cover, 2-hop cover, and 3-hop cover approaches. We also addressed how to support distance-aware queries such as to find the shortest distance between two nodes in a large directed graph using the 2-hop cover, and how to support graph pattern matching using the existing graph-based coding schema. As future work, it becomes important how to use the graph-based coding schema to support more real large graph-based applications.

References

[1] R. Agrawal, A. Borgida, and H. V. Jagadish. Efficient management of transitive relationships in large data and knowledge bases. In *Proceedings of the 1989 ACM SIGMOD international conference on Management of data (SIGMOD 1989)*, 1989.

[2] K. Anyanwu and A. Sheth. ρ-queries: enabling querying for semantic associations on the semantic web. In *Proceedings of the 12th international conference on World Wide Web (WWW 2003)*, 2003.

[3] B. Berendt and M. Spiliopoulou. Analysis of navigation behaviour in web sites integrating multiple information systems. *The VLDB Journal*, 9(1), 2000.

[4] R. Bramandia, J. Cheng, B. Choi, and J. X. Yu. Updating recursive XML views without transitive closure. To appear in *VLDB J.*, 2009.

[5] R. Bramandia, B. Choi, and W. K. Ng. On incremental maintenance of 2-hop labeling of graphs. In *Proceedings of the 17th international conference on World Wide Web (WWW 2008)*, 2008.

[6] D. Brickley and R. V. Guha. Resource Description Framework (RDF) Schema Specification 1.0. W3C Recommendation, 2000.

[7] N. Bruno, N. Koudas, and D. Srivastava. Holistic twig joins: optimal XML pattern matching. In *Proceedings of the 2002 ACM SIGMOD international conference on Management of data (SIGMOD 2002)*, 2002.

[8] L. Chen, A. Gupta, and M. E. Kurul. Stack-based algorithms for pattern matching on dags. In *Proceedings of the 31nd international conference on Very large data bases (VLDB 2005)*, 2005.

[9] Y. Chen and Y. Chen. An efficient algorithm for answering graph reachability queries. In *Proceedings of the 24th International Conference on Data Engineering (ICDE 2008)*, 2008.

[10] J. Cheng and J. X. Yu. On-line exact shortest distance query processing. In *Proceedings of the 12th International Conference on Extending Database Technology (EDBT 2009)*, 2009.

[11] J. Cheng, J. X. Yu, and B. Ding. Cost-based query optimization for multi reachability joins. In *Proceedings of the 12th International Conference on Database Systems for Advanced Applications (DASFAA 2007)*, 2007.

[12] J. Cheng, J. X. Yu, B. Ding, P. S. Yu, and H. Wang. Fast graph pattern matching. In *Proceedings of the 24th International Conference on Data Engineering (ICDE 2008)*.

[13] J. Cheng, J. X. Yu, X. Lin, H. Wang, and P. S. Yu. Fast computation of reachability labeling for large graphs. In *Proceedings of the 10th International Conference on Extending Database Technology (EDBT 2006)*, 2006.

[14] J. Cheng, J. X. Yu, X. Lin, H. Wang, and P. S. Yu. Fast computing reachability labelings for large graphs with high compression rate. In *Proceedings of the 11th International Conference on Extending Database Technology (EDBT 2008)*, 2008.

[15] J. Cheng, J. X. Yu, and N. Tang. Fast reachability query processing. In *Proceedings of the 11th International Conference on Database Systems for Advanced Applications (DASFAA 2006)*, 2006.

[16] Y. J. Chu and T. H. Liu. On the shortest arborescence of a directed graph. *Science Sinica*, 14:1396–1400, 1965.

[17] E. Cohen, E. Halperin, H. Kaplan, and U. Zwick. Reachability and distance queries via 2-hop labels. In *Proceedings of the 13th annual ACM-SIAM symposium on Discrete algorithms (SODA 2002)*, 2002.

[18] T. H. Cormen, C. E. Leiserson, R. L. Rivest, and C. Stein. *Introduction to algorithms*. MIT Press, 2001.

[19] S. DeRose, E. Maler, and D. Orchard. XML linking language (XLink) version 1.0. 2001.

[20] S. DeRose, E. Maler, and D. Orchard. XML pointer language (XPointer) version 1.0. 2001.

[21] J. Edmonds. Optimum branchings. *J. Research of the National Bureau of Standards*, 71B:233–240, 1967.

[22] M. Fernandez, D. Florescu, A. Levy, and D. Suciu. A query language for a web-site management system. *SIGMOD Rec.*, 26(3), 1997.

[23] H. He, H. Wang, J. Yang, and P. S. Yu. Compact reachability labeling for graph-structured data. In *Proceedings of the 2005 ACM CIKM International Conference on Information and Knowledge Management (CIKM 2005)*, pages 594–601, 2005.

[24] H. V. Jagadish. A compression technique to materialize transitive closure. *ACM Trans. Database Syst.*, 15(4):558–598, 1990.

[25] R. Jin, Y. Xiang, N. Ruan, and D. Fuhry. 3-HOP: A high-compression indexing scheme for reachability query. In *Proceedings of the 2009 ACM SIGMOD international conference on Management of data (SIGMOD 2009)*, 2009.

[26] R. Jin, Y. Xiang, N. Ruan, and H. Wang. Efficiently answering reachability queries on very large directed graphs. In *Proceedings of the 2008 ACM SIGMOD international conference on Management of data (SIGMOD 2008)*, 2008.

[27] D. S. Johnson. Approximation algorithms for combinatorial problems. In *Proceedings of the 5th annual ACM symposium on Theory of computing (STOC 1973)*, 1973.

[28] L. Roditty and U. Zwick. A fully dynamic reachability algorithm for directed graphs with an almost linear update time. In *Proceedings of the 36 annual ACM symposium on Theory of computing (STOC 2004)*, 2004.

[29] R. Schenkel, A. Theobald, and G. Weikum. Hopi: An efficient connection index for complex XML document collections. In *Proceedings of the 9th International Conference on Extending Database Technology (EDBT 2004)*, 2004.

[30] R. Schenkel, A. Theobald, and G. Weikum. Efficient creation and incremental maintenance of the HOPI index for complex XML document collections. In *Proceedings of the 21th International Conference on Data Engineering (ICDE 2005)*, 2005.

[31] K. Simon. An improved algorithm for transitive closure on acyclic digraphs. *Theor. Comput. Sci.*, 58(1-3):325–346, 1988.

[32] S. Trißl and U. Leser. Fast and practical indexing and querying of very large graphs. In *Proceedings of the 2007 ACM SIGMOD international conference on Management of data (SIGMOD 2007)*, 2007.

[33] J. van Helden, A. Naim, R. Mancuso, , M. Eldridge, L. Wernisch, D. Gilbert, and S. Wodak. Reresenting and analysing molecular and cellular function using the computer. *Journal of Biological Chemistry*, 381(9-10), 2000.

[34] H. Wang, H. He, J. Yang, P. S. Yu, and J. X. Yu. Dual labeling: Answering graph reachability queries in constant time. In *Proceedings of the 22th International Conference on Data Engineering (ICDE 2006)*, 2006.

[35] H. Wang, J. Li, J. Luo, and H. Gao. Hash-base subgraph query processing method for graph-structured XML documents. *Proceedings VLDB Endowment*, 1(1), 2008.

[36] H. Wang, W. Wang, X. Lin, and J. Li. Labeling scheme and structural joins for graph-structured XML data. In *Proceedings of the 7th Asia-Pacific Web Conference on Web Technologies Research and Development (APWeb 2005)*, 2005.

Chapter 7

EXACT AND INEXACT GRAPH MATCHING: METHODOLOGY AND APPLICATIONS

Kaspar Riesen

Institute of Computer Science and Applied Mathematics, University of Bern
Neubrückstrasse 10, CH-3012 Bern, Switzerland

riesen@iam.unibe.ch

Xiaoyi Jiang

Department of Mathematics and Computer Science, University of Münster
Einsteinstrasse 62, D-48149 Münster, Germany

xjiang@math.uni-muenster.de

Horst Bunke

Institute of Computer Science and Applied Mathematics, University of Bern
Neubrückstrasse 10, CH-3012 Bern, Switzerland

bunke@iam.unibe.ch

Abstract   Graphs provide us with a powerful and flexible representation formalism which can be employed in various fields of intelligent information processing. The process of evaluating the similarity of graphs is referred to as graph matching. Two approaches to this task exist, viz. exact and inexact graph matching. The former approach aims at finding a strict correspondence between two graphs to be matched, while the latter is able to cope with errors and measures the difference of two graphs in a broader sense. The present chapter reviews some fundamental concepts of both paradigms and shows two recent applications of graph matching in the fields of information retrieval and pattern recognition.

Keywords: Exact and Inexact Graph Matching, Graph Edit Distance, Information Retrieval by means of Graph Matching, Graph Embedding via Graph Matching

1.   Introduction

After many years of research, the fields of pattern recognition, machine learning and data mining have reached a high level of maturity [4]. Powerful methods for classification, clustering, information retrieval, and other tasks have become available. However, the vast majority of these approaches rely on object representations given in terms of feature vectors. Such object representations have a number of useful properties. For instance, the dissimilarity, or distance, of two objects can be easily computed by means of the Euclidean distance. Moreover, a large number of well-established methods for data mining, information retrieval, and related tasks in intelligent information processing are available. Recently, however, a growing interest in graph-based object representation can be observed [16]. Graphs are powerful and universal data structures able to explicitly model networks of relationships between substructures of a given object. Thereby, the size as well as the complexity of a graph can be adopted to the size and complexity of a particular object (in contrast to vectorial approaches where the number of features has to be fixed beforehand).

Yet, after the initial enthusiasm induced by the "smartness" and flexibility of graph representations in the late seventies, a number of problems became evident. First, working with graphs is unequally more challenging than working with feature vectors, as even basic mathematic operations cannot be defined in a standard way, but must be provided depending on the specific application. Hence, almost none of the common methods for data mining, machine learning, or pattern recognition can be applied to graphs without significant modifications.

Second, graphs suffer from of their own flexibility. For instance, computing the distances of a pair of objects, which is an important task in many areas, is linear in the number of data items in the case where vectors are employed. The same task for graphs, however, is much more complex, since one cannot simply compare the sets of nodes and edges, which are generally unordered and of different size. More formally, when computing graph dissimilarity or similarity one has to identify common parts of the graphs by considering all of their subgraphs. Regarding that there are $O(2^n)$ subgraphs of a graph with n nodes, the inherent difficulty of graph comparisons becomes obvious.

Despite adverse mathematical and computational conditions in the graph domain, various procedures for evaluating proximity, i.e. similarity or dissimilarity, of graphs have been proposed in the literature [15]. The process of evaluating the similarity of two graphs is commonly referred to as *graph matching*. The overall aim of graph matching is to find a correspondence between the nodes and edges of two graphs that satisfies some, more or less, stringent constraints. That is, by means of the graph matching process similar substructures in one graph are mapped to similar substructures in the other graph. Based on this matching, a dissimilarity or similarity score can eventually be computed indicating the proximity of two graphs.

Graph matching has been the topic of numerous studies in computer science over the last decades. Roughly speaking, there are two categories of tasks in graph matching, viz. *exact matching* and *inexact matching*. In the former case, for a matching to be successful, it is required that a strict correspondence is found between the two graphs being matched, or at least among their subparts. In the latter approach this requirement is substantially relaxed, since also matchings between completely non-identical graphs are possible. That is, inexact matching algorithms are endowed with a certain tolerance to errors and noise, enabling them to detect similarities in a more general way than the exact matching approach. Therefore, inexact graph matching is also referred to as *error-tolerant graph matching*.

For an extensive review of graph matching methods and applications, the reader is referred to [15]. In this chapter, basic notations and definitions are introduced (Sect. 2) and an overview of standard techniques for exact as well as error-tolerant graph matching is given (Sect. 3 and 4). In Sect. 3, dissimilarity models derived from graph isomorphism, subgraph isomorphism, and maximum common subgraph are discussed for exact graph matching. In Sect. 4, inexact graph matching and in particular the paradigm of edit distance applied to graphs is discussed. Finally, two recent applications of graph matching are reviewed. First, in Sect. 5 an algorithmic framework for information retrieval based on graph matching is described. This approach is based on both exact and inexact graph matching procedures and aims at querying large database graphs. Secondly, a graph embedding procedure based on graph matching is reviewed in Sect. 6. This framework aims at an explicit embedding of graphs in real vector spaces, which establishes access to the rich repository of algorithmic tools for classification, clustering, regression, and other tasks, originally developed for vectorial representations.

2. Basic Notations

Various definitions for graphs can be found in the literature, depending upon the considered application. It turns out that the definition given below is sufficiently flexible for a large variety of tasks.

Definition 7.1 (Graph). *Let L_V and L_E be a finite or infinite label alphabet for nodes and edges, respectively. A graph g is a four-tuple $g = (V, E, \mu, \nu)$, where*

- V *is the finite set of nodes,*

- $E \subseteq V \times V$ *is the set of edges,*

- $\mu : V \to L_V$ *is the node labeling function, and*

Figure 7.1. Different kinds of graphs: (a) undirected and unlabeled, (b) directed and unlabeled, (c) undirected with labeled nodes (different shades of gray refer to different labels), (d) directed with labeled nodes and edges.

- $\nu : E \rightarrow L_E$ *is the edge labeling function.*

The number of nodes of a graph g is denoted by $|g|$, while $\mathcal{G}$ represents the set of all graphs over the label alphabets L_V and L_E.

Definition 7.1 allows us to handle arbitrarily structured graphs with unconstrained labeling functions. For example, the labels for both nodes and edges can be given by the set of integers $L = \{1, 2, 3, \ldots\}$, the vector space $L = \mathbb{R}^n$, or a set of symbolic labels $L = \{\alpha, \beta, \gamma, \ldots\}$. Given that the nodes and/or the edges are labeled, the graphs are referred to as *labeled graphs*. *Unlabeled graphs* are obtained as a special case by assigning the same label ε to all nodes and edges, i.e. $L_V = L_E = \{\varepsilon\}$.

Edges are given by pairs of nodes (u, v), where $u \in V$ denotes the source node and $v \in V$ the target node of a directed edge. Commonly, the two nodes u and v connected by an edge (u, v) are referred to as *adjacent*. A graph is termed *complete* if all pairs of nodes are adjacent. *Directed graphs* directly correspond to the definition above. In addition, the class of *undirected graphs* can be modeled by inserting a reverse edge $(v, u) \in E$ for each edge $(u, v) \in E$ with identical labels, i.e. $\nu(u, v) = \nu(v, u)$. In Fig. 7.1 some graphs (directed/undirected, labeled/unlabeled) are shown.

Definition 7.2 (Subgraph). *Let* $g_1 = (V_1, E_1, \mu_1, \nu_1)$ *and* $g_2 = (V_2, E_2, \mu_2, \nu_2)$ *be graphs. Graph* g_1 *is a subgraph of* g_2, *denoted by* $g_1 \subseteq g_2$, *if*

(1) $V_1 \subseteq V_2$,

(2) $E_1 \subseteq E_2$,

(3) $\mu_1(u) = \mu_2(u)$ *for all* $u \in V_1$, *and*

(4) $\nu_1(e) = \nu_2(e)$ *for all* $e \in E_1$.

By replacing condition *(2)* in Definition 7.2 by the more stringent condition

(2') $E_1 = E_2 \cap V_1 \times V_1$,

g_1 becomes an *induced subgraph* of g_2. If g_2 is a subgraph of g_1, graph g_1 is called a *supergraph* of g_2.

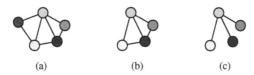

Figure 7.2. Graph (b) is an induced subgraph of (a), and graph (c) is a non-induced subgraph of (a).

Obviously, a subgraph g_1 is obtained from a graph g_2 by removing some nodes and their incident, as well as possibly some additional, edges from g_2. For g_1 to be an induced subgraph of g_2, some nodes and only their incident edges are removed from g_2, i.e. no additional edge removal is allowed. Fig. 7.2(b) and 7.2(c) show an induced and a non-induced subgraph of the graph in Fig. 7.2(a), respectively.

3. Exact Graph Matching

The aim in exact graph matching is to determine whether two graphs, or at least part of them, are identical in terms of structure and labels. A common approach to describe the structure of a graph is to define the *adjacency matrix* $\mathbf{A} = (a_{ij})_{n \times n}$ of graph $g = (V, E, \mu, \nu)$ ($|g| = n$). In this matrix the entry a_{ij} is equal to 1 if there is an edge $(v_i, v_j) \in E$ connecting the i-th node $v_i \in V$ with the $j - th$ node $v_j \in V$, and 0 otherwise.

Generally, for the nodes (and also the edges) of a graph there is no unique canonical order. Thus, for a single graph with n nodes, $n!$ different adjacency matrices exist, since there are $n!$ possibilities to order the nodes of g. Consequently, for checking two graphs for structural identity, we cannot simply compare their adjacency matrices. The identity of two graphs g_1 and g_2 is commonly established by defining a function, termed graph isomorphism, that maps g_1 to g_2.

Definition 7.3 (Graph Isomorphism). *Let us consider two graphs denoted by* $g_1 = (V_1, E_1, \mu_1, \nu_1)$ *and* $g_2 = (V_2, E_2, \mu_2, \nu_2)$ *respectively. A graph isomorphism is a bijective function* $f : V_1 \rightarrow V_2$ *satisfying*

(1) $\mu_1(u) = \mu_2(f(u))$ *for all nodes* $u \in V_1$

(2) for each edge $e_1 = (u, v) \in E_1$, *there exists an edge*

$$e_2 = (f(u), f(v)) \in E_2$$

such that $\nu_1(e_1) = \nu_2(e_2)$

(3) for each edge $e_2 = (u, v) \in E_2$, *there exists an edge*

$$e_1 = (f^{-1}(u), f^{-1}(v)) \in E_1$$

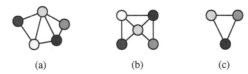

Figure 7.3. Graph (b) is isomorphic to (a), and graph (c) is isomorphic to a subgraph of (a). Node attributes are indicated by different shades of gray.

such that $\nu_1(e_1) = \nu_2(e_2)$

Two graphs are called isomorphic if there exists an isomorphism between them.

Obviously, isomorphic graphs are identical in both structure and labels. That is, a one-to-one correspondence between each node of the first graph and each node of the second graph has to be found such that the edge structure is preserved and node and edge labels are consistent.

Unfortunately, no polynomial runtime algorithm is known for the problem of graph isomorphism [25]. That is, in the worst case, the computational complexity of any of the available algorithms for graph isomorphism is exponential in the number of nodes of the two graphs. However, since most scenarios encountered in practice are often different from the worst case, and furthermore, the labels of both nodes and edges very often help to substantially reduce the complexity of the search, the actual computation time can still be manageable. Polynomial algorithms for graph isomorphism have been developed for special kinds of graphs, such as trees [1], ordered graphs [38], planar graphs [34], bounded-valence graphs [45], and graphs with unique node labels [18].

Standard procedures for testing graphs for isomorphism are based on tree search techniques with backtracking. The basic idea is that a partial node matching, which assigns nodes from the two graphs to each other, is iteratively expanded by adding new node-to-node correspondences. This expansion is repeated until either the edge structure constraint is violated or node or edge labels are inconsistent. In this case a backtracking procedure is initiated, i.e. the last node mappings are iteratively undone until a partial node mapping is found for which an alternative extension is possible. Obviously, if there is no further possibility for expanding the partial node matching without violating the constraints, the algorithm terminates indicating that there is no isomorphism between the considered graphs. Conversely, finding a complete node-to-node correspondence without violating any of the structure or label constraints proves that the investigated graphs are isomorphic. In Fig. 7.3 (a) and (b) two isomorphic graphs are shown.

A well known, and despite its age still very popular, algorithm implementing the idea of a tree search with backtracking for graph isomorphism is described in [89]. A more recent algorithm for graph isomorphism, also based on the idea of tree search, is the VF algorithm and its successor VF2 [17]. Here the basic tree search algorithm is endowed with an efficiently computable heuristic which substantially reduces the search time. In [43] the tree search method for isomorphism is sped up by means of another heuristic derived from *Constraint Satisfaction*. Other algorithms for exact graph matching, which are not based on tree search techniques, are *Nauty* [50], and decision tree based techniques [51], to name just two examples. The reader is referred to [15] for an exhaustive list of exact graph matching algorithms developed since 1973.

Closely related to graph isomorphism is subgraph isomorphism, which can be seen as a concept describing subgraph equality. A subgraph isomorphism is a weaker form of matching in terms of requiring only that an isomorphism holds between a graph g_1 and a subgraph of g_2. Intuitively, subgraph isomorphism is the problem to detect if a smaller graph is identically present in a larger graph. In Fig. 7.3 (a) and (c), an example of subgraph isomorphism is given.

Definition 7.4 (Subgraph Isomorphism). *Let* $g_1 = (V_1, E_1, \mu_1, \nu_1)$ *and* $g_2 = (V_2, E_2, \mu_2, \nu_2)$ *be graphs. An injective function* $f : V_1 \to V_2$ *from* g_1 *to* g_2 *is a subgraph isomorphism if there exists a subgraph* $g \subseteq g_2$ *such that* f *is a graph isomorphism between* g_1 *and* g.

The tree search based algorithms for graph isomorphism [17, 43, 89], as well as the decision tree based techniques [51], can also be applied to the subgraph isomorphism problem. In contrast with the problem of graph isomorphism, subgraph isomorphism is known to be NP-complete [25]. As a matter of fact, subgraph isomorphism is a harder problem than graph isomorphism as one has not only to check whether a permutation of g_1 is identical to g_2, but we have to decide whether g_1 is isomorphic to any of the subgraphs of g_2 with equal size as g_1.

The process of graph matching primarily aims at identifying corresponding substructures in the two graphs under consideration. Through the graph matching procedure an associated similarity or dissimilarity score can be easily inferred. In view of this, graph isomorphism as well as subgraph isomorphism provide us with a basic similarity measure, which is 1 (maximum similarity) for (sub)graph isomorphic, and 0 (minimum similarity) for non-isomorphic graphs. Hence, two graphs must be completely identical, or the smaller graph must be identically contained in the other graph, to be deemed similar. Consequently, the applicability of this graph similarity measure is rather limited. Consider a case where most, but not all, nodes and edges in two graphs are identical. The rigid concept of (sub)graph isomorphism fails in such a situation in the sense of considering the two graphs to be totally dissimilar. Due to this observation, the formal concept of the largest common part of two graphs is established.

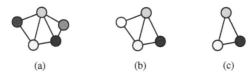

(a)   (b)   (c)

Figure 7.4. Graph (c) is a maximum common subgraph of graph (a) and (b).

Definition 7.5 (Maximum common subgraph). *Let* $g_1 = (V_1, E_1, \mu_1, \nu_1)$ *and* $g_2 = (V_2, E_2, \mu_2, \nu_2)$ *be graphs. A common subgraph of* g_1 *and* g_2, $cs(g_1, g_2)$, *is a graph* $g = (V, E, \mu, \nu)$ *such that there exist subgraph isomorphisms from* g *to* g_1 *and from* g *to* g_2. *We call* g *a maximum common subgraph of* g_1 *and* g_2, $mcs(g_1, g_2)$, *if there exists no other common subgraph of* g_1 *and* g_2 *that has more nodes than* g.

A maximum common subgraph of two graphs represents the maximal part of both graphs that is identical in terms of structure and labels. In Fig. 7.4(c) the maximum common subgraph is shown for the two graphs in Fig. 7.4(a) and (b). Note that, in general, the maximum common subgraph is not uniquely defined, that is, there may be more than one common subgraph with a maximal number of nodes. A standard approach to computing maximum common subgraphs is based on solving the maximum clique problem in an association graph [44, 49]. The association graph of two graphs represents the whole set of possible node-to-node mappings that preserve the edge structure and labels of both graphs. Finding a maximum clique in the association graph, that is, a fully connected maximal subgraph, is equivalent to finding a maximum common subgraph. In [10] the reader can find an experimental comparison of algorithms for maximum common subgraph computation on randomly connected graphs.

Graph dissimilarity measures can be derived from the maximum common subgraph of two graphs. Intuitively speaking, the larger a maximum common subgraph of two graphs is, the more similar are the two graphs. For instance, in [12] such a distance measure is introduced, defined by

$$d_{MCS}(g_1, g_2) = 1 - \frac{|mcs(g_1, g_2)|}{\max\{|g_1|, |g_2|\}} \qquad (7.1)$$

Note that, whereas the maximum common subgraph of two graphs is not uniquely defined, the d_{MCS} distance is. If two graphs are isomorphic, their d_{MCS} distance is 0; on the other hand, if two graphs have no part in common, their d_{MCS} distance is 1. It has been shown that d_{MCS} is a metric and produces a value in $[0, 1]$.

A second distance measure which has been proposed in [94], based on the idea of graph union, is

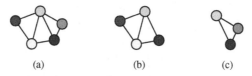

Figure 7.5. Graph (a) is a minimum common supergraph of graph (b) and (c).

$$d_{WGU}(g_1, g_2) = 1 - \frac{|mcs(g_1, g_2)|}{|g_1| + |g_2| - |mcs(g_1, g_2)|}$$

By "graph union" it is meant that the denominator represents the size of the union of the two graphs in the set-theoretic sense. This distance measure behaves similarly to d_{MCS}. The motivation of using graph union in the denominator is to allow for changes in the smaller graph to exert some influence on the distance measure, which does not happen with d_{MCS}. This measure was also demonstrated to be a metric and creates distance values in $[0, 1]$.

A similar distance measure [7] which is not normalized to the interval $[0, 1]$ is:

$$d_{UGU}(g_1, g_2) = |g_1| + |g_2| - 2 \cdot |mcs(g_1, g_2)|$$

Fernandez and Valiente [21] have proposed a distance measure based on both the maximum common subgraph and the minimum common supergraph

$$d_{MMCS}(g_1, g_2) = |MCS(g_1, g_2)| - |mcs(g_1, g_2)|$$

where $MCS(g_1, g_2)$ is the minimum common supergraph of graphs g_1 and g_2, which is the complimentary concept of minimum common subgraph.

Definition 7.6 (Minimum common supergraph). *Let* $g_1 = (V_1, E_1, \mu_1, \nu_1)$ *and* $g_2 = (V_2, E_2, \mu_2, \nu_2)$ *be graphs. A common supergraph of* g_1 *and* g_2, *$CS(g_1, g_2)$, is a graph* $g = (V, E, \mu, \nu)$ *such that there exist subgraph isomorphisms from* g_1 *to* g *and from* g_2 *to* g. *We call* g *a minimum common supergraph of* g_1 *and* g_2, *$MCS(g_1, g_2)$, if there exists no other common supergraph of* g_1 *and* g_2 *that has less nodes than* g.

In Fig. 7.5(a) the minimum common supergraph of the graphs in Fig. 7.5(b) and (c) is given. The computation of the minimum common supergraph can be reduced to the problem of computing a maximum common subgraph [11].

The concept that drives the distance measure above is that the maximum common subgraph provides a "lower bound" on the similarity of two graphs, while the minimum supergraph is an "upper bound". If two graphs are identical, then both their maximum common subgraph and minimum common supergraph are the same as the original graphs and $|g_1| = |g_2| = |MCS(g_1, g_2)| = |mcs(g_1, g_2)|$, which leads to $d_{MMCS}(g_1, g_2) = 0$. As the graphs become more dissimilar, the size of the maximum common subgraph decreases, while the size of the minimum supergraph increases. This in turn leads to increasing values of $d_{MMCS}(g_1, g_2)$. For two graphs with an empty maximum common subgraph, the distance will become $|MCS(g_1, g_2)| = |g_1| + |g_2|$. The distance $d_{MMCS}(g_1, g_2)$ has also been shown to be a metric, but it does not produce values normalized to the interval $[0, 1]$, unlike d_{MCS} or d_{WGU}. We can also create a version of this distance measure which is normalized to $[0, 1]$ as follows:

$$d_{MMCSN}(g_1, g_2) = 1 - \frac{|mcs(g_1, g_2)|}{|MCS(g_1, g_2)|}$$

Note that, because of $|MCS(g_1, g_2)| = |g_1| + |g_2| - |mcs(g_1, g_2)|$, d_{UGU} and d_{MMCS} are identical. The same is true for d_{WGU} and d_{MMCSN}.

The main advantage of exact graph matching methods is their stringent definition and solid mathematical foundation. This advantage may turn into a disadvantage, however, because in exact graph matching for finding two graphs g_1 and g_2 to be similar, it is required that a significant part of the topology together with the corresponding node and edge labels in g_1 and g_2 have to be identical. In fact, this constraint is too rigid in some applications. For this reason, a large number of error-tolerant, or inexact, graph matching methods have been proposed, dealing with a more general graph matching problem than the one of (sub)graph isomorphism.

4.   Inexact Graph Matching

Due to the intrinsic variability of the patterns under consideration and the noise resulting from the graph extraction process, it cannot be expected that two graphs representing the same class of objects are completely, or at least to a large part, identical in their structure. Moreover, if the node or edge label alphabet L is used to describe non-discrete properties of the underlying patterns, e.g. $L \subseteq \mathbb{R}^n$, it is most probable that the actual graphs differ somewhat from their ideal model. Obviously, such noise crucially hampers the applicability of exact graph matching techniques, and consequently exact graph matching is rarely used in real-world applications.

In order to overcome this drawback, it is advisable to endow the graph matching framework with a certain tolerance to errors. That is, the matching process must be able to accommodate the differences of the graphs by relaxing –to some extent– the underlying constraints. In the first part of this section the concept of graph edit distance is introduced to exemplarily illustrate the paradigm of inexact graph matching. In the second part, several other approaches to inexact graph matching are briefly discussed.

Figure 7.6. A possible edit path between graph g_1 and graph g_2 (node labels are represented by different shades of gray).

4.1   Graph Edit Distance

Graph edit distance [8, 71] offers an intuitive way to integrate error-tolerance into the graph matching process and is applicable to virtually all types of graphs. Originally, edit distance has been developed for string matching [93] and a considerable amount of variants and extensions to the edit distance have been proposed for strings and graphs. The key idea is to model structural variation by edit operations reflecting modifications in structure and labeling. A standard set of edit operations is given by *insertions*, *deletions*, and *substitutions* of both nodes and edges. Note that other edit operations, such as *merging* and *splitting* of nodes [2], can be useful in certain applications. Given two graphs, the source graph g_1 and the target graph g_2, the idea of graph edit distance is to delete some nodes and edges from g_1, relabel (substitute) some of the remaining nodes and edges, and insert some nodes and edges in g_2, such that g_1 is finally transformed into g_2. A sequence of edit operations $e_1, \ldots, e_k$ that transform g_1 into g_2 is called an *edit path* between g_1 and g_2. In Fig. 7.6 an example of an edit path between two graphs g_1 and g_2 is given. This edit path consists of three edge deletions, one node deletion, one node insertion, two edge insertions, and two node substitutions.

Let $\Upsilon(g_1, g_2)$ denote the set of all possible edit paths between two graphs g_1 and g_2. Clearly, every edit path between two graphs g_1 and g_2 is a model describing the correspondences found between the graphs' substructures. That is, the nodes of g_1 are either deleted or uniquely substituted with a node in g_2, and analogously, the nodes in g_2 are either inserted or matched with a unique node in g_1. The same applies for the edges. In [58] the idea of fuzzy edit paths was reported where both nodes and edges can be simultaneously mapped to several nodes and edges. The optimal fuzzy edit path is then determined by means of quadratic programming.

To find the most suitable edit path out of $\Upsilon(g_1, g_2)$, one introduces a cost for each edit operation, measuring the strength of the corresponding operation. The idea of such a cost is to define whether or not an edit operation represents a strong modification of the graph. Clearly, between two similar graphs, there should exist an inexpensive edit path, representing low cost operations, while for dissimilar graphs an edit path with high costs is needed. Consequently, the *edit distance* of two graphs is defined by the minimum cost edit path between two graphs.

Definition 7.7 (Graph Edit Distance). *Let* $g_1 = (V_1, E_1, \mu_1, \nu_1)$ *be the source and* $g_2 = (V_2, E_2, \mu_2, \nu_2)$ *the target graph. The graph edit distance between* g_1 *and* g_2 *is defined by*

$$d(g_1, g_2) = \min_{(e_1, \ldots, e_k) \in \Upsilon(g_1, g_2)} \sum_{i=1}^{k} c(e_i),$$

where $\Upsilon(g_1, g_2)$ *denotes the set of edit paths transforming* g_1 *into* g_2, *and* c *denotes the cost function measuring the strength* $c(e)$ *of edit operation* e.

The definition of adequate and application-specific cost functions is a key task in edit distance based graph matching. Prior knowledge of the graphs' labels is often inevitable for graph edit distance to be a suitable proximity measure. This fact is often considered as one of the major drawbacks of graph edit distance. Yet, contrariwise, the possibility to parametrize graph edit distance by means of the cost function crucially amounts for the versatility of this dissimilarity model. That is, by means of graph edit distance it is possible to integrate domain specific knowledge about object similarity, if available, when defining the costs of the elementary edit operations. Furthermore, if in a particular case prior knowledge about the labels and their meaning is not available, automatic procedures for learning the edit costs from a set of sample graphs are available as well [55, 56].

The overall aim of the cost function is to favor weak distortions over strong modifications of the graph. Hence, the cost is defined with respect to the underlying node or edge labels, i.e. the cost $c(e)$ is a function depending on the edit operation e. Typically, for numerical node and edge labels the Euclidean distance can be used to model the cost of a particular substitution operation on the graphs. For deletions and insertions of both nodes and edges, often a constant cost τ_{node}/τ_{edge} is assigned. We refer to this cost function as *Euclidean Cost Function*.

The Euclidean cost function defines substitution costs proportional to the Euclidean distance of two respective labels. The basic intuition behind this approach is that the further away two labels are, the stronger is the distortion associated with the corresponding substitution. Note that any node substitution having a higher cost than $2 \cdot \tau_{node}$ will be replaced by a composition of a deletion and an insertion of the involved nodes (the same accounts for the edges). This behavior reflects the basic intuition that substitutions should be favored over deletions and insertions to a certain degree.

Optimal algorithms for computing the edit distance of graphs g_1 and g_2 are typically based on combinatorial search procedures that explore the space of all possible mappings of the nodes and edges of g_1 to the nodes and edges of g_2 [8]. A major drawback of those procedures is their computational complexity, which is exponential in the number of nodes of the involved graphs.

Consequently, the application of optimal algorithms for edit distance computations is limited to graphs of rather small size in practice.

To render graph edit distance computation less computationally demanding, a number of suboptimal methods have been proposed. In some approaches, the basic idea is to perform a local search to solve the graph matching problem, that is, to optimize local criteria instead of global, or optimal ones [57, 80]. In [40], a linear programming method for computing the edit distance of graphs with unlabeled edges is proposed. The method can be used to derive lower and upper edit distance bounds in polynomial time. Two fast but suboptimal algorithms for graph edit distance computation are proposed in [59]. The authors propose simple variants of a standard edit distance algorithm that make the computation substantially faster. In [20] another suboptimal method has been proposed. The basic idea is to decompose graphs into sets of subgraphs. These subgraphs consist of a node and its adjacent nodes and edges. The graph matching problem is then reduced to the problem of finding a match between the sets of subgraphs. In [67] a method somewhat similar to the method described in [20] is proposed. However, while the optimal correspondence between local substructures is found by dynamic programming in [20], a bipartite matching procedure [53] is employed in [67].

4.2   Other Inexact Graph Matching Techniques

Several other important classes of error-tolerant graph matching algorithms have been proposed. Among others, algorithms based on Artificial Neural Networks, Relaxation Labeling, Spectral Decompositions, and Graph Kernels have been reported.

Artificial Neural Networks.   One class of error-tolerant graph matching methods employs *artificial neural networks*. In two seminal papers [24, 81] it is shown that neural networks can be used to classify directed acyclic graphs. The algorithms are based on an energy minimization framework, and use some kind of Hopfield network [84]. Hopfield networks consist of a set of neurons connected by synapses such that, upon activation of the network, the neuron output is fed back into the network. By means of an iterative learning procedure the given energy criterion is minimized. Similar to the approach of relaxation labeling (see below), compatibility coefficients are used to evaluate whether two nodes or edges constitute a successful match.

In [83] the optimization procedure is stabilized by means of a Potts MFT network. In [85] a self-organizing Hopfield network is introduced that learns most of the network parameters and eliminates the need for specifying them a priori. In [52, 72] the graph neural network is crucially extended such that also undirected and acyclic graphs can be processed. The general idea is to represent the nodes of a graph in an encoding network. In this encoding network local transition functions and local output functions are employed, expressing the dependency of a node on its neighborhood and describing how the output is produced, respectively. As both functions are implemented by feedforward neural networks, the encoding network can be interpreted as a recurrent neural network.

Further examples of graph matching based on artificial neural networks can be found in [37, 73, 101]

Relaxation Labeling.   Another class of error-tolerant graph matching methods employs *relaxation labeling techniques*. The basic idea of this particular approach is to formulate the graph matching problem as a labeling problem. Each node of one graph is to be assigned to one label out of a discrete set of possible labels, specifying a matching node of the other graph. During the matching process, Gaussian probability distributions are used to model compatibility coefficients measuring how suitable each candidate label is. The initial labeling, which is based on the node attributes, node connectivity, and other information available, is then refined in an iterative procedure until a sufficiently accurate labeling, i.e. a matching of two graphs, is found. Based on the pioneering work presented in [22], the idea of relaxation labeling has been refined in several contributions. In [30, 41] the probabilistic framework for relaxation labeling is endowed with a theoretical foundation. The main drawback of the initial formulation of this technique, viz. the fact that node and edge labels are used only in the initialization of the matching process, is overcome in [14]. A significant extension of the framework is introduced in [97] where a Bayesian consistency measure is adapted to derive a graph distance. In [35] this method is further improved by taking also edge labels into account in the evaluation of the consistency measure. The concept of Bayesian graph edit distance, which in fact builds up on the idea of probabilistic relaxation, is presented in [54]. The concept has also been successfully applied to special kinds of graphs, such as trees [87].

Spectral Methods.   *Spectral methods* build a further class of graph matching procedures [13, 47, 70, 78, 90, 98]. The general idea of this approach is based on the following observation. The eigenvalues and the eigenvectors of the adjacency or Laplacian matrix of a graph are invariant with respect to node permutation. Hence, if two graphs are isomorphic, their structural matrices will have the same eigendecomposition. The converse, i.e. deducing from the equality of eigendecompositions to graph isomorphism, is not true in general. However, by representing the underlying graphs by means of the eigendecomposition of their structural matrix, the matching process of the graphs can be conducted on some features derived from their eigendecomposition. The main problem of spectral methods is that they are rather sensitive towards structural errors, such as missing or spurious nodes. Moreover, most of these methods are purely structural, in the sense that they are only applicable to unlabeled graphs, or they allow only severely constrained label alphabets.

Graph Kernel.   Kernel methods were originally developed for vectorial representations, but the kernel framework can be extended to graphs in a very natural way. A number of *graph kernels* have been designed for graph matching [26, 57]. A seminal contribution is the work on convolution kernels, which provides a general framework for dealing with complex objects that consist of simpler parts [32, 95]. Convolution kernels infer the similarity of complex objects from the similarity of their parts.

A second class of graph kernels is based on the analysis of random walks in graphs. These kernels measure the similarity of two graphs by the number of random walks in both graphs that have all or some labels in common [5, 27]. In [27] an important result is reported. It is shown that the number of matching walks in two graphs can be computed by means of the product graph of two graphs, without the need to explicitly enumerate the walks. In order to handle continuous labels the random walk kernel has been extended in [5]. This extension allows one to also take non-identically labeled walks into account.

A third class of graph kernels is given by diffusion kernels. The kernels of this class are defined with respect to a base similarity measure which is used to construct a valid kernel matrix [42, 79, 92]. This base similarity measure only needs to satisfy the condition of symmetry and can be defined for any kind of objects.

Miscellaneous Methods.   Several other error-tolerant graph matching methods have been proposed in the literature, for instance, graph matching based on the Expectation Maximization algorithm [46], on replicator equations [61], and on graduated assignment [28]. Random walks in graphs [29, 69], approximate least-squares and interpolation theory algorithms [91], and random graphs [99] have also been employed for error-tolerant graph matching.

5.   Graph Matching for Data Mining and Information Retrieval

The use of graphs and graph matching has become a promising approach in data mining and related areas [16]. In fact, querying graph databases has a long tradition and dates back to the time when the first algorithms for subgraph isomorphism detection became available. Yet, the use of conventional subgraph isomorphism in graph based data mining implicates severe limitations. First of all, the underlying database graph often includes a rather large number of attributes, some of which might be irrelevant for a particular query. The second

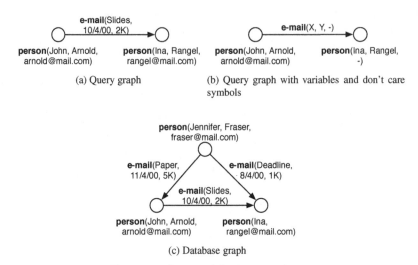

(a) Query graph

(b) Query graph with variables and don't care
symbols

(c) Database graph

Figure 7.7. Query and database graphs.

restriction arises from the limited answer format provided by conventional subgraph isomorphism which is only able to check whether or not a query graph
is embedded in a larger database graph. Thirdly, subgraph isomorphism in its
original mode does not allow constraints that may be imposed on the attributes
of a query to model restrictions or dependencies.

The generalized subgraph isomorphism retrieval procedure described in [6]
overcomes these three restrictions. First, the approach offers the possibility to
mask out attributes in queries. To this end, *don't care* values are introduced for
attributes that are irrelevant. Secondly, to make the retrieval of more specific
information from the database graph possible than just a binary decision **yes**
or **no**, *variables* are used. By means of these variables, one is able to retrieve
values of specific attributes from the database graph. Thirdly, the concept of
constrained variables, for example, variables that can assume only values from
a certain interval, allows one to define more specific queries.

The approach to knowledge mining and information retrieval proposed
in [6] is based on the idea of specifying a query by means of a query graph,
which can be used to extract information from a large database graph. In contrast with Definition 7.1, the graphs employed are defined in a more general
way. Rather than using just a single label, each node in a graph is labeled by
a type and some attributes. The same accounts for the edges. In Fig. 7.7 (a)
an example of a query graph is shown. In this illustration nodes are of the
type *person* and labeled with the person's first and second name, and e-mail
address. Edges are of the type *e-mail* and labeled with the e-mail's subject, the
date, and the size. Note that in general there may occur nodes as well as edges
of different type in the same graph.

Query graphs are more general than common graphs in the sense that don't care symbol and variables may occur as the values of attributes on the nodes and edges. The purpose of the variables is to define those attributes whose values are to be returned as an answer to a query (we will come back to this point later). In Fig. 7.7 (b) an example of a query graph with variables (X, Y) and don't care symbols $(-)$ is given. According to this query, we are particularly interested in the subject (X) and the date (Y) of an e-mail sent from John Arnold to Ina Rangel. As we do not care about the size of the e-mail and we do not know the e-mail address of Ina Rangel, two don't care symbols are used. Variables may also occur in a query because they may be used to express constraints on one or several attribute values. A constraint on a set of variables occurring in a query graph is a condition on one or several variables that evaluates to **true** or **false** if we assign a concrete attribute value to each variable. For instance, the query in Fig. 7.7 (b) can be augmented by the constraint that the e-mail in question was sent between October 1 and October 3 (formally $9/31/00 < Y < 10/4/00$).

Once the query graph has been constructed by the user, it is matched against a database graph. The process of matching a query graph to a database graph essentially means that we want to find out whether there exists a subgraph isomorphism from the query to the database graph. Obviously, as the query graph may include don't care symbols and variables, we need a more general notion of subgraph isomorphism than the one provided in Definition 7.4. Such a generalized subgraph isomorphism between a query and a database graph is referred to as a *match*, i.e., if a query graph q matches a database graph G, we call the injective function f a match between q and G. Note that for given q and G and a given set of constraints over the variables in q, there can be zero, one, or more than one matches.

For a match we require each edge of the query graph being included in the database graph. A node, u, can be mapped, via injective function f, only to a node of the same type. If the (type, attribute)-pair of a node u of the query graph includes an attribute value x_i, then it is required that the same value occur at the corresponding position in the (type, attribute)-pair of the node $f(u)$ in the database graph. Don't care symbols occurring in the (type, attribute)-pair of a node u will match any attribute value at the corresponding position in the (type, attribute)-pair of node $f(u)$. Similarly, unconstrained variables match any attribute value at their corresponding position in $f(u)$. In case there exist constraints on a variable in the query graph, the attribute values at the corresponding positions in $f(u)$ must satisfy these constraints.

By means of variables we indicate which attribute values are to be returned by our knowledge mining system as an answer to a query. Therefore, the answer to a query can be **no**, if there is no such structure as the query graph contained as a substructure in the database graph, or **yes** if the query graph exists (at least once) as a substructure in the database graph and the query graph does not contain any answer variables. In the case where answer variables are defined in the query graph and one or several matches are found an individual answer is generated for each match f_j. An answer is of the form $X_1 = x'_1, \ldots, X_n = x'_n$ where $X_1, \ldots, X_n$ are the answer variables occurring in the query and x'_i are the values of the attributes in the database graph that correspond to the variables X_i under match f_j. Obviously, there is a match between the query graph in Fig. 7.7 (b) and the database graph in Fig. 7.7 (c). Hence, the variables are linked by $X =$ Slides and $Y = 10/4/00$.

The proposed system described so far does not return any information from the database graph whenever no match is found. However, in some cases this behavior may be undesirable. Let us consider, for instance, a query graph that contains spurious attribute values or edges which do not occur in the underlying database graph. The graph matching framework presented so far merely returns the answer **no** as it finds no match in the database graph. However, we can easily endow the graph isomorphism framework with a certain tolerance to errors. To this end one can use graph edit distance. In cases when no perfect match of the query graph to the database graph is possible, the query is minimally modified such that a match becomes possible. The well-founded possibility of augmenting the data mining framework with some tolerance to errors definitely accounts for the power of this particular procedure based on graph matching.

In [6] an algorithmic procedure is described for finding matches between a query q and a database graph G. This procedure checks two given graphs, q and G, whether there exists a match from q to G by constructing all possible mappings $f: V_1 \to V_2$. This matching algorithm is of exponential complexity. However, as the underlying query graphs are typically limited in size and due to the fact that the attributes and constraints limit the potential search space for a match significantly, the computational complexity of this algorithm is usually still manageable, as shown in the experiments reported in [6].

For applications where large query graphs occur a novel approximate approach for querying graph databases has been introduced in [86]. This algorithm proceeds as follows. First, a number of important nodes from the query graph are selected. The importance of the nodes can be measured, for instance, by their degree. Using the label, the degree, and information about a node's local neighborhood, the most important nodes are matched against the database graph nodes. Clearly, by means of this procedure each node from the query graph may be mapped to several database nodes and vice versa. Given a quality criterion for the individual node mappings, a bipartite optimization procedure can be applied resulting in a one-to-one correspondence between query nodes and database nodes. The node pairs returned by the bipartite matching procedure serve us as *anchor points* of the complete matching. Based on these anchor points, the initial graph match is iteratively extended. For each node that has already been mapped to a database node, its nearby nodes (nodes that are at most two hops away) are tried to be mapped to database nodes. This extension is repeated until no more nodes can be added to the match. Clearly, in contrast with the method described in [6] this procedure is suboptimal in the sense of finding subgraphs in the database graph that are similar, but not necessarily equal, to the query graph. In exchange, a graph matching framework applicable to very large query graphs (hundreds to thousands of nodes and edges) is established.

6. Vector Space Embeddings of Graphs via Graph Matching

Classification and clustering of objects are common tasks in intelligent information processing. Classification refers to the process of assigning an unknown input object to one out of a given set of classes, while clustering refers to the process of dividing a set of given objects into homogeneous groups. A vast number of algorithms for classification [19] and clustering [100] have been proposed in the literature. Almost all of these algorithms have been designed for object representations given in terms of feature vectors. This means that there exists a severe lack of algorithmic tools for graph classification and clustering. This lack is mainly due to the fact that some of the basic operations needed in classification as well as clustering are not available for graphs. In other words, while it is possible to define graph dissimilarity measures via specific graph matching procedures, this is often not sufficient for standard algorithms in intelligent information processing. In fact, graph distance based pattern recognition is basically limited to nearest-neighbor classification and k-medians clustering [57].

A promising direction to overcome this severe limitation is graph embedding into vector spaces. Basically, such an embedding of graphs establishes access to the rich repository of algorithmic tools developed for vectorial representations. In [47], for instance, features derived from the eigendecomposition of graphs are studied. Another idea deals with string edit distance applied to the eigensystem of graphs [96]. This procedure results in distances between graphs which are used to embed the graphs into a vector space by means of multidimensional scaling. In [98] the authors turn to the spectral decomposition of the Laplacian matrix of a graph. They show how the elements of the spectral matrix of the Laplacian can be used to construct symmetric polynomials. In order to encode graphs as vectors, the coefficients of these polynomials are used as graph features. Another approach for graph embedding has been proposed in [70]. The authors use the relationship between the Laplace-

Beltrami operator and the graph Laplacian to embed a graph in a Riemannian manifold.

The present section considers a new class of graph embedding procedures which are based on dissimilarity representation and graph matching. Originally the idea was proposed in [60] in order to map feature vectors into dissimilarity spaces. Later it was generalized to string based object representation [82] and to the domain of graphs [62]. Graphs from a given problem domain are mapped to vector spaces by computing the distance to some predefined prototype graphs. The resulting distances can be used as a vectorial representation of the considered graph.

Formally, assume we have a set of sample graphs, $T = \{g, \ldots, g_N\}$ from some graph domain $\mathcal{G}$ and an arbitrary graph dissimilarity measure $d : \mathcal{G} \times \mathcal{G} \to \mathbb{R}$. Note that T can be any kind of graph set. However, for the sake of convenience we define T as a training set of given graphs. After selecting a set of prototypical graphs $\mathcal{P} \subseteq T$, we compute the dissimilarity of a given input graph g to each prototype graph $p_i \in \mathcal{P}$. Note that g can be an element of T or any other graph set $\mathcal{S}$. Given n prototypes, i.e. $\mathcal{P} = \{p_1, \ldots, p_n\}$, this procedure leads to n dissimilarities, $d_1 = d(g, p_1), \ldots, d_n = d(g, p_n)$, which can be arranged in an n-dimensional vector $(d_1, \ldots, d_n)$.

Definition 7.8 (Graph Embedding). *Let us assume a graph domain $\mathcal{G}$ is given. If $T = \{g, \ldots, g_N\} \subseteq \mathcal{G}$ is a training set with N graphs and $\mathcal{P} = \{p_1, \ldots, p_n\} \subseteq T$ is a prototype set with n graphs, the mapping*

$$\varphi_n^{\mathcal{P}} : \mathcal{G} \to \mathbb{R}^n$$

is defined as the function

$$\varphi_n^{\mathcal{P}}(g) = (d(g, p_1), \ldots, d(g, p_n)),$$

where $d(g, p_i)$ is any graph dissimilarity measure between graph g and the i-th prototype graph.

Obviously, by means of this definition we obtain a vector space where each axis corresponds to a prototype graph $p_i \in \mathcal{P}$ and the coordinate values of an embedded graph g are the distances of g to the elements in $\mathcal{P}$. In this way we can transform any graph g from the training set T as well as any other graph set $\mathcal{S}$ (for instance a validation or a test set of a classification problem), into a vector of real numbers. In [65] this procedure is further generalized towards Lipschitz embeddings [33]. Rather than singleton reference sets (i.e. prototypes $p_1, \ldots, p_n$), sets of prototypes $\mathcal{P}_1, \ldots, \mathcal{P}_n$ are used for embedding the graphs via dissimilarities.

The embedding procedure proposed in [62] makes use of graph edit distance. Note, however, that any other graph dissimilarity measure can be used as well. Yet, using graph edit distance allows us to deal with a large class of graphs (directed, undirected, unlabeled, node and/or edge labels from any finite or infinite domain). Furthermore, a high degree of robustness against various graph distortions can be expected. Hence, in contrast with other graph embedding techniques, where sometimes restrictions on the type of underlying graph are imposed (e.g. [47, 70, 98]), this approach is distinguished by a high degree of flexibility in the graph definition. Since the computation of graph edit distance is exponential in the number of nodes for general graphs, the complexity of this graph embedding is exponential as well. However, as mentioned in Sect. 4, there exist efficient approximation algorithms for graph edit distance computation with cubic time complexity (e.g. the procedure described in [67]). Consequently, given n predefined prototypes the embedding of one particular graph is established by means of n distance computations with polynomial time.

Dissimilarity embeddings are closely related to kernel methods [75, 77]. In the kernel approach objects are described by means of pairwise kernel functions, while in the dissimilarity approach they are described by pairwise dissimilarities. However, there is one fundamental difference between kernels and dissimilarity embeddings. In the former method, the kernel values are interpreted as dot products in some implicitly existing feature space. By means of kernel machines, the underlying algorithm is eventually carried out in this kernel feature space. In the latter approach, the set of dissimilarities is interpreted as a novel vectorial description of the object under consideration. Hence, no implicit feature space, but an explicit dissimilarity space is obtained.

Obviously, the embedding paradigm established by mapping $\varphi_n^{\mathcal{P}} : \mathcal{G} \to \mathbb{R}^n$ constitutes a foundation for a novel class of graph kernels. One can define a valid graph kernel κ based on the graph embedding by computing the standard dot product of two graph maps in the resulting vector space. Formally,

$$\kappa_{\langle\rangle}(g_1, g_2) = \langle \varphi_n^{\mathcal{P}}(g_1), \varphi_n^{\mathcal{P}}(g_2) \rangle \ .$$

Note that this approach is very similar to the empirical kernel map described in [88] where general similarity measures are turned into kernel functions. Of course, not only the standard dot product can be used but any valid kernel function defined for vectors. For instance an RBF kernel function

$$\kappa_{RBF}(g_1, g_2) = \exp\left(-\gamma \|\varphi_n^{\mathcal{P}}(g_1) - \varphi_n^{\mathcal{P}}(g_2)\|^2\right)$$

with $\gamma > 0$ can thus be applied to graph maps.

The selection of the n prototypes $\mathcal{P} = \{p_1, \ldots, p_n\}$ is a critical issue since not only the prototypes $p_i \in \mathcal{P}$ themselves but also their number n affect the resulting graph mapping $\varphi_n^{\mathcal{P}}(\cdot)$ and thus the performance of the corresponding pattern recognition algorithm. A good selection of n prototypes seems to be crucial to succeed with the classification or clustering algorithm in the embedding vector space. A first and very simple idea might be to use all available training graphs from $\mathcal{T}$ as prototypes. Yet, two severe shortcomings arise with such a plain approach. First, the dimensionality of the resulting vector space is equal to the size N of the training set $\mathcal{T}$. Consequently, if the training set is large, the dimensionality of the feature vectors will be high, which possibly leads to overfitting effects and compromises computational efficiency. Secondly, the presence of similar prototypes as well as outlier graphs in the training set $\mathcal{T}$ is most likely. Therefore, redundant, noisy, or irrelevant information will be captured in the graph maps which in turn may harm the performance of the underlying algorithms.

The selection of prototypes for graph embedding has been addressed in various papers [62, 64, 66, 68]. In [62], for instance, a number of *prototype selection methods* are discussed. These selection strategies use some heuristics based on the underlying dissimilarities in the original graph domain. The basic idea of these approaches is to select prototypes from $\mathcal{T}$ that reflect the distribution of the training set $\mathcal{T}$ or cover a predefined region of $\mathcal{T}$ in the best possible way.

A severe shortcoming of such heuristic prototype selection strategies is that the dimensionality of the embedding space has to be determined by the user. In other words, the number of prototypes to be selected by a certain prototype selection algorithm has to be experimentally defined by means of the target algorithm on a validation set. In order to overcome this limitation, in [68], various *prototype reduction schemes* [3] are adopted for the task of graph embedding. In contrast with the heuristic prototype selection strategies, with these procedures the number of prototypes n, i.e. the resulting dimensionality of the vector space, is defined by an algorithmic procedure.

Another solution to the problem of noisy and redundant vectors with too high dimensionality is offered by the following procedure. Rather than selecting the prototypes beforehand, the embedding is carried out first and then the problem of prototype selection is reduced to a feature subset selection problem. That is, for graph embedding all available elements from the training set are used as prototypes, i.e. we define $\mathcal{P} = \mathcal{T}$. Next, a huge number of different feature selection strategies [23, 36, 39] can be applied to the resulting large scale vectors eliminating redundancies and noise, finding good features, and reducing the dimensionality. In [66], for instance, principal component analysis (PCA) [39] and Fisher linear discriminant analysis (LDA) [23] are applied to the vector space embedded graphs. Rather than traditional PCA, in [64], kernel PCA [76] is used for feature transformation.

Regardless of the strategy actually employed for the task of prototype selection, it has been experimentally shown that the general graph embedding procedure proposed in [62] has great potential. Its performance in various graph classification and clustering problems was evaluated and compared to alternative methods, including various graph kernels [62–66]. The data sets used in the experimental evaluation are publicly available[1].

7.   Conclusions

Due to the ability of graphs to represent properties of entities and binary relations at the same time, a growing interest in graph-based object representation in intelligent information processing can be observed. In the fields of bioinformatics and chemoinformatics, for instance, graph based representations have been intensively used [5, 48]. Another field of research where graphs have been studied with emerging interest is that of web content mining [74]. Image classification is a further area of research where graph based representation draws the attention [31]. Finally, we like to mention computer network analysis, where graphs have been used to detect network anomalies and predict abnormal events [9].

The concept of similarity or dissimilarity is an important issue in many application domains. In case where graphs are employed as representation formalism, various procedures for evaluating proximity, i.e. similarity or dissimilarity, of graphs have been proposed [15]. The process of evaluating the similarity of two graphs is commonly referred to as graph matching. Graph matching has successfully been applied to various problems in pattern recognition, computer vision, machine learning, data mining, and related fields.

In the case of exact graph matching, the graph extraction process is assumed to be structurally flawless, i.e. the conversion of the underlying data into graphs always proceeds without errors. Otherwise, if distortions are present, graph and subgraph isomorphism detection are rather unsuitable, which seriously restricts the applicability of exact graph matching algorithms.

Inexact methods, sometimes also referred to as error-tolerant methods, are characterized by their ability to cope with errors, or non-corresponding parts, in terms of structure and labels of graphs. Hence, in order for two graphs to be positively matched, they need not be identical at all, but only similar. The notion of graph similarity depends on the error-tolerant matching method that is to be applied.

In this chapter we have given an overview of both exact and inexact graph matching. The emphasis has been on the fundamental concepts and on two recent applications. In the first application, it is shown how the concept of subgraph isomorphism can be extended, such that a powerful and flexible information retrieval framework is established. This framework can be used to retrieve information from large database graphs by means of query graphs. In

---

[1](www.iam.unibe.ch/fki/databases/iam-graph-database)

a further application it is shown how graphs can be embedded in vector spaces by means of dissimilarities derived from graph edit distance or some other dissimilarity measure. The crucial benefit of such a graph embedding is that it instantly makes available all algorithmic tools originally developed for vectorial object descriptions.

References

[1] A.V. Aho, J.E. Hopcroft, and J.D. Ullman. *The Design and Analysis of Computer Algorithms*. Addison Wesley, 1974.

[2] R. Ambauen, S. Fischer, and H. Bunke. Graph edit distance with node splitting and merging and its application to diatom identification. In E. Hancock and M. Vento, editors, *Proc. 4th Int. Workshop on Graph Based Representations in Pattern Recognition*, LNCS 2726, pages 95–106. Springer, 2003.

[3] J.C. Bezdek and L. Kuncheva. Nearest prototype classifier designs: An experimental study. *Int. Journal of Intelligent Systems*, 16(12):1445–1473, 2001.

[4] C. Bishop. *Pattern Recognition and Machine Learning*. Springer, 2008.

[5] K. Borgwardt, C. Ong, S. Schönauer, S. Vishwanathan, A. Smola, and H.-P. Kriegel. Protein function prediction via graph kernels. *Bioinformatics*, 21(1):47–56, 2005.

[6] A. Brügger, H. Bunke, P. Dickinson, and K Riesen. Generalized graph matching for data mining and information retrieval. In P. Perner, editor, *Advances in Data Mining. Medical Applications, E-Commerce, Marketing, and Theoretical Aspects*, LNCS 5077, pages 298–312. Springer, 2008.

[7] H. Bunke. On a relation between graph edit distance and maximum common subgraph. *Pattern Recognition Letters*, 18:689–694, 1997.

[8] H. Bunke and G. Allermann. Inexact graph matching for structural pattern recognition. *Pattern Recognition Letters*, 1:245–253, 1983.

[9] H. Bunke, P.J. Dickinson, M. Kraetzl, and W.D. Wallis. *A Graph-Theoretic Approach to Enterprise Network Dynamics*, volume 24 of *Progress in Computer Science and Applied Logic (PCS)*. Birkhauser, 2007.

[10] H. Bunke, P. Foggia, C. Guidobaldi, C. Sansone, and M. Vento. A comparison of algorithms for maximum common subgraph on randomly connected graphs. In T. Caelli, A. Amin, R. Duin, M. Kamel, and D. de Ridder, editors, *Structural, Syntactic, and Statistical Pattern Recognition*, pages 85–106. Springer, 2002. LNCS 2396.

[11] H. Bunke, X. Jiang, and A. Kandel. On the minimum common supergraph of two graphs. *Computing*, 65(1):13–25, 2000.

[12] H. Bunke and K. Shearer. A graph distance metric based on the maximal common subgraph. *Pattern Recognition Letters*, 19(3):255–259, 1998.

[13] T. Caelli and S. Kosinov. Inexact graph matching using eigen-subspace projection clustering. *Int. Journal of Pattern Recognition and Artificial Intelligence*, 18(3):329–355, 2004.

[14] W.J. Christmas, J. Kittler, and M. Petrou. Structural matching in computer vision using probabilistic relaxation. *IEEE Transactions on Pattern Analysis and Machine Intelligence*, 17(8):749–764, 1995.

[15] D. Conte, P. Foggia, C. Sansone, and M. Vento. Thirty years of graph matching in pattern recognition. *Int. Journal of Pattern Recognition and Artificial Intelligence*, 18(3):265–298, 2004.

[16] D. Cook and L. Holder, editors. *Mining Graph Data*. Wiley-Interscience, 2007.

[17] L.P. Cordella, P. Foggia, C. Sansone, and M. Vento. A (sub)graph isomorphism algorithm for matching large graphs. *IEEE Trans. on Pattern Analysis and Machine Intelligence*, 26(20):1367–1372, 2004.

[18] P.J. Dickinson, H. Bunke, A. Dadej, and M. Kraetzl. Matching graphs with unique node labels. *Pattern Analysis and Applications*, 7(3):243–254, 2004.

[19] R. Duda, P. Hart, and D. Stork. *Pattern Classification*. Wiley-Interscience, 2nd edition, 2000.

[20] M.A. Eshera and K.S. Fu. A graph distance measure for image analysis. *IEEE Transactions on Systems, Man, and Cybernetics (Part B)*, 14(3):398–408, 1984.

[21] M.-L. Fernandez and G. Valiente. A graph distance metric combining maximum common subgraph and minimum common supergraph. *Pattern Recognition Letters*, 22(6–7):753–758, 2001.

[22] M.A. Fischler and R.A. Elschlager. The representation and matching of pictorial structures. *IEEE Trans. on Computers*, 22(1):67–92, 1973.

[23] R.A. Fisher. The statistical utilization of multiple measurements. In *Annals of Eugenics*, volume 8, pages 376–386, 1938.

[24] P. Frasconi, M. Gori, and A. Sperduti. A general framework for adaptive processing of data structures. *IEEE Transactions on Neural Networks*, 9(5):768–786, 1998.

[25] M.R. Garey and D.S. Johnson. *Computers and Intractability: A Guide to the Theory of NP-Completeness*. Freeman and Co., 1979.

[26] T. Gartner. *Kernels for Structured Data*. World Scientific, 2008.

[27] T. Gartner, P. Flach, and S. Wrobel. On graph kernels: Hardness results and efficient alternatives. In B. Schölkopf and M. Warmuth, editors, *Proc. 16th Annual Conf. on Learning Theory*, pages 129–143, 2003.

[28] S. Gold and A. Rangarajan. A graduated assignment algorithm for graph matching. *IEEE Transactions on Pattern Analysis and Machine Intelligence*, 18(4):377–388, 1996.

[29] M. Gori, M. Maggini, and L. Sarti. Exact and approximate graph matching using random walks. *IEEE Transactions on Pattern Analysis and Machine Intelligence*, 27(7):1100–1111, 2005.

[30] E.R. Hancock and J. Kittler. Discrete relaxation. *Pattern Recognition*, 23(7):711–733, 1990.

[31] Z. Harchaoui and F. Bach. Image classification with segmentation graph kernels. In *IEEE Conference on Computer Vision and Pattern Recognition*, pages 1–8, 2007.

[32] D. Haussler. Convolution kernels on discrete structures. Technical Report UCSC-CRL-99-10, University of California, Santa Cruz, 1999.

[33] G. Hjaltason and H. Samet. Properties of embedding methods for similarity searching in metric spaces. *IEEE Trans. on Pattern Analysis ans Machine Intelligence*, 25(5):530–549, 2003.

[34] J.E. Hopcroft and J. Wong. Linear time algorithm for isomorphism of planar graphs. In *Proc. 6th Annual ACM Symposium on Theory of Computing*, pages 172–184, 1974.

[35] B. Huet and E.R. Hancock. Shape recognition from large image libraries by inexact graph matching. *Pattern Recognition Letters*, 20(11–13):1259–1269, 1999.

[36] A. Jain and D. Zongker. Feature selection: Evaluation, application, and small sample performance. *IEEE Trans. on Pattern Analysis and Machine Intelligence*, 19(2):153–158, 1997.

[37] B. Jain and F. Wysotzki. Automorphism partitioning with neural networks. *Neural Processing Letters*, 17(2):205–215, 2003.

[38] X. Jiang and H. Bunke. Optimal quadratic-time isomorphism of ordered graphs. *Pattern Recognition*, 32(17):1273–1283, 1999.

[39] I. Jolliffe. *Principal Component Analysis*. Springer, 1986.

[40] D. Justice and A. Hero. A binary linear programming formulation of the graph edit distance. *IEEE Trans. on Pattern Analysis ans Machine Intelligence*, 28(8):1200–1214, 2006.

[41] J. Kittler and E.R. Hancock. Combining evidence in probabilistic relaxation. *Int. Journal of Pattern Recognition and Art. Intelligence*, 3(1):29–51, 1989.

[42] R.I. Kondor and J. Lafferty. Diffusion kernels on graphs and other discrete input spaces. In *Proc. 19th Int. Conf. on Machine Learning*, pages 315–322, 2002.

[43] J. Larrosa and G. Valiente. Constraint satisfaction algorithms for graph pattern matching. *Mathematical Structures in Computer Science*, 12(4):403–422, 2002.

[44] G. Levi. A note on the derivation of maximal common subgraphs of two directed or undirected graphs. *Calcolo*, 9:341–354, 1972.

[45] E.M. Luks. Isomorphism of graphs of bounded valence can be tested in polynomial time. *Journal of Computer and Systems Sciences*, 25:42–65, 1982.

[46] B. Luo and E. Hancock. Structural graph matching using the EM algorithm and singular value decomposition. *IEEE Transactions on Pattern Analysis and Machine Intelligence*, 23(10):1120–1136, 2001.

[47] B. Luo, R. Wilson, and E.R. Hancock. Spectral embedding of graphs. *Pattern Recognition*, 36(10):2213–2223, 2003.

[48] P. Mahe, N. Ueda, and T. Akutsu. Graph kernels for molecular structures – activity relationship analysis with support vector machines. *Journal of Chemical Information and Modeling*, 45(4):939–951, 2005.

[49] J.J. McGregor. Backtrack search algorithms and the maximal common subgraph problem. *Software Practice and Experience*, 12:23–34, 1982.

[50] B.D. McKay. Practical graph isomorphism. *Congressus Numerantium*, 30:45–87, 1981.

[51] B.T. Messmer and H. Bunke. A decision tree approach to graph and subgraph isomorphism detection. *Pattern Recognition*, 32:1979–1998, 1008.

[52] A. Micheli. Neural network for graphs: A contextual constructive approach. *IEEE Transactions on Neural Networks*, 20(3):498–511, 2009.

[53] J. Munkres. Algorithms for the assignment and transportation problems. In *Journal of the Society for Industrial and Applied Mathematics*, volume 5, pages 32–38, March 1957.

[54] R. Myers, R.C. Wilson, and E.R. Hancock. Bayesian graph edit distance. *IEEE Transactions on Pattern Analysis and Machine Intelligence*, 22(6):628–635, 2000.

[55] M. Neuhaus and H. Bunke. Self-organizing maps for learning the edit costs in graph matching. *IEEE Transactions on Systems, Man, and Cybernetics (Part B)*, 35(3):503–514, 2005.

[56] M. Neuhaus and H. Bunke. Automatic learning of cost functions for graph edit distance. *Information Sciences*, 177(1):239–247, 2007.

[57] M. Neuhaus and H. Bunke. *Bridging the Gap Between Graph Edit Distance and Kernel Machines*. World Scientific, 2007.

[58] M. Neuhaus and H. Bunke. A quadratic programming approach to the graph edit distance problem. In F. Escolano and M. Vento, editors, *Proc.*

6th Int. Workshop on Graph Based Representations in Pattern Recognition, LNCS 4538, pages 92–102, 2007.

[59] M. Neuhaus, K. Riesen, and H. Bunke. Fast suboptimal algorithms for the computation of graph edit distance. In Dit-Yan Yeung, J.T. Kwok, A. Fred, F. Roli, and D. de Ridder, editors, *Proc. 11.th int. Workshop on Strucural and Syntactic Pattern Recognition*, LNCS 4109, pages 163–172. Springer, 2006.

[60] E. Pekalska and R. Duin. *The Dissimilarity Representation for Pattern Recognition: Foundations and Applications.* World Scientific, 2005.

[61] M. Pelillo. Replicator equations, maximal cliques, and graph isomorphism. *Neural Computation*, 11(8):1933–1955, 1999.

[62] K. Riesen and H. Bunke. Graph classification based on vector space embedding. *Int. Journal of Pattern Recognition and Artificial Intelligence*, 2008. accepted for publication.

[63] K. Riesen and H. Bunke. Kernel k-means clustering applied to vector space embeddings of graphs. In L. Prevost, S. Marinai, and F. Schwenker, editors, *Proc. 3rd IAPR Workshop Artificial Neural Networks in Pattern Recognition*, LNAI 5064, pages 24–35. Springer, 2008.

[64] K. Riesen and H. Bunke. Non-linear transformations of vector space embedded graphs. In A. Juan-Ciscar and G. Sanchez-Albaladejo, editors, *Pattern Recognition in Information Systems*, pages 173–186, 2008.

[65] K. Riesen and H. Bunke. On Lipschitz embeddings of graphs. In I. Lovrek, R.J. Howlett, and L.C. Jain, editors, *Proc. 12th International Conference, Knowledge-Based Intelligent Information and Engineering Systems, Part I*, LNAI 5177, pages 131–140. Springer, 2008.

[66] K. Riesen and H. Bunke. Reducing the dimensionality of dissimilarity space embedding graph kernels. *Engineering Applications of Artificial Intelligence*, 22(1):48–56, 2008.

[67] K. Riesen and H. Bunke. Approximate graph edit distance computation by means of bipartite graph matching. *Image and Vision Computing*, 27(4):950–959, 2009.

[68] K. Riesen and H. Bunke. Dissimilarity based vector space embedding of graphs using prototype reduction schemes. Accepted for publication in Machine Learning and Data Mining in Pattern Recognition, 2009.

[69] A. Robles-Kelly and E.R. Hancock. String edit distance, random walks and graph matching. *Int. Journal of Pattern Recognition and Artificial Intelligence*, 18(3):315–327, 2004.

[70] A. Robles-Kelly and E.R. Hancock. A Riemannian approach to graph embedding. *Pattern Recognition*, 40:1024–1056, 2007.

[71] A. Sanfeliu and K.S. Fu. A distance measure between attributed relational graphs for pattern recognition. *IEEE Transactions on Systems, Man, and Cybernetics (Part B)*, 13(3):353–363, 1983.

[72] F. Scarselli, M. Gori, A.C. Tsoi, M. Hagenbuchner, and G. Monfardini. The graph neural network model. *IEEE Transactions on Neural Networks*, 20(1):61–80, 2009.

[73] K. Schadler and F. Wysotzki. Comparing structures using a Hopfield-style neural network. *Applied Intelligence*, 11:15–30, 1999.

[74] A. Schenker, H. Bunke, M. Last, and A. Kandel. *Graph-Theoretic Techniques for Web Content Mining*. World Scientific, 2005.

[75] B. Schelkopf and A. Smola. *Learning with Kernels*. MIT Press, 2002.

[76] B. Schelkopf, A. Smola, and K.-R. Muller. Nonlinear component analysis as a kernel eigenvalue problem. *Neural Computation*, 10:1299–1319, 1998.

[77] J. Shawe-Taylor and N. Cristianini. *Kernel Methods for Pattern Analysis*. Cambridge University Press, 2004.

[78] A. Shokoufandeh, D. Macrini, S. Dickinson, K. Siddiqi, and S.W. Zucker. Indexing hierarchical structures using graph spectra. *IEEE Transactions on Pattern Analysis and Machine Intelligence*, 27(7):1125–1140, 2005.

[79] A. Smola and R. Kondor. Kernels and regularization on graphs. In *Proc. 16th. Int. Conf. on Comptuational Learning Theory*, pages 144–158, 2003.

[80] S. Sorlin and C. Solnon. Reactive tabu search for measuring graph similarity. In L. Brun and M. Vento, editors, *Proc. 5th Int. Worksho on Graph-based Representations in Pattern Recognition*, LNCS 3434, pages 172–182. Springer, 2005.

[81] A. Sperduti and A. Starita. Supervised neural networks for the classification of structures. *IEEE Transactions on Neural Networks*, 8(3):714–735, 1997.

[82] B. Spillmann, M. Neuhaus, H. Bunke, E. Pekalska, and R. Duin. Transforming strings to vector spaces using prototype selection. In Dit-Yan Yeung, J.T. Kwok, A. Fred, F. Roli, and D. de Ridder, editors, *Proc. 11.th int. Workshop on Strucural and Syntactic Pattern Recognition*, LNCS 4109, pages 287–296. Springer, 2006.

[83] P.N. Suganthan, E.K. Teoh, and D.P. Mital. Pattern recognition by graph matching using the potts MFT neural networks. *Pattern Recognition*, 28(7):997–1009, 1995.

[84] P.N. Suganthan, E.K. Teoh, and D.P. Mital. Pattern recognition by homomorphic graph matching using Hopfield neural networks. *Image Vision Computing*, 13(1):45–60, 1995.

[85] P.N. Suganthan, E.K. Teoh, and D.P. Mital. Self-organizing Hopfield network for attributed relational graph matching. *Image Vision Computing*, 13(1):61–73, 1995.

[86] Y. Tian and J.M. Patel. Tale: A tool for approximate large graph matching. In *IEEE 24th International Conference on Data Engineering*, pages 963–972, 2008.

[87] A. Torsello and E. Hancock. Computing approximate tree edit distance using relaxation labeling. *Pattern Recognition Letters*, 24(8):1089–1097, 2003.

[88] K. Tsuda. Support vector classification with asymmetric kernel function. In M. Verleysen, editor, *Proc. 7th European Symposium on Artifical Neural Netweorks*, pages 183–188, 1999.

[89] J.R. Ullmann. An algorithm for subgraph isomorphism. *Journal of the Association for Computing Machinery*, 23(1):31–42, 1976.

[90] S. Umeyama. An eigendecomposition approach to weighted graph matching problems. *IEEE Transactions on Pattern Analysis and Machine Intelligence*, 10(5):695–703, 1988.

[91] M.A. van Wyk, T.S. Durrani, and B.J. van Wyk. A RKHS interpolator-based graph matching algorithm. *IEEE Transactions on Pattern Analysis and Machine Intelligence*, 24(7):988–995, 2003.

[92] J.-P. Vert and M. Kanehisa. Graph-driven features extraction from microarray data using diffusion kernels and kernel CCA. In *Advances in Neural Information Processing Systems*, volume 15, pages 1425–1432. MIT Press, 2003.

[93] R.A. Wagner and M.J. Fischer. The string-to-string correction problem. *Journal of the Association for Computing Machinery*, 21(1):168–173, 1974.

[94] W.D. Wallis, P. Shoubridge, M. Kraetzl, and D. Ray. Graph distances using graph union. *Pattern Recognition Letters*, 22(6):701–704, 2001.

[95] C. Watkins. Dynamic alignment kernels. In A. Smola, P.L. Bartlett, B. Schølkopf, and D. Schuurmans, editors, *Advances in Large Margin Classifiers*, pages 39–50. MIT Press, 2000.

[96] R. Wilson and E.R. Hancock. Levenshtein distance for graph spectral features. In J. Kittler, M. Petrou, and M. Nixon, editors, *Proc. 17th Int. Conf. on Pattern Recognition*, volume 2, pages 489–492, 2004.

[97] R.C. Wilson and E. Hancock. Structural matching by discrete relaxation. *IEEE Transactions on Pattern Analysis and Machine Intelligence*, 19(6):634–648, 1997.

[98] R.C. Wilson, E.R. Hancock, and B. Luo. Pattern vectors from algebraic graph theory. *IEEE Trans. on Pattern Analysis ans Machine Intelligence*, 27(7):1112–1124, 2005.

[99] A.K.C. Wong and M. You. Entropy and distance of random graphs with application to structural pattern recognition. *IEEE Transactions on Pattern Analysis and Machine Intelligence*, 7(5):599–609, 1985.

[100] R. Xu and D. Wunsch. Survey of graph clustering algorithms. *IEEE Transactions on Neural Networks*, 16(3):645–678, 2005.

[101] Y. Yao, G.L. Marcialis, M. Pontil, P. Frasconi, and F. Roli. Combining flat and structured representations for fingerprint classification with recursive neural networks and support vector machines. *Pattern Recognition*, 36(2):397–406, 2003.

Chapter 8

A SURVEY OF ALGORITHMS FOR KEYWORD SEARCH ON GRAPH DATA

Haixun Wang

Microsoft Research Asia
Beijing, China 100190

haixunw@microsoft.com

Charu C. Aggarwal

IBM T. J. Watson Research Center
Hawthorne, NY 10532

charu@us.ibm.com

Abstract   In this chapter, we survey methods that perform keyword search on graph data. Keyword search provides a simple but user-friendly interface to retrieve information from complicated data structures. Since many real life datasets are represented by trees and graphs, keyword search has become an attractive mechanism for data of a variety of types. In this survey, we discuss methods of keyword search on schema graphs, which are abstract representation for XML data and relational data, and methods of keyword search on schema-free graphs. In our discussion, we focus on three major challenges of keyword search on graphs. First, what is the semantics of keyword search on graphs, or, what qualifies as an answer to a keyword search; second, what constitutes a good answer, or, how to rank the answers; third, how to perform keyword search efficiently. We also discuss some unresolved challenges and propose some new research directions on this topic.

Keywords: Keyword Search, Information Retrieval, Graph Structured Data, Semi-Structured Data

C.C. Aggarwal and H. Wang (eds.), *Managing and Mining Graph Data*,
Advances in Database Systems 40, DOI 10.1007/978-1-4419-6045-0_8,
© Springer Science+Business Media, LLC 2010

1.   Introduction

Keyword search is the *de facto* information retrieval mechanism for data on the World Wide Web. It also proves to be an effective mechanism for querying semi-structured and structured data, because of its user-friendly query interface. In this survey, we focus on keyword search problems for XML documents (semi-structured data), relational databases (structured data), and all kinds of schema-free graph data.

Recently, query processing over graph-structured data has attracted increasing attention, as myriads of applications are driven by and producing graph-structured data [14]. For example, in semantic web, two major W3C standards, RDF and OWL, conform to node-labeled and edge-labeled graph models. In bioinformatics, many well-known projects, e.g., BioCyc (http://biocyc.org), build graph-structured databases. In social network analysis, much interest centers around all kinds of personal interconnections. In other applications, raw data might not be graph-structured at the first glance, but there are many implicit connections among data items; restoring these connections often allows more effective and intuitive querying. For example, a number of projects [1, 18, 3, 26, 8] enable keyword search over relational databases. In personal information management (PIM) systems [10, 5], objects such as emails, documents, and photos are interwoven into a graph using manually or automatically established connections among them. The list of examples of graph-structured data goes on.

For data with relational and XML schema, specific query languages, such as SQL and XQuery, have been developed for information retrieval. In order to query such data, the user must master a complex query language and understand the underlying data schema. In relational databases, information about an object is often scattered in multiple tables due to normalization considerations, and in XML datasets, the schema are often complicated and embedded XML structures often create a lot of difficulty to express queries that are forced to traverse tree structures. Furthermore, many applications work on graph-structured data with no obvious, well-structured schema, so the option of information retrieval based on query languages is not applicable.

Both relational databases and XML databases can be viewed as graphs. Specifically, XML datasets can be regarded as graphs when IDREF/ID links are taken into consideration, and a relational database can be regarded as a data graph that has tuples and keywords as nodes. In the data graph, for example, two tuples are connected by an edge if they can be joined using a foreign key; a tuple and a keyword are connected if the tuple contains the keyword. Thus, traditional graph search algorithms, which extract features (e.g., paths [27], frequent-patterns [30], sequences [20]) from graph data, and convert queries into searches over feature spaces, can be used for such data.

However, traditional graph search methods usually focus more on the structure of the graph rather than the semantic content of the graph. In XML and relational data graphs, nodes contain keywords, and sometimes nodes and edges are labeled. The problem of keyword search requires us to determine a group of densely linked nodes in the graph, which may satisfy a particular keyword-based query. Thus, the keyword search problem makes use of *both* the content and the linkage structure. These two sources of information actually re-enforce each other, and improve the overall quality of the results. This makes keyword search a more preferred information retrieval method. Keyword search allows users to query the databases quickly, with no need to know the schema of the respective databases. In addition, keyword search can help discover unexpected answers that are often difficult to obtain via rigid-format SQL queries. It is for these reasons that keyword search over tree- and graph-structured data has attracted much attention [1, 18, 3, 6, 13, 16, 2, 28, 21, 26, 24, 8].

Keyword search over graph data presents many challenges. The first question we must answer is that, what constitutes an answer to a keyword. For information retrieval on the Web, answers are simply Web documents that contain the keywords. In our case, the entire dataset is considered as a single graph, so the algorithms must work on a finer granularity and decide what subgraphs are qualified as answers. Furthermore, since many subgraphs may satisfy a query, we must design ranking strategies to find top answers. The definition of answers and the design of their ranking strategies must satisfy users' intention. For example, several papers [16, 2, 12, 26] adopt IR-style answer-tree ranking strategies to enhance semantics of answers. Finally, a major challenge for keyword search over graph data is query efficiency, which to a large extent hinges on the semantics of the query and the ranking strategy. For instance, some ranking strategies score an answer by the sum of edge weights. In this case, finding the top-ranked answer is equivalent to the group Steiner tree problem [9], which is NP-hard. Thus, finding the exact top k answers is inherently difficult. To improve search efficiency, many systems, such as BANKS [3], propose ways to reduce the search space. As another example, BLINKS [14] avoids the inherent difficulty of the group Steiner tree problem by proposing an alternative scoring mechanism, which lowers complexity and enables effective indexing and pruning.

Before we delve into the details of various keyword search problems for graph data, we briefly summarize the scope of this survey chapter. We classify algorithms we survey into three categories based on the schema constraints in the underlying graph data.

- **Keyword Search on XML Data:**

  Keyword search on XML data [11, 6, 13, 23, 25] is a simpler problem than on schema-free graphs. They are basically constrained to tree structures, where each node only has a single incoming path. This property provides great optimization opportunities [28]. Connectivity information can also be efficiently encoded and indexed. For example, in XRank [13], the Dewey inverted list is used to index paths so that a keyword query can be evaluated without tree traversal.

- **Keyword Search over Relational Databases:**

    Keyword search on relational databases [1, 3, 18, 16, 26] has attracted much interest. Conceptually, a database is viewed as a labeled graph where tuples in different tables are treated as nodes connected via foreign-key relationships. Note that a graph constructed this way usually has a regular structure because schema restricts node connections. Different from the graph-search approach in BANKS [3], DBXplorer [1] and DISCOVER [18] construct join expressions and evaluate them, relying heavily on the database schema and query processing techniques in RDBMS.

- **Keyword Search on Graphs:** A great deal of work on keyword querying of structured and semi-structured data has been proposed in recent years. Well known algorithms includes the backward expanding search [3], bidirectional search [21], dynamic programming techniques DPBF [8], and BLINKS [14]. Recently, work that extend keyword search to graphs on external memory has been proposed [7].

This rest of the chapter is organized as follows. We first discuss keyword search methods for schema graphs. In Section 2 we focus on keyword search for XML data, and in Section 3, we focus on keyword search for relational data. In Section 4, we introduce several algorithms for keyword search on schema-free graphs. Section 5 contains a discussion of future directions and the conclusion.

2.   Keyword Search on XML Data

Sophisticated query languages such as XQuery have been developed for querying XML documents. Although XQuery can express many queries precisely and effectively, it is by no means a user-friendly interface for accessing XML data: users must master a complex query language, and in order to use it, they must have a full understanding of the schema of the underlying XML data. Keyword search, on the other hand, offers a simple and user-friendly interface. Furthermore, the tree structure of XML data gives nice semantics to the query and enables efficient query processing.

2.1   Query Semantics

In the most basic form, as in XRank [13] and many other systems, a keyword search query consists of n keywords: $Q = \{k_1, \cdots, k_n\}$. XSEarch [6] extends the syntax to allow users to specify which keywords *must* appear in a satisfying document, and which *may* or *may not* appear (although the appearance of such keywords is desirable, as indicated by the ranking function).

Syntax aside, one important question is, what qualifies as an answer to a keyword search query? In information retrieval, we simply return documents that contain all the keywords. For keyword search on an XML document, we want to return *meaningful* snippets of the document that contains the keywords. One interpretation of *meaningful* is to find the *smallest* subtrees that contain all the keywords.

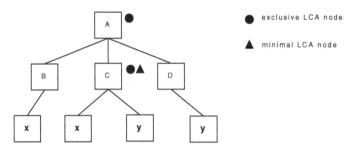

Figure 8.1. Query Semantics for Keyword Search $Q = \{x, y\}$ on XML Data

Specifically, for each keyword k_i, let L_i be the list of nodes in the XML document that contain keyword k_i. Clearly, subtrees formed by at least one node from each $L_i, i = 1, \cdots, n$ contain all the keywords. Thus, an answer to the query can be represented by $lca(n_1, \cdots, n_n)$, the lowest common ancestor (LCA) of nodes $n_1, \cdots, n_n$ where $n_i \in L_i$. In other words, answering the query is equivalent to finding:

$$LCA(k_1, \cdots, k_n) = \{lca(n_1, \cdots, n_n) | n_1 \in L_1, \cdots, n_n \in L_n\}$$

Moreover, we are only interested in the "smallest" answer, that is,

$$
\begin{aligned}
SLCA(k_1, \cdots, k_n) = \{v \mid v \in LCA(k_1, \cdots, k_n) \;\wedge \\
\forall v' \in LCA(k_1, \cdots, k_n), v \not\prec v'\}
\end{aligned}
\tag{8.1}
$$

where $\prec$ denotes the ancestor relationship between two nodes in an XML document. As an example, in Figure 8.1, we assume the keyword query is $Q = \{x, y\}$. We have $C \in SLCA(x, y)$ while $A \in LCA(x, y)$ but $A \notin SLCA(x, y)$.

Several algorithms including [28, 17, 29] are based on the SLCA semantics. However, SLCA is by no means the only meaningful semantics for keyword search on XML documents. Consider Figure 8.1 again. If we remove node C and the two keyword nodes under C, the remaining tree is still an answer to the query. Clearly, this answer is independent of the answer $C \in SLCA(x,y)$, yet it is not represented by the SLCA semantics.

XRank [13], for example, adopts different query semantics for keyword search. The set of answers to a query $Q = \{k_1, \cdots, k_n\}$ is defined as:

$$
\begin{aligned}
ELCA(k_1, \cdots, k_n) = \{v \mid \forall k_i \; \exists c \;\; & c \text{ is a child node of } v \; \wedge \\
& \nexists c' \in LCA(k_1, \cdots, k_n) \text{ and } c \prec c' \wedge \quad (8.2) \\
& c \text{ contains } k_i \text{ directly or indirectly} \}
\end{aligned}
$$

$ELCA(k_1, \cdots, k_n)$ contains the set of nodes that contain at least one occurrence of all of the query keywords, after excluding the sub-nodes that already contain all of the query keywords. Clearly, in Figure 8.1, we have $A \in ELCA(k_1, \cdots, k_n)$. More generally, we have

$$SLCA(k_1, \cdots, k_n) \subseteq ELCA(k_1, \cdots, k_n) \subseteq LCA(k_1, \cdots, k_n)$$

Query semantics has a direct impact on the complexity of query processing. For example, answering a keyword query according to the ELCA query semantics is more computationally challenging than according to the SLCA query semantics. In the latter, the moment we know a node l has a child c that contains all the keywords, we can immediately determine that node l is not an SLCA node. However, we cannot determine that l is not an ELCA node because l may contain keyword instances that are not under c and are not under any node that contains all keywords [28, 29].

2.2   Answer Ranking

It is clear that according to the lowest common ancestor (LCA) query semantics, potentially many answers will be returned for a keyword query. It is also easy to see that, due to the difference of the nested XML structure where the keywords are embedded, not all answers are equal. Thus, it is important to devise a mechanism to rank the answers based on their relevance to the query. In other words, for every given answer tree T containing all the keywords, we want to assign a numerical score to T. Many approaches for keyword search on XML data, including XRank [13] and XSEarch [6], present a ranking method.

To decide which answer is more desirable for a keyword query, we note several properties that we would like a ranking mechanism to take into consideration:

1 *Result specificity.* More specific answers should be ranked higher than less specific answers. The SLCA and ELCA semantics already exclude certain answers based on result specificity. Still, this criterion can be further used to rank satisfying answers in both semantics.

2 *Semantic-based keyword proximity.* Keywords in an answer should appear close to each other. Furthermore, such closeness must reflect the semantic distance as prescribed by the XML embedded structure. Example 8.1 demonstrates this need.

3 *Hyperlink Awareness.* LCA-based semantics largely ignore the hyperlinks in XML documents. The ranking mechanism should take hyperlinks into consideration when computing nodes' authority or prestige as well as keyword proximity.

The ranking mechanism used by XRank [13] is based on an adaptation of *PageRank* [4]. For each element v in the XML document, XRank defines *ElemRank(v)* as v's objective importance, and *ElemRank(v)* is computed using the underlying embedded structure in a way similar to *PageRank*. The difference is that *ElemRank* is defined at node granularity, while *PageRank* at document granularity. Furthermore, *ElemRank* looks into the nested structure of XML, which offers richer semantics than the hyperlinks among documents do.

Given a path in an XML document $v_0, v_1, \cdots, v_t, v_{t+1}$, where v_{t+1} directly contains a keyword k, and v_{i+1} is a child node of v_i, for $i = 0, \cdots, t$, XRank defines the rank of v_i as:

$$r(v_i, k) = ElemRank(v_t) \times decay^{t-i}$$

where *decay* is a value in the range of 0 to 1. Intuitively, the rank of v_i with respect to a keyword k is *ElemRank(v_t)* scaled appropriately to account for the specificity of the result, where v_t is the parent element of the value node v_{t+1} that directly contains the keyword k. By scaling down *ElemRank(v_t)*, XRank ensures that less specific results get lower ranks. Furthermore, from node v_i, there may exist multiple paths leading to multiple occurrences of keyword k. Thus, the rank of v_i with respect to k should be a combination of the ranks for all occurrences. XRank uses $\hat{r}(v, k)$ to denote the rank of node v with respect to keyword k:

$$\hat{r}(v, k) = f(r_1, r_2, \cdots, r_m)$$

where $r_1, \cdots, r_m$ are the ranks computed for each occurrence of k (using the above formula), and f is a combination function (e.g., sum or max). Finally, the overall ranking of a node v with respect to a query Q which contains n keywords $k_1, \cdots, k_n$ is defined as:

$$R(v, Q) = \left(\sum_{1 \leq i \leq n} \hat{r}(v, k_i) \right) \times p(v, k_1, k_2, \cdots, k_n) \qquad (8.3)$$

Here, the overall ranking $R(v, Q)$ is the sum of the ranks with respect to keywords in Q, multiplied by a measure of keyword proximity $p(v, k_1, k_2, \cdots, k_n)$, which ranges from 0 (keywords are very far apart) to 1 (keywords occur right next to each other). A simple proximity function is the one that is inversely proportional to the size of the smallest text window that contains occurrences of all keywords $k_1, k_2, \cdots, k_n$. Clearly, such a proximity function may not be optimal as it ignores the structure where the keywords are embedded, or in other words, it is not a semantic-based proximity measure.

Eq 8.3 depends on function $ElemRank()$, which measures the importance of XML elements bases on the underlying hyperlinked structure. $ElemRank$ is a global measure and is not related to specific queries. XRank [13] defines $ElemRank()$ by adapting PageRank:

$$PageRank(v) = \frac{1-d}{N} + d \times \sum_{(u,v) \in E} \frac{PageRank(u)}{N_u} \qquad (8.4)$$

where N is the total number of documents, and N_u is the number of out-going hyperlinks from document u. Clearly, $PageRank(v)$ is a combination of two probabilities: i) $\frac{1}{N}$, which is the probability of reaching v by a random walk on the entire web, and ii) $\frac{PageRank(u)}{N_u}$, which is the probability of reaching v by following a link on web page u.

Clearly, a link from page u to page v propagates "importance" from u to v. To adapt PageRank for our purpose, we must first decide what constitutes a "link" among elements in XML documents. Unlike HTML documents on the Web, there are three types of links within an XML document: importance can propagate through a hyperlink from one element to the element it points to; it can propagate from an element to its sub-element (containment relationship); and it can also propagate from a sub-element to its parent element. XRank [13] models each of the three relationships in defining $ElemRank()$:

$$ElemRank(v) = \frac{1 - d_1 - d_2 - d_3}{N_e} +$$
$$d_1 \times \sum_{(u,v) \in HE} \frac{ElemRank(u)}{N_h(u)} +$$
$$d_2 \times \sum_{(u,v) \in CE} \frac{ElemRank(u)}{N_c(u)} + \qquad (8.5)$$
$$d_3 \times \sum_{(u,v) \in CE^{-1}} ElemRank(u)$$

where N_e is the total number of XML elements, $N_c(u)$ is the number of sub-elements of u, and $E = HE \cup CE \cup CE^{-1}$ are edges in the XML document, where HE is the set of hyperlink edges, CE the set of containment edges, and CE^{-1} the set of reverse containment edges.

As we have mentioned, the notion of keyword proximity in XRank is quite primitive. The proximity measure $p(v, k_1, \cdots, k_n)$ in Eq 8.3 is defined to be inversely proportional to the size of the smallest text window that contains all the keywords. However, this does not guarantee that such an answer is always the most meaningful.

Example 8.1. *Semantic-based keyword proximity*

```
<proceedings>
   <inproceedings>
       <author>Moshe Y. Vardi</author>
       <title>Querying Logical Databases</title>
   </inproceedings>
   <inproceedings>
       <author>Victor Vianu</author>
       <title>A Web Odyssey: From Codd to XML</title>
   </inproceedings>
</proceedings>
```

For instance, given a keyword query "Logical Databases Vianu", the above XML snippet [6] will be regarded as a good answer by XRank, since all keywords occur in a small text window. But it is easy to see that the keywords do not appear in the same context: "Logical Databases" appears in one paper's title and "Vianu" is part of the name of another paper's author. This can hardly be an ideal response to the query. To address this problem, XSEarch [6] proposes a semantic-based keyword proximity measure that takes into account the nested structure of XML documents.

XSEarch defines an *interconnected* relationship. Let n and n' be two nodes in a tree structure T. Let $|n, n'$ denote the tree consisting of the paths from the lowerest common ancestor of n and n' to n and n'. The nodes n and n' are *interconnected* if one of the following conditions holds:

- $T_{|n,n'}$ does not contain two distinct nodes with the same label, or

- the only two distinct nodes in $T_{|n,n'}$ with the same label are n and n'.

As we can see, the element that matches keywords "Logical Databases" and the element that matches keyword "Vianu" in the previous example are not interconnected, because the answer tree contains two distinct nodes with the same label "inproceedings". XSEarch requires that all pairs of matched elements in the answer set are interconnected, and XSEarch proposes an all-pairs index to efficiently check the connectivity between the nodes.

In addition to using a more sophisticated keyword proximity measure, XSEarch [6] also adopts a *tfidf* based ranking mechanism. Unlike standard information retrieval techniques that compute *tfidf* at document level, XSEarch computes the weight of keywords at a lower granularity, i.e., at the level of the leaf nodes of a document. The term frequency of keyword k in a leaf node n_l is defined as:

$$tf(k, n_l) = \frac{occ(k, n_l)}{max\{occ(k', n_l)|k' \in words(n_l)\}}$$

where $occ(k, n_l)$ denotes the number of occurrences of k in n_l. Similar to the standard tf formula, it gives a larger weight to frequent keywords in sparse nodes. XSEarch also defines the inverse leaf frequency (ilf):

$$ilf(k) = \log\left(1 + \frac{|N|}{|\{n' \in N|k \in words(n')|\}}\right)$$

where N is the set of all leaf nodes in the corpus. Intuitively, $ilf(k)$ is the logarithm of the inverse leaf frequency of k, i.e., the number of leaves in the corpus over the number of leaves that contain k. The weight of each keyword $w(k, n_l)$ is a normalized version of the value $tfilf(k, n_l)$, which is defined as $tf(k, n_l) \times ilf(k)$.

With the $tfilf$ measure, XSEarch uses the standard vector space model to determine how well an answer satisfies a query. The measure of similarity between a query Q and an answer N is the sum of the cosine distances between the vectors associated with the nodes in N and the vectors associated with the terms that they match in Q [6].

2.3   Algorithms for LCA-based Keyword Search

Search engines endeavor to speed up the query: find the documents where word X occurs. A word level inverted list is used for this purpose. For each word X, the inverted list stores the id of the documents that contain the word X. Keyword search over XML documents operates at a finer granularity, but still we can use an inverted list based approach: For each keyword, we store all the elements that either directly contain the keyword, or contain the keyword through their descendents. Then, given a query $Q = \{k_1, \cdots, k_n\}$, we find common elements in all of the n inverted lists corresponding to k_1 through k_n. These common elements are potential root nodes of the answer trees.

This na"ve approach, however, may incur significant cost of time and space as it ignores the ancestor-descendant relationships among elements in the XML document. Clearly, for each smallest LCA that satisfies the query, the algorithm will produce all of its ancestors, which may likely be pruned according to the query semantics. Furthermore, the na"ve approach also incurs significant storage overhead, as each inverted list not only contains the XML element that directly contains the keyword, but also all of its ancestors [13].

Several algorithms have been proposed to improve the na"ve approach. Most systems for keyword search over XML documents [13, 25, 28, 19, 17, 29] are based on the notion of lowest common ancestors (LCAs) or its variations. XRank [13], for example, uses the ELCA semantics. XRank proposes two core algorithms, DIL (Dewey Inverted List) and RDIL (Ranked Dewey Inverted List). As RDIL is basically DIL integrated with ranking, due to space considerations, we focus on DIL in this section.

The DIL algorithm encodes ancestor-descendant relationships into the element IDs stored in the inverted list. Consider the tree representation of an XML document, where the root of the XML tree is assigned number 0, and sibling nodes are assigned sequential numbers $0, 1, 2, \cdots, i$. The Dewey ID of a node n is the concatenation of the numbers assigned to the nodes on the path from the root to n. Unlike the na"ve algorithm, in XRank, the inverted list for a keyword k contains only the Dewey IDs of nodes that *directly* contain k. This reduces much of the space overhead of the na"ve approach. From their Dewey IDs, we can easily figure out the ancestor-descendant relationships between two nodes: node A is an ancestor of node B iff the Dewey ID of node A is a prefix of that of node B.

Given a query $Q = \{k_1, \cdots, k_n\}$, the DIL algorithm makes a single pass over the n inverted lists corresponding to k_1 through k_n. The goal is to sort-merge the n inverted lists to find the ELCA answers of the query. However, since only nodes that directly contain the keywords are stored in the inverted lists, the standard sort-merge algorithm cannot be used. Nevertheless, the ancestor-descendant relationships have been encoded in the Dewey ID, which enables the DIL algorithm to derive the common ancestors from the Dewey IDs of nodes in the lists. More specifically, as each prefix of a node's Dewey ID is the Dewey ID of the node's ancestor, computing the longest common prefix will compute the ID of the lowest ancestor that contains the query keywords. In XRank, the inverted lists are sorted on the Dewey ID, which means all the common ancestors are clustered together. Hence, this computation can be done in a single pass over the n inverted lists. The complexity of the DIL algorithm is thus $O(nd|S|)$ where $|S|$ is the size of the largest inverted list for keyword $k_1, \cdots, k_n$ and d is the depth of the tree.

More recent approaches seek to further improve the performance of XRank [13]. Both the DIL and the RDIL algorithms in XRank need to perform a full scan of the inverted lists for every keyword in the query. However, certain keywords may be very frequent in the underlying XML documents. These keywords correspond to long inverted lists that become the bottleneck in query processing. XKSearch [28], which adopts the SLCA semantics for keyword search, is proposed to address the problem. XKSearch makes an observation that, in contrast to the general LCA semantics, the number of SLCAs is bounded by the length of the inverted list that corresponds to the least frequent keyword. The key intuition of XKSearch is that, given two keywords w_1 and w_2 and a node v that contains keyword w_1, there is no need to inspect the whole inverted list of keyword w_2 in order to find all possible answers. Instead, we only have to find the *left match* and the *right match* of the list of w_2, where the left (right) match is the node with the greatest (least) id that is smaller (greater) than or equal to the id of v. Thus, instead of scanning the inverted lists, XKSearch performs an indexed search on the lists. This enables XKSearch to reduce the number of disk accesses to $O(n|S_{min}|)$, where n is the number of the keywords in the query, and S_{min} is the length of the inverted list that corresponds to the least frequent keyword in the query (XKSearch assumes a B-tree disk-based structure where non-leaf nodes of the B-Tree are cached in memory). Clearly, this approach is meaningful only if at least one of the query keywords has very low frequency.

3.   Keyword Search on Relational Data

A tremendous amount of data resides in relational databases but is reachable via SQL only. To provide the data to users and applications that do not have the knowledge of the schema, much recent work has explored the possibility of using keyword search to access relational databases [1, 18, 3, 16, 21, 2]. In this section, we discuss the challenges and methods of implementing this new query interface.

3.1   Query Semantics

Enabling keyword search in relational databases without requiring the knowledge of the schema is a challenging task. Keyword search in traditional information retrieval (IR) is on the document level. Specifically, given a query $Q = \{k_1, \cdots, k_n\}$, we employ techniques such as the inverted lists to find documents that contain the keywords. Then, our question is, what is relational database's counterpart of IR's notion of "documents"?

It turns out that there is no straightforward mapping. In a relational schema designed according to the normalization principle, a logical unit of information is often disassembled into a set of entities and relationships. Thus, a relational database's notion of "document" can only be obtained by joining multiple tables.

Naturally, the next question is, can we enumerate all possible joins in a database? In Figure 8.2, as an example (borrowed from [1]), we show all potential joins among database tables $\{T_1, T_2, \cdots, T_5\}$. Here, a node represents a table. If a foreign key in table T_i references table T_j, an edge is created between T_i and T_j. Thus, any connected subgraph represents a potential join.

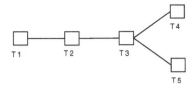

Figure 8.2. Schema Graph

Given a query $Q = \{k_1, \cdots, k_n\}$, a possible query semantics is to check all potential joins (subgraphs) and see if there exists a row in the join results that contains all the keywords in Q.

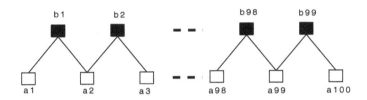

Figure 8.3. The size of the join tree is only bounded by the data Size

However, Figure 8.2 does not show the possibility of self-joins, i.e., a table may contain a foreign key that references the table itself. More generally, the schema graph may contain a cycle, which involves one or more tables. In this case, the size of the join is only bounded by the data size [18]. We demonstrates this issue with a self-join in Figure 8.3, where the self-join is on a table containing tuples (a_i, b_j), and the tuple (a_1, b_1) can be connected with tuple (a_{100}, b_{99}) by repeated self-joins. Thus, the join tree in Figure 8.3 satisfies keyword query $Q = \{a_1, a_{100}\}$. Clearly, the size of the join is only bounded by the number of tuples in the table. Such query semantics is hard to implement in practice. To mitigate this vulnerability, we change the semantics by introducing a parameter K to limit the size of the join we search for answers. In the above example, the result of (a_1, a_{100}) is only returned if K is as large as 100.

3.2 DBXplorer and DISCOVER

DBXplorer [1] and DISCOVER [18] are the most well known systems that support keyword search in relational databases. While implementing the query semantics discussed before, these approaches also focus on how to leverage the physical database design (e.g., the availability of indexes on various database columns) for building compact data structures critical for efficient keyword search over relational databases.

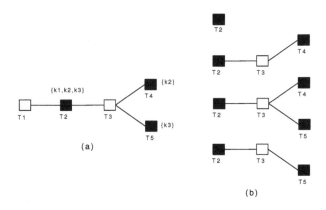

Figure 8.4. Keyword matching and join trees enumeration

Traditional information retrieval techniques use inverted lists to efficiently identify documents that contain the keywords in the query. In the same spirit, DBXplorer maintains a symbol table, which identifies columns in database tables that contain the keywords. Assuming index is available on the column, then given the keyword, we can efficiently find the rows that contain the keyword. If index is not available on a column, then the symbol table needs to map keywords to rows in the database tables directly.

Figure 8.4 shows an example. Assume the query contains three keywords $Q = \{k_1, k_2, k_3\}$. From the symbol table, we find tables/columns that contain one or more keywords in the query, and these tables are represented by black nodes in the Figure: k_1, k_2, k_3 all occur in T_2 (in different columns), k_2 occurs in T_4, and k_3 occurs in T_5. Then, DBXplorer enumerates the four possible join trees, which are shown in Figure 8.4(b). Each join tree is then mapped to a single SQL statement that joins the tables as specified in the tree, and selects those rows that contain all the keywords. Note that DBXplorer does not consider solutions that include two tuples from the same relation, or the query semantics required for problems shown in Figure 8.3.

DISCOVER [18] is similar to DBXplorer in the sense that it also finds all join trees (called candidate networks in DISCOVER) by constructing join expressions. For each candidate join tree, an SQL statement is generated. The trees may have many common components, that is, the generated SQL statements have many common join structures. An optimal execution plan seeks to maximize the reuse of common subexpressions. DISCOVER shows that the task of finding the optimal execution plan is NP-complete. DISCOVER introduces a greedy algorithm that provides near-optimal plan execution time cost. Given a set of join trees, in each step, it chooses the join m between two base tables or intermediate results that maximizes the quantity $\frac{frequency^a}{\log^b(size)}$, where $frequency$ is the number of occurences of m in the join trees, $size$ is the estimated number of tuples of m and a, b are constants. The $frequency^a$ term of the quantity maximizes the reusability of the intermediate results, while the $log^b(size)$ minimizes the size of the intermediate results that are computed first.

DBXplorer and DISCOVER use very simple ranking strategy: the answers are ranked in ascending order of the number of joins involved in the tuple trees; the reasoning being that joins involving many tables are harder to comprehend. Thus, all tuple trees consisting of a single tuple are ranked ahead of all tuples trees with joins. Furthermore, when two tuple trees have the same number of joins, their ranks are determined arbitrarily. BANKS [3] (see Section 4) combines two types of information in a tuple tree to compute a score for ranking: a weight (similar to PageRank for web pages) of each tuple, and a weight of each edge in the tuple tree that measures how related the two tuples are. Hristidis et al. [16] propose a strategy that applies IR-style ranking methods into the computation of ranking scores in a straightforward manner.

4. Keyword Search on Schema-Free Graphs

Graphs formed by relational and XML data are confined by their schemas, which not only limit the search space of keyword query, but also help shape the query semantics. For instance, many keyword search algorithms for XML data are based on the lowest common ancestor (LCA) semantics, which is only meaningful for tree structures. Challenges for keyword search on graph data are two-fold: what is the appropriate query semantics, and how to design efficient algorithms to find the solutions.

4.1 Query Semantics and Answer Ranking

Let the query consist of n keywords $Q = \{k_1, k_2, \cdots, k_n\}$. For each keyword k_i in the query, let S_i be the set of nodes that match the keyword k_i. The goal is to define what is a qualified answer to Q, and the score of the answer.

As we know, the semantics of keyword search over XML data is largely defined by the tree structure, as most approaches are based on the lowest common ancestor (LCA) semantics. Many algorithms for keyword search over graphs try to use similar semantics. But in order to do that, the answer must first form trees embedded in the graph. In many graph search algorithms, including BANKS [3], the bidirectional algorithm [21], and BLINKS [14], a response or an answer to a keyword query is a minimal rooted tree T embedded in the graph that contains at least one node from each S_i.

We need a measure for the "goodness" of each answer. An answer tree T is good if it is meaningful to the query, and the meaning of T lies in the tree structure, or more specifically, how the keyword nodes are connected through paths in T. In [3, 21], their goodness measure tries to decompose T into edges and nodes, score the edges and nodes separately, and combine the scores. Specifically, each edge has a pre-defined weight, and default to 1. Given an answer tree T, for each keyword k_i, we use $s(T, k_i)$ to represent the sum of the edge weights on the path from the root of T to the leaf containing keyword k_i. Thus, the aggregated edge score is $E = \sum_i^n s(T, k_i)$. The nodes, on the other hand, are scored by their global importance or prestige, which is usually based on PageRank [4] random walk. Let N denote the aggregated score of nodes that contain keywords. The combined score of an answer tree is given by $s(T) = EN^\lambda$ where λ helps adjust the importance of edge and node scores [3, 21].

Query semantics and ranking strategies used in BLINKS [14] are similar to those of BANKS [14] and the bidirectional search [21]. But instead of using a measure such as $S(T) = EN^\lambda$ to find top-K answers, BLINKS requires that each of the top-K answer has a different root node, or in other words, for all answer trees rooted at the same node, only the one with the highest score is considered for top-K. This semantics guards against the case where a "hub" pointing to many nodes containing query keywords becomes the root for a huge number of answers. These answers overlap and each carries very little additional information from the rest. Given an answer (which is the best, or one of the best, at its root), users can always choose to further examine other answers with this root [14].

Unlike most keyword search on graph data approaches [3, 21, 14], ObjectRank [2] does not return answer trees or subgraphs containing keywords in the query, instead, for ObjectRank, an answer is simply a node that has high authority on the keywords in the query. Hence, a node that does not even contain a particular keyword in the query may still qualify as an answer as long as enough authority on that keyword has flown into that node (Imagine a node that represents a paper which does not contain keyword *OLAP*, but many important papers that contain keyword *OLAP* reference that paper, which makes it an authority on the topic of *OLAP*). To control the flow of authority in the graph, ObjectRank models *labeled* graphs: Each node u has a label $\lambda(u)$ and contains a set of keywords, and each edge e from u to v has a label $\lambda(e)$ that represents a relationship between u and v. For example, a node may be labeled as a *paper*, or a *movie*, and it contains keywords that describe the paper or the movie; a directed edge from a paper node to another paper node may have a label *cites*, etc. A keyword that a node contains directly gives the node certain authority on that keyword, and the authority flows to other nodes through edges connecting them. The amount or the rate of the outflow of authority from keyword nodes to other nodes is determined by the types of the edges which represent different semantic connections.

4.2 Graph Exploration by Backward Search

Many keyword search algorithms try to find trees embedded in the graph so that similar query semantics for keyword search over XML data can be used. Thus, the problem is how to construct an embedded tree from keyword nodes in the graph. In the absence of any index that can provide graph connectivity information beyond a single hop, BANKS [3] answers a keyword query by exploring the graph starting from the nodes containing at least one query keyword – such nodes can be identified easily through an inverted-list index. This approach naturally leads to a *backward search* algorithm, which works as follows.

1 At any point during the backward search, let E_i denote the set of nodes that we know can reach query keyword k_i; we call E_i the *cluster* for k_i.

2 Initially, E_i starts out as the set of nodes O_i that directly contain k_i; we call this initial set the *cluster origin* and its member nodes *keyword nodes*.

3 In each search step, we choose an incoming edge to one of previously visited nodes (say v), and then follow that edge *backward* to visit its source node (say u); any E_i containing v now expands to include u as well. Once a node is visited, all its incoming edges become known to the search and available for choice by a future step.

4 We have discovered an answer root x if, for each cluster E_i, either $x \in E_i$ or x has an edge to some node in E_i.

BANKS uses the following two strategies for choosing what nodes to visit next. For convenience, we define the distance from a node n to a set of nodes N to be the shortest distance from n to any node in N.

1 *Equi-distance expansion in each cluster*: This strategy decides which node to visit for expanding a keyword. Intuitively, the algorithm expands a cluster by visiting nodes in order of increasing distance from the cluster origin. Formally, the node u to visit next for cluster E_i (by following edge $u \to v$ backward, for some $v \in E_i$) is the node with the shortest distance (among all nodes not in E_i) to O_i.

2 *Distance-balanced expansion across clusters*: This strategy decides the frontier of which keyword will be expanded. Intuitively, the algorithm attempts to balance the distance between each cluster's origin to its frontier across all clusters. Specifically, let (u, E_i) be the node-cluster pair such that $u \notin E_i$ and the distance from u to O_i is the shortest possible. The cluster to expand next is E_i.

He et al. [14] investigated the optimality of the above two strategies introduced by BANKS [3]. They proved the following result with regard to the first strategy, *equi-distance expansion of each cluster* (the complete proof can be found in [15]):

Theorem 8.2. *An optimal backward search algorithm must follow the strategy of equi-distance expansion in each cluster.*

However, the investigation [14] also showed that the second strategy, *distance-balanced expansion across clusters*, is not optimal and may lead to poor performance on certain graphs. Figure 8.5 shows one such example. Suppose that $\{k_1\}$ and $\{k_2\}$ are the two cluster origins. There are many nodes that can reach k_1 through edges with a small weight (1), but only one edge into k_2 with a large weight (100). With distance-balanced expansion across clusters, we would not expand the k_2 cluster along this edge until we have visited all nodes within distance 100 to k_1. It would have been unnecessary to visit many of these nodes had the algorithm chosen to expand the k_2 cluster earlier.

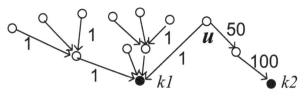

Figure 8.5. Distance-balanced expansion across clusters may perform poorly.

4.3   Graph Exploration by Bidirectional Search

To address the problem shown in Figure 8.5, Kacholia et al. [21] proposed a *bidirectional search* algorithm, which has the option of exploring the graph by following forward edges as well. The rationale is that, for example, in Figure 8.5, if the algorithm is allowed to explore forward from node u towards k_2, we can identify u as an answer root much faster.

To control the order of expansion, the bidirectional search algorithm prioritizes nodes by heuristic *activation factors* (roughly speaking, PageRank with decay), which intuitively estimate how likely nodes can be roots of answer trees. In the bidirectional search algorithm, nodes matching keywords are added to the iterator with an initial activation factor computed as:

$$a_{u,i} = \frac{nodePrestige(u)}{|S_i|}, \forall u \in S_i \qquad (8.6)$$

where S_i is the set of nodes that match keyword i. Thus, nodes of high prestige will have a higher priority for expansion. But if a keyword matches a large number of nodes, the nodes will have a lower priority. The activation factor is spreaded from keyword nodes to other nodes. Each node v spreads a fraction μ of the received activation to its neighbours, and retains the remaining $1 - \mu$ fraction.

As a result, keyword search in Figure 8.5 can be performed more efficiently. The bidirectional search will start from the keyword nodes (dark solid nodes). Since keyword node k_1 has a large fanout, all the nodes pointing to k_1 (including node u) will receive a small amount of activation. On the other hand, the node pointing to k_2 will receive most of the activation of k_2, which then spreads to node u. Thus, node u becomes the most activated node, which happens to be the root of the answer tree.

While this strategy is shown to perform well in multiple scenarios, it is difficult to provide any worst-case performance guarantee. The reason is that activation factors are heuristic measures derived from general graph topology and parts of the graph already visited. They do not accurately reflect the likelihood of reaching keyword nodes through an unexplored region of the graph within a reasonable distance. In other words, without additional connectivity information, forward expansion may be just as aimless as backward expansion [14].

4.4   Index-based Graph Exploration – the BLINKS Algorithm

The effectiveness of forward and backward expansions hinges on the structure of the graph and the distribution of keywords in the graph. However, both forward and backward expansions explore the graph link by link, which means the search algorithms do not have knowledge of either the structure of the graph nor the distribution of keywords in the graph. If we create an index structure to store the keyword reachability information in advance, we can avoid aimless exploration on the graph and improve the performance of keyword search. BLINKS [14] is designed based on this intuition.

BLINKS makes two contributions: First, it proposes a new, *cost-balanced* strategy for controlling expansion across clusters, with a provable bound on its worst-case performance. Second, it uses indexing to support forward jumps in search. Indexing enables it to determine whether a node can reach a keyword and what the shortest distance is, thereby eliminating the uncertainty and inefficiency of step-by-step forward expansion.

Cost-balanced expansion across clusters.   Intuitively, BLINKS attempts to balance the number of accessed nodes (i.e., the search cost) for expanding each cluster. Formally, the cluster E_i to expand next is the cluster with the smallest cardinality.

This strategy is intended to be combined with the equi-distance strategy for expansion within clusters: First, BLINKS chooses the smallest cluster to expand, then it chooses the node with the shortest distance to this cluster's origin to expand.

To establish the optimality of an algorithm A employing these two expansion strategies, let us consider an optimal "oracle" backward search algorithm P. As shown in Theorem 8.2, P must also do equi-distance expansion within each cluster. The additional assumption here is that P "magically" knows the right amount of expansion for each cluster such that the total number of nodes visited by P is minimized. Obviously, P is better than the best practical backward search algorithm we can hope for. Although A does not have the advantage of the oracle algorithm, BLINKS gives the following theorem (the complete proof can be found in [15]) which shows that A is m-optimal, where m is the number of query keywords. Since most queries in practice contain very few keywords, the cost of A is usually within a constant factor of the optimal algorithm.

Theorem 8.3. *The number of nodes accessed by A is no more than m times the number of nodes accessed by P, where m is the number of query keywords.*

Index-based Forward Jump. The BLINKS algorithm [14] leverages the new search strategy (*equi-distance* plus *cost-balanced* expansions) as well as indexing to achieve good query performance. The index structure consists of two parts.

- **Keyword-node lists L_{KN}.** BLINKS pre-computes, for each keyword, the shortest distances from every node to the keyword (or, more precisely, to any node containing this keyword) in the data graph. For a keyword w, $L_{KN}(w)$ denotes the list of nodes that can reach keyword w, and these nodes are ordered by their distances to w. In addition to other information used for reconstructing the answer, each entry in the list has two fields $(dist, node)$, where $dist$ is the shortest distance between $node$ and a node containing w.

- **Node-keywordmap M_{NK}.** BLINKS pre-computes, for each node u, the shortest graph distance from u to every keyword, and organize this information in a hash table. Given a node u and a keyword w, $M_{NK}(u, w)$ returns the shortest distance from u to w, or ∞ if u cannot reach any node that contains w. In fact, the information in M_{NK} can be derived from L_{KN}. The purpose of introducing M_{NK} is to reduce the linear time search over L_{KN} for the shortest distance between u and w to $O(1)$ time search over M_{NK}.

The search algorithm can be regarded as index-assisted backward and forward expansion. Given a keyword query $Q = \{k_1, \cdots, k_n\}$, for backward expansion, BLINKS uses a cursor to traverse each keyword-node list $L_{KN}(k_i)$. By construction, the list gives the equi-distance expansion order in each cluster. Across clusters, BLINKS picks a cursor to expand next in a round-robin manner, which implements cost-balanced expansion among clusters. These two together ensure optimal backward search. For forward expansion, BLINKS uses the node-keyword map M_{NK} in a direct fashion. Whenever BLINKS visits a node, it looks up its distance to other keywords. Using this information, it can immediately determine if the root of an answer is found.

The index L_{KN} and M_{NK} are defined over the entire graph. Each of them contains as many as $N \times K$ entries, where N is the number of nodes, and K is the number of distinct keywords in the graph. In many applications, K is on the same scale as the number of nodes, so the space complexity of the index comes to $O(N^2)$, which is clearly infeasible for large graphs. To solve this problem, BLINKS partitions the graph into multiple blocks, and the L_{KN} and M_{NK} index for each block, as well as an additional index structure to assist graph exploration across blocks.

4.5   The ObjectRank Algorithm

Instead of returning sub-graphs that contain all the keywords, ObjectRank [2] applies authority-based ranking to keyword search on labeled graphs, and returns nodes having high authority with respect to all keywords. To certain extent, ObjectRank is similar to BLINKS [14], whose query semantics prescribes that all top-K answer trees have different root nodes. Still, BLINKS returns sub-graphs as answers.

Recall that the bidirectional search algorithm [21] assigns activation factors to nodes in the graph to guide keyword search. Activation factors originate at nodes containing the keywords and propagate to other nodes. For each keyword node u, its activation factor is weighted by $nodePrestige(u)$ (Eq. 8.6), which reflects the importance or authority of node u. Kacholia et al. [21] did not elaborate on how to derive $nodePrestige(u)$. Furthermore, since graph edges in [21] are all the same, to spread the activation factor from a node u, it simply divides u's activation factor by u's fanout.

Similar to the activation factor, in ObjectRank [2], authority originates at nodes containing the keywords and flows to other nodes. Furthermore, nodes and edges in the graphs are labeled, giving graph connections semantics that controls the amount or the rate of the authority flow between two nodes.

Specifically, ObjectRank assumes a labeled graph G is associated with some predetermined schema information. The schema information decides the rate of authority transfer from a node labeled u_G, through an edge labeled e_G, and to a node labeled v_G. For example, authority transfers at a fixed rate from a *person* to a *paper* through an edge labeled *authoring*, and at another fixed rate from a *paper* to a *person* through an edge labeled *authoring*. The two rates are potentially different, indicating that authority may flow at a different rate backward and forward. The schema information, or the rate of authority transfer, is determined by domain experts, or by a trial and error process.

To compute node authority with regard to every keyword, ObjectRank computes the following:

- **Rates of authority transfer through graph edges.** For every edge $e = (u \rightarrow v)$, ObjectRank creates a forward authority transfer edge $e^f = (u \rightarrow v)$ and a backward authority transfer edge $e^b = (v \rightarrow u)$. Specifically, the authority transfer edges e^f and e^b are annotated with rates $\alpha(e^f)$ and $\alpha(e^b)$:

$$\alpha(e^f) = \begin{cases} \frac{\alpha(e_G^f)}{OutDeg(u,e_G^f)} & \text{if } OutDeg(u,e_G^f) > 0 \\ 0 & \text{if } OutDeg(u,e_G^f) = 0 \end{cases} \quad (8.7)$$

where $\alpha(e_G^f)$ denotes the fixed authority transfer rate given by the schema, and $OutDeg(u,e_G^f)$ denotes the number of outgoing nodes from u, of type e_G^f. The authority transfer rate $\alpha(e^b)$ is defined similarly.

- **Node authorities.** ObjectRank can be regarded as an extension to PageRank [4]. For each node v, ObjectRank assigns a global authority $ObjectRank^G(v)$ that is independent of the keyword query. The global $ObjectRank^G$ is calculated using the random surfer model, which is similar to PageRank. In addition, for each keyword w and each node v, ObjectRank integrates authority transfer rates in Eq 8.7 with PageRank to calculate a keyword-specific ranking $ObjectRank^w(v)$:

$$ObjectRank^w(v) = d \times \sum_{e=(u\rightarrow v) or (v\rightarrow u)} \alpha(e) \times ObjectRank^w(u) +$$

$$+\frac{1-d}{|S(w)|}$$
$$(8.8)$$

where $S(w)$ is s the set of nodes that contain the keyword w, and d is the damping factor that determines the portion of ObjectRank that a node transfers to its neighbours as opposed to keeping to itself [4]. The final ranking of a node v is the combination combination of $ObjectRank^G(v)$ and $ObjectRank^w(v)$.

5. Conclusions and Future Research

The work surveyed in this chapter include various approaches for keyword search for XML data, relational databases, and schema-free graphs. Because of the underlying graph structure, keyword search over graph data is much more complex than keyword search over documents. The challenges have three aspects, namely, how to define intuitive query semantics for keyword search over graphs, how to design meaningful ranking strategies for answers, and how to devise efficient algorithms that implement the semantics and the ranking strategies.

There are many remaining challenges in the area of keyword search over graphs. One area that is of particular importance is how to provide a semantic search engine for graph data. The graph is the best representation we have for complex information such as human knowledge, social and cultural dynamics, etc. Currently, keyword-oriented search merely provides best-effort heuristics to find relevant "needles" in this humongous "haystack". Some recent work, for example, NAGA [22], has looked into the possibility of creating a semantic search engine. However, NAGA is not keyword-based, which introduces complexity for posing a query. Another important challenge is that the size of the graph is often significantly larger than memory. Many graph keyword search algorithms [3, 21, 14] are memory-based, which means they cannot handle graphs such as the English Wikipedia that has over 30 million edges. Some reacent work, such as [7], organizes graphs into different levels of granularity, and supports keyword search on disk-based graphs.

References

[1] S. Agrawal, S. Chaudhuri, and G. Das. DBXplorer: A system for keyword-based search over relational databases. In *ICDE*, 2002.

[2] A. Balmin, V. Hristidis, and Y. Papakonstantinou. ObjectRank: Authority-based keyword search in databases. In *VLDB*, pages 564–575, 2004.

[3] G. Bhalotia, C. Nakhe, A. Hulgeri, S. Chakrabarti, and S. Sudarshan. Keyword searching and browsing in databases using BANKS. In *ICDE*, 2002.

[4] S. Brin and L. Page. The anatomy of a large-scale hypertextual Web search engine. *Computer networks and ISDN systems*, 30(1-7):107–117, 1998.

[5] Y. Cai, X. Dong, A. Halevy, J. Liu, and J. Madhavan. Personal information management with SEMEX. In *SIGMOD*, 2005.

[6] S. Cohen, J. Mamou, Y. Kanza, and Y. Sagiv. XSEarch: A semantic search engine for XML. In *VLDB*, 2003.

[7] Bhavana Bharat Dalvi, Meghana Kshirsagar, and S. Sudarshan. Keyword search on external memory data graphs. In *VLDB*, pages 1189–1204, 2008.

[8] B. Ding, J. X. Yu, S. Wang, L. Qing, X. Zhang, and X. Lin. Finding top-k min-cost connected trees in databases. In *ICDE*, 2007.

[9] S. E. Dreyfus and R. A. Wagner. The Steiner problem in graphs. *Networks*, 1:195–207, 1972.

[10] S. Dumais, E. Cutrell, JJ Cadiz, G. Jancke, R. Sarin, and D. C. Robbins. Stuff i've seen: a system for personal information retrieval and re-use. In *SIGIR*, 2003.

[11] D. Florescu, D. Kossmann, and I. Manolescu. Integrating keyword search into XML query processing. *Comput. Networks*, 33(1-6):119–135, 2000.

[12] J. Graupmann, R. Schenkel, and G. Weikum. The spheresearch engine for unified ranked retrieval of heterogeneous XML and web documents. In *VLDB*, pages 529–540, 2005.

[13] L. Guo, F. Shao, C. Botev, and J. Shanmugasundaram. XRANK: ranked keyword search over XML documents. In *SIGMOD*, pages 16–27, 2003.

[14] H. He, H. Wang, J. Yang, and P. S. Yu. BLINKS: Ranked keyword searches on graphs. In *SIGMOD*, 2007.

[15] H. He, H. Wang, J. Yang, and P. S. Yu. BLINKS: Ranked keyword searches on graphs. Technical report, Duke CS Department, 2007.

[16] V. Hristidis, L. Gravano, and Y. Papakonstantinou. Efficient IR-style keyword search over relational databases. In *VLDB*, pages 850–861, 2003.

[17] V. Hristidis, N. Koudas, Y. Papakonstantinou, and D. Srivastava. Keyword proximity search in XML trees. *IEEE Transactions on Knowledge and Data Engineering*, 18(4):525–539, 2006.

[18] V. Hristidis and Y. Papakonstantinou. Discover: Keyword search in relational databases. In *VLDB*, 2002.

[19] V. Hristidis, Y. Papakonstantinou, and A. Balmin. Keyword proximity search on XML graphs. In *ICDE*, pages 367–378, 2003.

[20] Haoliang Jiang, Haixun Wang, Philip S. Yu, and Shuigeng Zhou. GString: A novel approach for efficient search in graph databases. In *ICDE*, 2007.

[21] V. Kacholia, S. Pandit, S. Chakrabarti, S. Sudarshan, R. Desai, and H. Karambelkar. Bidirectional expansion for keyword search on graph databases. In *VLDB*, 2005.

[22] G. Kasneci, F.M. Suchanek, G. Ifrim, M. Ramanath, and G. Weikum. Naga: Searching and ranking knowledge. In *ICDE*, pages 953–962, 2008.

[23] R. Kaushik, R. Krishnamurthy, J. F. Naughton, and R. Ramakrishnan. On the integration of structure indexes and inverted lists. In *SIGMOD*, pages 779–790, 2004.

[24] B. Kimelfeld and Y. Sagiv. Finding and approximating top-k answers in keyword proximity search. In *PODS*, pages 173–182, 2006.

[25] Yunyao Li, Cong Yu, and H. V. Jagadish. Schema-free XQuery. In *VLDB*, pages 72–83, 2004.

[26] F. Liu, C. T. Yu, W. Meng, and A. Chowdhury. Effective keyword search in relational databases. In *SIGMOD*, pages 563–574, 2006.

[27] Dennis Shasha, Jason T.L. Wang, and Rosalba Giugno. Algorithmics and applications of tree and graph searching. In *PODS*, pages 39–52, 2002.

[28] Y. Xu and Y. Papakonstantinou. Efficient keyword search for smallest LCAs in XML databases. In *SIGMOD*, 2005.

[29] Yu Xu and Yannis Papakonstantinou. Efficient LCA based keyword search in XML data. In *EDBT*, pages 535–546, New York, NY, USA, 2008. ACM.

[30] Xifeng Yan, Philip S. Yu, and Jiawei Han. Substructure similarity search in graph databases. In *SIGMOD*, pages 766–777, 2005.

Chapter 9

A SURVEY OF CLUSTERING ALGORITHMS FOR GRAPH DATA

Charu C. Aggarwal

IBM T. J. Watson Research Center
Hawthorne, NY 10532

charu@us.ibm.com

Haixun Wang

Microsoft Research Asia
Beijing, China 100190

haixunw@microsoft.com

Abstract   In this chapter, we will provide a survey of clustering algorithms for graph data. We will discuss the different categories of clustering algorithms and recent efforts to design clustering methods for various kinds of graphical data. Clustering algorithms are typically of two types. The first type consists of node clustering algorithms in which we attempt to determine dense regions of the graph based on edge behavior. The second type consists of structural clustering algorithms, in which we attempt to cluster the different graphs based on overall structural behavior. We will also discuss the applicability of the approach to other kinds of data such as semi-structured data, and the utility of graph mining algorithms to such representations.

Keywords: Graph Clustering, Dense Subgraph Discovery

1.   Introduction

Graph mining has been a popular area of research in recent years because of numerous applications in computational biology, software bug localization and computer networking. In addition, many new kinds of data such as semi-

C.C. Aggarwal and H. Wang (eds.), *Managing and Mining Graph Data*,
Advances in Database Systems 40, DOI 10.1007/978-1-4419-6045-0_9,
© Springer Science+Business Media, LLC 2010

structured data and XML [2] can typically be represented as graphs. In particular, XML data is a popular representation of different kinds of data sets. Since core graph-mining algorithms can be extended to this scenario, it follows that the extension of mining algorithms to graphs has tremendous applicability of a wide variety of data sets which are represented as semi-structured data. Many traditional algorithms such as clustering, classification, and frequent-pattern mining have been extended to the graph scenario. A detailed discussion of various kinds of graph mining algorithms may be found in [15].

In this chapter, we will study the clustering problem for the graph domain. The problem of clustering is defined as follows: For a given set of objects, we would like to divide it into groups of *similar objects*. The similarity between objects is typically defined with the use of a mathematical objective function. This problem is useful in a number of practical applications such as marketing, customer-segmentation, and data summarization. The problem of clustering is extremely important in a number of important data domains. A detailed description of clustering algorithms may be found in [24].

Clustering algorithms have significant applications in a variety of graph scenarios such as congestion detection, facility location, and XML data integration [28]. The graph clustering problems are typically defined into two categories:

- **Node Clustering Algorithms:** Node-clustering algorithms are generalizations of multi-dimensional clustering algorithms in which we use functions of the multi-dimensional data points in order to define the distances. In the case of graph clustering algorithms, we associate numerical values with the edges. These numerical values need not satisfy traditional properties of distance functions such as the triangle inequality. We use these distance values in order to create clusters of nodes. We note that the numerical value associated with a given node may either be a distance value or a similarity value. Correspondingly, the objective function associated with the partitioning may either be minimized or maximized respectively. We note that the problem of minimizing the inter-cluster similarity for a fixed number of clusters essentially reduces to the problem of *graph partitioning* or the *minimum multi-way cut problem*. This is also referred to the problem of mining dense graphs and pseudo-cliques. Recently, the problem has also been studied in the database literature as that of *quasi-clique determination*. In this problem, we determine groups of nodes which are "almost cliques". In other words, an edge exists between any pair of nodes in the set with high probability. A closely related problem is that of determining *shingles* [5, 22]. Shingles are defined as those sub-graphs which have a large number of common links. This is particularly useful for massive graphs which contain a large number of nodes. In such cases, a min-hash ap-

proach [5] can be used in order to summarize the structural behavior of
the underlying graph.

- **Graph Clustering Algorithms:** In this case, we have a (possibly large)
  number of graphs which need to be clustered based on their underlying
  structural behavior. This problem is challenging because of the need to
  match the structures of the underlying graphs, and use these structures
  for clustering purposes. Such algorithms are discussed both in the context of classical graph data sets as well as semi-structured data. In the
  case of semi-structured data, the problem arises in the context of a large
  number of documents which need to be clustered on the basis of the underlying structure and attributes. It has been shown in [2] that the use of
  the underlying document structure leads to significantly more effective
  algorithms.

This chapter is organized as follows. In the next section, we will discuss a
variety of node clustering algorithms. Methods for clustering multiple graphs
and XML records are discussed in section 3. Section 4 discusses numerous
applications of graph clustering algorithms. Section 5 contains the conclusions
and summary.

2.   Node Clustering Algorithms

A number of algorithms for graph node clustering are discussed in [19]. In
[19], the graph clustering problem is related to the minimum cut and graph
partitioning problems. In this case, it is assumed that the underlying graphs
have weights on the edges. It is desired to partition the graph in such a way
so as to minimize the weights of the edges across the partitions. In general,
we would like to partition the graph into k groups of nodes. However, since
the special case $k = 2$ is efficiently solvable, we would like to first provide a
special discussion for this case. This version is polynomially solvable, since it
is the mathematical dual of the maximum flow problem. This problem is also
referred to as the *minimum-cut problem.*

2.1   The Minimum Cut Problem

The simplest case is the 2-way minimum cut problem, in which we wish to
partition the graph into two clusters, so as to minimize the weight of the edges
across the partitions. This version of the problem is efficiently solvable, and
can be resolved by use of the *maximum flow problem* [4].

The minimum-cut problem is defined as follows. Consider a graph $G =
(N, A)$ with node set N and edge set A. The node set N contains the source
s and sink t. Each edge $(i, j) \in A$ has a weight associated with it which is
denoted by u_{ij}. We note that the edges may be either undirected or directed,

though the undirected case is often much more relevant for connectivity applications. We would like to partition the node set N into two groups S and $N - S$. The set of edges such that one end lies in S and the other lies in $N - S$ is denoted by $C(S, N - S)$. We would like to partition the node set N into two sets S and $N - S$, such that the sum of the weights in $C(S, N - S)$ is minimized. In other words, we would like to minimize $\sum_{(i,j) \in C(S,N-S)} u_{ij}$. This is the unrestricted version of the minimum-cut problem. We will examine two variations of the minimum-cut problem:

- We wish to determine the global minimum s-t cut with no restrictions on the membership of nodes to different partitions.

- We wish to determine the minimum s-t cut, in which one partition contains the source node s and the other partition contains the sink node t.

It is easy to see that the former problem can be solved by using repeated applications of the latter algorithm. By fixing s and choosing different values of the sink t, it can be shown that the global minimum-cut may be effectively determined.

It turns out that the maximum flow problem is the mathematical dual of the minimum cut problem. In the maximum-flow problem, we assume that the weight u_{ij} is a capacity of the edge (i, j). Each edge is allowed to have a *flow* x_{ij} which is at most equal to the capacity u_{ij}. Each node other than the source s and sink t is assumed to satisfy the *flow conservation property*. In other words, for each node $i \in N$ we have:

$$\sum_{j:(i,j) \in A} x_{ij} = \sum_{j:(j,i) \in A} x_{ji} \qquad (9.1)$$

We would like to maximize the total flow originating from the source and reaching the sink t, subject to the above constraints. The maximum flow problem is solved with the use of a variety of *augmenting-path* and *preflow push* algorithms [4]. In augmenting-path methods, we pick a path from s to t which has current unused capacity, and increase the flow on this path, such that at least one edge on this path is filled to capacity. We repeat this process, until no path with unfilled capacity exists from source s to sink t. Many different variations of this technique exist in terms of the choice of path used in order to augment the flow from source s to the sink t. Example, include the shortest-paths or maximum-capacity augmenting paths. Different choices of augmenting-paths will typically lead to different trade-offs in running time. These trade-offs are discussed in [4]. In general, the two-way cut problem can be solved quite efficiently in polynomial time with these different methods. It can be shown that the minimum-cut may be determined by determining all nodes S which are reachable from s by some path of unfilled capacity. We note that S will not contain the sink node t at maximum flow, since the sink is not reachable from the source with the use of a path of unfilled capacity. The set $C(S, N - S)$ is the minimum s-t cut. Every edge in this set is saturated, and the total flow across the cut is essentially equal to the s-t maximum flow. We can then determine the global minimum cut by fixing the source s, and varying the sink node t. The minimum cut over all these different possibilities will provide us with the global minimum-cut value. A particularly important variant of this method is the shortest augmenting-path approach. In this approach we always augment the maximum amount of flow from the source to sink along the corresponding shortest path. It can be shown that for a network containing n nodes, and m edges, the shortest path is guaranteed to increase by at least one after $O(m)$ augmentations. Since the shortest path cannot be larger than n, it follows that the maximum number of augmentations is $O(n \cdot m)$. It is possible to implement each augmentation in $O(\log(n))$ time with the use of dynamic data structures. This implies that the overall technique requires at most $O(n \cdot m \cdot \log(n))$ time.

A second class of algorithms which are often used in order to solve the maximum flow problem are preflow push algorithms, which do not maintain the flow conservation constraints in their intermediate solutions. Rather, an excess flow is maintained at each node, and we try to push as much of this flow as possible along any edge on the shortest path from the source to sink. A detailed discussion of preflow push methods is beyond the scope of this chapter, and may be found in [4]. Most maximum flow methods require at least $\Omega(n \cdot m)$ time, where n is the number of nodes, and m is the number of edges.

A closely related problem to the minimum s-t cur problem is that of determining a *global minimum cut* in an undirected graph. This particular case is more efficient than that of finding the s-t minimum cut. One way of determining a minimum cut is by using a contraction-based edge-sampling approach. While the previous technique is applicable to both the directed and undirected version of the problem, the contraction-based approach is applicable only to the undirected version of the problem. Furthermore, the contraction-based approach is applicable only for the case in which the weight of each edge is $u_{ij} = 1$. While the method can easily be extended to the weighted version by varying the edge-sampling probability, the polynomial running time bounds discussed in [37] do not apply to this case. The contraction approach is a probabilistic technique in which we successively sample edges in order to collapse nodes into larger sets of nodes. By successively sampling different sequences of edges and picking the optimum value [37], it is possible to determine a global minimum cut. The broad idea of the contraction-based approach is as follows. We pick an edge randomly in the graph, and contract its two end points into a single node. We remove all self-loops which are created as a result of the contraction. We may also create some parallel edges, which are allowed to remain, since they influence the sampling probability[1] of contractions. The process of contraction is repeated until we are left with two nodes. We note that each of this pair of "super-nodes" corresponds to a set of nodes in the original data. These two sets of nodes provide us with the final minimum cut. We note that the minimum cut will survive in this approach, if none of the edges in the minimum cut are sampled during the contraction. An immediate observation is that cuts with smaller number of edges are more likely to survive using this approach. This is because the edges in cuts which contain a large number of edges are much more likely to be sampled. One of the key observations in [37] is the following:

Lemma 9.1. *When a graph containing n nodes is contracted to t nodes, the probability that the minimum-cut survives during the contraction is given by $O(t^2/n^2)$.*

Proof: Let the minimum-cut have k edges. Then, each vertex must have degree at least k, and therefore the graph must contain at least $n \cdot k/2$ edges. Then, the probability that the minimum cut survives the first contraction is given by $1 - k/(\#\text{Edges}) \geq 1 - 2/n$. This relationship is derived by substituting the lower bound of $n \cdot k/2$ for the number of edges. Similarly, in the second round of contractions, the probability of survival is given by $1 - 2/(n-1)$. Therefore, the overall probability p_s of survival is given by:

$$p_s = \Pi_{i=0}^{n-t-1}(1 - 2/(n - i)) = \frac{t \cdot (t - 1)}{n \cdot (n - 1)} \tag{9.2}$$

This provides the result.   □

Thus, if we contract to two nodes, the probability of the survival of the minimum cut is $2/(n \cdot (n - 1))$. By repeating the process $n \cdot (n - 1)/2$ times, we can show that the probability that the minimum-cut survives is given by at least $1 - 1/e$. If we further scale up by a constant factor $C > 1$, we can show that the probability of survival is given by $1 - (1/e)^C$. By picking $C = \log(1/\delta)$, we can assure that the cut survives with probability at least $1 - \delta$, where $\delta << 1$. The logarithmic relationship assures that we can determine minimum cuts with very high probability at a small additional cost. An additional implication of Lemma 9.1 is that the total number of *distinct minimum cuts* is bounded above by $n \cdot (n-1)/2$. This is because the probability of the survival of *any particular* minimum cut is at least $2/(n \cdot (n - 1))$, and the probability of the survival of *any* minimum cut cannot be greater than 1.

---

[1]Alternatively, we may replace parallel edges by a single edge of weight which is equal to the number of parallel edges. We use this weight in order to bias the sampling process.

Another observation is that the probability of survival of the minimum cut in the first iteration is the largest, and it reduces in successive iterations. For example, in the first iteration, the probability of survival is $1 - (2/n)$, but the probability of survival in the last iteration is only $1/3$. Thus, most of the errors are caused in the last few iterations. This is particularly reflected in the cumulative error across many iterations, since the probability of maintaining the correct cut on contracting down to t nodes is t^2/n^2, whereas the probability of maintaining the correct cut in the remaining contractions is $1/t^2$.

Therefore, a natural solution is to use a two-phase approach. In the first phase, we do not contract down to 2 nodes, but we contract down to t nodes. The probability of maintaining the correct cut by the use of this approach is at least $\Omega(t^2/n^2)$. Therefore, $O(n^2/t^2)$ contractions are required in order to reduce the graph to t nodes. Since each contraction requires $O(n)$ time, the running time of the first phase is given by $O(n^3/t^2)$. In the second phase, we use a standard maximum flow based method in order to determine the minimum cut. This maximum flow problem needs to be repeated t times for a fixed source and different sinks. However, the base graph on which this is performed is much smaller, and contains only $O(t)$ nodes. Each maximum flow problem requires $O(t^3)$ time by using the method discussed in [8], and therefore the total time for all t problems is given by $O(t^4)$. Therefore, the total running time is given by $O(n^3/t^2 + t^4)$. By picking $t = \sqrt{n}$, we can obtain a running time of $O(n^2)$. Thus, by using a two-phase approach, it is possible to obtain a much better running time, than by using a single-phase contraction approach. The key idea behind this improvement is that since most of the error probability is concentrated in the last contractions, it is better to stop the contraction process when the the underlying graph is "small enough", and then use conventional algorithms in order to determine the minimum cut. This combination approach is theoretically more efficient than any other known algorithm.

2.2 Multi-way Graph Partitioning

The *multi-way graph partitioning problem* is significantly more difficult, and is NP-hard [21]. In this case, we wish to partition a graph into $k > 2$ components, so that the total weight of the edges whose ends lie in different partitions is minimized. A well known technique for graph partitioning is the Kerninghan-Lin algorithm [26]. This classical algorithm is based on a hill-climbing (or more generally neighborhood-search technique) for determining the optimal graph partitioning. Initially, we start off with a random cut of the graph. In each iteration, we exchange a pair of vertices in two partitions, to see if the overall cut value is reduced. In the event that the cut value is reduced, then the interchange is performed. Otherwise, we pick another pair of vertices in order to perform the interchange. This process is repeated until we converge to a optimal solution. We note that this optimum may not be a global optimum, but may only be a local optimum of the underlying data. The main variation in different versions of the Kerninghan-Lin algorithm is the policy which is used for performing the interchanges on the vertices. Some examples of strategies which may be used in order to perform the interchange are as follows:

- We randomly pick a pair of vertices and perform the interchange, if it improves the underlying solution quality.

- We test all possible vertex-pair interchanges (or a sample of possible interchanges), and pick the interchange which improves the solution by the greatest amount.

- A k-interchange is one in which a sequence of k interchanges are performed at one time. We can test any k-interchange and perform it, if it improves the underlying solution quality.

- We can pick the optimal k-interchange from a sample of possibilities.

We note that the use of more sophisticated strategies allows a better improvement in the objective function for each interchange, but also requires more time for each interchange. For example, the determination of an optimal k-interchange requires much more time than a straightforward interchange. This is a natural tradeoff which may work out differently depending upon the nature of the application at hand. Furthermore, the choice of the policy also affects the likelihood of getting stuck at a local optimum. For example, the use of k-interchange techniques are far less likely to result in local optimum for larger values of k. In fact, by choosing the best interchange across all possible values of k, it is possible to ensure that a global optimum is always reached. On the other hand, it because increasingly difficult to implement the algorithm efficiently with increasing value of k. This is because the time-complexity of the interchange increases exponentially with the value of k. A detailed survey on different methods for optimal graph partitioning may be found in [18].

2.3   Conventional Generalizations and Network Structure Indices

Two well known (and related) techniques for clustering in the context of multi-dimensional data [24] are the k-medoid and k-means algorithms. In the k-medoid algorithm (for multi-dimensional data), we sample a small number of points from the original data as *seeds* and assign every other data point from the clusters to the closest of these seeds. The closeness may be defined based on a user-defined objective function. The objective function for the clustering is defined as the sum of the corresponding distances of data points to the corresponding seeds. In the next iteration, the algorithm interchanges one of the seeds for another randomly selected seed from the data, and checks if the quality of the objective function improves upon performing the interchange. If this is indeed the case, then the interchange is accepted. Otherwise, we do not accept the interchange and try another sample interchange. This process is repeated, until the objective function does not improve over a pre-defined number of interchanges. A closely related method is the k-means method. The main difference with the k-medoid method is that we do not use representative points from the original data after the first iteration of picking the original seeds. In subsequent iterations, we use the centroid of each cluster as the seed set for the next iteration. This process is repeated until the cluster membership stabilizes.

A method has been proposed in [35], which uses characteristics of both[2] the k-means and k-medoids algorithms. As in the case of the conventional partitioning algorithms, it picks k graph nodes as seeds. The main differences from the conventional algorithms are in terms of computation of distances (for assignment purposes), and in determination of subsequent seeds. A natural distance function for graphs is the *geodesic distance*, or the smallest number of hops between a pair of nodes. In order to determine the seed set for the next iteration, we compute the *local closeness centrality* [20] for each cluster, and use the corresponding node as the sample seed. Thus, while this algorithm continues to use seeds from the original data set (as in the k-medoids algorithm), it uses intuitive ideas from the k-means algorithms in order to determine the identity of these seeds.

There are some subtle challenges in the use of the graphical versions of distance based clustering algorithms. One challenge is that since distances are integers, it is possible for data points to be equidistant to several seeds. While ties can be resolved by randomly selecting one of the best assignments, this may result in clusterings which do not converge. In order to handle this instability, a more relaxed threshold is imposed on the number of medoids which may change from iteration to iteration. Specifically, a clustering is considered stable, when the change between iterations is below a certain threshold (say 1 to 3%).

Another challenge is that the computation of geodesic distances can be very challenging. The computational complexity of the all-pairs shortest paths algorithm can be $O(n^3)$, where n is the number of nodes. Even pre-storage of all-pairs shortest paths can require $O(n^2)$ time. This is computationally not feasible in most practical scenarios, especially when the underlying graphs are large. Even the space-requirement can be infeasible for very large graphs may

---

[2]In [35], the method has been proposed as a generalization of the k-medoid algorithm. However, it actually uses characteristics of both the k-means and k-medoid algorithms, since it uses centrality notions in determination of subsequent seeds.

not be practical. In order to handle such cases, the method in [36] uses the concept of *network-structure indices*, which can summarize the behavior of the network by using *randomized division into zones*.

In this case, the graph is divided into multiple zones. The set of zones form a connected, mutually exclusive and exhaustive partitioning of the graph. The partitioning of the graph into zones is accomplished with the use of a *competitive flooding algorithm*. In this algorithm, we start off with randomly selected seeds which are labeled by zone identification, and randomly select some unlabeled neighbor of a currently labeled node, and add a label which is matching with its current value. This approach is repeated until all nodes have been labeled. We note that while this approach is extremely fast, it may sometimes result in zones which do not reflect locality well. In order to deal with this situation, we use *multiple sets of randomly selected partitions*. Each of these partitions is considered a dimension. Note that when we use multiple such random partitions, each node becomes distinguishable from other nodes by virtue of its membership.

The distance between a node i and a zone containing node j is denoted as $ZoneDistance(i, zone(j))$, and is defined as the shortest path between node i and any node in zone j. The distance between i and j along a *particular zone partitioning* (or dimension) is approximated as $ZoneDistance(i, zone(j)) + ZoneDistance(j, zone(i))$. This value is then averaged over all the sets of randomized partitions in order to provide better robustness. It has been shown in [36] that this approach seems to approximate pairwise distances quite well. The key observation is that the value of $ZoneDistance(i, zone(j))$ can be pre-computed in $n \cdot q$ space, where q is the number of zones. For a small number of zones, this is quite efficient. Upon using r different sets of partitions, the overall space requirement is $n \cdot q \cdot r$, which is much smaller than the $\Omega(n^2)$ space-requirement of all-pairs computation, for typical values of q and r as suggested in [35].

2.4   The Girvan-Newman Algorithm

The Girvan-Newman algorithm [23] is a divisive clustering algorithm, which is based on the concept of *edge betweenness centrality*. Betweenness centrality attempts to identify edges which form *critical bridges* between different connected components, and delete them, until a natural set of clusters remains. Formally, betweenness centrality is defined as the proportion of shortest paths between nodes which pass through a certain edge. Therefore, for a given edge e, we define the betweenness centrality $B(e)$ as follows:

$$B(e) = \frac{NumConstrainedPaths(e, i, j)}{NumShortPaths(i, j)} \qquad (9.3)$$

Here $NumConstrainedPaths(e, i, j)$ refers to the number of (global) short-est paths between i and j which pass through e, and $NumShortPaths(i, j)$ refers to the number of shortest paths between i and j. Note that the value of $NumConstrainedPaths)(e, i, j)$ may be 0 if none of the shortest paths between i and j contain e. The algorithm ranks the edges by order of their betweenness, and and deletes the edge with the highest score. The between-ness coefficients are recomputed, and the process is repeated. The set of con-nected components after repeated deletion form the natural clusters. A variety of termination-criteria (eg. fixing the number of connected components) can be used in conjunction with the algorithm.

A key issue is the efficient determination of edge-betweenness centrality. The number of paths between any pair of nodes can be exponentially large, and it would seem that the computation of the betweenness measure would be a key bottleneck. It has been shown in [36], that the network structure index can also be used in order to estimate edge-betweenness centrality effectively by pairwise node sampling.

2.5   The Spectral Clustering Method

Eigenvector techniques are often used in multi-dimensional data in order to determine the underlying correlation structure in the data. It is natural to question as to whether such techniques can also be used for the more general case of graph data. It turns out that this is indeed possible with the use of a method called *spectral clustering*.

In the spectral clustering method, we make use of the node-node adjacency matrix of the graph. For a graph containing n nodes, let us assume that we have a $n \times n$ adjacency matrix, in which the entry (i, j) correspond to the weight of the edge between the nodes i and j. This essentially corresponds to the similar-ity between nodes i and j. This entry is denoted by w_{ij}, and the corresponding matrix is denoted by W. This matrix is assumed to be symmetric, since we are working with undirected graphs. Therefore, we assume that $w_{ij} = w_{ji}$ for any pair (i, j). All diagonal entries of the matrix W are assumed to be 0. As discussed earlier, the aim of any node partitioning algorithm is minimize (a function of) the weights across the partitions. The spectral clustering method constructs this minimization function in terms of the matrix structure of the adjacency matrix, and another matrix which is referred to as the *degree matrix*.

The*degree matrix* D is simply a diagonal matrix, in which all entries are zero except for the diagonal values. The diagonal entry d_{ii} is equal to the sum of the weights of the incident edges. In other words, the entry d_{ij} is defined as follows:

$$d_{ij} = \begin{matrix} \sum_{j=1}^{n} w_{ij} & i = j \\ 0 & i \neq j \end{matrix}$$

We formally define the *Laplacian Matrix* as follows:

Definition 9.2. (Laplacian Matrix) *The Laplacian Matrix L is defined by subtracting the weighted adjacency matrix from the degree matrix. In other words, we have:*

$$L = D - W \tag{9.4}$$

This matrix encodes the structural behavior of the graph effectively and its eigenvector behavior can be used in order to determine the important clusters in the underlying graph structure. We can be shown that the Laplacian matrix L is positive semi-definite i.e., for any n-dimensional row vector $f = [f_1 \ldots f_n]$ we have $f \cdot L \cdot f^T \geq 0$. This can be easily shown by expressing L in terms of its constituent entries which are a function of the corresponding weights w_{ij}. Upon expansion, it can be shown that:

$$f \cdot L \cdot f^T = (1/2) \cdot \sum_{i=1}^{n} \sum_{j=1}^{n} w_{ij} \cdot (f_i - f_j)^2 \tag{9.5}$$

We summarize as follows.

Lemma 9.3. *The Laplacian matrix L is positive semi-definite. Specifically, for any n-dimensional row vector $f = [f_1 \ldots f_n]$, we have:*

$$f \cdot L \cdot f^T = (1/2) \cdot \sum_{i=1}^{n} \sum_{j=1}^{n} w_{ij} \cdot (f_i - f_j)^2$$

At this point, let us examine some *interpretations* of the vector f in terms of the underlying graph partitioning. Let us consider the case in which each f_i is drawn from the set $\{0, 1\}$, and this determines a two-way partition by labeling each node either 0 or 1. The particular partition to which the node i belongs is defined by the corresponding label. Note that the expansion of the expression $f \cdot L \cdot f^T$ from Lemma 9.3 simply represents the sum of the weights of the edges across the partition defined by f. Thus, the determination of an appropriate value of f for which the function $f \cdot L \cdot f^T$ is minimized also provides us with a good node partitioning. Unfortunately, it is not easy to determine the *discrete values* of f which determine this optimum partitioning. Nevertheless, we will see later in this section that even when we restrict f to real values, this provides us with the intuition necessary to create an effective partitioning.

An immediate observation is that the indicator vector $f = [1 \ldots 1]$ is an eigenvector with a corresponding eigenvalue of 0. We note that $f = [1 \ldots 1]$ must be an eigenvector, since L is positive semi-definite and $f \cdot L \cdot f^T$ can be 0 only for eigenvectors with 0 eigenvalues. This observation can be generalized further in order to determine the number of connected components in the graph. We make the following observation.

Lemma 9.4. *The number of (linearly independent) eigenvectors with zero eigenvalues for the Laplacian matrix L is equal to the number of connected components in the underlying graph.*

Proof: Without loss of generality, we can order the vertices corresponding to the particular connected component that they belong to. In this case, the Laplacian matrix takes on the following *block form*, which is illustrated below for the case of three connected components.

$$L = \begin{matrix} L_1 & 0 & 0 \\ 0 & L_2 & 0 \\ 0 & 0 & L_3 \end{matrix}$$

Each of the blocks L_1, L_2 and L_3 is a Laplacian itself of the corresponding component. Therefore, the corresponding indicator vector for that component is an eigenvector with corresponding eigenvalue 0. The result follows. ☐

We observe that connected components are the most obvious examples of clusters in the graph. Therefore, the determination of eigenvectors corresponding to zero eigenvalues provides us information about this (relatively rudimentary set) of clusters. Broadly speaking, it may not be possible to glean such clean membership behavior from the other eigenvectors. One of the problems is that other than this particular rudimentary set of eigenvectors (which correspond to the connected components), the vector components of the other eigenvectors are drawn from the real domain rather than the discrete $\{0, 1\}$ domain. Nevertheless, because of the nature of the natural interpretation of $f \cdot L \cdot f^T$ in terms of the weights of the edges across nodes with very differing values of f_i, it is natural to cluster together nodes for which the values of f_i are as similar as possible across any particular eigenvector on the average. This provides us with the intuition necessary to define an effective spectral clustering algorithm, which partitions the data set into k clusters for any arbitrary value of k. The algorithm is as follows:

- Determine the k eigenvectors with the smallest eigenvalues. Note that each eigenvector has as many components as the number of nodes. Let the component of the jth eigenvector for the ith node be denoted by p_{ij}.

- Create a new data set with as many records as the number of nodes. The ith record in this data set corresponds to the ith node, and has k components. The record for this node is simply the eigenvector components for that node, which are denoted by $p_{i1} \dots p_{ik}$.

- Since we would like to cluster nodes with similar eigenvector components, we use any conventional clustering algorithm (e.g. k-means) in order to create k clusters from this data set. Note that the main focus of the approach was to create a *transformation* of a structural clustering algorithm into a more conventional multi-dimensional clustering algorithm, which is easy to solve. The particular choice of the multi-dimensional clustering algorithm is orthogonal to the broad spectral approach.

The above algorithm provides a broad framework for the spectral clustering algorithm. The input parameter for the above algorithm is the number of clusters k. In practice, a number of variations are possible in order to tune the quality of the clusters which are found. Some examples are as follows:

- It is not necessary to use the same number of eigenvectors as the input parameter for the number of clusters. In general, one should use at least as many eigenvectors as the number of clusters to be created. However, the exact number of eigenvectors to be used in order to get the optimum results may vary with the particular data set. This can be known only with experimentation.

- There are other ways of creating *normalized* Laplacian matrices which can provide more effective results in some situations. Some classic examples of such Laplacian matrices in terms of the adjacency matrix W, degree matrix D and the identity matrix I are defined as follows:

$$L^A = I - D^{-(1/2)} \cdot W \cdot D^{-(1/2)}$$
$$L^B = I - D^{-1} \cdot W$$

More details on the different methods which can be used for effective spectral graph clustering may be found in [9].

2.6 Determining Quasi-Cliques

A different way of determining massive graphs in the underlying data is that of determining *quasi-cliques*. This technique is different from many other partitioning algorithms, in that it focuses on definitions which maximize edge densities *within a partition*, rather than minimizing edge densities across partitions. A clique is a graph in which every pair of nodes are connected by an edge. A quasi-clique is a relaxation on this concept, and is defined by imposing a lower bound on the degree of each vertex in the given set of nodes. Specifically, we define a γ-quasiclique is as follows:

Definition 9.5. *A k-graph ($k \geq 1$) G is a γ-quasiclique if the degree of each node in the corresponding sub-graph of vertices is at least $\gamma \cdot k$.*

The value of γ always lies in the range $(0, 1]$. We note that by choosing $\gamma = 1$, this definition reverts to that of standard cliques. Choosing lower values of γ allows for the relaxations which are more true in the case of real applications. This is because we rarely encounter complete cliques in real applications, and at least some edges within a dense subgraph would always be missing. A vertex is said to be critical, if its degree in the corresponding subgraph is the smallest integer which is at least equal to $\gamma \cdot k$.

The earliest piece of work on this problem is from [1] The work in [1] uses a greedy randomized adaptive search algorithm GRASP, to find a quasi-clique with the maximum size. A closely related problem is that of finding finding *frequently occurring cliques* in *multiple data sets*. In other words, when multiple graphs are obtained from different data sets, some dense subgraphs occur frequently together in the different data sets. Such graphs help in determining *important dense patterns of behavior in different data sources*. Such techniques find applicability in mining important patterns in graphical representations of customers. The techniques are also helpful in mining cross-graph quasi-cliques in gene expression data. A description of the application of the technique to the problem of gene-expression data may be found in [33]. An efficient algorithm for determining cross graph quasi-cliques was proposed in [32]. The main restriction of the work proposed in [32] is that the support threshold for the algorithms is assumed to be 100%. This restriction has been relaxed in subsequent work [43]. The work in [43] examines the problem of mining frequent closed quasi-cliques from a graph database with arbitrary support thresholds. In [31] a multi-graph version of the quasi-clique problem was explored. However, instead of finding the complete set of quasi-cliques in the graph, they proposed an approximation algorithm to cover all the vertices in the graph with a minimum number of p-quasi-complete subgraphs. Thus, this technique is more suited for summarization of the overall graph with a smaller number of densely connected subgraphs.

2.7 The Case of Massive Graphs

A closely related problem is that of dense subgraph determination in massive graphs. This problem is frequently encountered in large graph data sets. For example, the problem of determining large subgraphs of web graphs was studied in [5, 22]. A min-hash approach was first used in [5] in order to determine syntactically related clusters. This paper also introduces the advantages of using a min-hash approach in the context of graph clustering. Subsequently, the approach was generalized to the case of large dense graphs with the use of recursive application of the basic min-hash algorithm.

The broad idea in the min-hash approach is to represent the outlinks of a particular node as sets. Two nodes are considered similar, if they share many outlinks. Thus, consider a node A with an outlink set S_A and a node B with outlink set S_B. Then the similarity between the two nodes is defined by the *Jaccard coefficient*, which is defined as $\frac{S_A \cap S_B}{S_A \cup S_B}$. We note that explicit enumeration of all the edges in order to compute this can be computationally inefficient. Rather, a *min-hash approach* is used in order to perform the estimation. This *min-hash approach* is as follows. We sort the universe of nodes in a random order. For any set of nodes in random sorted order, we determine the first node $First(A)$ for which an outlink exists from A to $First(A)$. We also determine the first node $First(B)$ for which an outlink exists from B to $First(B)$. It can be shown that the Jaccard coefficient is an unbiased estimate of the probability that $First(A)$ and $First(B)$ are the same node. By repeating this process over different permutations over the universe of nodes, it is possible to accurately estimate the Jaccard coefficient. This is done by using a constant number of permutations c of the node order. The actual permutations are implemented by associated c different randomized hash values with each node. This creates c sets of hash values of size n. The sort-order for any particular set of hash-values defines the corresponding permutation order. For each such permutation, we store the minimum node index of the outlink set. Thus, for each node, there are c such minimum indices. This means that, for each node, a fingerprint of size c can be constructed. By comparing the fingerprints of two nodes, the Jaccard coefficient can be estimated. This approach can be further generalized with the use of every s element set contained entirely with S_A and S_B. Thus, the above description is the special case when s is set to 1. By using different values of s and c, it is possible to design an algorithm which distinguishes between two sets that are above or below a certain threshold of similarity.

The overall technique in [22] first generates a set of c shingles of size s for each node. The process of generating the c shingles is extremely straightforward. Each node is processed independently. We use the min-wise hash function approach in order to generate subsets of size s from the outlinks at each node. This results in c subsets for each node. Thus, for each node, we have a set of c shingles. Thus, if the graph contains a total of n nodes, the total size of this shingle fingerprint is $n \times c \times sp$, where sp is the space required for each shingle. Typically sp will be $O(s)$, since each shingle contains s nodes. For each distinct shingle thus created, we can create a list of nodes which contain it. In general, we would like to determine groups of shingles which contain a large number of common nodes. In order to do so, the method in [22] performs a second-order shingling in which the meta-shingles are created from the shingles. Thus, this further compresses the graph in a data structure of size $c \times c$. This is essentially a constant size data structure. We note that this group of meta-shingles have the the property that they contain a large number of common nodes. The dense subgraphs can then be extracted from these meta-shingles. More details on this approach may be found in [22].

The min-hash approach is frequently used for graphs which are extremely large and cannot be easily processed by conventional quasi-clique mining algorithms. Since the min-hash approach summarizes the massive graph in a small amount of space, it is particularly useful in leveraging the small space representation for a variety of query-processing techniques. Examples of such applications include the web graph and social networks. In the case of web graphs, we desire to determine closely connected clusters of web pages with similar content. The related problem in social networks is that of finding closely related communities. The min-hash approach discussed in [5, 22] precisely helps us achieve this goal, because we can process the summarized min-hash structure in a variety of ways in order to extract the important communities from the summarized structure. More details of this approach may be found in [5, 22].

3. Clustering Graphs as Objects

In this section, we will discuss the problem of clustering *entire graphs* in a *multi-graph database*, rather than examining the node clustering problem within a single graph. Such situations are often encountered in the context of XML data, since each XML document can be regarded as a structural record, and it may be necessary to create clusters from a large number of such objects. We note that XML data is quite similar to graph data in terms of how the data is organized structurally. The attribute values can be treated as graph labels and the corresponding semi-structural relationships as the edges. In has been shown in [2, 10, 28, 29] that this structural behavior can be leveraged in order to create effective clusters.

3.1 Extending Classical Algorithms to Structural Data

Since we are examining entre graphs in this version of the clustering problem, the problem simply boils down to that of clustering arbitrary *objects*, where the objects in this case have structural characteristics. Many of the conventional algorithms discussed in [24] (such as k-means type partitional algorithms and hierarchical algorithms can be extended to the case of graph data. The main changes required in order to extend these algorithms are as follows:

- Most of the underlying classical algorithms typically use some form of distance function in order to measure similarity. Therefore, we need appropriate measures in order to define similarity (or distances) between structural objects.

- Many of the classical algorithms (such as k-means) use *representative objects* such as centroids in critical intermediate steps. While this is straightforward in the case of multi-dimensional objects, it is much more challenging in the case of graph objects. Therefore, appropriate methods need to be designed in order to create representative objects. Furthermore, in some cases it may be difficult to create representatives in terms of single objects. We will see is that it is often more robust to use *representative summaries* of the underlying objects.

There are two main classes of conventional techniques, which have been extended to the case of structural objects. These techniques are as follows:

- **Structural Distance-based Approach:** This approach computes structural distances between documents and uses them in order to compute clusters of documents. One of the earliest work on clustering tree structured data is the *XClust algorithm* [28], which was designed to cluster XML schemas in order for efficient integration of large numbers of Document Type Definitions (DTDs) of XML sources. It adopts the agglomerative hierarchical clustering method which starts with clusters of single DTDs and gradually merges the two most similar clusters into one larger cluster. The similarity between two DTDs is based on their element similarity, which can be computed according to the semantics, structure, and context information of the elements in the corresponding DTDs. One of the shortcomings of the XClust algorithm is that it does not make full use of the structure information of the DTDs, which is quite important in the context of clustering tree-like structures. The method in [7] computes similarity measures based on the structural edit-distance between documents. This edit-distance is used in order to compute the distances between clusters of documents.

  S-GRACE is hierarchical clustering algorithm [29]. In [29], an XML document is converted to a structure graph (or s-graph), and the distance between two XML documents is defined according to the number of the common element-subelement relationships, which can capture better structural similarity relationships than the tree edit distance in some cases [29].

- **Structural Summary Based Approach:** In many cases, it is possible to create summaries from the underlying documents. These summaries are used for creating groups of documents which are similar to these summaries. The first summary-based approach for clustering XML documents was presented in [10]. In [10], the XML documents are modeled as rooted ordered labeled trees. A framework for clustering XML documents by using structural summaries of trees is presented. The aim is to improve algorithmic efficiency without compromising cluster quality.

A second approach for clustering XML documents is presented in [2]. This technique is a partition-based algorithm. The primary idea in this approach is to use frequent-pattern mining algorithms in order to determine the summaries of frequent structures in the data. The technique uses a k-means type approach in which each cluster center comprises a set of frequent patterns which are local to the partition for that cluster. The frequent patterns are mined using the documents assigned to a cluster center in the last iteration. The documents are then further reassigned to a cluster center based on the average similarity between the document and the newly created cluster centers from the local frequent patterns. In each iteration the document-assignment and the mined frequent patterns are iteratively re-assigned, until the cluster centers and document partitions converge to a final state. It has been shown in [2] that such a structural summary based approach is significantly superior to a similarity function based approach as presented in [7]. The method of also superior to the structural approach in [10] because of its use of more robust representations of the underlying structural summaries.

Since the most recent algorithm is the structural summary method discussed in [2], we will discuss this in more detail in the next section.

3.2   The XProj Approach

In this section, we will present XProj, which is a summary-based approach for clustering of XML documents. The pseudo-code for clustering of XML documents is illustrated in Figure 9.1. The primary approach is to use a sub-structural modification of a partition based approach in which the clusters of documents are built around groups of representative sub-structures. Thus, instead of a single representative of a partition-based algorithm, we use a *sub-structural set representative* for the structural clustering algorithm. Initially, the document set $\mathcal{D}$ is randomly divided into k partitions with equal size, and the sets of sub-structure representatives are generated by mining frequent sub-structures of size l from these partitions. In each iteration, the sub-structural representatives (of a particular size, and a particular support level) of a given partition are the frequent structures from that partition. These structural representatives are used to partition the document collection and vice-versa. We note that this can be a potentially expensive operation because of the determination of frequent substructures; in the next section, we will illustrate an interesting way to speed it up. In order to actually partition the document collection, we calculate the number of nodes in a document which are covered by each sub-structural set representative. A larger coverage corresponds to a greater level of similarity. The aim of this approach is that the algorithm will determine the most important *localized sub-structures* over time. This

Algorithm XProj(Document Set: $\mathcal{D}$, Minimum Support:
min_sup, Structural Size: l, NumClusters: k)
begin
  Initialize representative sets $\mathcal{S}_1 \ldots \mathcal{S}_k$;
  while ($convergence criterion$ =false)
  begin
  Assign each document $D \in \mathcal{D}$ to one of the sets in
      $\{\mathcal{S}_1 \ldots \mathcal{S}_k\}$ using coverage based similarity criterion;
      /* Let the corresponding document partitions be
          denoted by $\mathcal{M}_1 \ldots \mathcal{M}_k$; */
  Compute the freq. substructures of size l from each
      set $\mathcal{M}_i$ using sequential transformation paradigm;
  if ($|\mathcal{M}_i| \times min\_sup) \geq 1$
      set $\mathcal{S}_i$ to frequent substructures of size l from $\mathcal{M}_i$;
      /* If ($|\mathcal{M}_i| \times min\_sup) < 1$, $\mathcal{S}_i$ remains unchanged; */
  end;
end

Figure 9.1. The Sub-structural Clustering Algorithm (High Level Description)

is analogous to the projected clustering approach which determines the most
important localized projections over time. Once the partitions have been computed, we use them to re-compute the representative sets. These re-computed
representative sets are defined as the frequent sub-structures of size l from
each partition. Thus, the representative set S_i is defined as the substructural
set from the partition $\mathcal{M}_i$ which has size l, and which has absolute support
no less than ($|\mathcal{M}_i| \times min\_sup$). Thus, the newly defined representative set
S_i also corresponds to the local structures which are defined from the partition $\mathcal{M}_i$. Note that if the partition $\mathcal{M}_i$ contains too few documents such that
($|\mathcal{M}_i| \times min\_sup$) < 1, the representative set S_i remains unchanged.

Another interesting observation is that the similarity function between a
document and a given representative set is defined by the number of nodes
in the document which are covered by that set. This makes the similarity function more sensitive to the underlying projections in the document structures.
This leads to more robust similarity calculations in most circumstances.

In order to ensure termination, we need to design a convergence criterion.
One useful criterion is based on the increase of the average sub-structural
self-similarity over the k partitions of documents. Let the partitions of documents with respect to the current iteration be $\mathcal{M}_1 \ldots \mathcal{M}_k$, and their corresponding frequent sub-structures of size l be $\mathcal{S}_1 \ldots \mathcal{S}_k$ respectively. Then,
the average sub-structural self-similarity at the end of the current iteration

is $\Phi = \sum_{i=1}^{k} \Delta(\mathcal{M}_i, \mathcal{S}_i)/k$. Similarly, let the average sub-structural self-similarity at the end of the the previous iteration be Φ'. In the beginning of the next iteration, the algorithm computes the increase of the average sub-structural self-similarity, $\Phi - \Phi'$, and checks if it is smaller than a user-specified threshold ϵ. If not, the algorithm proceeds with another iteration. Otherwise, the algorithm terminates. In addition, an upper bound on the number of iterations is imposed. This is done in order to effectively handle situations in which the threshold ϵ is chosen to be too small. Two further issues need to be implemented in order to effectively use the underlying algorithm:

- We need to determine effective methods for determining the similarity between a given document, and a group of other documents. Techniques for computing the similarity are discussed in [2].

- We need to determine frequent structural patterns in the underlying documents. This can be a huge challenge in many applications, especially since structural data is far more challenging to mine than transactional data. It has been shown in [2], how sequential pattern mining algorithms can be adapted to the case of structural data. The broad idea is to *flatten out* the tree structure into a sequential pattern by using a pre-order traversal. Then the clustering is performed on the resulting sequential patterns. It has been shown [2] that such an approach is able to retain most of the structural information in the data, while introducing some spurious relations. The overall approach has been shown in [2] to be experimentally quite effective.

It has been shown in [2], that this method is far more effective than competing techniques such as those discussed in [10, 29].

4. Applications of Graph Clustering Algorithms

Graph clustering algorithms find numerous applications in the literature. As discussed in this chapter, graph mining algorithms fall into the categories of node clustering and more generally object-based clustering algorithms. Object-based clustering algorithms are similar to general clustering algorithms in the literature, except that we use the underlying graphs as records rather than standard multi-dimensional attributes. Such algorithms are useful in a number of data domains such as molecular biology, chemical graphs, and XML data. In general, any data domain which can represent the underlying records in terms of compact graphs can benefit from such algorithms.

Node clustering algorithms can be used for a variety of real applications such as facility location. These algorithms can also be used for clustering with arbitrary distance functions between groups of objects. These algorithms are more general than those used for clustering records with the use of multi-dimensional distance functions.

Node clustering algorithms are closely related to the problem of graph partitioning. These methods are particularly useful for applications which need to determine dense regions of the graphs. The determination of dense regions of the graph is closely related to the problem of graph summarization and dimensionality reduction. The process of dimensionality reduction on graphs can be used in order to represent them in a small space, so that they can be used effectively for indexing and retrieval. Furthermore, compressed graphs can be used in a variety of applications in which it is desirable to use the summary behavior in order to *estimate* the approximate structural properties of the network. These estimates can then be subsequently refined for more exact results at a later stage. Some specific applications for which clustering algorithms may be leveraged are as follows:

4.1   Community Detection in Web Applications and Social Networks

Many web applications and social networks can be typically represented as massive graphs. For example, the structure of the web is itself a graph [22, 30, 34], in which nodes represent web pages, and hyperlinks represent the edges of this graph. Similarly social networks are graphs in which nodes represent the members of the social network, and the friendship relationship between members represent the corresponding links. Node clustering algorithms are a natural fit for community detection in massive graphs. The communities have natural interpretations in the context of a variety of web applications:

- For the case of web applications such as web sites, communities typically refer to communities of closely linked pages. Such communities are typically linked because of common material in terms of topic, or similar interests in terms of readership.

- For the case of social networks, communities refer to groups of members who may know each other very well, and may therefore be closely linked with one another. This is useful in determining important associations in the underlying social network.

- Blogging communities often behave like social networks, and contain links between related blogs. The techniques discussed in this chapter are also useful for determining the closely related blogs with the use of community detection methods.

Many of the node clustering applications discussed in this chapter are used in the context of social networks [22, 30, 34]. The min-hash approach [5, 22] is commonly used when the underlying graph is massive in nature, such as that in the case of the web. This is because the min-hash approach is able to summarize the graph in a very small amount of space. This is very useful for practical applications in which it may be possible to represent the entire graph on disk. For example, the size of the web graph is so large, that it may not even be possible to store it on disk without the use of add-ons onto standard desktop hardware. Such situations lead to further constraints during the mining process, which are handled quite well by min-hash style approaches. This is because the min-hash summary is of extremely small size compared to the size of the graph itself. This compressed representation can even be maintained in main memory and used to determine the underlying communities in the network directly. It has been shown in [5, 22], that such an approach is able to determine communities of very high quality.

4.2 Telecommunication Networks

Large telecommunication companies may have millions of customers who may make billions of phone calls to one another over a period of time. In this case, the individual phone numbers may be represented as node, and phone calls may be represented as edges. In such cases, it may be desirable to determine groups of customers who call each other frequently. This information can be very useful for target marketing purposes. Furthermore, we note that the graphs in a tele-communication network are represented in the form of *edge streams*, since the edges may be received continuously over time. These result in even greater challenges from the point of view of analysis, since the edges cannot be explicitly stored on disk. The methods discussed in [22] are particularly useful in such scenarios.

4.3 Email Analysis

An interesting application in the context of the Enron crisis was to determine important email interactions between groups of Enron employees. In this case, the individuals are represented as nodes, and the emails sent between them are represented as edges. Node clustering algorithms are very useful in order to isolate dense email interactions between different groups of customers. This approach can be used for a variety of intelligence applications such as that of determining suspicious communities in groups of interactions.

5. Conclusions and Future Research

In this chapter, we presented a review of the commonly known algorithms for clustering graph data. The problem of clustering graphs has been widely studied in the literature, because of its application to a variety of data mining and data management problems. Graph clustering algorithms are of two types:

- **Node Clustering Algorithms:** In this case, we attempt to partition the graph into groups of clusters, so that each cluster contains groups of nodes which are densely connected. These densely connected groups of nodes may often provide significant information about how the entities in the underlying graph are inter-connected with one another.

- **Graph Clustering Algorithms:** In this case, we have complete graphs available, and we wish to determine the clusters with the use of the structural information in the underlying graphs. Such cases are often encountered in the case of XML data, which are commonly encountered in many real domains.

We provided an overview of the different clustering algorithms available, and the tradeoffs with the use of different methods. The major challenges that remain in the area of graph clustering are as follows:

- **Clustering Massive Data Sets:** In some cases, the data sets containing the graphs may be so large that they may be held only on disk. For example, if we have a dense graph containing 10^7 nodes, then the number of edges may be as high as 10^{13}. In such cases, it may not even be possible to store the graph effectively on disk. In cases in which the graph can be stored on disk, it is critical that the algorithm should be designed in order to take the disk-resident behavior of the underlying data into account. This is especially challenging in the case of graph data sets, because the structural behavior of the graph interferes with our ability to process the edges sequentially for many applications. In cases in which the graph is too large to store on disk, it is essential to design summary structures which can effectively store the underlying structural behavior of the graph. This stored summary can then be used effectively for graph clustering algorithms.

- **Clustering Graph Streams:** In this case, we have large graphs which are received as edge streams. Such graphs are more challenging, since a given edge cannot be processed more than once during the computation process. In such cases, summary structures need to be designed in order to facilitate an effective clustering process. These summary structures may be utilized in order to determine effective clusters in the underlying data. This approach is similar to the case discussed above in which the size of the graph is too large to store on disk.

In addition, techniques need to be designed for interfacing clustering algorithms with traditional database management techniques. In order to achieve this goal, effective representations and query languages need to be designed for graph data. This is a new and emerging area of research, and can be leveraged upon in order to further improve the effectiveness of graph algorithms.

References

[1] J. Abello, M. G. Resende, S. Sudarsky, Massive quasi-clique detection. *Proceedings of the 5th Latin American Symposium on Theoretical Informatics (LATIN)*, pp. 598-612, 2002.

[2] C. Aggarwal, N. Ta, J. Feng, J. Wang, M. J. Zaki. XProj: A Framework for Projected Structural Clustering of XML Documents, *KDD Conference*, 2007.

[3] R. Agrawal, A. Borgida, H.V. Jagadish. Efficient Maintenance of transitive relationships in large data and knowledge bases, *ACM SIGMOD Conference*, 1989.

[4] R. Ahuja, J. Orlin, T. Magnanti. Network Flows: Theory, Algorithms, and Applications, *Prentice Hall*, Englewood Cliffs, NJ, 1992.

[5] A. Z. Broder, M. Charikar, A. Frieze, and M. Mitzenmacher, Syntactic clustering of the web, *WWW Conference, Computer Networks*, 29(8–13):1157–1166, 1997.

[6] D. Chakrabarti, Y. Zhan, C. Faloutsos R-MAT: A Recursive Model for Graph Mining. *SDM Conference*, 2004.

[7] S.S. Chawathe. Comparing Hierachical data in external memory. *Very Large Data Bases Conference*, 1999.

[8] J. Cheriyan, T. Hagerup, K. Melhorn An $O(n^3)$-time maximum-flow algorithm, *SIAM Journal on Computing*, Volume 25 , Issue 6, pp. 1144 – 1170, 1996.

[9] F. Chung,. Spectral graph theory. *Washington: Conference Board of the Mathematical Sciences*, 1997.

[10] T. Dalamagas, T. Cheng, K. Winkel, T. Sellis. Clustering XML Documents Using Structural Summaries. Information Systems, Elsevier, January 2005.

[11] J. Cheng, J. Xu Yu, X. Lin, H. Wang, and P. S. Yu, Fast Computing Reachability Labelings for Large Graphs with High Compression Rate, *EDBT Conference*, 2008.

[12] J. Cheng, J. Xu Yu, X. Lin, H. Wang, and P. S. Yu, Fast Computation of Reachability Labelings in Large Graphs, *EDBT Conference*, 2006.

[13] E. Cohen. Size-estimation framework with applications to transitive closure and reachability, *Journal of Computer and System Sciences*, v.55 n.3, p.441-453, Dec. 1997.

[14] E. Cohen, E. Halperin, H. Kaplan, and U. Zwick, Reachability and distance queries via 2-hop labels, *ACM Symposium on Discrete Algorithms*, 2002.

[15] D. Cook, L. Holder, Mining Graph Data, *John Wiley & Sons Inc*, 2007.

[16] E. W. Dijkstra, A note on two problems in connection with graphs. *Numerische Mathematik*, 1 (1959), S. 269-271.

[17] M. Faloutsos, P. Faloutsos, C. Faloutsos, On Power Law Relationships of the Internet Topology. *SIGCOMM Conference*, 1999.

[18] P.-O. Fjallstrom, Algorithms for Graph Partitioning: A Survey, Linkoping Electronic Articles in Computer and Information Science Vol 3, no 10, 1998.

[19] G. Flake, R. Tarjan, M. Tsioutsiouliklis. Graph Clustering and Minimum Cut Trees, *Internet Mathematics*, 1(4), 385–408, 2003.

[20] I. Freeman. Centrality in Social Networks, *Social Networks*, 1, 215–239, 1979.

[21] M. S. Garey, D. S. Johnson. Computers and Intractability: A Guide to the Theory of NP-completeness,*W. H. Freeman*, 1979.

[22] D. Gibson, R. Kumar, A. Tomkins, Discovering Large Dense Subgraphs in Massive Graphs, *VLDB Conference*, 2005.

[23] M. Girvan, M. Newman. Community Structure in Social and Biological Networks, *Proceedings of the National Academy of Science*, 99, 7821–7826, 2002.

[24] A. Jain and R. Dubes, *Algorithms for Clustering Data*, Prentice Hall, New Jersey, 1998.

[25] H. Kashima, K. Tsuda, A. Inokuchi. Marginalized Kernels between Labeled Graphs, *ICML*, 2003.

[26] B.W. Kernighan, S. Lin. An efficient heuristic procedure for partitioning graphs, *Bell System Tech. Journal*, vol. 49, Feb. 1970, pp. 291-307.

[27] T. Kudo, E. Maeda, Y. Matsumoto. An Application of Boosting to Graph Classification, *NIPS Conf.* 2004.

[28] M. Lee, W. Hsu, L. Yang, X. Yang. XClust: Clustering XML Schemas for Effective Integration. *ACM Conference on Information and Knowledge Management*, 2002

[29] W. Lian, D.W. Cheung, N. Mamoulis, S. Yiu. An Efficient and Scalable Algorithm for Clustering XML Documents by Structure, *IEEE Transactions on Knowledge and Data Engineering*, Vol 16, No. 1, 2004.

[30] R. Kumar, P Raghavan, S. Rajagopalan, D. Sivakumar, A. Tomkins, E. Upfal. The Web as a Graph. *ACM PODS Conference*, 2000.

[31] M. Matsuda et al. Classifying molecular sequences using a linkage graph with their pairwise similarities. *Theoretical Computer Science*, 210(2):305-325, 1999.

[32] J. Pei, D. Jiang, A. Zhang. On Mining Cross-Graph Quasi-Cliques, *ACM KDD Conference*, 2005.

[33] J. Pei, D. Jiang, A. Zhang. Mining Cross-Graph Quasi-Cliques in Gene Expression and Protein Interaction Data, *ICDE Conference*, 2005.

[34] S. Raghavan, H. Garcia-Molina. Representing web graphs. *ICDE Conference*, pages 405-416, 2003.

[35] M. Rattigan, M. Maier, D. Jensen: Graph Clustering with Network Sructure Indices. *ICML*, 2007.

[36] M. Rattigan, M. Maier, D. Jensen: Using structure indices for approximation of network properties. *ACM KDD Conference*, 2006.

[37] A. A. Tsay, W. S. Lovejoy, David R. Karger, Random Sampling in Cut, Flow, and Network Design Problems, *Mathematics of Operations Research*, 24(2):383-413, 1999.

[38] H. Wang, H. He, J. Yang, J. Xu-Yu, P. Yu. Dual Labeling: Answering Graph Reachability Queries in Constant Time. *ICDE Conference*, 2006.

[39] X. Yan, J. Han. CloseGraph: Mining Closed Frequent Graph Patterns, *ACM KDD Conference*, 2003.

[40] X. Yan, H. Cheng, J. Han, and P. S. Yu, Mining Significant Graph Patterns by Scalable Leap Search, *SIGMOD Conference*, 2008.

[41] X. Yan, P. S. Yu, and J. Han, Graph Indexing: A Frequent Structure-based Approach, *SIGMOD Conference*, 2004.

[42] M. J. Zaki, C. C. Aggarwal. XRules: An Effective Structural Classifier for XML Data, *KDD Conference*, 2003.

[43] Z. Zeng, J. Wang, L. Zhou, G. Karypis, Out-of-core Coherent Closed Quasi-Clique Mining from Large Dense Graph Databases, *ACM Transactions on Database Systems*, Vol 31(2), 2007.

Chapter 10

A SURVEY OF ALGORITHMS FOR DENSE SUBGRAPH DISCOVERY

Victor E. Lee

Department of Computer Science
Kent State University
Kent, OH 44242
vlee@cs.kent.edu

Ning Ruan

Department of Computer Science
Kent State University
Kent, OH 44242
nruan@cs.kent.edu

Ruoming Jin

Department of Computer Science
Kent State University
Kent, OH 44242
jin@cs.kent.edu

Charu Aggarwal

IBM T.J. Watson Research Center
Yorktown Heights, NY 10598
charu@us.ibm.com

Abstract   In this chapter, we present a survey of algorithms for dense subgraph discovery. The problem of dense subgraph discovery is closely related to clustering though the two problems also have a number of differences. For example, the problem of clustering is largely concerned with that of finding a fixed partition in the data, whereas the problem of dense subgraph discovery defines these dense components in a much more flexible way. The problem of dense subgraph discovery

C.C. Aggarwal and H. Wang (eds.), *Managing and Mining Graph Data*,
Advances in Database Systems 40, DOI 10.1007/978-1-4419-6045-0_10,
© Springer Science+Business Media, LLC 2010

may wither be defined over single or multiple graphs. We explore both cases. In the latter case, the problem is also closely related to the problem of the frequent subgraph discovery. This chapter will discuss and organize the literature on this topic effectively in order to make it much more accessible to the reader.

Keywords: Dense subgraph discovery, graph clustering

1.   Introduction

In almost any network, density is an indication of importance. Just as someone reading a road map is interesting in knowing the location of the larger cities and towns, investigators who seek information from abstract graphs are often interested in the dense components of the graph. Depending on what properties are being modeled by the graph's vertices and edges, dense regions may indicate high degrees of interaction, mutual similarity and hence collective characteristics, attractive forces, favorable environments, or critical mass.

From a theoretical perspective, dense regions have many interesting properties. Dense components naturally have small diameters (worst case shortest path between any two members). Routing within these components is rapid. A simple strategy also exists for global routing. If most vertices belong to a dense component, only a few selected inter-hub links are needed to have a short average distance between any two arbitrary vertices in the entire network. Commercial airlines employ this hub-based routing scheme. Dense regions are also robust, in the sense that many connections can be broken without splitting the component. A less well-known but equally important property of dense subgraphs comes from percolation theory. If a graph is sufficiently dense, or equivalently, if messages are forwarded from one node to its neighbors with higher than a certain probability, then there is very high probability of propagating a message across the diameter of the graph [20]. This fact is useful in everything from epidemiology to marketing.

Not all graphs have dense components, however. A sparse graph may have few or none. In order to understand this issue, we first need to define a formal notion of the words 'dense' and 'sparse'. We will address this issue shortly. A uniform graph is either entirely dense or not dense at all. Uniform graphs, however, are rare, usually limited to either small or artificially created ones. Due to the usefulness of dense components, it is generally accepted that their existence is the rule rather than the exception in nature and in human-planned networks [39].

Dense components have been identified in and have enhanced understanding of many types of networks; among the best-known are social networks [53, 44], the World Wide Web [30, 17, 11], financial markets [5], and biological systems [26]. Much of the early motivation, research, and nomenclature regarding dense components was in the field of social network analysis. Even before the advent of computers, sociologists turned to graph theory to formulate models for the concept of social cohesion. Clique, K-core, K-plex, and K-club are metrics originally devised to measure social cohesiveness [53]. It is not surprising that we also see dense components in the World Wide Web. In many ways, the Web is simply a virtual implementation of traditional direct human-human social networks.

Today, the natural sciences, the social sciences, and technological fields are all using network and graph analysis methods to better understand complex systems. Dense component discovery and analysis is one important aspect of network analysis. Therefore, readers from many different backgrounds will benefit from understanding more about the characteristics of dense components and some of the methods used to uncover them.

In the next section, we outline the graph terminology and define the fundamental measures of density to be used in the rest of the chapter. Section 3 categorizes the algorithmic approaches and presents representative implementations in more detail. Section 4 expands the topic to consider frequently-occurring dense components in a set of graphs. Section 5 provides examples of how these techniques have been applied in various scientific fields. Section 6 concludes the chapter with a look to the future.

2. Types of Dense Components

Different applications find different definitions of *dense component* to be useful. In this section, we outline the many ways to define a dense component, categorizing them by their important features. Understanding these features of the various types of components are valuable for deciding which type of component to pursue.

2.1 Absolute vs. Relative Density

We can divide density definitions into two classes, absolute density and relative density. An absolute density measure establishes rules and parameter values for what constitutes a dense component, independent of what is outside the component. For example, we could say that we are only interested in cliques, fully-connected subgraphs of maximum density. Absolute density measures take the form of relaxations of the pure clique measure.

On the other hand, a relative density measure has no preset level for what is sufficiently dense. It compares the density of one region to another, with the goal of finding the densest regions. To establish the boundaries of components, a metric typically looks to maximize the difference between intra-component connectedness and inter-component connectedness. Often but not necessarily, relative density techniques look for a user-defined number k densest regions. The alert reader may have noticed that relative density discovery is closely related to clustering and in fact shares many features with it.

Since this book contains another chapter dedicated to graph clustering, we will focus our attention on absolute density measures. However, we will have more so say about the relationship between clustering and density at the end of this section.

2.2   Graph Terminology

Let $G(V, E)$ be a graph with $|V|$ vertices and $|E|$ edges. If the edges are weighted, then $w(u)$ is the weight of edge u. We treat unweighted graphs as the special case where all weights are equal to 1. Let S and T be subsets of V. For an undirected graph, $E(S)$ is the set of induced edges on S: $E(S) = \{(u, v) \in E \,|u, v \in S\}$. Then, H_S is the induced subgraph $(S, E(S))$. Similarly, $E(S, T)$ designates the set of edges from S to T. $H_{S,T}$ is the induced subgraph $(S, T, E(S, T))$. Note that S and T are not necessarily disjoint from each other. If $S \cap T = \emptyset$, $H_{S,T}$ is a bipartite graph. If S and T are not disjoint (possibly $S = T = V$), this notation can be used to represent a directed graph.

A dense component is a maximal induced subgraph which also satisfies some density constraint. A component H_S is maximal if no other subgraph of G which is a superset of H_S would satisfy the density constraints. Table 10.1 defines some basic graph concepts and measures that we will use to define density metrics.

Table 10.1. Graph Terminology

Symbol	Description
$G(V, E)$	graph with vertex set V and edge set E
H_S	subgraph with vertex set S and edge set $E(S)$
$H_{S,T}$	subgraph with vertex set $S \cup T$ and edge set $E(S, T)$
$w(u)$	weight of edge u
$N_G(u)$	neighbor set of vertex u in G: $\{v| (u, v) \in E\}$
$N_S(u)$	only those neighbors of vertex u that are in S: $\{v| (u, v) \in S\}$
$\delta_G(u)$	(weighted) degree of u in $G : \sum_{v \in N_G(u)} w(v)$
$\delta_S(u)$	(weighted) degree of u in $S : \sum_{v \in N_S(u)} w(v)$
$d_G(u, v)$	shortest (weighted) path from u to v traversing any edges in G
$d_S(u, v)$	shortest (weighted) path from u to v traversing only edges in $E(S)$

We now formally define the **density of S**, $den(S)$, as the ratio of the total weight of edges in $E(S)$ to the number of possible edges among $|S|$ vertices. If the graph is unweighted, then the numerator is simply the number of actual edges, and the maximum possible density is 1. If the graph is weighted, the maximum density is unbounded. The number of possible edges in an undirected graph of size n is $\binom{n}{2} = n(n-1)/2$. We give the formulas for an undirected graph; the formulas for a directed graph lack the factor of 2.

$$den(S) = \frac{2|E(S)|}{|S|(|S|-1)}$$

$$den_W(S) = \frac{2\sum_{u,v \in S} w(u,v)}{|S|(|S|-1)}$$

Some authors define density as the ratio of the number of edges to the number of *vertices*: $\frac{|E|}{|V|}$. We will refer to this as **average degree of S**.

Another important metric is the **diameter of S**, $diam(S)$. Since we have given two different distance measures, d_S and d_G, we accordingly offer two different diameter measures. The first is the standard one, in which we consider only paths within S. The second permits paths to stray outside S, if it offers a shorter path.

$$diam(S) = max\{d_S(u,v)| \ u,v \in S\}$$
$$diam_G(S) = max\{d_G(u,v)| \ u,v \in S\}$$

2.3   Definitions of Dense Components

We now present a collection of measures that have been used to define dense components in the literature (Table 10.2). To focus on the fundamentals, we assume unweighted graphs. In a sense, all dense components are either cliques, which represent the ideal, or some relaxation of the ideal. There relaxations fall into three categories: density, degree, and distance. Each relaxation can be quantified as either a percentage factor or a subtractive amount. While most of there definitions are widely-recognized standards, the name *quasi-clique* has been applied to any relaxation, with different authors giving different formal definitions. Abello [1] defined the term in terms of overall edge density, without any constraint on individual vertices. This offers considerable flexibility in the component topology. Several other authors [36, 32, 33] have opted to define quasi-clique in terms of minimum degree of each vertex. Li et al. [32] provide a brief overview and comparison of quasi-cliques. In our table, when the authorship of a specific metric can be traced, it is given. Our list is not exhaustive; however, the majority of definitions can be reduced to some combination of density, degree, and diameter.

Note that in unweighted graphs, cliques have a density of 1. Density-based quasi-cliques are only defined for unweighted graphs. We use the term *Kd-clique* instead of Mokken's original name *K-clique*, because K-clique is already defined in the mathematics and computer science communities to mean a clique with k vertices.

Table 10.2. Types of Dense Components

Component	Reference	Formal definition	Description
Clique		$\exists(i,j), i \neq j \in S$	Every vertex connects to every other vertex in S.
Quasi-Clique (density-based)	[1]	$den(S) \geq \gamma$	S has at least $\gamma|S|(|S|-1)/2$ edges. Density may be imbalanced within S.
Quasi-Clique (degree-based)	[36]	$\delta_S(u) \geq \gamma * (k-1)$	Each vertex has γ percent of the possible connections to other vertices. Local degree satisfies a minimum. Compare to K-core and K-plex.
K-core	[45]	$\delta_S(u) \geq k$	Every vertex connects to at least k other vertices in S. A clique is a $(k\text{-}1)$-core.
K-plex	[46]	$\delta_S(u) \geq |S| - k$	Each vertex is missing no more than $k-1$ edges to its neighbors. A clique is a 1-plex.
Kd-clique	[34]	$diam_G(S) \leq k$	The shortest path from any vertex to any other vertex is not more than k. An ordinary clique is a 1d-clique. Paths may go outside S.
K-club	[37]	$diam(S) \leq k$	The shortest path from any vertex to any other vertex is not more than k. Paths may not go outside S. Therefore, every K-club is a K-clique.

Figure 10.1, a superset of an illustration from Wasserman and Faust [53], demonstrates each of the dense components that we have defined above.

- Cliques: $\{1,2,3\}$ and $\{2,3,4\}$
- 0.8-Quasi-clique: $\{1,2,3,4\}$ (includes $5/6 > 0.83$ of possible edges)
- 2-Core: $\{1,2,3,4,5,6,7\}$
- 3-Core: none
- 2-Plex: $\{1,2,3,4\}$ (vertices 1 and 3 are missing one edge each)
- 2d-Cliques: $\{1,2,3,4,5,6\}$ and $\{2,3,4,5,6,7\}$ (In the first component, 5 connects to 6 via 7, which need not be a member of the component)
- 2-Clubs: $\{1,2,3,4,5\}$, $\{1,2,3,4,6\}$, and $\{2,3,5,6,7\}$

2.4   Dense Component Selection

When mining for dense components in a graph, a few additional questions must be addressed:

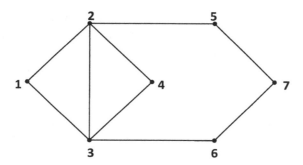

Figure 10.1. Example Graph to Illustrate Component Types

1 **Minimum size** σ: What is the minimum number of vertices in a dense component S? I.e., $|S| \geq \sigma$.

2 **All or top-N?**: One of the following criteria should be applied.

- Select all components which meet the size, density, degree, and distance constraints.

- Select the N highest ranking components that meet the minimum constraints. A ranking function must be established. This can be as simple as one of the same metrics used for minimum constraints (size, density, degree, distance, etc.) or a linear combination of them.

- Select the N highest ranking components, with no minimum constraints.

3 **Overlap**: May two components share vertices?

2.5 Relationship between Clusters and Dense Components

The measures described above set an absolute standard for what constitutes a dense component. Another approach is to find the most dense components on a relative basis. This is the domain of clustering. It may seem that clustering, a thoroughly-studied topic in data mining with many excellent methodologies, would provide a solution to dense component discovery. However, clustering is a very broad term. Readers interested in a survey on clustering may wish to consult either Jain, Murty, and Flynn [24] or Berkhin [8]. In the data mining community, *clustering* refers to the task of assigning similar or nearby items to the same group while assigning dissimilar/distant items to different groups. In most clustering algorithms, similarity is a relative concept; therefore it is potentially suitable for relative density measures. However, not all clustering algorithms are based on density, and not all types of dense components can be discovered with clustering algorithms.

Partitioning refers to one class of clustering problem, where the objective is to assign every item to exactly one group. A k-partitioning requires the result to have k groups. K-partitioning is not a good approach for identifying absolute dense components, because the objectives are at odds. Consider the well-known k-Means algorithm applied to a uniform graph. It will generate k partitions, because it must. However, the partitioning is arbitrary, changing as the seed centroids change.

In *hierarchical clustering*, we construct a tree of clusters. Conceptually, as well as in actual implementation, this can be either agglomerative (bottom-up), where the closest clusters are merged together to form a parent cluster, or divisive (top-down), where a cluster is subdivided into relatively distant child clusters. In basic greedy agglomerative clustering, the process starts by grouping together the two closest items. The pair are now treated as a single item, and the process is repeated. Here, pairwise distance is the density measure, and the algorithm seeks to group together the densest pair. If we use divisive clustering, we can choose to stop subdividing after finding k leaf clusters. A drawback of both hierarchical clustering and partitioning is that they do not allow for a separate "non-dense" partition. Even sparse regions are forced to belong to some cluster, so they are lumped together with their closest denser cores.

Spectral clustering describes a graph as a adjacency matrix W, from which is derived the Laplacian matrix $L = D - W$(unnormalized) or $L = I - D^{1/2}WD^{-1/2}$(normalized), where D is the diagonal matrix featuring each vertex's degree. The eigenvectors of L can be used as cluster centroids, with the corresponding eigenvalues giving an indication of the cut size between clusters. Since we want minimum cut size, the smallest eigenvalues are chosen first. This ranking of clusters is an appealing feature for dense component discovery.

None of these clustering methods, however, are suited for an absolute density criterion. Nor can they handle overlapping clusters. Therefore, some but not all clustering criteria are dense component criteria. Most clustering methods are suitable for relative dense component discovery, excluding k-partitioning methods.

3. Algorithms for Detecting Dense Components in a Single Graph

In this section, we explore algorithmic approaches for finding dense components. First we look at basic exact algorithms for finding cliques and quasi-cliques and comment on their time complexity. Because the clique problem is NP-hard, we then consider some more time efficient solutions. The algorithms can be categorized as follows: Exact enumeration (Section 3.1), Fast Heuristic Enumeration (Section 3.2), and Bounded Approximation Algorithms (Section 3.3). We review some recent works related to dense component discovery, concentrating on the details of several well-received algorithms.

The following table (Table 10.3) gives an overview of the major algorithmic approaches and lists the representative examples we consider in this chapter.

Table 10.3. Overview of Dense Component Algorithms

Algorithm Type	Component Type	Example	Comments
Enumeration	Clique	[12]	
	Biclique	[35]	
	Quasi-clique	[33]	min. degree for each vertex
	Quasi-biclique	[47]	
	k-core	[7]	
Fast Heuristic Enumeration	Maximal biclique	[30]	nonoverlapping
	Quasi-clique/biclique	[13]	spectral analysis
	Relative density	[18]	shingling
	Maximal quasi-biclique	[32]	balanced noise tolerance
	Quasi-clique, k-core	[52]	pruned search; visual results with upper-bounded estimates
Bounded Approximation	Max. average degree	[14]	undirected graph: 2-approx. directed graph: 2+ϵ-approx.
	Densest subgraph, $n \geq k$	[4]	1/3-approx.
	Subgraph of known density θ	[3]	finds subgraph with density $\Omega(\theta/\log \Delta)$

3.1 Exact Enumeration Approach

The most natural way to discover dense components in a graph is to enumerate all possible subsets of vertices and to check if some of them satisfy the definition of dense components. In the following, we investigate some algorithms for discovering dense components by explicit enumeration.

Enumeration Approach.   Finding maximal cliques in a graph may be straightforward, but it is time-consuming. The clique decision problem, deciding whether a graph of size n has a clique of size at least k, is one of Karp's 21 NP-Complete problems [28]. It is easy to show that the clique optimization problem, finding a largest clique in a graph, is also NP-Complete, because the optimization and decision problems each can be reduced in polynomial time to the other. Our goal is to enumerate all cliques. Moon and Moser showed that a graph may contain up to $3^{n/3}$ maximal cliques [38]. Therefore, even for modest-sized graphs, it is important to find the most effective algorithm.

One well-known enumeration algorithm for generating cliques was proposed by Bron and Kerbosch [12]. This algorithm utilizes the branch-and-bound technique in order to prune branches which are unable to generate a clique. The basic idea is to extend a subset of vertices, until the clique is maximal, by adding a vertex from a candidate set but not in a exclusion set. Let C be the set of vertices which already form a clique, $Cand$ be the set of vertices which may potentially be used for extending C, and $NCand$ be the set of vertices which are not allowed to be candidates for C. $N(v)$ are the neighbors of vertex v. Initially, C and $NCand$ are empty, and $Cand$ contains all vertices in the graph. Given C, $Cand$ and $NCand$, we describe the Bron-Kerbosch algorithm below. The authors experimentally observed $O(3.14^n)$, but did not prove their theoretical performance.

Algorithm 6 CliqueEnumeration(C,$Cand$,$NCand$)

---

  if $Cand = \emptyset$ and $NCand = \emptyset$ **then**
    output the clique induced by vertices C;
  else
    for all $v_i \in Cand$ **do**
      $Cand \leftarrow Cand \setminus \{v_i\}$;
      call $CliqueEnumeration(C \cup \{v_i\}, Cand \cap N(v_i), NCand \cap N(v_i))$;
      $NCand \leftarrow NCand \cup \{v_i\}$;
    end for
  end if

Makino et al. [35] proposed new algorithms making full use of efficient matrix multiplication to enumerate all maximal cliques in a general graph or bicliques in a bipartite graph. They developed different algorithms for different types of graphs (general graph, bipartite, dense, and sparse). In particular, for a sparse graph such that the degree of each vertex is bounded by $\Delta \ll |V|$, an algorithm with $O(|V||E|)$ preprocessing time, $O(\Delta^4)$ time delay (i.e, the bound of running time between two consecutive outputs) and $O(|V| + |E|)$ space is developed to enumerate all maximal cliques. Experimental results demonstrate good performance for sparse graphs.

Quasi-clique Enumeration. Compared to exact cliques, quasi-cliques provide both more flexibility of the components being sought as well as more opportunities for pruning the search space. However, the time complexity generally remains NP-complete. The *Quick* algorithm, introduced in [33], provided an illustrative example. The authors studied the problem of mining maximal degree-based quasi-cliques with size at least min_size and degree of each vertex at least $\lceil \gamma(|V| - 1) \rceil$. The *Quick* algorithm integrates some novel pruning techniques based on degree of vertices with a traditional depth-first search framework to prune the unqualified vertices as soon as possible. Those pruning techniques also can be combined with other existing algorithms to achieve the goal of mining maximal quasi-cliques.

They employ these established pruning techniques based on diameter, minimum size threshold, and vertex degree. Let $N_k^G(v) = \{u | dist^G(u, v) \leq k\}$ be the set of vertices that are within a distance of k from vertex v, $indeg^X(u)$ denotes the number of vertices in X that are adjacent to u, and $exdeg^X(u)$ represents the number of vertices in $cand\_exts(X)$ that are adjacent to u. All vertices are sorted in lexicographic order, then $cand\_exts(X)$ is the set of vertices after the last vertex in X which can be used to extend X. For the pruning technique based on graph diameter, the vertices which are not in $\cap_{v \in X} N_k^G(v)$ can be removed from $cand\_exts(X)$. Considering the minimum size threshold, the vertices whose degree is less than $\lceil \gamma(min\_size - 1) \rceil$ should be removed.

In addition, they introduce five new pruning techniques. The first two techniques consider the lower and upper bound of the number of vertices that can be used to extend current X. The first pruning technique is based on the upper bound of the number of vertices that can be added to X concurrently to form a γ-quasi-clique. In other words, given a vertex set X, the maximum number of vertices in $cand\_exts(X)$ that can be added into X is bounded by the minimal degree of the vertices in X; The second one is based on the lower bound of the number of vertices that can be added to X concurrently to form a γ-quasi-clique. The third technique is based on critical vertices. If we can find some critical vertices of X, then all vertices in $cand\_exts(X)$ and adjacent to critical vertices are added into X. Technique 4 is based on cover vertex u which maximizes the size of $C_X(u) = cand\_exts(X) \cap N^G(u) \cap (\cap_{v \in X \wedge (u,v) \ni E} N^G(v))$.

Lemma 10.1. *[33] Let X be a vertex set and u be a vertex in $cand\_exts(X)$ such that $indeg^X(u) \geq \lceil \gamma \times |X| \rceil$. If for any vertex $v \in X$ such that $(u, v) \in E$, we have $indeg^X(v) \geq \lceil \gamma \times |X| \rceil$, then for any vertex set Y such that $G(Y)$ is a γ-quasi-clique and $Y \subseteq (X \cup (cand\_exts(X) \cap N^G(u) \cap (\cap_{v \in X \wedge (u,v) \ni E} N^G(v))))$, $G(Y)$ cannot be a maximal γ-quasi-clique.*

From the above lemma, we can prune the $C_X(u)$ of cover vertex u from $cand\_exts(X)$ to reduce the search space. The last technique, the so-called lookahead technique, is to check if $X \cup cand\_exts(X)$ is γ-quasi-clique. If so, we do not need to extend X anymore and reduce some computational cost. See Algorithm $Quick$ above.

Algorithm 7 Quick($X, cand\_exts(X), \gamma, min\_size$)

---

find the cover vertex u of X and sort vertices in $cand\_exts(X)$;
for all $v \in cand\_exts(X) - C_X(u)$ **do**
    apply minimum size constraint on $|X| + |cand\_exts(X)|$;
    apply lookahead technique (technique 5) to prune search space;
    remove the vertices that are not in $N_k^G(v)$;
    $Y \leftarrow X \cup \{u\}$;
    calculate the upper bound and lower bound of the number vertices to be added to Y in order to form γ-quasi-clique;
    recursively prune unqualified vertices (techniques 1,2);
    identify critical vertices of Y and apply pruning (technique 3);
    apply existing pruning techniques to further reduce the search space;
end for
return γ-quasi-cliques;

K-Core Enumeration.   For k-cores, we are happily able to escape NP-complete time complexity; greedy algorithms with polynomial time exist. Batagelj et al. [7] developed a efficient algorithm running in $O(m)$ time, based on the following observation: given a graph $G = (V, E)$, if we recursively eliminate the vertices with degree less than k and their incident edges, the resulting graph is a k-core. The algorithm is quite simple and can be considered as a variant of [29]. This algorithm attempts to assign each vertex with a core number to which it belongs. At the beginning, the algorithm places all vertices in a priority queue based on minimim degree. For each iteration, we eliminate the first vertex v (i.e, the vertex with lowest degree) from the queue. After then, we assign the degree of v as its core number. Considering v's neighbors whose degrees are greater than that of v, we decrease their degrees by one and reorder the remaining vertices in the queue. We repeat such procedure until the queue is empty. Finally, we output the k-cores based on their assigned core numbers.

3.2   Heuristic Approach

As mentioned before, it is impractical to exactly enumerate all maximal cliques, especially for some real applications like protein-protein interaction networks which have a very large number of vertices. In this case, fast heuristic methods are available to address this problem. These methods are able to efficiently identify some dense components, but they cannot guarantee to discover all dense components.

Shingling Technique. Gibson et al. [18] propose an new algorithm based on shingling for discovering large dense bipartite subgraphs in massive graphs. In this paper, a dense bipartite subgraph is considered a cohesive group of vertices which share many common neighbors. Since this algorithm utilizes the shingling technique to convert each dense component with arbitrary size into shingles with constant size, it is very efficient and practical for single large graphs and can be easily extended for streaming graph data.

We first provide some basic knowledge related to the shingling technique. Shingling was firstly introduced in [11] and has been widely used to estimate the similarity of web pages, as defined by a particular feature extraction scheme. In this work, shingling is applied to generate two constant-size fingerprints for two different subsets A and B from set S of a universe U of elements, such that the similarity of A and B can be computed easily by comparing fingerprints of A and B, respectively. Assuming π is a random permutation of the elements in the ordered universe U which contains A and B, the probability that the smallest element of A and B is the same, is equal to the Jaccard coefficient. That is,

$$Pr[\pi^{-1}(min_{a \in A}\{\pi(a)\}) = \pi^{-1}(min_{b \in B}\{\pi(b)\})] = \frac{|A \cap B|}{|A \cup B|}$$

Given a constant number c of permutations $\pi_1, \cdots, \pi_c$ of U, we generate a fingerprinting vector whose i-th element is $min_{a \in A}\pi_i(a)$. The similarity between A and B is estimated by the number of positions which have the same element with respect to their corresponding fingerprint vectors. Furthermore, we can generalize this approach by considering every s-element subset of entire set instead of the subset with only one element. Then the similarity of two sets A and B can be measured by the fraction of these s-element subsets that appear in both. This actually is an agreement measure used in information retrieval. We say each s-element subset is a *shingle*. Thus this feature extraction approach is named the (s, c) shingling algorithm. Given a n-element set $A = \{a_i, 0 \le i \le n\}$ where each element a_i is a string, the (s, c) shingling algorithm tries to extract c shingles such that the length of each shingle is exact s. We start from converting each string a_i into a integer x_i by a hashing function. Following that, given two random integer vectors R, S with size c, we generate a n-element temporary set $Y = \{y_i, 0 \le i \le n\}$ where each element $y_i = R_j \times x_i + S_j$. Then the s smallest elements of Y are selected and concatenated together to form a new string y. Finally, we apply a hash function on string y to get one shingle. We repeat such procedure c times in order to generate c shingles.

Remember that our goal is to discover dense bipartite subgraphs such that vertices in one side share some common neighbors in another side. Figure 10.2 illustrates a simple scenario in a web community where each web page

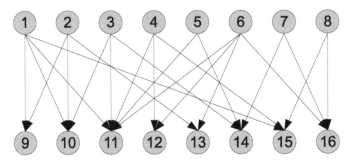

Figure 10.2. Simple example of web graph

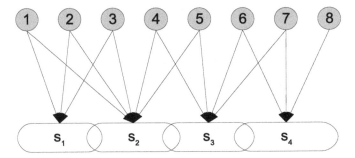

Figure 10.3. Illustrative example of shingles

in the upper part links to some other web pages in the lower part. We can describe each upper web page (vertex) by the list of lower web pages to which it
links. In order to put some vertices into the same group, we have to measure
the similarity of the vertices which denotes to what extent they share common
neighbors. With the help of shingling, for each vertex in the upper part, we can
generate constant-size shingles to describe its outlinks (i.e, its neighbors in the
lower part). As shown in Figure 10.3, the outlinks to the lower part are converted to shingles s_1, s_2, s_3, s_4. Since the size of shingles can be significantly
smaller than the original data, much computational cost can be saved in terms
of time and space.

In the paper, Gibson et al. repeatedly employ the shingling algorithm for
converting dense component into constant-size shingles. The algorithm is a
two-step procedure. Step 1 is recursive shingling, where the goal is to exact
some subsets of vertices where the vertices in each subset share many common neighbors. Figure 10.4 illustrates the recursive shingling process for a
graph ($\Gamma(V)$ is the outlinks of vertices V). After the first shingling process,
for each vertex $v \in V$, its outlinks $\Gamma(v)$ are converted into a constant size of
first-level shingles v'. Then we can transpose the mapping relation E_0 to E_1 so
that each shingle in v' corresponds to a set of vertices which share this shingle.
In other words, a new bipartite graph is constructed where each vertex in one

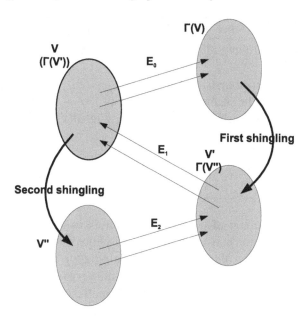

Figure 10.4. Recursive Shingling Step

part represents one shingle, and each vertex in another part is the original vertex. If there is a edge from shingle v' to vertex v, v' is one of the shingles for
v's outlinks generated by shingling. From now on, V is considered as $\Gamma(V')$.
Following the same procedure, we apply shingling on V' and $\Gamma(V')$. After
the second shingling process, V is converted into a constant-size V'', so-called
second-level shingles. Similar to the transposition in the first shingling process, we transpose E_1 to E_2 and obtain many pairs $< v'', \Gamma(v'') >$ where v''
is second-level shingles and $\Gamma(v'')$ are all the first-level shingles that share a
second-level shingle. Step 2 is clustering, where the aim is to merge first-level
shingles which share some second-level shingles. Essentially, merges a number of biclique subsets into one dense component. Specifically, given all pairs
$< v'', \Gamma(v'') >$, a traditional algorithm, namely $UnionFind$, is used to merge
some first-level shingles in $\Gamma(V'')$ such that any two first-level shingles at least
share one second-level shingle. To the end, we map the clustering results back
to the vertices of the original graph and generate one dense bipartite subgraph
for each cluster. The entire algorithm is presented in Algorithm *DiscoverDens-
eSubgraph*.

GRASP Algorithm.   As mentioned in Table 10.2, Abello et al. [1] were
one of the first to formally define quasi-dense components, namely γ-cliques,
and to investigate their discovery. They utilize a existing framework known
as a Greedy Randomized Adaptive Search Procedure (GRASP). Their paper
makes two major contributions. First, they propose a novel evaluation measure

---

Algorithm 8 DiscoverDenseSubgraph(c_1, s_1, c_2, s_2)

---

apply recursive shingling algorithms to obtain first- and second-level shingles;
let $S = < s, \Gamma(s) >$ be first-level shingles;
let $T = < t, \Gamma(t) >$ be second-level shingles;
apply clustering approach to get the clustering result $\mathcal{C}$ in terms of first-level shingles;
for all $C \in \mathcal{C}$ **do**
    output $\cup_{s \in C} \Gamma(s)$ as a dense subgraph;
end for

---

on potential improvement of adding a new vertex to a current quasi-clique. This measure enables the construction of quasi-cliques incrementally. Second, a semi-external memory algorithm incorporating edge pruning and external breath first search traversal is introduced to handle very large graphs. The basic idea is to decompose a large graph into several small components, then process each of them using GRASP. In the following, we concentrate our efforts on discussing the first point and its usage in GRASP. Interested readers can refer to [1] for the details of the second algorithm.

GRASP is a multi-start iterative process, with two steps per iteration, initial construction and local optimization. The initial construction step aims to produce a feasible solution for subsequent processing. For local optimization, we examine the neighborhood of the current solution in terms of the solution space, and try to find a better local solution. A comprehensive survey of the GRASP approach can be found in [41]. In this paper, Abello et al. proposed a incremental algorithm to build a maximal γ-clique, which serves as the initial feasible solution in GRASP. Before we move to the algorithm, we first define the potential of a vertex set R as

$$\phi(R) = |E(R)| - \gamma \binom{|R|}{2}$$

and the potential of R with respect to a disjoint vertices set S to be

$$\phi_S(R) = \phi(S \cup R)$$

Furthermore, considering a graph $G = (V, E)$ and a γ-clique induced by vertices set $S \subset V$, we call a vertex $x \in (V \setminus S)$ a δ-vertex with respect to S if and only if the graph induced by $S \cup \{x\}$ is a γ-clique. Then, the set of γ-vertices with respect to S is denoted as $\mathcal{N}_\gamma(S)$. Given this, the incremental algorithm tries to add a *good* vertex in $\mathcal{N}_\gamma(S)$ into S. To facilitate our discussion, a potential difference of a vertex $y \in \mathcal{N}_\gamma(S) \setminus \{x\}$ is defined to be

$$\delta_{S,x}(y) = \phi_{S \cup \{x\}}(\{y\}) - \phi_S(\{y\})$$

The above equation can also expressed as

$$\delta_{S,x}(y) = deg(x)|_S + deg(y)|_{\{x\}} - \gamma(|S| + 1)$$

where $deg(x)|_S$ is the degree of x in the graph induced by vertex set S. This equation implies that the potential of y which is a γ-neighbor of x does not decrease when x is included in S. Here the γ-neighbors of vertex x are the neighbors of x with $deg(x)|_S$ greater than $\gamma|S|$. The total effect caused by adding vertex x to current γ-clique S is

$$\Delta_{S,x} = \sum_{y \in \mathcal{N}_\gamma(S) \setminus \{x\}} \delta_{S,x}(y) = |\mathcal{N}_\gamma(\{x\})| + |\mathcal{N}_\gamma(S)|(deg(x)|_S - \gamma(|S| + 1))$$

We see that the vertices with a large number of γ-neighbors and high degree with respect to S are preferred to be selected. A greedy algorithm to build a maximal γ-clique is outlined in Algorithm *DiscoverMaximalQuasi-Clique*. The time complexity of this algorithm is $O(|S||V|^2)$, where S the vertex set used to induce a maximal γ-clique.

Algorithm 9 DiscoverMaximalQuasi-clique(V, E, γ)

$\gamma^* \leftarrow 1, S^* \leftarrow \emptyset$;
select a vertex $x \in V$ and add into S^*;
while $\gamma^* \geq \gamma$ **do**
  $S \leftarrow S^*$;
  if $\mathcal{N}_{\gamma^*}(S) \neq \emptyset$ **then**
    select $x \in \mathcal{N}_{\gamma^*}(S)$;
  else
    if $\mathcal{N}(S) \setminus S = \emptyset$ **then**
      return S;
    end if
    select $x \in \mathcal{N}(S) \setminus S$;
  end if
  $S^* \leftarrow S \cup \{x\}$;
  $\gamma^* \leftarrow 2|E(S^*)|/(|S^*|(|S^*| - 1))$;
end while
return S;

Then applying GRASP, a local search procedure tries to improve the generated maximal γ-clique. Generally speaking, given current γ-clique induced by vertex set S, this procedure attempts to substitute two vertices within S with one vertex outside S in order to improve aforementioned $\Delta_{S,x}$. GRASP guarantees to obtain a local optimum.

Visualization of Dense Components.   Wang et al. [52] combine theoretical bounds, a greedy heuristic for graph traversal, and visual cues to develop a mining technique for clique, quasi-clique, and k-core components. Their approach is named CSV for Cohesive Subgraph Visualization. Figure 10.5 shows a representative plot and how it is interpreted.

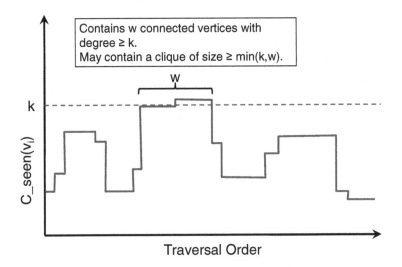

Figure 10.5. Example of CSV Plot

A key measure in CSV is co-cluster size $CC(v, x)$, meaning the (estimated) size of the largest clique containing both vertices v and x. Then, $C(v) = max\{CC(v, x), \forall x \in N(v)\}$.

At the top level of abstraction, the algorithm is not difficult. We maintain a priority queue of vertices observed so far, sorted by $C(v)$ value. We traverse the graph and draw a density plot by iterating the following steps:

    1 Remove the top vertex from the queue, making this the current vertex v.

    2 Plot v.

    3 Add v's neighbors to the priority queue.

Now for some details. If this is the i-th iteration, plot the point $(i, C_{seen}(v_i))$, where $C_{seen}(v_i)$ is the largest value of $C(v_i)$ observed so far. We say "seen so far" because we may not have observed all of v neighbors yet, and even when we have, we are only estimating clique sizes. Next, some neighbors of v may already be in the queue. In this case, update their C values and reprioritize. Due to the estimation method described below, the new estimate is no worse that the previous one.

Since an exact determination of $CC(v, x)$ is computationally expensive, CSV takes several steps to efficiently find a good estimate of the actual clique size. First, to reduce the clique search space, the graph's vertices and edges are pre-processed to map them to a multi-dimensional space. A certain number of vertices are selected as pivot points. Then each vertex is mapped to a vector: $v \rightarrow M(v) = \{d(v, p_1), \cdots, d(v, p_p)\}$, where $d(v, p_i)$ is the shortest distance in the graph from v to pivot p_i. The authors prove that all the vertices of a clique map to the same unit cell, so we can search for cliques by searching individual cells.

Second, CSV further prunes the vertices within each occupied cell. Do the following for each vertex v in each occupied cell: For each neighbor x of v, identify the set of vertices Y which connect to both v and x. Construct the induced subgraph $S(v, x, Y)$. If there is a clique, it must be a subgraph of S. Sort Y by decreasing order of degree in S. To be in a k-clique, a vertex must have degree $\geq k - 1$. Consequently, we step through the sorted Y list and eliminate the remainder when the threshold $\delta_S(y_i) < i - 1$ is reached. The size of the remaining list is an upper bound estimate for $C(v)$ and $CC(v, x)$. With relatively minor modification, the same general approach can be used for quasi-cliques and k-cores.

The slowest step in CSV is searching the cells for pseudo-cliques, with overall time complexity $O(|V|^2 log|V|2^d)$. This becomes exponential when the graph is a single large clique. However, when tested on two real-life datasets, DBLP co-authorship and SMD stock market networks, $d << |V|$, so performance is polynomial.

Other Heuristic Approaches.   We give a brief overview of three additional heuristic approaches. Li et al. [32] studied the problem of discovering dense bipartite subgraphs with so-called balanced noise tolerance, meaning that each vertex in one part is allowed no more than a certain number or a certain percentage of missing edges to the other part. This definition can avoid the density skew found within density-based quasi-cliques. Li et al. observed that their type of maximal quasi-biclique cannot be trivially expanded from traditional maximal bicliques. Some useful properties such as bounded closure and the fixed point property are utilized to develop an efficient algorithm, $\mu - CompleteQB$, for discovering maximal quasi-bicliques with balanced noise tolerance. Given a bipartite graph, the algorithm looks for maximal quasi-bicliques where the number of vertices in each part exceeds a specified value $ms \geq \mu$. Two cases are considered. If $ms \geq 2\mu$, the problem is converted into the problem to find exact maximal μ-quasi bicliques that has been well discussed in [47]. On the other hand, if $ms < 2\mu$, a depth-first search for μ-tolerance maximal quasi-bicliques whose vertex size is between ms and 2μ is conducted to achieve the goal.

A spectral analysis method [13] is used to uncover the functionality of a certain dense component. To begin, the similarity matrix for a protein-protein interaction network is defined, and the corresponding eigenvalues and eigenvectors are calculated. In particular, each eigenvector with positive eigenvalue is identified as a quasi-clique, while each eigenvector with negative eigenvalue is considered a quasi-biclique. Given these dense components, a statistical test based on p-value is applied to measure whether a dense component is enriched with proteins from a particular category more than would be expected by chance. Simply speaking, the statistical test ensures that the existence of each dense component is significant with respect to a specific protein category. If so, that dense component annotated with the corresponding protein functionality.

Kumar et al. [30] focus on enumerating emerging communities which have little or no representation in newsgroups or commercial web directories. They define an (i, j) biclique, where the number of vertices in each part are i and j, respectively, to be the *core* of interested communities. Therefore, this paper aims to extract a non-overlapping maximal set of *cores* for interested communities. A stream-based algorithm combining a set of pruning techniques is presented to process huge raw web data and eventually generate the appropriate cores. Some open problems like how to automatically extract semantic information and organize them into a useful structure are also discussed.

3.3   Exact and Approximation Algorithms for Discovering Densest Components

In this section, we focus on the problem of finding the densest components, i.e., the quasi-cliques with the highest values of *gamma*. We first look at exact solutions, utilizing max-flow/min-cut related algorithms. To reach faster performance, we then consider several greedy approximation algorithms that guarantee. These bounded-approximation algorithms are able to efficiently handle the large graphs and obtain guaranteed reasonable results.

Exact Solution for Discovering Densest Subgraph.   We first consider density of a graph defined as its average degree. Using this definition, Goldberg [19] showed that the problem of finding the densest subgraph can be exactly reduced to a sequence of max-flow/min-cut problems. Given a value g, algorithm constructs a network and finds a min-cut on it. The resulting sizes tell us whether there is a subgraph with density at least g. Given a graph G

with n vertices and m edges, the construction of its corresponding cut network are as follows:

1. Add two vertices source s and sink t to undirected G;

2. Replace each undirected edge with two directed edges with capacity 1 such that each endpoint is the source and target of those two edges, respectively;

3. Add directed edges with capacity m from s to all vertices in G, and add directed edges with capacity $m + 2g - d_i$ from all vertices in G to t, where d_i is the degree of vertex v_i in the original graph.

We apply the max-flow/min-cut algorithm to decompose the vertices of the new network into two non-overlapping sets S and T, such that $s \in S$ and $t \in T$. Let $V_s = S \setminus \{s\}$. Goldberg proved that there exists a subgraph with density at least g if $V_s \neq \emptyset$. The following theorem formally presents this result:

Theorem 10.2. *Given S and T which are generated by the algorithm for max-flow min-cut problem, if $V_s \neq \emptyset$, then there is no subgraph with density D such that $g \leq D$. If $V_s = \emptyset$, then there exists a subgraph with density D such that $g \geq D$.*

The remaining issue is to enumerate all possible values of density and apply the max-flow/min-cut algorithm for each value. Goldberg observed that the difference between any two subgraphs is no more than $\frac{1}{n(n-1)}$. Combined with binary search, this observation provides a effective stop criteria to reduce the search space. The sketch of the entire algorithm is outlined in Algorithm *FindDensestSubgraph*.

Greedy Approximation Algorithm with Bound.   In [14], Charikar describes exact and greedy approximation algorithms to discover subgraphs which can maximize two different notions of density, one for undirected graphs and one for directed graphs. The density notion utilized for undirected graphs is the average degree of the subgraph, such that density $f(S)$ of the subset S is $\frac{|E(S)|}{|S|}$. For directed graphs, the criteria first proposed by Kannan and Vinay [27] is applied. That is, given two subsets of vertices $S \subseteq V$ and $T \subseteq V$, the density of subgraph $H_{S,T}$ is defined as $d(S, T) = \frac{|E(S,T)|}{\sqrt{|S||T|}}$. Here, S and T are not necessarily disjoint. This paper studies the optimization problem of discovering a subgraph H_s induced by a subset S with maximum $f(S)$ or $H_{S,T}$ induced by two subsets S and T with maximum $d(S, T)$, respectively.

The author shows that finding a subgraph H_S in undirected graph with maximum $f(S)$ is equivalent to solving the following linear programming (LP) problem:

Algorithm 10 FindDensestSubgraph(G)

$mind \leftarrow 0; maxd \leftarrow m$;
$V_s \leftarrow \emptyset$;
while $maxd - mind \geq \frac{1}{n(n-1)}$ **do**
  $g \leftarrow \frac{maxd+mind}{2}$;
  Construct new network as we have mentioned;
  Generate S and T utilizing max-flow min-cut algorithm;
  if $S = \{s\}$ **then**
    $maxd \leftarrow g$;
  else
    $mind \leftarrow g$;
    $V_s \leftarrow S - \{s\}$;
  end if
end while
return subgraph induced by V_s;

---

(1) $max \sum_{ij} x_{ij}$
(2) $\forall ij \in E \; x_{ij} \leq y_i$
(3) $\forall ij \in E \; x_{ij} \leq y_j$
(4) $\sum_i y_i \leq 1$
(5) $x_{ij}, y_i \geq 0$

From a graph viewpoint, we assign each vertex v_i with weight $\sum_j x_{ij}$, and $min(y_i, y_j)$ is the threshold for the weight of all edges (v_i, v_j) incident to vertex v_i. Then x_{ij} can be considered as the weight of edge (v_i, v_j) which vertex v_i distributes. Weights are normalized so that the sum of threshold for edges incident to vertex v_i, $\sum_i y_i$, is bounded by 1. In this sense, finding the optimal solution of $\sum_{ij} x_{ij}$ is equivalent to finding a set of edges such that the weights of their incident vertices mostly distribute to them. Charikar shows that the optimality of the above LP problem is exactly equivalent to discovering the densest subgraph in a undirected graph.

Intuitively, the complexity of this LP problem depends highly on the number of edges and vertices in the graph (i.e., the number of inequality constraints in LP). It is impractical for large graphs. Therefore, Charikar proposes an efficient greedy algorithm and proves that this algorithm produces a 2-approximation for $f(G)$. This greedy algorithm is a simple variant of [29]. Let S is a subset of V and H_S is its induced subgraph with density $f(H_S)$. Given this, we outline this greedy algorithm as follows:

1 Let S be the subset of vertices, initialized as V;

2 Let H_S be the subgraph induced by vertices S;

3 For each iteration, eliminate the vertex with lowest degree in H_S from S and recompute its density;

4 For each iteration, measure the density of H_S and record it as a candidate for densest component

Similar techniques are also applied to finding the densest subgraph in a directed graph. The greedy algorithm for directed graphs takes $O(m + n)$ time. According to the analysis, Charikar claims that we have to run the greedy algorithm for $O(\frac{\log n}{\epsilon})$ values of c in order to get a $2 + \epsilon$ approximation, where $c = |S|/|T|$ and S, T are two subset of vertices in the graph.

A variant of this approach is presented in [25]. Jin et al. developed an approximation algorithm for discovering the densest subgraph by introducing a new notion of *rank subgraph*. The rank subgraph can be defined as follows:

Definition 10.3. *(Rank Subgraph) [25]. Given an undirected graph $G = (V, E)$ and a positive integer d, we remove all vertices with degree less than d and their incident edges from G. Repeat this procedure until no vertex can be eliminated and form a new graph G_d. Each vertex in G_d is adjacent to at least d vertices in G_d. If G_d has no vertices, it is denoted $G_\emptyset$. Given this, construct a subgraph sequence $G \supseteq G_1 \supseteq G_2 \cdots \supseteq G_l \supset G_{l+1} = G_\emptyset$, where $G_l \neq G_\emptyset$ and contains at least $l + 1$ vertices. Define l as the rank of the graph G, and G_l as the rank subgraph of G.*

Lemma 10.4. *Given an undirected graph G, let G_s be the densest subgraph of G with density $d(G_s)$ and G_l be its rank subgraph with density $d(G_l)$. Then, the density of G_l is no less than half of the density of G_s:*

$$d(G_l) \geq \frac{d(G_s)}{2}$$

The above lemma implies that we can use the rank subgraph G_l with highest rank of G to approximate its densest subgraph. This technique is utilized to derive a efficient search algorithm for finding densest subgraphs from a sequence of bipartite graphs. The interested reader can refer to [25] for details.

Other Approximation Algorithms.   Anderson et al. [4] consider the problem of discovering dense subgraphs with lower bound or upper bound of size. Three problems including **dalks**, **damks** and **dks** are formulated. In detail, **dalks** is the abbreviation for Densest-At-Least-K subgraph problem aiming at extracting an induced subgraph with highest average degree among all subgraphs with at least k vertices. Similarly, **damks** looks for the Densest At-Most-K subgraph and **dks** seeks the densest subgraph with exactly k vertices. Clearly, both **dalks** and **damks** are relaxed versions of **dks**. Anderson et al. show that **daks** is approximately as hard as **dks** which has been proven to be NP-Complete. More importantly, an effective 1/3-approximation algorithm based on core decomposition of a graph is proposed for **dalks**. This algorithm runs in $O(m + n)$ and $O(m + n \log n)$ time for unweighted and weighted graphs, respectively.

We describe the algorithm for **dalks** as follows. Given a graph $G = (V, E)$ with n vertices and a lower bound of size k, let H_i be the subgraph induced by i vertices. At the beginning, i is initialized with n and H_i is the original graph G. Then, we remove the vertex v_i with minimum weighted degree from H_i to form H_{i-1}. Next, we update its corresponding total weight $W(H_{i-1})$ and density $d(H_{i-1})$. We repeat this procedure and get a sequence of subgraphs $H_n, H_{n-1}, \cdots, H_1$. Finally, we choose the subgraph H_k with maximal density $d(H_k)$ as the resulting dense component.

Anderson [3] develops a local search algorithm to find a dense bipartite subgraph near a specified starting vertex in a bipartite graph. Specifically, for any bipartite subgraph with K vertices and density θ (the definition of density is identical to the definition in [27]), the proposed algorithm guarantees to generate a subgraph with density $\Omega(\theta/\log \Delta)$ near any starting vertex v where Δ is the maximum degree in the graph. The time complexity of this algorithm is $O(\Delta K^2)$ which is independent of the size of graph, and thus has potential to be scaled for large graphs.

4.   Frequent Dense Components

The dense component discovery problem can be extended to consider a dataset consisting of a set of graphs $D = \{G_1, \cdots, G_n\}$. In this case, we have two criteria for components: they must be dense and they must occur frequently. The density requirement can be any of our earlier criteria. The frequency requirement says that a component satisfies a minumum *support* threshold; that is, it appears in at least a certain number of graphs. Obviously, if we say that we find the same component in different graphs, there must be a correspondence of vertices from one graph to another. If the graphs have exactly the same vertex sets, then we call this a relation graph set.

Many authors have considered the broader problem of frequent pattern mining in graphs [50, 23, 31]; however, not until recently has there been a clear focus on patterns defined and restricting by density. Several recent papers have looked into discovery methods for frequent dense subgraphs. We take a more detailed look at some of these papers.

4.1   Frequent Patterns with Density Constraints

One approach is to impose a density constraint on the patterns discovered by frequent pattern mining. In [55], Yan et al. use the minumum cut clustering criterion: a component must have an edge cut less than or equal to k. Note that this is equivalent to a k-core criterion. Furthermore, each frequent pattern must be closed, meaning it does not have any supergraph with the same support level. They develop two approaches, pattern growth and pattern reduction. In pattern growth, begin with a small subgraph (possibly a single vertex) that satisfies both the frequency and density requirements but may not be closed. The algorithm incrementally adds adjacent edges until the pattern is closed. In pattern reduction, initialize the working set P_1 to be the first graph G_1. Update the working set by intersecting its edge set with the edges of the next graph:

$$P_i = P_{i-1} \cap G_I = (V, E(P_{i-1}) \cap E(G_I))$$

This removes any edges that do not appear in both input graphs. Decompose P_i into k-core subgraphs. Recursively call pattern reduction for each dense subgraph. Record the dense subgraphs that survive enough intersections to be considered frequent.

The greedy removal of edges at each iteration quickly reduces the working set size, leading to fast execution time. The trade-off is that we prune away edges that might have contributed to a frequent dense component. The consequence of edge intersection is that we only find components whose edges happen to appear in the first $min\_support$ graphs. Therefore, a useful heuristic would be to order the graphs by decreasing overall density. In [55], they find that pattern reduction works better when targeting high connectivity but a low support threshold. Conversely, pattern growth works better when targeting high support but only modest connectivity.

4.2   Dense Components with Frequency Constraint

Hu et al. [22] take a different perspective, providing a simple meta-algorithm on top of an existing dense component algorithm. From the input graphs, which must be a relation graph set, they derive two new graphs, the Summary Graph and the Second-Order Graph. The Summary Graph is $\hat{G} = (V, \hat{E})$, where an edge exists if it appears in at least k graphs in D. For the Second-Order Graph, we transform each edge in D into a vertex, giving us $F = (V \times V, E_F)$. An edge joins two vertices in F (equivalent to two edges in G) if they have similar support patterns in D. An edge's support pattern is represented as the n-dimensional vector of weights in each graph: $\boldsymbol{w}(e) = \{w_{G_1}(e), \cdots, w_{G_n}(e)\}$. Then, a similarity measure such as Euclidean distance can be used to determine whether two vertices in F should be connected.

Given these two secondary graphs, the problem is quite simple to state: find *coherent dense subgraphs*, where a subgraph S qualifies if its vertices form a dense component in $\hat{G}$ and if its edges form a dense component in F. Density in $\hat{G}$ means that the component's edges occur frequently, when considering the whole relation graph set D. Density in F ensures that these frequent edges are coherent, that is, they tend to appear in the same graphs.

To efficiently find dense subgraphs, Hu uses a modified version of Hartuv and Shamir's HCS mincut algorithm [21]. Because Hu's approach converts any n graphs into only 2 graphs, it scales well with the number of graphs. A drawback, however, is the potentially large size of the second-order graph. The worst case would occur when all n graphs are identical. Since all edge support vectors would be identical, the second order graph would become a clique of size $|E|$ with $O(|E|^2)$ edges.

4.3   Enumerating Cross-Graph Quasi-Cliques

Pei et al. [40] consider the problem of finding so-called cross-graph quasi-cliques, CGQC for short. They use the balanced quasi-clique definition. Given a set of graphs $D = \{G_1, \cdots, G_n\}$ on the same set of vertices U, corresponding parameters $\gamma_1, \cdots, \gamma_n$ for the completeness of vertex connectivity, and a minimum component size min_S, they seek to find all subsets of vertices of cardinality $\geq min_S$ such that when each subset is induced upon graph G_i, it will form a maximal γ_i-quasi-clique.

A complete enumeration is $\#P$-Complete. Therefore, they derive several graph-theoretical pruning methods that will typically reduce the execution time. They employ a *set enumeration tree* [43] to list all possible subsets of

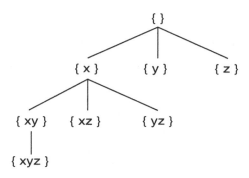

Figure 10.6. The Set Enumeration Tree for {x,y,z}

vertices, while taking advantage of some tree-based concepts, such as depth-first search and sub-tree pruning. An example of a set enumeration tree is shown in Figure 10.6. Below is a brief listing of some of the graph and tree properties they utilize to prune the set of candidate components, followed by the main algorithm, called *Crochet*.

1. Given γ and graph size n, there exist upper bounds on the graph diameter $diam(G)$. For example, $diam(G) \leq n - 1$ if $\gamma > \frac{1}{n-1}$.
2. Define $N^k(u) =$ vertices within a distance k of u.
3. Reducing vertices: If $\delta(u) < \gamma_i(min_S - 1)$ or $|N^k(u)| < (min_S - 1)$, then u cannot be in a CGQC.
4. Candidate projection: when traversing the tree, a child cannot be in a CGQC if it does not satisfy its parent's neighbor distance bounds $N_{G_i}^{k_i}$.
5. Subtree pruning: apply various rules on min_S, redundancy, monotonicity.

5. Applications of Dense Component Analysis

In financial and economic analysis, dense components represent entities that are highly correlated. For example, Boginski et al. define a market graph, where each vertex is a financial instrument, and two vertices are connected if their behaviors (say, price change over time) are highly correlated [9, 10]. A dense component then indicates a set of instruments whose members are well-correlated to one another. This information is valuable both for understanding market dynamics and for predicting the behavior of individual instruments. Density can also indicate strength and robustness. Du et al. [15] identify cliques in a financial grid space to assist in discovering price-value motifs. Some researchers have employed bipartite and multipartite networks. Sim et al. [47] correlates stocks to financial ratios using quasi-bicliques. Alkemade

---

Algorithm 11 Crochet($G_1, G_2, \gamma_1, \gamma_2, min_s$)

1: **for all** graph G_i **do**
2:     construct *set enumeration tree* for all possible vertex subsets of G_i;
3:     $k_i \leftarrow$ upper bound diameter of complete γ_i-quasi-complete graph in G_i;
4: **end for**
5: apply Vertex and Edge Reduction to G_1 and G_2;
6: **for all** $v \in V(G_1)$, using DFS and highest-degree-child-first order **do**
7:     *recursive-mine* $(\{v\}, G_1, G_2)$;
8: **end for**
9:
10: **Function** *recursive-mine*(X, G_1, G_2); {returns TRUE if still seeking quasi-cliques in this branch}
11: $G_i \leftarrow G_i(P), P = \{u | u \in \cap_{v \in X, i=1,2} N_{G_i}^{k_i}(v)\}$ {Candidate Projection}
12: $G_i \leftarrow G_i(P(X))$;
13: apply Vertex Reduction;
14: **if** a Subtree Pruning condition applies **then return** FALSE;
15: *continue* $\leftarrow$ FALSE;
16: **for all** $v \in P(X) \backslash X$, using DFS and highest-degree-child-first order **do**
17:     *continue* $\leftarrow$ *continue* $\lor$ *recursive-mine* $(X \cup \{v\}, G_1, G_2)$;
18: **end for**
19: **if** (not *continue*) $\land$ ($G_i(X)$ is a γ_i-quasi-complete graph) **then**
20:     output X;
21:     **return** TRUE;
22: **else**
23:     **return** *continue*;
24: **end if**

---

et al. [2] finds edge density in a tripartite graph of producers, consumers, and intermediaries to be an important factor in the dynamics of commerce.

In the first decade of the 21st century, the field that perhaps has shown the greatest interest and benefitted the most from dense component analysis is biology. Molecular and systems biologists have formulated many types of networks: signal transduction and gene regulation networks, protein interaction networks, metabolic networks, phylogenetic networks, and ecological networks. [26].

Proteins are so numerous that even simple organisms such as *Saccharomyces cerevisiae*, a budding yeast, are believed to have over 6000 [51]. Understanding the function and interrelationships of each one is a daunting task. Fortunately, there is some organization among the proteins. Dense components in protein-protein interaction networks have been shown to correlate to functional units [49, 42, 54, 13, 6]. Finding these modules and complexes helps to explain metabolic processes and to annotate proteins whose functions are as yet unknown.

Gene expression faces similar challenges. Microarray experiments can record which of the thousands of genes in a genome are expressed under a set of test conditions and over time. By compiling the expression results from several trials and experiments, a network can be constructed. Clustering the genes into dense groups can be used to identify not only healthy functional classes, but also the expression pattern for genetic diseases [48].

Proteins interact with genes by activating and regulating gene transcription and translation. Density in a protein-gene bipartite graph suggests which protein groups or complexes operate on which genes. Everett et al. [16] have extended this to a tripartite protein-gene-tissue graph.

Other biological systems are also being modeled as networks. Ecological networks, famous for food chains and food webs, are receiving new attention as more data becomes available for analysis and as the effects of climate change become more apparent.

Today, the natural sciences, the social sciences, and technological fields are all using network and graph analysis methods to better understand complex systems. Dense component discovery and analysis is one important aspect of network analysis. Therefore, readers from many different backgrounds will benefit from understanding more about the characteristics of dense components and some of the methods used to uncover them.

6. Conclusions and Future Research

In this chapter, we presented a survey of algorithms for dense subgraph discovery. This problem has been studied in the classical literature in the context of the problem of graph partitioning. Subsequently, a number of techniques have been designed for quasi-clique detection, as well as shingling approaches for dense subgraph discovery. Many of the recent applications are designed in the contexts of the web, social, communication and biological networks. These networks have a number of properties, in that they are *massive* and often *dynamic* in nature. This leads to a number of interesting problems for future research:

- In many large scale applications, the data is often disk-resident. This leads to issues involving efficient processing of the underlying network. This is because it is not possible to perform random access of the edges in a disk-resident networks.

- In applications such as the web and social networks, the *domain* of the underlying graph may be massive. In many web, telecommunication, biological and social networks, we may have millions of nodes in the underlying graph. Consequently, the number of edges may range in the trillions. This may lead to storage issues, since the number of distinct edges may not even be possible to store effectively on many desktop machines.

- A number of recent applications may lead to the streaming scenario in which the edges in the graph are received incrementally over time at a fast speed. This is the case in many large telecommunication and social networks. In such cases, it may be extremely challenging to analyze the underlying graph in real time to determine dense patterns.

The area of dense graph mining in massive graphs is still relatively unexplored and represents a fertile area of future research for a number of different applications.

References

[1] J. Abello, M. G. C. Resende, and S. Sudarsky. Massive quasi-clique detection. In *LATIN '02: Proc. 5th Latin American Symposium on Theoretical Informatics*, pages 598–612. Springer-Verlag, 2002.

[2] F. Alkemade, H. A. La Poutre, and H. A. Amman. An agent-based evolutionary trade network simulation. In A. Nagurney, editor, *Innovations in Financial and Economic Networks (New Dimensions in Networks)*, chapter 11, pages 237–255. Edward Elgar Publishing, 2004.

[3] R. Andersen. A local algorithm for finding dense subgraphs. In *SODA '08: Proc. 19th ACM-SIAM Symp. on Discrete Algorithms*, pages 1003–1009. Society for Industrial and Applied Mathematics, 2008.

[4] R. Andersen and K. Chellapilla. Finding dense subgraphs with size bounds. In *WAW '09: Proc. 6th Intl. Workshop on Algorithms and Models for the Web-Graph*, pages 25–37. Springer-Verlag, 2009.

[5] Anna Nagurney, ed. *Innovations in Financial and Economic Networks (New Dimensions in Networks)*. Edward Elgar Publishing, 2004.

[6] G. Bader and C. Hogue. An automated method for finding molecular complexes in large protein interaction networks. *BMC Bioinformatics*, 4(1):2, 2003.

[7] V. Batagelj and M. Zaversnik. An o(m) algorithm for cores decomposition of networks. *CoRR (Computing Research Repository)*, cs.DS/0310049, 2003.

[8] P. Berkhin. Survey of clustering data mining techniques. In C. N. Jacob Kogan and M. Teboulle, editors, *Grouping Multidimensional Data*, chapter 2, pages 25–71. Springer Berlin Heidelberg, 2006.

[9] V. Boginski, S. Butenko, and P. M. Pardalos. On structural properties of the market graph. In A. Nagurney, editor, *Innovations in Financial and Economic Networks (New Dimensions in Networks)*, chapter 2, pages 29–45. Edward Elgar Publishing, 2004.

[10] V. Boginski, S. Butenko, and P. M. Pardalos. Mining market data: A network approach. *Computers and Operations Research*, 33(11):3171–3184, 2006.

[11] A. Z. Broder, S. C. Glassman, M. S. Manasse, and G. Zweig. Syntactic clustering of the web. *Comput. Netw. ISDN Syst.*, 29(8-13):1157–1166, 1997.

[12] C. Bron and J. Kerbosch. Algorithm 457: finding all cliques of an undirected graph. *Commun. ACM*, 16(9):575–577, 1973.

[13] D. Bu, Y. Zhao, L. Cai, H. Xue, and X. Z. andH. Lu. Topological structure analysis of the protein-protein interaction network in budding yeast. *Nucl. Acids Res.*, 31(9):2443–2450, 2003.

[14] M. Charikar. Greedy approximation algorithms for finding dense components in a graph. In *APPROX '00: Proc. 3rd Intl. Workshop on Approximation Algoritms for Combinatorial Optimization*, volume 1913, pages 84–95. Springer, 2000.

[15] X. Du, J. H. Thornton, R. Jin, L. Ding, and V. E. Lee. Migration motif: A spatial-temporal pattern mining approach for financial markets. In *KDD '09: Proc. 15th ACM SIGKDD Intl. Conf. on Knowledge Discovery and Data Mining*. ACM, 2009.

[16] L. Everett, L.-S. Wang, and S. Hannenhalli. Dense subgraph computation via stochastic search: application to detect transcriptional modules. *Bioinformatics*, 22(14), July 2006.

[17] G. W. Flake, S. Lawrence, and C. L. Giles. Efficient identification of web communities. In *KDD'00: Proc. 6th ACM SIGKDD Intl. Conf. on Knowledge Discovery and Data Mining*, pages 150 – 160, 2000.

[18] D. Gibson, R. Kumar, and A. Tomkins. Discovering large dense subgraphs in massive graphs. In *VLDB '05: Proc. 31st Intl. Conf. on Very Large Data Bases*, pages 721–732. ACM, 2005.

[19] A. V. Goldberg. Finding a maximum density subgraph. Technical report, UC Berkeley, 1984.

[20] G. Grimmett. *Precolation*. Springer Verlag, 2nd edition, 1999.

[21] E. Hartuv and R. Shamir. A clustering algorithm based on graph connectivity. *Inf. Process. Lett.*, 76(4-6):175–181, 2000.

[22] H. Hu, X. Yan, Y. H. 0003, J. Han, and X. J. Zhou. Mining coherent dense subgraphs across massive biological networks for functional discovery. In *ISMB (Supplement of Bioinformatics)*, pages 213–221, 2005.

[23] A. Inokuchi, T. Washio, and H. Motoda. An apriori-based algorithm for mining frequent substructures from graph data. In *PKDD '00: Proc. 4th European Conf. on Principles of Data Mining and Knowledge Discovery*, pages 13–23, 2000.

[24] A. K. Jain, M. N. Murty, and P. J. Flynn. Data clustering: a review. *ACM Comput. Surv.*, 31(3):264–323, 1999.

[25] R. Jin, Y. Xiang, N. Ruan, and D. Fuhry. 3-hop: A high-compression indexing scheme for reachability query. In *SIGMOD '09: Proc. ACM SIGMOD Intl. Conf. on Management of Data*. ACM, 2009.

[26] B. H. Junker and F. Schreiber. *Analysis of Biological Networks*. Wiley-Interscience, 2008.

[27] R. Kannan and V. Vinay. Analyzing the structure of large graphs. manuscript, August 1999.

[28] R. M. Karp. Reducibility among combinatorial problems. In R. E. Miller and J. W. Thatcher, editors, *Complexity of Computer Computations*, pages 85–103. Plenum, New York, 1972.

[29] G. Kortsarz and D. Peleg. Generating sparse 2-spanners. *J. Algorithms*, 17(2):222–236, 1994.

[30] R. Kumar, P. Raghavan, S. Rajagopalan, and A. Tomkins. Trawling the web for emerging cyber-communities. *Computer Networks*, 31(11-16):1481–1493, 1999.

[31] M. Kuramochi and G. Karypis. Frequent subgraph discovery. In *ICDM '01: Proc. IEEE Intl. Conf. on Data Mining*, pages 313–320. IEEE Computer Society, 2001.

[32] J. Li, K. Sim, G. Liu, and L. Wong. Maximal quasi-bicliques with balanced noise tolerance: Concepts and co-clustering applications. In *SDM '08: Proc. SIAM Intl. Conf. on Data Mining*, pages 72–83. SIAM, 2008.

[33] G. Liu and L. Wong. Effective pruning techniques for mining quasi-cliques. In W. Daelemans, B. Goethals, and K. Morik, editors, *ECML/PKDD (2)*, volume 5212 of *Lecture Notes in Computer Science*, pages 33–49. Springer, 2008.

[34] R. Luce. Connectivity and generalized cliques in sociometric group structure. *Psychometrika*, 15(2):169–190, 1950.

[35] K. Makino and T. Uno. New algorithms for enumerating all maximal cliques. *Algorithm Theory - SWAT 2004*, pages 260–272, 2004.

[36] H. Matsuda, T. Ishihara, and A. Hashimoto. Classifying molecular sequences using a linkage graph with their pairwise similarities. *Theor. Comput. Sci.*, 210(2):305–325, 1999.

[37] R. Mokken. Cliques, clubs and clans. *Quality and Quantity*, 13(2):161–173, 1979.

[38] J. W. Moon and L. Moser. On cliques in graphs. *Israel Journal of Mathematics*, 3:23–28, 1965.

[39] M. E. J. Newman. The structure and function of complex networks. *SIAM REVIEW*, 45:167–256, 2003.

[40] J. Pei, D. Jiang, and A. Zhang. On mining cross-graph quasi-cliques. In *KDD'05: Proc. 11th ACM SIGKDD Intl. Conf. on Knowledge Discovery and Data Mining*, pages 228–238. ACM, 2005.

[41] L. Pitsoulis and M. Resende. Greedy randomized adaptive search procedures. In P. Pardalos and M. Resende, editors, *Handbook of Applied Optimization*, pages 168–181. Oxford University Press, 2002.

[42] N. Przulj, D. Wigle, and I. Jurisica. Functional topology in a network of protein interactions. *Bioinformatics*, 20(3):340–348, 2004.

[43] R. Rymon. Search through systematic set enumeration. In *Proc. Third Intl. Conf. on Knowledge Representation and Reasoning*, 1992.

[44] J. P. Scott. *Social Network Analysis: A Handbook*. Sage Publications Ltd., 2nd edition, 2000.

[45] S. B. Seidman. Network structure and minimum degree. *Social Networks*, 5(3):269–287, 1983.

[46] S. B. Seidman and B. Foster. A graph theoretic generalization of the clique concept. *J. Math. Soc.*, 6(1):139–154, 1978.

[47] K. Sim, J. Li, V. Gopalkrishnan, and G. Liu. Mining maximal quasi-bicliques to co-cluster stocks and financial ratios for value investment. In *ICDM '06: Proc. 6th Intl. Conf. on Data Mining*, pages 1059–1063. IEEE Computer Society, 2006.

[48] D. K. Slonim. From patterns to pathways: gene expression data analysis comes of age. *Nature Genetics*, 32:502–508, 2002.

[49] V. Spirin and L. Mirny. Protein complexes and functional modules in molecular networks. *Proc. Natl. Academy of Sci.*, 100(21):1123–1128, 2003.

[50] Y. Takahashi, Y. Sato, H. Suzuki, and S.-i. Sasaki. Recognition of largest common structural fragment among a variety of chemical structures. *Analytical Sciences*, 3(1):23–28, 1987.

[51] P. Uetz, L. Giot, G. Cagney, T. A. Mansfield, R. S. Judson, J. R. Knight, D. Lockshon, V. Narayan, M. Srinivasan, P. Pochart, A. Qureshi-Emili, Y. Li, B. Godwin, D. Conover, T. Kalbfleisch, G. Vijayadamodar, M. Yang, M. Johnston, S. Fields, and J. M. Rothberg. A comprehensive analysis of protein-protein interactions in saccharomyces cerevisiae. *Nature*, 403:623–631, 2000.

[52] N. Wang, S. Parthasarathy, K.-L. Tan, and A. K. H. Tung. Csv: visualizing and mining cohesive subgraphs. In *SIGMOD '08: Proc. ACM SIGMOD Intl. Conf. on Management of Data*, pages 445–458. ACM, 2008.

[53] S. Wasserman and K. Faust. *Social Network Analysis: Methods and Applications*. Cambridge University Press, 1994.

[54] S. Wuchty and E. Almaas. Peeling the yeast interaction network. *Proteomics*, 5(2):444–449, 2205.

[55] X. Yan, X. J. Zhou, and J. Han. Mining closed relational graphs with connectivity constraints. In *KDD '05: Proc. 11th ACM SIGKDD Intl. Conf. on Knowledge Discovery in Data Mining*, pages 324–333. ACM, 2005.

Chapter 11

GRAPH CLASSIFICATION

Koji Tsuda

Computational Biology Research Center, National Institute of Advanced Industrial Science and Technology (AIST)
Tokyo, Japan
koji.tsuda@aist.go.jp

Hiroto Saigo

Max Planck Institute for Informatics
Saarbrucken, Germany
hiroto@mpi-inf.mpg.de

Abstract   Supervised learning on graphs is a central subject in graph data processing. In graph classification and regression, we assume that the target values of a certain number of graphs or a certain part of a graph are available as a training dataset, and our goal is to derive the target values of other graphs or the remaining part of the graph. In drug discovery applications, for example, a graph and its target value correspond to a chemical compound and its chemical activity. In this chapter, we review state-of-the-art methods of graph classification. In particular, we focus on two representative methods, graph kernels and graph boosting, and we present other methods in relation to the two methods. We describe the strengths and weaknesses of different graph classification methods and recent efforts to overcome the challenges.

Keywords: graph classification, graph mining, graph kernels, graph boosting

1.   Introduction

Graphs are general and powerful data structures that can be used to represent diverse kinds of objects. Much of the real world data is represented not

C.C. Aggarwal and H. Wang (eds.), *Managing and Mining Graph Data,*
Advances in Database Systems 40, DOI 10.1007/978-1-4419-6045-0_11,
© Springer Science+Business Media, LLC 2010

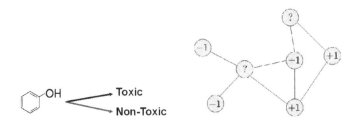

Figure 11.1. Graph classification and label propagation.

as vectors, but as graphs (including sequences and trees, which are specialized graphs). Examples include biological sequences, semi-structured texts such as HTML and XML, chemical compounds, RNA secondary structures, API call graphs, etc. The topic of graph data processing is not new. Over the last three decades, there have been continuous efforts in developing new methods for processing graph data. Recently we have seen a surge of interest in this topic, fueled partly by new technical advances, for example, development of graph kernels [21] and graph mining [52] techniques, and partly by demands from new applications, for example, chemical informatics. In fact, chemical informatics is one of the most prominent fields that deal with large repositories of graph data. For example, NCBI's PubChem has millions of chemical compounds that are naturally represented as molecular graphs. Also, many different kinds of chemical activity data are available, which provides a huge test-bed for graph classification methods.

This chapter aims at giving an overview of existing graph classification methods. The term "graph classification" can mean two different tasks. The first task is to build a model to predict the class label of a whole graph (Figure 11.1, left). The second task is to predict the class labels of nodes in a large graph (Figure 11.1, right). For clarity, we used the term to represent the first task, and we call the second task "label propagation"[6]. This chapter mainly deals with graph classification, but we will provide a short review of label propagation in Section 5.

Graph classification tasks can either be unsupervised or supervised. Unsupervised methods classify graphs into a certain number of categories by similarity [47, 46]. In supervised classification, a classification model is constructed by learning from training data. In the training data, each graph (e.g., a chemical compound) has a target value or a class label (e.g., biochemical activity). Supervised methods are more fundamental from a technical point of view, because unsupervised learning problems can be solved by supervised methods via probabilistic modeling of latent class labels [46]. In this chapter, we focus on two supervised methods for graph classification: graph kernels and graph boosting [40], which are similarity- and feature-based respectively. The two

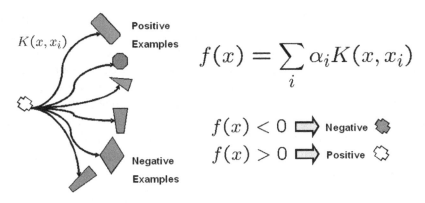

$$f(x) = \sum_i \alpha_i K(x, x_i)$$

$$f(x) < 0 \implies \text{Negative}$$
$$f(x) > 0 \implies \text{Positive}$$

Figure 11.2. Prediction rules of kernel methods.

methods differ in many aspects, and a characterization of the difference of these two methods would be helpful in characterizing other methods.

Kernel methods, such as support vector machines, construct a prediction rule based on a similarity function between two objects [42]. Similarity functions which satisfy a mathematical condition called positive definiteness are called *kernel functions*. For example, in Figure 11.2, the similarity between two objects is represented by a *kernel function* $K(x, x')$. The prediction function $f(x)$ is a linear combination of x's similarities to each training example $K(x, x_i)$, $i = 1, \ldots, n$. In order to apply kernel methods to graph data, it is necessary to define a kernel function for graphs that can measure the similarity between two graphs. It is natural to use the number of shared substructures in two graphs as a similarity measure. However, the enumeration of subgraphs of a given graph is NP-hard [12]. Therefore, one needs to use simpler substructures such as paths and trees. Graph kernels [21] are based on the weighted counts of common paths. A clever recursive algorithm is employed to compute the similarity without total enumeration of substructures.

One obvious drawback of graph kernels is that it is not clear which substructures have the biggest contribution to classification. For a new graph classified by similarity, it is not always possible to know which part of the compound is essential in classification. In many chemical applications, the users are interested not only in accurate prediction of biochemical activities, but also in the mechanism creating the activities. This interpretation problem motivates us to reexamine the approach of subgraph enumeration. Recently, frequent subgraph enumeration algorithms such as AGM [18], Gaston [33] and gSpan [52] have been proposed. They can enumerate all the subgraph patterns that appear more than m times in a graph database. The threshold m is called *minimum support*. Frequent subgraph patterns are determined by branch-and-bound search in a tree shaped search space (Figure 11.7). The computational time crucially depends on the minimum support parameter. For larger values of the support parameter, the search tree can be pruned earlier. For chemical compound datasets, it is easy to mine tens of thousands of graphs on a commodity desktop computer, if the minimum support is reasonably high (e.g., 10% of the number of graphs). However, it is known that, to achieve the best accuracy, the minimum support has to be set to a small value (e.g., smaller than 1%) [51, 23, 16]. In such a setting, the graph mining becomes prohibitively inefficient, because the algorithm creates millions of patterns. This also makes subsequent processing very expensive. Graph boosting [40] progressively constructs the prediction rule in an iterative fashion, and in each iteration only a few informative subgraphs are discovered. In comparison to the na"ve method of using frequent mining and support vector machines, the graph mining routine has to be invoked multiple times. However, an additional search tree pruning condition can speed up each call, and the overall time is shorter than the na"ve method.

The rest of this chapter is organized as follows. In Section 2, we will explain graph kernels, and review its recent extensions for graph classification. In Section 3, we will discuss graph boosting and other methods based on explicit substructure mining. Applications of graph classification methods are reviewed in Section 4. Section 5 briefly presents the label propagation techniques. We conclude the chapter in Section 6.

2.   Graph Kernels

We consider a graph kernel as a similarity measure for two graphs whose nodes and edges are labeled (Figure 11.3). In this section, we present the most fundamental kernel called the marginalized graph kernel [21], which is based on graph paths. Recently, different versions of graph kernels have been proposed using different substructures. Examples include cyclic paths [17] and trees [29].

The proposed graph kernel is based on the idea of random walking. For the labeled graph shown in Figure 11.3a, a label sequence is produced by traversing the graph. A representative example is as follows:

$$(A, c, C, b, A, a, B), \tag{2.1}$$

The vertex labels A, B, C, D and the edge labels a, b, c, d appear alternately. By repeating random walks with random initial and end points, it is possible to obtain the probabilities for all possible walks (Figure 11.3b). The essential idea of the graph kernel is to derive a similarity measure of two graphs by comparing their probability tables. It is computationally infeasible to perform all possible random walks. Therefore, we employ a recursive algorithm which can estimate the underlying probabilities. The node and edge labels are either

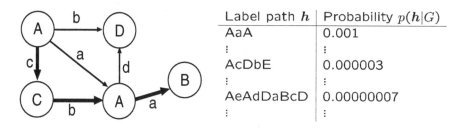

Label path h	Probability $p(h|G)$
AaA	0.001
⋮	⋮
AcDbE	0.000003
⋮	⋮
AeAdDaBcD	0.00000007
⋮	⋮

Figure 11.3. (a) An example of labeled graphs. Vertices and edges are labeled by uppercase and lowercase letters, respectively. By traversing along the bold edges, the label sequence (2.1) is produced. (b) By repeating random walks, one can construct a list of probabilities.

discrete symbols or vectors. In the latter case, it is necessary to define node kernels and edge kernels to specify the similarity of vectors.

Before describing technical details, we formally define a labeled graph. Let Σ_V denote the set of vertex labels, and Σ_E the set of edge labels. Let $\mathcal{X}$ be a finite nonempty set of vertices, v be a function $v : \mathcal{X} \to \Sigma_V$. Let $\mathcal{L}$ be a set of vertex pairs that denote edges, and e be a function $e : \mathcal{L} \to \Sigma_E$. (We assume that there are no multiple edges from one vertex to another.) Then $G = (\mathcal{X}, v, \mathcal{L}, e)$ is a labeled graph with directed edges. Our task is to construct a kernel function $k(G, G')$ between two labeled graphs G and G'.

2.1 Random Walks on Graphs

We extract features (labeled sequences) from a graph G by performing random walks. At the first step, we sample a node $x_1 \in \mathcal{X}$ from an initial probability distribution $p_s(x_1)$. Subsequently, at the ith step, the next vertex $x_i \in \mathcal{X}$ is sampled subject to a transition probability $p_t(x_i|x_{i-1})$, or the random walk ends at node x_{i-1} with probability $p_q(x_{i-1})$. In other words, at the ith step, we have:

$$\sum_{k=1}^{|\mathcal{X}|} p_t(x_k|x_{i-1}) + p_q(x_{i-1}) = 1 \qquad (2.2)$$

that is, at each step, the probabilities of transitions and termination sum to 1.

When we do not have any prior knowledge, we can set the initial probability distribution p_s to be the uniform distribution, the transition probability p_t to be a uniform distribution over the vertices adjacent to the current vertex, and the termination probability p_q to be a small constant probability.

From the random walk, we obtain a sequence of vertices called a *path*:

$$\mathbf{x} = (x_1, x_2, \ldots, x_\ell), \qquad (2.3)$$

where ℓ is the length of $\mathbf{x}$ (possibly infinite). The final probability of obtaining path $\mathbf{x}$ is the product of the probabilities that the path starts with x_1, transits

342 MANAGING AND MINING GRAPH DATA

from x_{i-1} to x_i for each i, and finally terminates with x_l:

$$p(\mathbf{x}|G) = p_s(x_1) \prod_{i=2}^{\ell} p_t(x_i|x_{i-1}) p_q(x_\ell).$$

Let us define a *label sequence* as sequence of alternating vertex labels and edge labels:

$$\mathbf{h} = (h_1, h_2, \ldots, h_{2\ell-1}) \in (\Sigma_V \Sigma_E)^{\ell-1} \Sigma_V.$$

Associated with a path $\mathbf{x}$, we obtain a label sequence

$$\mathbf{h_x} = (v_{x_1}, e_{x_1, x_2}, v_{x_2}, e_{x_2, x_3}, \ldots, v_{x_\ell}).$$

which is a sequence of alternating vertex and edge labels. Since multiple vertices (edges) may have the same label, multiple paths may map to one label sequence. The probability of obtaining a label sequence $\mathbf{h}$ is thus the sum of the probabilities of each path that emits $\mathbf{h}$. This can be expressed as

$$p(\mathbf{h}|G) = \sum_{\mathbf{x}} \delta(\mathbf{h} = \mathbf{h_x}) \cdot \left(p_s(x_1) \prod_{i=2}^{\ell} p_t(x_i|x_{i-1}) p_q(x_\ell) \right),$$

where δ is a function that returns 1 if its argument holds, 0 otherwise.

2.2 Label Sequence Kernel

We now define a kernel k_z between two label sequences $\mathbf{h}$ and $\mathbf{h'}$. The sequence kernel is defined based on kernels for vertex labels and edge labels.

We assume two kernel functions, $k_v(v, v')$ and $k_e(e, e')$, are readily defined between vertex labels and edge labels. We constrain both kernels to be non-negative[1]. An example of a vertex label kernel is the identity kernel, that is, the kernel return 1 if the two labels are the same, 0 otherwise. It can be expressed as:

$$k_v(v, v') = \delta(v = v') \tag{2.4}$$

where $\delta(\cdot)$ is a function that returns 1 if its argument holds, and 0 otherwise. The above kernel (2.4) is for labels of discrete values. If the labels are defined in $\mathbb{R}$, then the Gaussian kernel can be used as a natural choice [42]:

$$k_v(v, v') = \exp(- \parallel v - v' \parallel^2 / 2\sigma^2), \tag{2.5}$$

Edge kernels can be defined in the same way as in (2.4) and (2.5).

Based on the vertex label and the edge label kernels, we defome the kernel for label sequences. If two sequences $\mathbf{h}$ and $\mathbf{h'}$ are of the same length, or

---

[1]This constraint will play an important role in proving the convergence of our kernel.

$\ell(\mathbf{h}) = \ell(\mathbf{h}')$, then the sequence kernel is defined as the product of the label kernels:

$$k_z(\mathbf{h}, \mathbf{h}') = k_v(h_1, h_1') \prod_{i=2}^{\ell} k_e(h_{2i-2}, h_{2i-2}') k_v(h_{2i-1}, h_{2i-1}'). \qquad (2.6)$$

If the two sequences are of different length, or $\ell(\mathbf{h}) \neq \ell(\mathbf{h}')$, then the sequence kernel returns 0, that is, $k_z(\mathbf{h}, \mathbf{h}') = 0$.

Finally, our label sequence kernel is defined as the expectation of k_z over all possible $\mathbf{h} \in G$ and $\mathbf{h}' \in G'$.

$$k(G, G') = \sum_{\mathbf{h}} \sum_{\mathbf{h}'} k_z(\mathbf{h}, \mathbf{h}') p(\mathbf{h}|G) p(\mathbf{h}'|G'). \qquad (2.7)$$

Here, $p(\mathbf{h}|G)p(\mathbf{h}'|G')$ is the probabilty that $\mathbf{h}$ and $\mathbf{h}'$ occur in G and G', respectively, and $k_z(\mathbf{h}, \mathbf{h}')$ is their similarity. This kernel is valid, as it is described as an inner product of two vectors $p(\mathbf{h}|G)$ and $p(\mathbf{h}'|G')$.

2.3 Efficient Computation of Label Sequence Kernels

The label sequence kernel (2.7) defined above can be expanded as follows:

$$
\begin{aligned}
k(G, G') = \sum_{\ell=1}^{\infty} \sum_{\mathbf{h}} \sum_{\mathbf{h}'} & k_v(h_1, h_1') \times \\
& \left(\prod_{i=2}^{\ell} k_e(h_{2i-2}, h_{2i-2}') k_v(h_{2i-1}, h_{2i-1}') \right) \times \\
& \left(\sum_{\mathbf{x}} \delta(\mathbf{h} = \mathbf{h_x}) \cdot \left(p_s(x_1) \prod_{i=2}^{\ell} p_t(x_i|x_{i-1}) p_q(x_\ell) \right) \right) \times \\
& \left(\sum_{\mathbf{x}'} \delta(\mathbf{h} = \mathbf{h_{x'}}) \cdot \left(p_s(x_1') \prod_{i=2}^{\ell} p_t(x_i'|x_{i-1}') p_q(x_\ell') \right) \right).
\end{aligned}
$$

The straightforward enumeration of all terms to compute the sum has a prohibitive computational cost. In particular, for cyclic graphs, it is infeasible to perform this computation in an enumerative way, because the possible length of a sequence spans from 1 to infinity. Nevertheless, there is an efficient method to compute this kernel as shown below. The method is based on the observation that the kernel has the following nested structure.

$$k(G, G') = \lim_{L \to \infty} \sum_{\ell=1}^{L} \quad (2.8)$$

$$\sum_{x_1, x_1'} s(x_1, x_1') \times$$

$$\left(\sum_{x_2, x_2'} t(x_2, x_2', x_1, x_1') \times \left(\sum_{x_3, x_3'} t(x_3, x_3', x_2, x_2') \times \right.\right.$$

$$\left.\left. \cdots \times \sum_{x_\ell, x_\ell'} t(x_\ell, x_\ell', x_{\ell-1}, x_{\ell-1}') q(x_\ell, x_\ell') \right) \cdots \right)$$

where

$$s(x_1, x_1') = p_s(x_1) p_s'(x_1') k_v(v_{x_1}, v_{x_1'}),$$

$$q(x_\ell, x_\ell') = p_q(x_\ell) p_q'(x_\ell')$$

$$t(x_i, x_i', x_{i-1}, x_{i-1}') = p_t(x_i | x_{i-1}) p_t'(x_i' | x_{i-1}') k_v(v_{x_i}, v_{x_i'}) k_e(e_{x_{i-1} x_i}, e_{x_{i-1}' x_i'})$$

Intuitively, (2.8) computes the expectation of the kernel function over all possible pairs of paths of the same length l. Consider one of such pairs: $(x_1, \cdots, x_\ell)$ in G and $(x_1', \cdots, x_\ell')$ in G'. Here, p_s, p_t, and p_q denote the initial, transition, and termination probability of nodes in graph G, and p_s', p_t', and p_q' denote the initial, transition, and termination probability of nodes in graph G'. Thus, $s(x_1, x_1')$ is the probability-weighted similarity of the first elements in the two paths, $q(x_\ell, x_\ell')$ is the probability that the two paths end with x_ℓ and x_ℓ', and $t(x_i, x_i', x_{i-1}, x_{i-1}')$ is the probability-weighted similarity of the ith node pair and edge pair in the two paths.

Acyclic Graphs.   Let us first consider the case of acyclic graphs. In an acyclic graph, if there is a directed path from vertex x_1 to x_2, then there is no directed path from vertex x_2 to x_1. It is well known that vertices of a directed, acyclic graph can be numbered in a topological order[2] such that every edge from a vertex numbered i to a vertex numbered j satisfies $i < j$ (see Figure 11.4).

Since there are no directed paths from vertex j to vertex i if $i < j$, we can employ dynamic programming to achieve our goal. Given that both G and G'

---

[2]Topological sorting of graph G can be done in $O(|\mathcal{X}| + |\mathcal{L}|)$ [7].

are directed acyclic graphs, we can rewrite (2.8) into the following:

$$k(G, G') = \sum_{x_1, x_1'} s(x_1, x_1')q(x_1, x_1') + \lim_{L \to \infty} \sum_{\ell=2}^{L} \sum_{x_1, x_1'} s(x_1, x_1') \times$$
$$\left(\sum_{x_2 > x_1, x_2' > x_1'} t(x_2, x_2', x_1, x_1') \left(\sum_{x_3 > x_2, x_3' > x_2'} t(x_3, x_3', x_2, x_2') \times \right. \right.$$
$$\left. \left. \left(\cdots \left(\sum_{x_\ell > x_{\ell-1}, x_\ell' > x_{\ell-1}'} t(x_\ell, x_\ell', x_{\ell-1}, x_{\ell-1}')q(x_\ell, x_\ell') \right) \right) \cdots \right) \right).$$
$$(2.9)$$

The first term corresponds to paths of length 1, and the second term corresponds to paths longer than 1. We define $r(\cdot, \cdot)$ as follows:

$$r(x_1, x_1') := q(x_1, x_1') + \lim_{L \to \infty} \sum_{\ell=2}^{L} \left(\sum_{x_2 > x_1, x_2' > x_1'} t(x_2, x_2', x_1, x_1') \times \right.$$
$$\left. \left(\cdots \left(\sum_{x_\ell > x_{\ell-1}, x_\ell' > x_{\ell-1}'} t(x_\ell, x_\ell', x_{\ell-1}, x_{\ell-1}')q(x_\ell, x_\ell') \right) \right) \cdots \right),$$
$$(2.10)$$

We can rewrite (2.9) as the follows:

$$k(G, G') = \sum_{x_1, x_1'} s(x_1, x_1')r(x_1, x_1').$$

The merit of defining (2.10) is that we can exploit the following recursive equation.

$$r(x_1, x_1') = q(x_1, x_1') + \sum_{j > x_1, j' > x_1'} t(j, j', x_1, x_1')r(j, j'). \qquad (2.11)$$

Since all vertices are topologically ordered, $r(x_1, x_1')$ can be efficiently computed by dynamic programming (Figure 11.5) for all x_1 and x_1'. The worst-case time complexity of computing $k(G, G')$ is $O(c \cdot c' \cdot |\mathcal{X}| \cdot |\mathcal{X}'|)$ where c and c' are the maximum out-degree of G and G', respectively.

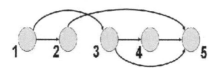

Figure 11.4. A topologically sorted directed acyclic graph. The label sequence kernel can be efficiently computed by dynamic programming running from right to left.

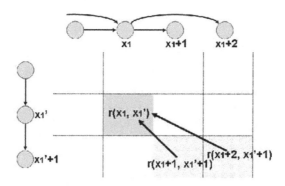

Figure 11.5. Recursion for computing $r(x_1, x_1')$ using recursive equation (2.11). $r(x_1, x_1')$ can be computed based on the precomputed values of $r(x_2, x_2')$, $x_2 > x_1$, $x_2' > x_1'$.

General Directed Graphs.   For cyclic graphs, nodes cannot be topologically sorted. This means that we cannot employ a one-pass dynamic programming algorithm for acyclic graphs. However, we can obtain a recursive form of the kernel like (2.11), and reduce the problem to solving a system of simultaneous linear equations.

Let us rewrite (2.8) as

$$k(G, G') = \lim_{L \to \infty} \sum_{\ell=1}^{L} \sum_{x_1, x_1'} s(x_1, x_1') r_\ell(x_1, x_1'), \qquad (2.12)$$

where

$$r_1(x_1, x_1') := q(x_1, x_1')$$

and

$$r_\ell(x_1, x_1') := \left(\sum_{x_2, x_2'} t(x_2, x_2', x_1, x_1') \left(\sum_{x_3, x_3'} t(x_3, x_3', x_2, x_2') \times \right. \right.$$
$$\left. \left. \left(\cdots \left(\sum_{x_\ell, x_\ell'} t(x_\ell, x_\ell', x_{\ell-1}, x_{\ell-1}') q(x_\ell, x_\ell') \right) \right) \cdots \right) \right)$$
$$\text{for } \ell \geq 2$$

Replacing the order of summation in (2.12), we have the following:

$$
\begin{aligned}
k(G, G') &= \sum_{x_1, x_1'} s(x_1, x_1') \lim_{L \to \infty} \sum_{\ell=1}^{L} r_\ell(x_1, x_1') \\
&= \sum_{x_1, x_1'} s(x_1, x_1') \lim_{L \to \infty} R_L(x_1, x_1'), \qquad (2.13)
\end{aligned}
$$

where

$$R_L(x_1, x_1') := \sum_{\ell=1}^{L} r_\ell(x_1, x_1').$$

Thus we need to compute $R_\infty(x_1, x_1')$ to obtain $k(G, G')$.

Now let us restate this problem in terms of linear system theory [38]. The following recursive relationship holds between r_k and r_{k-1} ($k \geq 2$):

$$r_k(x_1, x_1') = \sum_{i,j} t(i, j, x_1, x_1') r_{k-1}(i, j). \qquad (2.14)$$

Using (2.14), the recursive relationship for R_L also holds as follows:

$$
\begin{aligned}
R_L(x_1, x_1') &= r_1(x_1, x_1') + \sum_{k=2}^{L} r_k(x_1, x_1') \\
&= r_1(x_1, x_1') + \sum_{k=2}^{L} \sum_{i,j} t(i, j, x_1, x_1') r_{k-1}(i, j) \\
&= r_1(x_1, x_1') + \sum_{i,j} t(i, j, x_1, x_1') R_{L-1}(i, j). \quad (2.15)
\end{aligned}
$$

Thus, R_L can be perceived as a discrete-time linear system [38] evolving as the time L increases. Assuming that R_L converges (see [21] for the convergence condition), we have the following equilibrium equation:

$$
R_\infty(x_1, x_1') = r_1(x_1, x_1') + \sum_{i,j} t(i, j, x_1, x_1') R_\infty(i, j). \quad (2.16)
$$

Therefore, the computation of the kernel finally requires solving simultaneous linear equations (2.16) and substituting the solutions into (2.13).

Now let us restate the above discussion in the language of matrices. Let $\mathbf{s}$, $\mathbf{r}_1$, and $\mathbf{r}_\infty$ be $|\mathcal{X}| \cdot |\mathcal{X}'|$ dimensional vectors such that

$$
\begin{aligned}
\mathbf{s} &= (\cdots, s(i, j), \cdots)^\top \\
\mathbf{r}_1 &= (\cdots, r_1(i, j), \cdots)^\top \\
\mathbf{r}_\infty &= (\cdots, R_\infty(i, j), \cdots)^\top
\end{aligned}
$$

Let the transition probability matrix T be a $|\mathcal{X}||\mathcal{X}'| \times |\mathcal{X}||\mathcal{X}'|$ matrix,

$$
[T]_{(i,j),(k,l)} = t(i, j, k, l).
$$

Equation (2.13) can be rewritten as

$$
k(G, G') = \mathbf{r}_\infty^T \mathbf{s} \quad (2.17)
$$

Similarly, the recursive equation (2.16) is rewritten as

$$
\mathbf{r}_\infty = \mathbf{r}_1 + T \mathbf{r}_\infty.
$$

The solution of this equation is

$$
\mathbf{r}_\infty = (I - T)^{-1} \mathbf{r}_1.
$$

Finally, the matrix form of the kernel is

$$
k(G, G') = (I - T)^{-1} \mathbf{r}_1 \mathbf{s}. \quad (2.18)
$$

Computing the kernel requires solving a linear equation or inverting a matrix with $|\mathcal{X}||\mathcal{X}'| \times |\mathcal{X}||\mathcal{X}'|$ coefficients. However, the matrix $I - T$ is actually sparse because the number of non-zero elements of T is less than $c \cdot c' \cdot |\mathcal{X}| \cdot |\mathcal{X}'|$ where c and c' are the maximum out degree of G and G', respectively. Therefore, we can employ efficient numerical algorithms that exploit sparsity [3]. In our implementation, we employed a simple iterative method that updates R_L by using (2.15) until convergence starting from $R_1(x_1, x_1') = r_1(x_1, x_1')$.

2.4   Extensions

Vishwanathan et al. [50] proposed a fast way to compute the graph kernel based on the Sylvestor equation. Let A_X, A_Y and B denote $M \times M$, $N \times N$ and $M \times N$ matrices, respectively. They have used the following equation to speed up the computation.

$$(A_X \otimes A_Y)\text{vec}(B) = \text{vec}(A_X B A_Y)$$

where $\otimes$ corresponds to the Kronecker product (tensor product) and vec is the vectorization operator. The left hand side requires $O(M^2 N^2)$ time, while the right hand side requires only $O(MN(M + N))$ time. Notice that this trick ("vec-trick") has recently been used in link prediction tasks as well [20].

A random walk can trace the same edge back and forth many times ("tottering"), which could be harmful for similarity measurement. Mahe et al. [28] presented an extension of the kernel without tottering and applied it successfully to chemical informatics data.

3.   Graph Boosting

Frequent pattern mining techniques are important tools in data mining [14]. Its simplest form is the classic problem of itemset mining [1], where frequent subsets are enumerated from a series of sets. The original work on this topic is for transactional data, and since then, researchers have applied frequent pattern mining to other structured data such as sequences [35] and trees [2]. Every pattern mining method uses a search tree to systematically organize the patterns. For general graphs, there are technical difficulties about duplication: it is possible to generate the same graph with different paths of the search tree. Methods such as AGM [18] and gspan [52] solve this duplication problem by pruning the search nodes whenever duplicates are found.

The simplest way to apply such pattern mining techniques to graph classification is to build a binary feature vector based on the presence or absence of frequent patterns and apply an off-the-shelf classifier. Such methods are employed in a few chemical informatics papers [16, 23]. However, they are obviously suboptimal because frequent patterns are not necessarily useful for

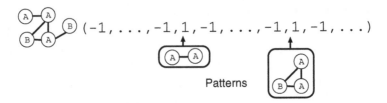

Figure 11.6. Feature space based on subgraph patterns. The feature vector consists of binary pattern indicators.

classification. In chemical data, patterns such as C-C or C-C-C are frequent, but have almost no significance.

To discuss pattern mining strategies for graph classification, let us first define the binary classification problem. The task is to learn a prediction rule from training examples $\{(G_i, y_i)\}_{i=1}^n$, where G_i is a training graph and $y_i \in \{+1, -1\}$ is its associated class label. Let $\mathcal{P}$ be the set of all patterns, i.e., the set of all subgraphs included in at least one training graph, and $d := |\mathcal{P}|$. Then, each graph G_i is encoded as a d-dimensional vector

$$x_{i,p} = \begin{cases} 1 & \text{if } p \subseteq G_i, \\ -1 & \text{otherwise,} \end{cases}$$

This feature space is illustrated in Figure 11.6.

Since the whole feature space is intractably large, we need to obtain a set of informative patterns without enumerating all patterns (i.e., discriminative pattern mining). This problem is close to feature selection in machine learning. The difference is that it is not allowed to scan all features. As in feature selection, we can consider the following three categories in discriminative pattern mining methods: filter, wrapper and embedded [24]. In filter methods, discriminative patterns are collected by a mining call before the learning algorithm is started. They employ a simple statistical criterion such as information gain [31]. In wrapper and embedded methods, the learning algorithm chooses features via minimization of a sparsity-inducing objective function. Typically, they have a high dimensional weight vector and most of these weights coverage to zero after optimization. In most cases, the sparsity is induced by L1-norm regularization [40]. The difference between wrapper and embedded methods are subtle, but wrapper methods tend to be based on heuristic ideas by reducing the features recursively (recursive feature elimination)[13]. Graph boosting is an embedded method, but to deal with graphs, we need to combine L1-norm regularization with graph mining.

3.1 Formulation of Graph Boosting

The name 'boosting' comes from the fact that linear program boosting (LP-Boost) is used as a fundamental computational framework. In chemical informatics experiments [40], it was shown that the accuracy of graph boosting is better than graph kernels. At the same time, key substructures are explicitly discovered.

Our prediction rule is a convex combination of binary indicators $x_{i,j}$, and has the form

$$f(\boldsymbol{x}_i) = \sum_{p \in \mathcal{P}} \beta_p \boldsymbol{x}_{i,p}, \tag{3.1}$$

where $\boldsymbol{\beta}$ is a $|\mathcal{P}|$-dimensional column vector such that $\sum_{p \in \mathcal{P}} \beta_p = 1$ and $\beta_p \geq 0$.

This is a linear discriminant function in an intractably large dimensional space. To obtain an interpretable rule, we need to obtain a *sparse* weight vector $\boldsymbol{\beta}$, where only a few weights are nonzero. In the following, we will present a linear programming approach for efficiently capturing such patterns. Our formulation is based on that of LPBoost [8], and the learning problem is represented as

$$\min_{\boldsymbol{\beta}} \quad \|\boldsymbol{\beta}\|_1 + \lambda \sum_{i=1}^{n} [1 - y_i f(\boldsymbol{x}_i)]_+, \tag{3.2}$$

where $\|x\|_1 = \sum_{i=1}^{n} |x_i|$ denotes the ℓ_1 norm of x, λ is a regularization parameter, and the subscript "+" indicates positive part. A soft-margin formulation of the above problem exists [8], and can be written as follows:

$$\min_{\boldsymbol{\beta}, \boldsymbol{\xi}, \rho} \quad -\rho + \lambda \sum_{i=1}^{n} \xi_i \tag{3.3}$$

$$\text{s.t.} \quad \boldsymbol{y}^\top \boldsymbol{X} \boldsymbol{\beta} + \xi_i \geq \rho, \quad \xi_i \geq 0, \quad i = 1, \dots, n \tag{3.4}$$

$$\sum_{p \in \mathcal{P}} \beta_p = 1, \quad \beta_p \geq 0,$$

where $\boldsymbol{\xi}$ are slack variables, ρ is the margin separating negative examples from positives, $\lambda = \frac{1}{\nu n}$, $\nu \in (0, 1)$ is a parameter controlling the cost of misclassification which has to be found using model selection techniques, such as cross-validation. It is known that the optimal solution has the following ν-property:

Theorem 11.1 ([36]). *Assume that the solution of (3.3) satisfies $\rho \geq 0$. The following statements hold:*

1 ν is an upper-bound of the fraction of margin errors, *i.e., the examples with*

$$\boldsymbol{y}^\top \boldsymbol{X} \boldsymbol{\beta} < \rho.$$

2 ν *is a lower-bound of the fraction of the examples such that*

$$y^\top X\beta < \rho.$$

Directly solving this optimization problem is intractable due to the large number of variables in β. So we solve the following *equivalent* dual problem instead.

$$\min_{u,v} \quad v \tag{3.5}$$

$$\text{s.t.} \quad \sum_{i=1}^{n} u_i y_i x_{i,p} \leq v, \ \forall p \in \mathcal{P} \tag{3.6}$$

$$\sum_{i=1}^{n} u_i = 1, \quad 0 \leq u_i \leq \lambda, \ i = 1, \ldots, n.$$

After solving the dual problem, the primal solution β is obtained from the Lagrange multipliers [8]. The dual problem has a limited number of variables, but a huge number of constraints. Such a linear program can be solved by the *column generation* technique [27]: Starting with an empty pattern set, the pattern whose corresponding constraint is violated the most is identified and added iteratively. Each time a pattern is added, the optimal solution is updated by solving the restricted dual problem. Denote by $u^{(k)}$, $v^{(k)}$ the optimal solution of the restricted problem at iteration $k = 0, 1, \ldots$, and denote by $\hat{X}^{(k)} \subseteq \mathcal{P}$ the set at iteration k. Initially, $\hat{X}^{(0)}$ is empty and $u_i^{(0)} = 1/n$. The restricted problem is defined by replacing the set of constraints (3.6) with

$$\sum_{i=1}^{n} u_i^{(k)} y_i x_{i,p} \leq v, \ \forall p \in \hat{X}^{(k)}.$$

The left hand side of the inequality is called as *gain* in boosting literature. After solving the problem, $\hat{X}^{(k)}$ is updated to $\hat{X}^{(k+1)}$ by adding a column. Several criteria have been proposed to select the new columns [10], but we adopt the most simple rule that is amenable to graph mining: We select the constraint with the largest gain.

$$p^* = \underset{p \in \mathcal{P}}{\operatorname{argmax}} \sum_{i=1}^{n} u_i^{(k)} y_i x_{i,p}. \tag{3.7}$$

The solution set is updated as $\hat{X}^{(k+1)} \leftarrow \hat{X}^{(k)} \cup X_{j^*}$. In the next section, we discuss how to efficiently find the largest gain in detail.

One of the big advantages of our method is that we have a stopping criterion that guarantees that the optimal solution is found: If there is no $p \in \mathcal{P}$ such

Tree of Substructures

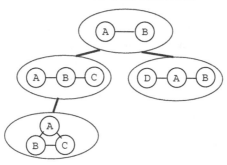

Figure 11.7. Schematic figure of the tree-shaped search space of graph patterns (i.e., the DFS code tree). To find the optimal pattern efficiently, the tree is systematically expanded by rightmost extensions.

that

$$\sum_{i=1}^{n} u_i^{(k)} y_i x_{i,p} > v^{(k)}, \tag{3.8}$$

then the current solution is the optimal dual solution. Empirically, the patterns found in the last few iterations have negligibly small weights. The number of iterations can be decreased by relaxing the condition as

$$\sum_{i=1}^{n} u_i^{(k)} y_i x_{i,p} > v^{(k)} + \epsilon, \tag{3.9}$$

Let us define the primal objective function as $V = -\rho + \lambda \sum_{i=1}^{n} \xi_i$. Due to the convex duality, we can guarantee that, for the solution obtained from the early termination (3.9), the objective satisfies $V \leq V^* + \epsilon$, where V^* is the optimal value with the exact termination (3.8) [8]. In our experiments, $\epsilon = 0.01$ is always used.

3.2 Optimal Pattern Search

Our search strategy is a branch-and-bound algorithm that requires a canonical search space in which a whole set of patterns are enumerated without duplication. As the search space, we adopt the DFS (depth first search) code tree [52]. The basic idea of the DFS code tree is to organize patterns as a tree, where a child node has a super graph of the parent's pattern (Figure 11.7). A pattern is represented as a text string called the DFS code. The patterns are enumerated by generating the tree from the root to leaves using a recursive algorithm. To avoid duplications, node generation is systematically done by rightmost extensions.

All embeddings of a pattern in the graphs $\{G_i\}_{i=1}^n$ are maintained in each node. If a pattern matches a graph in different ways, all such embeddings are stored. When a new pattern is created by adding an edge, it is not necessary to perform full isomorphism checks with respect to all graphs in the database. A new list of embeddings are made by extending the embeddings of the parent [52]. Technically, it is necessary to devise a data structure such that the embeddings are stored incrementally, because it takes a prohibitive amount of memory to keep all embeddings independently in each node. As mentioned in (3.7), our aim is to find the optimal hypothesis that maximizes the gain $g(p)$.

$$g(p) = \sum_{i=1}^n u_i^{(k)} y_i x_{i,p}. \tag{3.10}$$

For efficient search, it is important to minimize the size of the actual search space. To this aim, *tree pruning* is crucially important: Suppose the search tree is generated up to the pattern p and denote by g^* the maximum gain among the ones observed so far. If it is guaranteed that the gain of any super graph p' is not larger than g^*, we can avoid the generation of downstream nodes without losing the optimal pattern. We employ the following pruning condition.

Theorem 11.2. *[30, 26] Let us define*

$$\mu(p) = 2 \sum_{\{i|y_i=+1, p \subseteq G_i\}} u_i^{(k)} - \sum_{i=1}^n y_i u_i^{(k)}.$$

If the following condition is satisfied,

$$g^* > \mu(p), \tag{3.11}$$

the inequality $g(p') < g^*$ holds for any p' such that $p \subseteq p'$.

The gBoost algorithm is summarized in Algorithms 12 and 13.

3.3   Computational Experiments

In [40], it is shown that graph boosting performs better than graph kernels in classification accuracy in chemical compound datasets. The top 20 discriminative subgraphs for a mutagenicity dataset called CPDB are displayed in Figure 11.8. We found that the top 3 substructures with positive weights (0.0672,0.0656, 0.0577) correspond to known *toxicophores* [23]. They correspond to *aromatic amine, aliphatic halide,* and *three-membered heterocycle*, respectively. In addition, the patterns with weights 0.0431, 0.0412, 0.0411 and 0.0318 seem to be related to *polycyclic aromatic systems*. Only from this result, we cannot conclude that graph boosting is better in general data. However, since important chemical substructures cannot be represented in paths, it would be reasonable to say that subgraph features are better in chemical data.

Algorithm 12 gBoost algorithm: main part

1: $\hat{\boldsymbol{X}}^{(0)} = \emptyset, \boldsymbol{u}_i^{(0)} = 1/n, k = 0$
2: **loop**
3:     Find the optimal pattern p^* based on $\boldsymbol{u}^{(k)}$
4:     **if** termination condition (3.9) holds **then**
5:         break
6:     **end if**
7:     $\hat{\boldsymbol{X}} \leftarrow \hat{\boldsymbol{X}} \cup \boldsymbol{X}_{j^*}$
8:     Solve the restricted dual problem (3.5) to obtain $\boldsymbol{u}^{(k+1)}$
9:     $k = k + 1$
10: **end loop**

Algorithm 13 Finding the Optimal Pattern

1: **Procedure** OPTIMAL PATTERN
2: Global variables: g^*, p^*
3: $g^* = -\infty$
4: **for** $p \in$ DFS codes with single nodes **do**
5:     project(p)
6: **end for**
7: return p^*
8: **EndProcedure**
9:
10: **Function** PROJECT(p)
11: **if** p is not a minimum DFS code **then**
12:     return
13: **end if**
14: **if** pruning condition (3.11) holds **then**
15:     return
16: **end if**
17: **if** $g(p) > g^*$ **then**
18:     $g^* = g(p), p^* = p$
19: **end if**
20: **for** $p' \in$ rightmost extensions of p **do**
21:     project(p')
22: **end for**
23: **EndFunction**

3.4   Related Work

Graph algorithms can be designed based on existing statistical frameworks (i.e., mother algorithms). It allows us to use theoretical results and insights

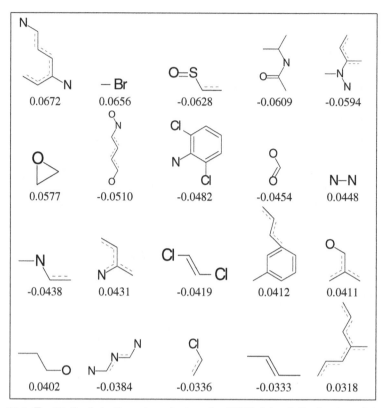

Figure 11.8. Top 20 discriminative subgraphs from the CPDB dataset. Each subgraph is shown with the corresponding weight, and ordered by the absolute value from the top left to the bottom right. H atom is omitted, and C atom is represented as a dot for simplicity. Aromatic bonds appeared in an open form are displayed by the combination of dashed and solid lines.

accumulated in the past studies. In graph boosting, we employed LPboost as a mother algorithm. It is possible to employ other algorithms such as partial least squares regression (PLS) [39] and least angle regression (LARS) [45].

When applied to ordinary vectorial data, partial least squares regression extracts a few orthogonal features and perform least squares regression in the projected space [37]. A PLS feature is a linear combination of original features, and it is often the case that correlated features are summarized into a PLS feature. Sometimes, the subgraph features chosen by graph boosting is not robust against bootstrapping or other data perturbations, whereas the classification accuracy is quite stable. It is due to strong correlation among features corresponding to similar subgraphs. The graph mining version of PLS, gPLS [39], solves this problem by summarizing similar subgraphs into each feature (Figure 11.9). Since only one graph mining call is required to construct each

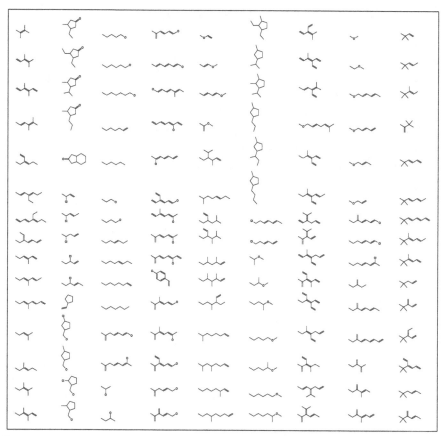

Figure 11.9. Patterns obtained by gPLS. Each column corresponds to the patterns of a PLS component.

feature, gPLS can build the classification rule more quickly than graph boosting.

In graph boosting, it is necessary to set the regularization parameter λ in (3.2). Typically it is determined by cross validation, but there is a different approach called "regularization path tracking". When $\lambda = 0$, the weight vector converges to the origin. As λ is increased continuously, the weight vector draws a piecewise linear path. Because of this property, one can track the whole path by repeating to jump to the next turning point. We combined the tracking with graph mining in [45]. In ordinary tracking, a feature is added or removed at each turning point. In our graph version, a subgraph to add or remove is found by a customized gSpan search.

The examples shown above were for supervised classification. For unsupervised clustering of graphs, the combinations with the EM algorithm [46] and the Dirichlet process [47] have been reported.

4.   Applications of Graph Classification

Borgwardt et al. [5] applied the graph kernel method to classify protein 3D structures. It outperformed classical alignment-based approaches. Karklin et al. [19] built a classifier for non-coding RNAs employing a graph representation of RNAs. Outside biology and chemistry, Harchaoui and Bach [15] applied graph kernels to image classification where each region corresponds to a node and their positional relationships are represented by edges.

Traditionally, graph mining methods are mainly used for small chemical compounds [28, 9]. However, new application areas are emerging. In image processing [34], geometric relationships between points are represented as edges. Software bug detection is an interesting area, where the relationships of APIs are represented as directed graphs and anomalous patterns are detected to identify bugs [11]. In natural language processing, the relationships between words are represented as a graph (e.g., predicate-argument structures) and key phrases are identified as subgraphs [26].

5.   Label Propagation

In the previous discussion, the term graph classification means classifying an entire graph. In many applications, we are interested in classifying the nodes. For example, in large-scale network analysis for social networks and biological networks, it is a central task to classify unlabeled nodes given a limited number of labeled nodes (Figure 11.1, right). In FaceBook, one can label people who responded to a certain advertisement as positive nodes, and people who did not respond as negative nodes. Based on these labeled nodes, our task is to predict other people's response to the advertisement.

In earlier studies, diffusion kernels are used in combination with support vector machines [25, 48]. The basic idea is to compute the closeness between two nodes in terms of commute time of random walks between the nodes. Though this approach gained popularity in the machine learning community, a significant drawback is that the derived kernel matrix is dense. For large networks, the diffusion kernel is not suitable because it takes $O(n^3)$ time and $O(n^2)$ memory. In contrast, label propagation methods use simpler computational strategies that exploit sparsity of the adjacency matrix [54, 53]. The label propagation method of Zhou et al.[53] is achieved by solving simultaneous linear equations with a sparse coefficient matrix. The time complexity is nearly linear to the number of non-zero entries of the coefficient matrix [49], which is much more efficient than the diffusion kernels. Due to its efficiency, label propagation is gaining popularity in applications with biological networks, where web servers should return the propagation result without much delay [32]. However, the classification performance is quite sensitive to methodological details. For example, Shin et al. pointed out that the introduction of directional propagation can increase the performance significantly [43]. Also, Mostafavi et al. [32] reported that their engineered version has outperformed the vanilla version [53]. Label propagation is still an active research field. Recent extensions include automatic combination of multiple networks [49, 22] and the introduction of probabilistic inference in label propagation [54, 44].

6. Concluding Remarks

We have covered the two different methods for graph classification. Graph kernel is a similarity measure between two graphs, while graph mining methods can derive characteristic subgraphs that can be used for any subsequent machine learning algorithms. We have the impression that so far graph kernels are more frequently applied. Probably it is due to the fact that graph kernels are easier to implement and currently used graph datasets are not so large. However, graph kernels are not suitable for very large data, because it takes $O(n^2)$ time to derive the kernel matrix of n training graphs, which is very hard to improve. Toward large scale data, graph mining methods seem more promising because it requires only $O(n)$ time. Nevertheless, there remains much to be done in graph mining methods. Existing methods such as gSpan enumerate all subgraphs satisfying a certain frequency-based criterion. However, it is often pointed out that, for graph classification, it is not always necessary to enumerate all subgraphs. Recently, Boley and Grosskreutz proposed a uniform sampling method of frequent itemsets [4]. Such theoretically guaranteed sampling procedures will certainly contribute to graph classification as well.

One fact that hinders the further popularity of graph mining methods is that it is not common to make the code public in the machine learning and data mining community. We have made several easy-to-use code available: SPIDER (http://www.kyb.tuebingen.mpg.de/bs/people/spider/) contains codes for graph kernels and the gBoost package contains codes for graph mining and boosting (http://www.kyb.mpg.de/bs/people/nowozin/gboost/).

References

[1] R. Agrawal and R. Srikant. Fast algorithms for mining association rules in large databases. In *Proc. VLDB 1994*, pages 487–499, 1994.

[2] T. Asai, K. Abe, S. Kawasoe, H. Arimura, H. Sakamoto, and S. Arikawa. Efficient substructure discovery from large semi-structured data. In *Proc 2nd SIAM Data Mining Conference (SDM)*, pages 158–174, 2002.

[3] R. Barrett, M. Berry, T. F. Chan, J. Demmel, J. Donato, J. Dongarra, V. Eijkhout, R. Pozo, C. Romine, and H. Van der Vorst. *Templates for the Solution of Linear Systems: Building Blocks for Iterative Methods, 2nd Edition.*

SIAM, Philadelphia, PA, 1994.

[4] M. Boley and H. Grosskreutz. A randomized approach for approximating the number of frequent sets. In *Proceedings of the 8th IEEE International Conference on Data Mining*, pages 43–52, 2008.

[5] K. M. Borgwardt, C. S. Ong, S. Schonauer, S. V. N. Vishwanathan, A. J. Smola, and H.-P. Kriegel. Protein function prediction via graph kernels. *Bioinformatics*, 21(suppl. 1):i47–i56, 2006.

[6] O. Chapelle, A. Zien, and B. Schelkopf, editors. *Semi-Supervised Learning*. MIT Press, Cambridge, MA, 2006.

[7] T. Cormen, C. Leiserson, and R. Rivest. Introduction to Algorithms. MIT Press and McGraw Hill, 1990.

[8] A. Demiriz, K.P. Bennet, and J. Shawe-Taylor. Linear programming boosting via column generation. *Machine Learning*, 46(1-3):225–254, 2002.

[9] M. Deshpande, M. Kuramochi, N. Wale, and G. Karypis. Frequent substructure-based approaches for classifying chemical compounds. *IEEE Trans. Knowl. Data Eng.*, 17(8):1036–1050, 2005.

[10] O. du Merle, D. Villeneuve, J. Desrosiers, and P. Hansen. Stabilized column generation. *Discrete Mathematics*, 194:229–237, 1999.

[11] F. Eichinger, K. Behm, and M. Huber. Mining edge-weighted call graphs to localise software bugs. In *Proceedings of the European Conference on Machine Learning and Principles and Practice of Knowledge Discovery in Databases (ECML PKDD)*, pages 333–348, 2008.

[12] T. Gartner, P. Flach, and S. Wrobel. On graph kernels: Hardness results and efficient alternatives. In *Proc. of the Sixteenth Annual Conference on Computational Learning Theory*, 2003.

[13] I. Guyon, J. Weston, S. Bahnhill, and V. Vapnik. Gene selection for cancer classification using support vector machines. *Machine Learning*, 46(1-3):389–422, 2002.

[14] J. Han and M. Kamber. *Data Mining: Concepts and Techniques*. Morgan Kaufmann, 2000.

[15] Z. Harchaoui and F. Bach. Image classification with segmentation graph kernels. In *2007 IEEE Computer Society Conference on Computer Vision and Pattern Recognition*. IEEE Computer Society, 2007.

[16] C. Helma, T. Cramer, S. Kramer, and L.D. Raedt. Data mining and machine learning techniques for the identification of mutagenicity inducing substructures and structure activity relationships of noncongeneric compounds. *J. Chem. Inf. Comput. Sci.*, 44:1402–1411, 2004.

[17] T. Horvath, T. Gartner, and S. Wrobel. Cyclic pattern kernels for predictive graph mining. In *Proceedings of the 10th ACM SIGKDD International*

Conference on Knowledge Discovery and Data Mining, pages 158–167, 2004.

[18] A. Inokuchi. Mining generalized substructures from a set of labeled graphs. In *Proceedings of the 4th IEEE Internatinal Conference on Data Mining*, pages 415–418. IEEE Computer Society, 2005.

[19] Y. Karklin, R.F. Meraz, and S.R. Holbrook. Classification of non-coding rna using graph representations of secondary structure. In *Pacific Symposium on Biocomputing*, pages 4–15, 2005.

[20] H. Kashima, T. Kato, Y. Yamanishi, M. Sugiyama, and K. Tsuda. Link propagation: A fast semi-supervised learning algorithm for link prediction. In *2009 SIAM Conference on Data Mining*, pages 1100–1111, 2009.

[21] H. Kashima, K. Tsuda, and A. Inokuchi. Marginalized kernels between labeled graphs. In *Proceedings of the 21st International Conference on Machine Learning*, pages 321–328. AAAI Press, 2003.

[22] T. Kato, H. Kashima, and M. Sugiyama. Robust label propagation on multiple networks. *IEEE Trans. Neural Networks*, 20(1):35–44, 2008.

[23] J. Kazius, S. Nijssen, J. Kok, T. Back, and A.P. Ijzerman. Substructure mining using elaborate chemical representation. *J. Chem. Inf. Model.*, 46:597–605, 2006.

[24] R. Kohavi and G. H. John. Wrappers for feature subset selection. *Artificial Intelligence*, 1-2:273–324, 1997.

[25] R. I. Kondor and J. Lafferty. Diffusion kernels on graphs and other discrete input. In *ICML 2002*, 2002.

[26] T. Kudo, E. Maeda, and Y. Matsumoto. An application of boosting to graph classification. In *Advances in Neural Information Processing Systems 17*, pages 729–736. MIT Press, 2005.

[27] D. G. Luenberger. *Optimization by Vector Space Methods*. Wiley, 1969.

[28] P. Mahe, N. Ueda, T. Akutsu, J.-L. Perret, and J.-P. Vert. Graph kernels for molecular structure - activity relationship analysis with support vector machines. *J. Chem. Inf. Model.*, 45:939–951, 2005.

[29] P. Mahe and J.P. Vert. Graph kernels based on tree patterns for molecules. *Machine Learning*, 75:3–35, 2009.

[30] S. Morishita. Computing optimal hypotheses efficiently for boosting. In *Discovery Science*, pages 471–481, 2001.

[31] S. Morishita and J. Sese. Traversing itemset lattices with statistical metric pruning. In *Proceedings of ACM SIGACT-SIGMOD-SIGART Symposium on Database Systems (PODS)*, pages 226–236, 2000.

[32] S. Mostafavi, D. Ray, D. Warde-Farley, C. Grouios, and Q. Morris. GeneMANIA: a real-time multiple association network integration algorithm for predicting gene function. *Genome Biology*, 9(Suppl. 1):S4, 2008.

[33] S. Nijssen and J.N. Kok. A quickstart in frequent structure mining can make a difference. In *Proceedings of the 10th ACM SIGKDD International Conference on Knowledge Discovery and Data Mining*, pages 647–652. ACM Press, 2004.

[34] S. Nowozin, K. Tsuda, T. Uno, T. Kudo, and G. Bakir. Weighted substructure mining for image analysis. In *IEEE Computer Society Conference on Computer Vision and Pattern Recognition (CVPR)*. IEEE Computer Society, 2007.

[35] J. Pei, J. Han, B. Mortazavi-asl, J. Wang, H. Pinto, Q. Chen, U. Dayal, and M. Hsu. Mining sequential patterns by pattern-growth: The prefixspan approach. *IEEE Transactions on Knowledge and Data Engineering*, 16(11):1424–1440, 2004.

[36] G. Ratsch, S. Mika, B. Schelkopf, and K.-R. Muller. Constructing boosting algorithms from SVMs: an application to one-class classification. *IEEE Trans. Patt. Anal. Mach. Intell.*, 24(9):1184–1199, 2002.

[37] R. Rosipal and N. Kramer. Overview and recent advances in partial least squares. In *Subspace, Latent Structure and Feature Selection Techniques*, pages 34–51. Springer, 2006.

[38] W.J. Rugh. *Linear System Theory*. Prentice Hall, 1995.

[39] H. Saigo, N. Kramer, and K. Tsuda. Partial least squares regression for graph mining. In *Proceedings of the 14th ACM SIGKDD International Conference on Knowledge Discovery and Data Mining*, pages 578–586, 2008.

[40] H. Saigo, S. Nowozin, T. Kadowaki, T. Kudo, and K. Tsuda. GBoost: A mathematical programming approach to graph classification and regression. *Machine Learning*, 2008.

[41] A. Sanfeliu and K.S. Fu. A distance measure between attributed relational graphs for pattern recognition. *IEEE Trans. Syst. Man Cybern.*, 13:353–362, 1983.

[42] B. Schelkopf and A. J. Smola. *Learning with Kernels: Support Vector Machines, Regularization, Optimization, and Beyond*. MIT Press, 2002.

[43] H. Shin, A.M. Lisewski, and O. Lichtarge. Graph sharpening plus graph integration: a synergy that improves protein functional classification. *Bioinformatics*, 23:3217–3224, 2007.

[44] A. Subramanya and J. Bilmes. Soft-supervised learning for text classification. In *Proceedings of the 2008 Conference on Empirical Methods in Natural Language Processing*, pages 1090–1099, 2008.

[45] K. Tsuda. Entire regularization paths for graph data. In *Proceedings of the 24th International Conference on Machine Learning*, pages 919–926, 2007.

[46] K. Tsuda and T. Kudo. Clustering graphs by weighted substructure mining. In *Proceedings of the 23rd International Conference on Machine Learning*, pages 953–960. ACM Press, 2006.

[47] K. Tsuda and K. Kurihara. Graph mining with variational dirichlet process mixture models. In *SIAM Conference on Data Mining (SDM)*, 2008.

[48] K. Tsuda and W.S. Noble. Learning kernels from biological networks by maximizing entropy. *Bioinformatics*, 20(Suppl. 1):i326–i333, 2004.

[49] K. Tsuda, H.J. Shin, and B. Schelkopf. Fast protein classification with multiple networks. *Bioinformatics*, 21(Suppl. 2):ii59–ii65, 2005.

[50] S.V.N. Vishwanathan, K.M. Borgwardt, and N.N. Schraudolph. Fast computation of graph kernels. In *Advances in Neural Information Processing Systems 19*, Cambridge, MA, 2006. MIT Press.

[51] N. Wale and G. Karypis. Comparison of descriptor spaces for chemical compound retrieval and classification. In *Proceedings of the 2006 IEEE International Conference on Data Mining*, pages 678–689, 2006.

[52] X. Yan and J. Han. gSpan: graph-based substructure pattern mining. In *Proceedings of the 2002 IEEE International Conference on Data Mining*, pages 721–724. IEEE Computer Society, 2002.

[53] D. Zhou, O. Bousquet, J. Weston, and B. Schelkopf. Learning with local and global consistency. In *Advances in Neural Information Processing Systems (NIPS) 16*, pages 321–328. MIT Press, 2004.

[54] X. Zhu, Z. Ghahramani, and J. Lafferty. Semi-supervised learning using gaussian fields and harmonic functions. In *Proc. of the Twentieth International Conference on Machine Learning (ICML)*, pages 912–919. AAAI Press, 2003.

Chapter 12

MINING GRAPH PATTERNS

Hong Cheng
Department of Systems Engineering and Engineering Management
Chinese University of Hong Kong
hcheng@se.cuhk.edu.hk

Xifeng Yan
Department of Computer Science
University of California at Santa Barbara
xyan@cs.ucsb.edu

Jiawei Han
Department of Computer Science
University of Illinois at Urbana-Champaign
hanj@cs.uiuc.edu

Abstract   Graph pattern mining becomes increasingly crucial to applications in a variety of domains including bioinformatics, cheminformatics, social network analysis, computer vision and multimedia. In this chapter, we first examine the existing frequent subgraph mining algorithms and discuss their computational bottleneck. Then we introduce recent studies on mining significant and representative subgraph patterns. These new mining algorithms represent the state-of-the-art graph mining techniques: they not only avoid the exponential size of mining result, but also improve the applicability of graph patterns significantly.

Keywords: Apriori, frequent subgraph, graph pattern, significant pattern, representative pattern

C.C. Aggarwal and H. Wang (eds.), *Managing and Mining Graph Data,*
Advances in Database Systems 40, DOI 10.1007/978-1-4419-6045-0_12,
© Springer Science+Business Media, LLC 2010

1.   Introduction

Frequent pattern mining has been a focused theme in data mining research for over a decade. Abundant literature has been dedicated to this research area and tremendous progress has been made, including efficient and scalable algorithms for frequent itemset mining, frequent sequential pattern mining, frequent subgraph mining, as well as their broad applications.

Frequent graph patterns are subgraphs that are found from a collection of graphs or a single massive graph with a frequency no less than a user-specified support threshold. Frequent subgraphs are useful at characterizing graph sets, discriminating different groups of graphs, classifying and clustering graphs, and building graph indices. Borgelt and Berthold [2] illustrated the discovery of active chemical structures in an HIV-screening dataset by contrasting the support of frequent graphs between different classes. Deshpande et al. [7] used frequent structures as features to classify chemical compounds. Huan et al. [13] successfully applied the frequent graph mining technique to study protein structural families. Frequent graph patterns were also used as indexing features by Yan et al. [35] to perform fast graph search. Their method outperforms the traditional path-based indexing approach significantly. Koyuturk et al. [18] proposed a method to detect frequent subgraphs in biological networks, where considerably large frequent sub-pathways in metabolic networks are observed.

In this chapter, we will first review the existing graph pattern mining methods and identify the combinatorial explosion problem in these methods – the graph pattern search space grows exponentially with the pattern size. It causes two serious problems: (1) the computational bottleneck, *i.e.*, it takes very long, or even forever, for the algorithms to complete the mining process, and (2) patterns' applicability, *i.e.*, the huge mining result set hinders the potential usage of graph patterns in many real-life applications. We will then introduce scalable graph pattern mining paradigms which mine *significant* subgraphs [19, 11, 27, 25, 31, 24] and *representative* subgraphs [10].

2.   Frequent Subgraph Mining

2.1   Problem Definition

The vertex set of a graph g is denoted by $V(g)$ and the edge set by $E(g)$. A label function, l, maps a vertex or an edge to a label. A graph g is a subgraph of another graph g' if there exists a subgraph isomorphism from g to g', denoted by $g \subseteq g'$. g' is called a supergraph of g.

Definition 12.1 (Subgraph Isomorphism). *For two labeled graphs g and g', a subgraph isomorphism is an injective function $f : V(g) \rightarrow V(g')$, s.t., (1), $\forall v \in V(g), l(v) = l'(f(v))$; and (2), $\forall (u, v) \in E(g), (f(u),$*

$f(v)) \in E(g')$ and $l(u, v) = l'(f(u), f(v))$, where l and l' are the labeling functions of g and g', respectively. f is called an embedding of g in g'.

Definition 12.2 (Frequent Graph). *Given a labeled graph dataset $D = \{G_1, G_2, \ldots, G_n\}$ and a subgraph g, the supporting graph set of g is $D_g = \{G_i | g \subseteq G_i, G_i \in D\}$. The support of g is $support(g) = \frac{|D_g|}{|D|}$. A frequent graph is a graph whose support is no less than a minimum support threshold, min_sup.*

An important property, called *anti-monotonicity*, is crucial to confine the search space of frequent subgraph mining.

Definition 12.3 (Anti-Monotonicity). *Anti-monotonicity means that a size-k subgraph is frequent only if all of its subgraphs are frequent.*

Many frequent graph pattern mining algorithms [12, 6, 16, 20, 28, 32, 2, 14, 15, 22, 21, 8, 3] have been proposed. Holder et al. [12] developed SUBDUE to do approximate graph pattern discovery based on minimum description length and background knowledge. Dehaspe et al. [6] applied inductive logic programming to predict chemical carcinogenicity by mining frequent subgraphs. Besides these studies, there are two basic approaches to the frequent subgraph mining problem: the Apriori-based approach and the pattern-growth approach.

2.2 Apriori-based Approach

Apriori-based frequent subgraph mining algorithms share similar characteristics with Apriori-based frequent itemset mining algorithms. The search for frequent subgraphs starts with small-size subgraphs, and proceeds in a bottom-up manner. At each iteration, the size of newly discovered frequent subgraphs is increased by one. These new subgraphs are generated by joining two similar but slightly different frequent subgraphs that were discovered already. The frequency of the newly formed graphs is then checked. The framework of Apriori-based methods is outlined in Algorithm 14.

Typical Apriori-based frequent subgraph mining algorithms include AGM by Inokuchi et al. [16], FSG by Kuramochi and Karypis [20], and an edge-disjoint path-join algorithm by Vanetik et al. [28].

The AGM algorithm uses a *vertex-based candidate generation* method that increases the subgraph size by one vertex in each iteration. Two size-$(k + 1)$ frequent subgraphs are joined only when the two graphs have the same size-k subgraph. Here, *graph size* means the number of vertices in a graph. The newly formed candidate includes the common size-k subgraph and the additional two vertices from the two size-$(k + 1)$ patterns. Figure 12.1 depicts the two subgraphs joined by two chains.

Algorithm 14 Apriori(D, *min_sup*, S_k)

Input: Graph dataset D, minimum support threshold *min_sup*,
      size-k frequent subgraphs S_k
Output: The set of size-$(k+1)$ frequent subgraphs S_{k+1}

1: $S_{k+1} \leftarrow \varnothing$;
2: **for** each frequent subgraph $g_i \in S_k$ **do**
3:    **for** each frequent subgraph $g_j \in S_k$ **do**
4:       **for** each size-$(k+1)$ graph g formed by joining g_i and g_j **do**
5:          **if** g is frequent in D and $g \notin S_{k+1}$ **then**
6:             insert g to S_{k+1};
7: **if** $S_{k+1} \neq \varnothing$ **then**
8:    call Apriori(D, *min_sup*, S_{k+1});
9: **return**;

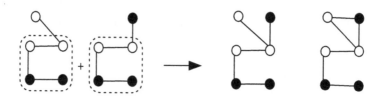

Figure 12.1. AGM: Two candidate patterns formed by two chains

The **FSG** algorithm adopts an *edge-based candidate generation* strategy that increases the subgraph size by one edge in each iteration. Two size-$(k+1)$ patterns are merged if and only if they share the same subgraph having k edges. In the *edge-disjoint path* method [28], graphs are classified by the number of disjoint paths they have, and two paths are edge-disjoint if they do not share any common edge. A subgraph pattern with $k+1$ disjoint paths is generated by joining subgraphs with k disjoint paths.

The **Apriori**-based algorithms mentioned above have considerable overhead when two size-k frequent subgraphs are joined to generate size-$(k+1)$ candidate patterns. In order to avoid this kind of overhead, non-**Apriori**-based algorithms were developed, most of which adopt the pattern-growth methodology, as discussed below.

2.3   Pattern-Growth Approach

Pattern-growth graph mining algorithms include **gSpan** by Yan and Han [32], **MoFa** by Borgelt and Berthold [2], **FFSM** by Huan et al. [14], **SPIN** by Huan et al. [15], and **Gaston** by Nijssen and Kok [22]. These algorithms are inspired by PrefixSpan [23], TreeMinerV [37], and FREQT [1] in mining sequences and trees, respectively.

The pattern-growth algorithm extends a frequent graph directly by adding a new edge, in every possible position. It does not perform expensive join operations. A potential problem with the edge extension is that the same graph can be discovered multiple times. The gSpan algorithm helps avoiding the discovery of duplicates by introducing a *right-most extension* technique, where the only extensions take place on the *right-most path* [32]. A right-most path for a given graph is the straight path from the starting vertex v_0 to the last vertex v_n, according to a depth-first search on the graph.

Besides the frequent subgraph mining algorithms, constraint-based subgraph mining algorithms have also been proposed. Mining closed graph patterns was studied by Yan and Han [33]. Mining coherent subgraphs was studied by Huan et al. [13]. Chi et al. proposed CMTreeMiner to mine closed and maximal frequent subtrees [5]. For relational graph mining, Yan et al. [36] developed two algorithms, CloseCut and Splat, to discover exact dense frequent subgraphs in a set of relational graphs. For large-scale graph database mining, a disk-based frequent graph mining method was introduced by Wang et al. [29]. Jin et al. [17] proposed an algorithm, TSMiner, for mining frequent large-scale structures (defined as topological structures) from graph datasets.

For a comprehensive introduction on basic graph pattern mining algorithms including Apriori-based and pattern-growth approaches, readers are referred to the survey written by Washio and Motoda [30] and Yan and Han [34].

2.4   Closed and Maximal Subgraphs

A major challenge in mining frequent subgraphs is that the mining process often generates a huge number of patterns. This is because if a subgraph is frequent, all of its subgraphs are frequent as well. A frequent graph pattern with n edges can potentially have 2^n frequent subgraphs, which is an exponential number. To overcome this problem, *closed subgraph mining* and *maximal subgraph mining* algorithms were proposed.

Definition 12.4 (Closed Subgraph). *A subgraph g is a closed subgraph in a graph set D if g is frequent in D and there exists no proper supergraph g' such that $g \subset g'$ and g' has the same support as g in D.*

Definition 12.5 (Maximal Subgraph). *A subgraph g is a maximal subgraph in a graph set D if g is frequent, and there exists no supergraph g' such that $g \subset g'$ and g' is frequent in D.*

The set of closed frequent subgraphs contains the complete information of frequent patterns; whereas the set of maximal subgraphs, though more compact, usually does not contain the complete support information regarding to its corresponding frequent sub-patterns. Close subgraph mining methods include CloseGraph [33]. Maximal subgraph mining methods include SPIN [15] and MARGIN [26].

2.5   Mining Subgraphs in a Single Graph

While most frequent subgraph mining algorithms assume the input graph data is a set of graphs $D = \{G_1, ..., G_n\}$, there are some studies [21, 8, 3] on mining graph patterns from a single large graph. Defining the support of a subgraph in a set of graphs is straightforward, which is the number of graphs in the database that contain the subgraph. However, it is much more difficult to find an appropriate support definition in a single large graph since multiple embeddings of a subgraph may have overlaps. If arbitrary overlaps between non-identical embeddings are allowed, the resulting support does not satisfy the anti-monotonicity property, which is essential for most frequent pattern mining algorithms. Therefore, [21, 8, 3] investigated appropriate support measures in a single graph.

Kuramochi and Karypis [21] proposed two efficient algorithms that can find frequent subgraphs within a large sparse graph. The first algorithm, called HSIGRAM, follows a horizontal approach and finds frequent subgraphs in a breadth-first fashion. The second algorithm, called VSIGRAM, follows a vertical approach and finds the frequent subgraphs in a depth-first fashion. For the support measure defined in [21], all possible occurrences φ of a pattern p in a graph g are calculated. An *overlap-graph* is constructed where each occurrence φ corresponds to a node and there is an edge between the nodes of φ and φ' if they overlap. This is called *simple overlap* as defined below.

Definition 12.6 (Simple Overlap). *Given a pattern $p = (V(p), E(p))$, a simple overlap of occurrences φ and φ' of pattern p exists if $\varphi(E(p)) \cap \varphi'(E(p)) \neq \varnothing$.*

The support of p is defined as the size of the maximum independent set (MIS) of the overlap-graph. A later study [8] proved that the MIS-support is anti-monotone.

Fiedler and Borgelt [8] suggested a definition that relies on the non-existence of equivalent ancestor embeddings in order to guarantee that the resulting support is anti-monotone. The support is called *harmful overlap support*. The basic idea of this measure is that some of the simple overlaps (in [21]) can be disregarded without harming the anti-monotonicity of the support measure. As in [21], an overlap graph is constructed and the support is defined as the size of the MIS. The major difference is the definition of the overlap.

Definition 12.7 (Harmful Overlap). *Given a pattern* $p = (V(p), E(p))$, *a harmful overlap of occurrences* φ *and* φ' *of pattern* p *exists if* $\exists v \in V(p) :$ $\varphi(v), \varphi'(v) \in \varphi(V(p)) \cap \varphi'(V(p))$.

Bringmann and Nijssen [3] examined the existing studies [21, 8] and identified the expensive operation of solving the MIS problem. They defined a new support measure.

Definition 12.8 (Minimum Image based Support). *Given a pattern* $p = (V(p), E(p))$, *the minimum image based support of* p *in* g *is defined as*

$$\sigma_\wedge(p, g) = \min_{v \in V(p)} |\{\varphi_i(v) : \varphi_i \text{ is an occurrence of } p \text{ in } g\}|.$$

It is based on the number of unique nodes in the graph g to which a node of the pattern p is mapped. This measure avoids the MIS computation. Therefore it is computationally less expensive and often closer to intuition than measures proposed in [21, 8].

By taking the node in p which is mapped to the least number of unique nodes in g, the anti-monotonicity of $\sigma_\wedge$ can be guaranteed. For the definition of support, several computational benefits could be identified: (1) instead of $O(n^2)$ potential overlaps, where n is the possibly exponential number of occurrences, the method only needs to maintain a set of vertices for every node in the pattern, which can be done in $O(n)$; (2) the method does not need to solve an NP complete MIS problem; and (3) it is not necessary to compute all occurrences: it is sufficient to determine for every pair of $v \in V(p)$ and $v' \in V(g)$ if there is one occurrence in which $\varphi(v) = v'$.

2.6   The Computational Bottleneck

Most graph mining methods follow the combinatorial pattern enumeration paradigm. In real world applications including bioinformatics and social network analysis, the complete enumeration of patterns is practically infeasible. It often turns out that the mining results, even those for closed graphs [33] or maximal graphs [15], are explosive in size.

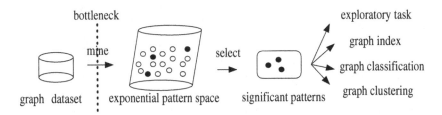

Figure 12.2. Graph Pattern Application Pipeline

Figure 12.2 depicts the pipeline of graph applications built on frequent subgraphs. In this pipeline, frequent subgraphs are mined first; then significant patterns are selected based on user-defined objective functions for different applications. Unfortunately, the potential of graph patterns is hindered by the limitation of this pipeline, due to a scalability issue. For instance, in order to find subgraphs with the highest statistical significance, one has to enumerate all the frequent subgraphs first, and then calculate their p-value one by one. Obviously, this two-step process is not scalable due to the following two reasons: (1) for many objective functions, the minimum frequency threshold has to be set very low so that none of significant patterns will be missed—a low-frequency threshold often means an exponential pattern set and an extremely slow mining process; and (2) there is a lot of redundancy in frequent subgraphs; most of them are not worth computing at all. When the complete mining results are prohibitively large, yet only the significant or representative ones are of real interest. It is inefficient to wait forever for the mining algorithm to finish and then apply post-processing to the huge mining result. In order to complete mining in a limited period of time, a user usually has to sacrifice patterns' quality. In short, the frequent subgraph mining step becomes the bottleneck of the whole pipeline in Figure 12.2.

In the following discussion, we will introduce recent graph pattern mining methods that overcome the scalability bottleneck. The first series of studies [19, 11, 27, 31, 25, 24] focus on mining the optimal or significant subgraphs according to user-specified objective functions in a timely fashion by accessing only a small subset of promising subgraphs. The second study [10] by Hasan et al. generates an orthogonal set of graph patterns that are representative. All these studies avoid generating the complete set of frequent subgraphs while presenting only a compact set of interesting subgraph patterns, thus solving the scalability and applicability issues.

3. Mining Significant Graph Patterns

3.1 Problem Definition

Given a graph database $D = \{G_1, ..., G_n\}$ and an objective function F, a general problem definition for mining significant graph patterns can be formulated in two different ways: (1) find all subgraphs g such that $F(g) \geq \delta$ where δ is a significance threshold; or (2) find a subgraph g^* such that $g^* = \operatorname{argmax}_g F(g)$. No matter which formulation or which objective function is used, an efficient mining algorithm shall find significant patterns directly without exhaustively generating the whole set of graph patterns. There are several algorithms [19, 11, 27, 31, 25, 24] proposed with different objective functions and pruning techniques. We are going to discuss four recent studies: gboost [19], gPLS [25], LEAP [31] and GraphSig [24].

3.2 gboost: A Branch-and-Bound Approach

Kudo et al. [19] presented an application of boosting for classifying labeled graphs, such as chemical compounds, natural language texts, *etc.* A weak classifier called decision stump uses a subgraph as a classification feature. Then a boosting algorithm repeatedly constructs multiple weak classifiers on weighted training instances. A gain function is designed to evaluate the quality of a decision stump, *i.e.*, how many weighted training instances can be correctly classified. Then the problem of finding the optimal decision stump in each iteration is formulated as mining an "optimal" subgraph pattern. gboost designs a branch-and-bound mining approach based on the gain function and integrates it into gSpan to search for the "optimal" subgraph pattern.

A Boosting Framework. gboost uses a simple classifier, *decision stump*, for prediction according to a single feature. The subgraph-based decision stump is defined as follows.

Definition 12.9 (Decision Stumps for Graphs). *Let* t *and* x *be labeled graphs and* $y \in \{\pm1\}$ *be a class label. A decision stump classifier for graphs is given by*

$$h_{\langle t,y \rangle}(x) = \begin{cases} y, & t \subseteq x \\ -y, & otherwise \end{cases}.$$

The decision stumps are trained to find a rule $\langle \hat{t}, \hat{y} \rangle$ that minimizes the error rate for the given training data $T = \{\langle \mathbf{x}_i, y_i \rangle\}_{i=1}^{L}$,

$$\langle \hat{t}, \hat{y} \rangle = \arg\min_{t \in \mathcal{F}, y \in \{\pm1\}} \frac{1}{L} \sum_{i=1}^{L} I(y_i \neq h_{\langle t,y \rangle}(\mathbf{x}_i))$$

$$= \arg\min_{t \in \mathcal{F}, y \in \{\pm1\}} \frac{1}{2L} \sum_{i=1}^{L} (1 - y_i h_{\langle t,y \rangle}(\mathbf{x}_i)), \quad (3.1)$$

where $\mathcal{F}$ is a set of candidate graphs or a feature set (*i.e.*, $\mathcal{F} = \bigcup_{i=1}^{L} \{t | t \subseteq \mathbf{x}_i\}$) and $I(\cdot)$ is the indicator function. The gain function for a rule $\langle t, y \rangle$ is defined as

$$gain(\langle t, y \rangle) = \sum_{i=1}^{L} y_i h_{\langle t,y \rangle}(\mathbf{x}_i). \quad (3.2)$$

Using the gain, the search problem in Eq.(3.1) becomes equivalent to the problem: $\langle \hat{t}, \hat{y} \rangle = \arg\max_{t \in \mathcal{F}, y \in \{\pm1\}} gain(\langle t, y \rangle)$. Then the gain function is used instead of error rate.

gboost applies AdaBoost [9] by repeatedly calling the decision stumps and finally produces a hypothesis f, which is a linear combination of K hypotheses produced by the decision stumps $f(\mathbf{x}) = sgn(\sum_{k=1}^{K} \alpha_k h_{\langle t_k, y_k \rangle}(\mathbf{x}))$. In the kth iteration, a decision stump is built with weights $\mathbf{d}^{(k)} = (d_1^{(k)}, ..., d_L^{(k)})$ on the training data, where $\sum_{i=1}^{L} d_i^{(k)} = 1$, $d_i^{(k)} \geq 0$. The weights are calculated to concentrate more on hard examples than easy ones. In the boosting framework, the gain function is redefined as

$$gain(\langle t, y \rangle) = \sum_{i=1}^{L} y_i d_i h_{\langle t, y \rangle}(\mathbf{x}_i). \tag{3.3}$$

A Branch-and-Bound Search Approach. According to the gain function in Eq.(3.3), the problem of finding the optimal rule $\langle \hat{t}, \hat{y} \rangle$ from the training dataset is defined as follows.

Problem 1 [Find Optimal Rule] Let $T = \{\langle \mathbf{x}_1, y_1, d_1 \rangle, ..., \langle \mathbf{x}_L, y_L, d_L \rangle\}$ be a training data set where $\mathbf{x}_i$ is a labeled graph, $y_i \in \{\pm 1\}$ is a class label associated with $\mathbf{x}_i$ and d_i ($\sum_{i=1}^{L} d_i = 1$, $d_i \geq 0$) is a normalized weight assigned to $\mathbf{x}_i$. Given T, find the optimal rule $\langle \hat{t}, \hat{y} \rangle$ that maximizes the gain, *i.e.*, $\langle \hat{t}, \hat{y} \rangle = \arg\max_{t \in \mathcal{F}, y \in \{\pm 1\}} y_i d_i h_{\langle t, y \rangle}$, where $\mathcal{F} = \bigcup_{i=1}^{L} \{t | t \subseteq \mathbf{x}_i\}$.

A naive method is to enumerate all subgraphs $\mathcal{F}$ and then calculate the gains for all subgraphs. However, this method is impractical since the number of subgraphs is exponential to their size. To avoid such exhaustive enumeration, the method to find the optimal rule is modeled as a branch-and-bound algorithm based on the upper bound of the gain function which is defined as follows.

Lemma 12.10 (Upper bound of the gain). *For any $t' \supseteq t$ and $y \in \{\pm 1\}$, the gain of $\langle t', y \rangle$ is bounded by $\mu(t)$ (i.e., $gain(\langle t', y \rangle) \leq \mu(t)$), where $\mu(t)$ is given by*

$$\mu(t) = max(2 \sum_{\{i | y_i = +1, t \subseteq x_i\}} d_i - \sum_{i=1}^{L} y_i \cdot d_i, 2 \sum_{\{i | y_i = -1, t \subseteq x_i\}} d_i + \sum_{i=1}^{L} y_i \cdot d_i). \tag{3.4}$$

Figure 12.3 depicts a graph pattern search tree where each node represents a graph. A graph g' is a child of another graph g if g' is a supergraph of g with one more edge. g' is also written as $g' = g \diamond e$, where e is the extra edge. In order to find an optimal rule, the branch-and-bound search estimates the upper bound of the gain function for all descendants below a node g. If it is smaller than the value of the best subgraph seen so far, it cuts the search branch of that node. Under the branch-and-bound search, a tighter upper bound is always preferred since it means faster pruning.

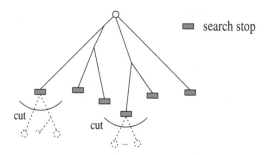

Figure 12.3. Branch-and-Bound Search

Algorithm 15 outlines the framework of branch-and-bound for searching the optimal graph pattern. In the initialization, all the subgraphs with one edge are enumerated first and these seed graphs are then iteratively extended to large subgraphs. Since the same graph could be grown in different ways, Line 5 checks whether it has been discovered before; if it has, then there is no need to grow it again. The optimal $gain(\langle \hat{t}, \hat{y} \rangle)$ discovered so far is maintained. If $\mu(t) \leq gain(\langle \hat{t}, \hat{y} \rangle)$, the branch of t can safely be pruned.

Algorithm 15 Branch-and-Bound

Input: Graph dataset D
Output: Optimal rule $\langle \hat{t}, \hat{y} \rangle$

1: $S = \{1\text{-edge graph}\}$;
2: $\langle \hat{t}, \hat{y} \rangle = \varnothing$; $gain(\langle \hat{t}, \hat{y} \rangle) = -\infty$;
3: **while** $S \neq \varnothing$ **do**
4:     choose t from S, $S = S \setminus \{t\}$;
5:     **if** t was examined **then**
6:         **continue**;
7:     **if** $gain(\langle t, y \rangle) > gain(\langle \hat{t}, \hat{y} \rangle)$ **then**
8:         $\langle \hat{t}, \hat{y} \rangle = \langle t, y \rangle$;
9:     **if** $\mu(t) \leq gain(\langle \hat{t}, \hat{y} \rangle)$ **then**
10:         **continue**;
11:     $S = S \cup \{t'|t' = t \diamond e\}$;
12: **return** $\langle \hat{t}, \hat{y} \rangle$;

3.3   gPLS: A Partial Least Squares Regression Approach

Saigo et al. [25] proposed gPLS, an iterative mining method based on partial least squares regression (PLS). To apply PLS to graph data, a sparse version of PLS is developed first and then it is combined with a weighted pattern mining algorithm. The mining algorithm is iteratively called with different weight vectors, creating one latent component per one mining call. Branch-and-bound search is integrated into graph mining with a designed gain function and a pruning condition. In this sense, gPLS is very similar to the branch-and-bound mining approach in gboost.

Partial Least Squares Regression.   This part is a brief introduction to partial least squares regression (PLS). Assume there are n training examples $(x_1, y_1), ..., (x_n, y_n)$. The output y_i is assumed to be centralized $\sum_i y_i = 0$. Denote by X the design matrix, where each row corresponds to x_i^T. The regression function of PLS is

$$f(x) = \sum_{i=1}^{m} \alpha_i w_i^T x,$$

where m is the pre-specified number of components that form a subset of the original space, and w_i are weight vectors that reduce the dimensionality of x, satisfying the following orthogonality condition,

$$w_i^T X^T X w_j = \begin{cases} 1 & (i = j) \\ 0 & (i \neq j) \end{cases}.$$

Basically w_i are learned in a greedy way first, then the coefficients α_i are obtained by least squares regression without any regularization. The solutions to α_i and w_i are

$$\alpha_i = \sum_{k=1}^{n} y_k w_i^T x_k, \tag{3.5}$$

and

$$w_i = \arg\max_w \frac{(\sum_{k=1}^{n} y_k w^T x_k)^2}{w^T w},$$

subject to $w^T X^T X w = 1$, $w^T X^T X w_j = 0$, $j = 1, ..., i - 1$.

Next we present an alternative derivation of PLS called *non-deflation sparse PLS*. Define the i-th latent component as $t_i = X w_i$ and T_{i-1} as the matrix of latent components obtained so far, $T_{i-1} = (t_1, ..., t_{i-1})$. The residual vector is computed by

$$r_i = (I - T_{i-1} T_{i-1}^T) y.$$

Then multiply it with X^T to obtain

$$v = \frac{1}{\eta} X^T (I - T_{i-1} T_{i-1}^T) y.$$

The non-deflation sparse PLS follows this idea.

In graph mining, it is useful to have sparse weight vectors w_i such that only a limited number of patterns are used for prediction. To this aim, we introduce the sparseness to the pre-weight vectors v_i as

$$v_{ij} = 0, \ \ if \ |v_{ij}| \le \epsilon, \ \ j = 1, .., d.$$

Due to the linear relationship between v_i and w_i, w_i becomes sparse as well. Then we can sort $|v_{ij}|$ in the descending order, take the top-k elements and set all the other elements to zero.

It is worthwhile to notice that the residual of regression up to the $(i - 1)$-th features,

$$r_{ik} = y_k - \sum_{j=1}^{i-1} \alpha_j w_j^T x_k, \tag{3.6}$$

is equal to the k-th element of r_i. It can be verified by substituting the definition of α_j in Eq.(3.5) into Eq.(3.6). So in the non-deflation algorithm, the pre-weight vector v is obtained as the direction that maximizes the covariance with residues. This observation highlights the resemblance of PLS and boosting algorithms.

Graph PLS: Branch-and-Bound Search. In this part, we discuss how to apply the non-deflation PLS algorithm to graph data. The set of training graphs is represented as $(G_1, y_1), ..., (G_n, y_n)$. Let $\mathcal{P}$ be the set of all patterns, then the feature vector of each graph G_i is encoded as a $|\mathcal{P}|$-dimensional vector x_i. Since $|\mathcal{P}|$ is a huge number, it is infeasible to keep the whole design matrix. So the method sets X as an empty matrix first, and grows the matrix as the iteration proceeds. In each iteration, it obtains the set of patterns p whose pre-weight $|v_{ip}|$ is above the threshold, which can be written as

$$P_i = \{p|| \sum_{j=1}^{n} r_{ij} x_{jp}| \ge \epsilon\}. \tag{3.7}$$

Then the design matrix is expanded to include newly introduced patterns. The pseudo code of **gPLS** is described in Algorithm 16.

The pattern search problem in Eq.(3.7) is exactly the same as the one solved in **gboost** through a branch-and-bound search. In this problem, the gain function is defined as $s(p) = |\sum_{j=1}^{n} r_{ij} x_{jp}|$. The pruning condition is described as follows.

Theorem 12.11. *Define $\tilde{y}_i = sgn(r_i)$. For any pattern p' such that $p \subseteq p'$, $s(p') < \epsilon$ holds if*

$$\max\{s^+(p), s^-(p)\} < \epsilon, \tag{3.8}$$

where

$$s^+(p) = 2 \sum_{\{i|\tilde{y}_i=+1, x_{i,j}=1\}} |r_i| - \sum_{i=1}^n r_i,$$

$$s^-(p) = 2 \sum_{\{i|\tilde{y}_i=-1, x_{i,j}=1\}} |r_i| + \sum_{i=1}^n r_i.$$

Algorithm 16 gPLS

Input: Training examples $(G_1, y_1), (G_2, y_2), ..., (G_n, y_n)$
Output: Weight vectors $w_i, i = 1, ..., m$

1: $r_1 = y, X = \varnothing$;
2: **for** $i = 1, ..., m$ **do**
3:     $P_i = \{p| |\sum_{j=1}^n r_{ij} x_{jp}| \geq \epsilon\}$;
4:     X_{P_i}: design matrix restricted to P_i;
5:     $X \leftarrow X \cup X_{P_i}$;
6:     $v_i = X^T r_i / \eta$;
7:     $w_i = v_i - \sum_{j=1}^{i-1} (w_j^T X^T X v_i) w_j$;
8:     $t_i = X w_i$;
9:     $r_{i+1} = r_i - (y^T t_i) t_i$;

3.4   LEAP: A Structural Leap Search Approach

Yan et al. [31] proposed an efficient algorithm which mines the most significant subgraph pattern with respect to an objective function. A major contribution of this study is the proposal of a general approach for significant graph pattern mining with non-monotonic objective functions. The mining strategy, called **LEAP** (Descending Leap Mine), explored two new mining concepts: (1) *structural leap search*, and (2) *frequency-descending mining*, both of which are related to specific properties in pattern search space. The same mining strategy can also be applied to searching other simpler structures such as itemsets, sequences and trees.

Structural Leap Search.   Figure 12.4 shows a search space of subgraph patterns. If we examine the search structure horizontally, we find that the subgraphs along the neighbor branches likely have similar compositions and frequencies, hence similar objective scores. Take the branches A and B as an example. Suppose A and B split on a common subgraph pattern g. Branch A

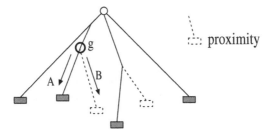

Figure 12.4. Structural Proximity

contains all the supergraphs of $g \diamond e$ and B contains all the supergraphs of g except those of $g \diamond e$. For a graph g' in branch B, let $g'' = g' \diamond e$ in branch A.

LEAP assumes each input graph is assigned either a positive or a negative label (*e.g.*, compounds active or inactive to a virus). One can divide the graph dataset into two subsets: a positive set D_+ and a negative set D_-. Let $p(g)$ and $q(g)$ be the frequency of a graph pattern g in positive graphs and negative graphs. Many objective functions can be represented as a function of p and q for a subgraph pattern g, as $F(g) = f(p(g), q(g))$.

If in a graph dataset, $g \diamond e$ and g often occur together, then g'' and g' might also occur together. Hence, likely $p(g'') \operatorname{sim} p(g')$ and $q(g'') \operatorname{sim} q(g')$, which means similar objective scores. This is resulted by the structural and embedding similarity between the starting structures $g \diamond e$ and g. We call it **structural proximity**: Neighbor branches in the pattern search tree exhibit strong similarity not only in pattern composition, but also in their embeddings in the graph datasets, thus having similar frequencies and objective scores. In summary, a conceptual claim can be drawn,

$$g' \operatorname{sim} g'' \Rightarrow F(g') \operatorname{sim} F(g''). \tag{3.9}$$

According to structural proximity, it seems reasonable to skip the whole search branch once its nearby branch is searched, since the best scores between neighbor branches are likely similar. Here, we would like to emphasize "likely" rather than "surely". Based on this intuition, if the branch A in Figure 12.4 has been searched, B could be "leaped over" if A and B branches satisfy some similarity criterion. The length of leap can be controlled by the frequency difference of two graphs g and $g \diamond e$. The leap condition is defined as follows.

Let $I(G, g, g \diamond e)$ be an indicator function of a graph G: $I(G, g, g \diamond e) = 1$, for any supergraph g' of g, if $g' \subseteq G$, $\exists g'' = g' \diamond e$ such that $g'' \subseteq G$; otherwise 0. When $I(G, g, g \diamond e) = 1$, it means if a supergraph g' of g has an embedding in G, there must be an embedding of $g' \diamond e$ in G. For a positive dataset D_+, let $D_+(g, g \diamond e) = \{G | I(G, g, g \diamond e) = 1, g \subseteq G, G \in D_+\}$. In $D_+(g, g \diamond e)$,

$g' \supset g$ and $g'' = g' \diamond e$ have the same frequency. Define $\Delta_+(g, g \diamond e)$ as follows,

$$\Delta_+(g, g \diamond e) = p(g) - \frac{|D_+(g, g \diamond e)|}{|D_+|}.$$

$\Delta_+(g, g \diamond e)$ is actually the maximum frequency difference that g' and g'' could have in D_+. If the difference is smaller than a threshold σ, then leap,

$$\frac{2\Delta_+(g, g \diamond e)}{p(g \diamond e) + p(g)} \le \sigma \text{ and } \frac{2\Delta_-(g, g \diamond e)}{q(g \diamond e) + q(g)} \le \sigma. \tag{3.10}$$

σ controls the leap length. The larger σ is, the faster the search is. Structural leap search will generate an optimal pattern candidate and reduce the need for thoroughly searching similar branches in the pattern search tree. Its goal is to help program search significantly distinct branches, and limit the chance of missing the most significant pattern.

Algorithm 17 Structural Leap Search: sLeap(D, σ, $g^\star$)

Input: Graph dataset D, difference threshold σ
Output: Optimal graph pattern candidate $g^\star$

1: $S = \{1 - \text{edge graph}\}$;
2: $g^\star = \varnothing$; $F(g^\star) = -\infty$;
3: **while** $S \neq \varnothing$ **do**
4: 　　$S = S \setminus \{g\}$;
5: 　　**if** g was examined **then**
6: 　　　　**continue**;
7: 　　**if** $\exists g \diamond e, g \diamond e \prec g, \frac{2\Delta_+(g, g \diamond e)}{p(g \diamond e) + p(g)} \le \sigma, \frac{2\Delta_-(g, g \diamond e)}{q(g \diamond e) + q(g)} \le \sigma$ **then**
8: 　　　　**continue**;
9: 　　**if** $F(g) > F(g^\star)$ **then**
10: 　　　　$g^\star = g$;
11: 　　**if** $\widehat{F}(g) \le F(g^\star)$ **then**
12: 　　　　**continue**;
13: 　　$S = S \cup \{g' | g' = g \diamond e\}$;
14: **return** $g^\star$;

Algorithm 17 outlines the pseudo code of structural leap search (sLeap). The leap condition is tested on Lines 7-8. Note that sLeap does not guarantee the optimality of result.

Frequency Descending Mining. Structural leap search takes advantages of the correlation between structural similarity and significance similarity. However, it does not exploit the possible relationship between patterns' frequency and patterns' objective scores. Existing solutions have to set the frequency threshold very low so that the optimal pattern will not be missed. Unfortunately, low-frequency threshold could generate a huge set of low-significance redundant patterns with long mining time.

Although most of objective functions are not correlated with frequency monotonically or anti-monotonically, they are not independent of each other. Cheng et al. [4] derived a frequency upper bound of discriminative measures such as information gain and Fisher score, showing a relationship between frequency and discriminative measures. According to this analytical result, if all frequent subgraphs are ranked in increasing order of their frequency, significant subgraph patterns are often in the high-end range, though their real frequency could vary dramatically across different datasets.

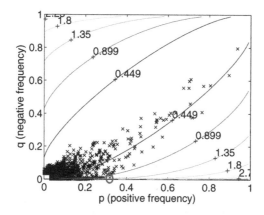

Figure 12.5. Frequency vs. G-test score

Figure 12.5 illustrates the relationship between frequency and G-test score for an AIDS Anti-viral dataset [31]. It is a contour plot displaying isolines of G-test score in two dimensions. The X axis is the frequency of a subgraph g in the positive dataset, *i.e.*, $p(g)$, while the Y axis is the frequency of the same subgraph in the negative dataset, $q(g)$. The curves depict G-test score. Left upper corner and right lower corner have the higher G-test scores. The "circle" marks the highest G-score subgraph discovered in this dataset. As one can see, its positive frequency is higher than most of subgraphs.

[Frequency Association]*Significant patterns often fall into the high-quantile of frequency.*

To profit from frequency association, an iterative frequency-descending mining method is proposed in [31]. Rather than performing mining with very low frequency, the method starts the mining process with high frequency threshold $\theta = 1.0$, calculates an optimal pattern candidate $g^\star$ whose frequency is at least θ, and then repeatedly lowers down θ to check whether $g^\star$ can be improved further. Here, the search leaps in the frequency domain, by leveling down the minimum frequency threshold exponentially.

Algorithm 18 Frequency-Descending Mine: fLeap(D, ε, $g^\star$)

Input: Graph dataset D, converging threshold ε
Output: Optimal graph pattern candidate $g^\star$

1: $\theta = 1.0$;
2: $g = \varnothing$; $F(g) = -\infty$;
3: **do**
4:      $g^\star = g$;
5:      g=fpmine(D, θ);
6:      $\theta = \theta/2$;
7: **while** $(F(g) - F(g^\star) \geq \varepsilon)$
8: **return** $g^\star = g$;

Algorithm 18 (fLeap) outlines the frequency-descending strategy. It starts with the highest frequency threshold, and then lowers the threshold down till the objective score of the best graph pattern converges. Line 5 executes a frequent subgraph mining routine, fpmine, which could be **FSG** [20], **gSpan** [32] *etc.* fpmine selects the most significant graph pattern g from the frequent subgraphs it mined. Line 6 implements a simple frequency descending method.

Descending Leap Mine.   With structural leap search and frequency-descending mining, a general mining pipeline is built for mining significant graph patterns in a complex graph dataset. It consists of three steps as follows.

Step 1. perform structural leap search with threshold $\theta = 1.0$, generate an optimal pattern candidate $g^\star$.

Step 2. repeat frequency-descending mining with structural leap search until the objective score of $g^\star$ converges.

Step 3. take the best score discovered so far; perform structural leap search again (leap length σ) without frequency threshold; output the discovered pattern.

3.5   GraphSig: A Feature Representation Approach

Ranu and Singh [24] proposed **GraphSig**, a scalable method to mine significant (measured by p-value) subgraphs based on a feature vector representation of graphs. The first step is to convert each graph into a set of feature vectors where each vector represents a region within the graph. Prior probabilities of features are computed empirically to evaluate statistical significance of patterns in the feature space. Following the analysis in the feature space, only a small portion of the exponential search space is accessed for further analysis. This enables the use of existing frequent subgraph mining techniques to mine significant patterns in a scalable manner even when they are infrequent. The major steps of GraphSig are described as follows.

Sliding Window across Graphs.   As the first step, random walk with restart (abbr. RWR) is performed on each node in a graph to simulate sliding a window across the graph. RWR simulates the trajectory of a random walker that starts from the target node and jumps from one node to a neighbor. Each neighbor has an equal probability of becoming the new station of the walker. At each jump, the feature traversed is updated which can either be an edge label or a node label. A restart probability α brings the walker back to the starting node within approximately $\frac{1}{\alpha}$ jumps. The random walk iterates till the feature distribution converges. As a result, RWR produces a continuous distribution of features for each node where a feature value lies in the range $[0, 1]$, which is further discretized into 10 bins. RWR can therefore be visualized as placing a window at each node of a graph and capturing a feature vector representation of the subgraph within it. A graph of m nodes is represented by m feature vectors. RWR inherently takes proximity of features into account and preserves more structural information than simply counting occurrence of features inside the window.

Calculating P-value of A Feature Vector.   To calculate p-value of a feature vector, we model the occurrence of a feature vector $\underline{x}$ in a feature vector space formulated by a random graph. The frequency distribution of a vector is generated using the prior probabilities of features obtained empirically. Given a feature vector $\underline{x} = [x_1, ..., x_n]$, the probability of $\underline{x}$ occurring in a random feature vector $\underline{y} = [y_1, ..., y_n]$ can be expressed as a joint probability

$$P(\underline{x}) = P(y_1 \geq x_1, ..., y_n \geq x_n). \tag{3.11}$$

To simplify the calculation, we assume independence of the features. As a result, Eq.(3.11) can be expressed as a product of the individual probabilities, where

$$P(\underline{x}) = \prod_{i=1}^{n} P(y_i \geq x_i). \tag{3.12}$$

Once $P(\underline{x})$ is known, the support of $\underline{x}$ in a database of random feature vectors can be modeled as a binomial distribution. To illustrate, a random vector can be viewed as a trial and $\underline{x}$ occurring in it as "success". A database consisting m feature vectors will involve m trials for $\underline{x}$. The support of $\underline{x}$ in the database is the number of successes. Therefore, the probability of $\underline{x}$ having a support μ is

$$P(\underline{x};\mu) = C_m^\mu P(\underline{x})^\mu (1 - P(\underline{x}))^{m-\mu}. \tag{3.13}$$

The probability distribution function (abbr. pdf) of $\underline{x}$ can be generated from Eq.(3.13) by varying μ in the range $[0, m]$. Therefore, given an observed support μ_0 of $\underline{x}$, its p-value can be calculated by measuring the area under the pdf in the range $[\mu_0, m]$, which is

$$p\text{-}value(x, \mu_0) = \sum_{i=\mu_0}^{m} P(\underline{x}; i). \tag{3.14}$$

Identifying Regions of Interest.   With the conversion of graphs into feature vectors, and a model to evaluate significance of a graph region in the feature space, the next step is to explore how the feature vectors can be analyzed to extract the significant regions. Based on the feature vector representation, the presence of a "common" sub-feature vector among a set of graphs points to a common subgraph. Similarly, the absence of a "common" sub-feature vector indicates the non-existence of any common subgraph. Mathematically, the *floor* of the feature vectors produces the "common" sub-feature vector.

Definition 12.12 (Floor of vectors). *The floor of a set of vectors $\{v_1, ..., v_m\}$ is a vector v_f where $v_{f_i} = min(v_{1_i}, ..., v_{m_i})$ for $i = 1, ..., n$, n is the number of dimensions of a vector. Ceiling of a set of vectors is defined analogously.*

The next step is to mine common sub-feature vectors that are also significant. Algorithm 19 presents the FVMine algorithm which explores closed sub-vectors in a bottom-up, depth-first manner. FVMine explores all possible common vectors satisfying the significance and support constraints.

With a model to measure the significance of a vector, and an algorithm to mine closed significant sub-feature vectors, we integrate them to build the significant graph mining framework. The idea is to mine significant sub-feature vectors and use them to locate similar regions which are significant. Algorithm 20 outlines the GraphSig algorithm.

The algorithm first converts each graph into a set of feature vectors and puts all vectors together in a single set D' (lines 3-4). D' is divided into sets, such that D'_a contains all vectors produced from RWR on a node labeled a. On each set D'_a, FVMine is performed with a user-specified support and p-value thresholds to retrieve the set of significant sub-feature vectors (line 7). Given that each sub-feature vector could describe a particular subgraph, the algorithm scans the database to identify the regions where the current sub-feature vector occurs. This involves finding all nodes labeled a and described by a feature vector such that the vector is a super-vector of the current sub-feature vector $\underline{v}$ (line 9). Then the algorithm isolates the subgraph centered

---

Algorithm 19 FVMine($\underline{x}$, S, b)

Input: Current sub-feature vector $\underline{x}$, supporting set S of $\underline{x}$,
  current starting position b
Output: The set of all significant sub-feature vectors A

1: **if** $p\text{-}value(\underline{x}) \leq maxPvalue$ **then**
2:   $A \leftarrow A + x$;
3:   **for** $i = b$ to m **do**
4:     $S' \leftarrow \{\underline{y} | y \in S, y_i > x_i\}$;
5:     **if** $|S'| < min\_sup$ **then**
6:       **continue**;
7:     $\underline{x}' = floor(S')$;
8:     **if** $\exists j < i$ such that $x'_j > x_j$ **then**
9:       **continue**;
10:     **if** $p\text{-}value(ceiling(S'), |S'|) \geq maxPvalue$ **then**
11:       **continue**;
12:     $FVMine(\underline{x}', S', i)$;

---

at each node by using a user-specified radius (line 12). This produces a set of subgraphs for each significant sub-feature vector. Next, maximal subgraph mining is performed with a high frequency threshold since it is expected that all of graphs in the set contain a common subgraph (line 13). The last step also prunes out false positives where dissimilar subgraphs are grouped into a set due to the vector representation. For the absence of a common subgraph, when frequent subgraph mining is performed on the set, no frequent subgraph will be produced and as a result the set is filtered out.

4.   Mining Representative Orthogonal Graphs

In this section we will discuss **ORIGAMI**, an algorithm proposed by Hasan et al. [10], which mines a set of α-orthogonal, β-representative graph patterns. Intuitively, two graph patterns are α-orthogonal if their similarity is bounded by a threshold α. A graph pattern is a β-representative of another pattern if their similarity is at least β. The orthogonality constraint ensures that the resulting pattern set has controlled redundancy. For a given α, more than one set of graph patterns qualify as an α-orthogonal set. Besides redundancy control, representativeness is another desired property, *i.e.*, for every frequent graph pattern not reported in the α-orthogonal set, we want to find a representative of this pattern with a high similarity in the α-orthogonal set.

The set of representative orthogonal graph patterns is a compact summary of the complete set of frequent subgraphs. Given user specified thresholds $\alpha, \beta \in$

Algorithm 20 GraphSig(D, min_sup, $maxPvalue$)

Input: Graph dataset D, support threshold min_sup,
      p-value threshold $maxPvalue$
Output: The set of all significant sub-feature vectors A

1: $D' \leftarrow \varnothing$;
2: $A \leftarrow \varnothing$;
3: **for** each $g \in D$ **do**
4:     $D' \leftarrow D' + RWR(g)$;
5: **for** each node label a in D **do**
6:     $D'_a \leftarrow \{v|v \in D', label(v) = a\}$;
7:     $S \leftarrow FVMine(floor(D'_a), D'_a, 1)$;
8:     **for** each vector $\underline{v} \in S$ **do**
9:         $V \leftarrow \{u|u\ is\ a\ node\ of\ label\ a, \underline{v} \subseteq vector(u)\}$;
10:        $E \leftarrow \varnothing$;
11:       **for** each node $u \in V$ **do**
12:           $E \leftarrow E + CutGraph(u, radius)$;
13:           $A \leftarrow A + Maximal\_FSM(E, freq)$;

$[0, 1]$, the goal is to mine an α-orthogonal, β-representative graph pattern set that minimizes the set of unrepresented patterns.

4.1   Problem Definition

Given a collection of graphs D and a similarity threshold $\alpha \in [0, 1]$, a subset of graphs $\mathcal{R} \subseteq D$ is α-orthogonal with respect to D iff for any $G_a, G_b \in \mathcal{R}$, $sim(G_a, G_b) \leq \alpha$ and for any $G_i \in D \backslash \mathcal{R}$ there exists a $G_j \in \mathcal{R}$, $sim(G_i, G_j) > \alpha$.

Given a collection of graphs D, an α-orthogonal set $\mathcal{R} \subseteq D$ and a similarity threshold $\beta \in [0, 1]$, $\mathcal{R}$ represents a graph $G \in D$, provided that there exists some $G_a \in \mathcal{R}$, such that $sim(G_a, G) \geq \beta$. Let $\Upsilon(\mathcal{R}, D) = \{G|G \in D\ s.t.\ \exists G_a \in \mathcal{R}, sim(G_a, G) \geq \beta\}$, then $\mathcal{R}$ is a β-representative set for $\Upsilon(\mathcal{R}, D)$.

Given D and $\mathcal{R}$, the residue set of $\mathcal{R}$ is the set of unrepresented patterns in D, denoted as $\triangle(\mathcal{R}, D) = D \backslash \{\mathcal{R} \cup \Upsilon(\mathcal{R}, D)\}$.

The problem defined in [10] is to find the α-orthogonal, β-representative set for the set of all maximal frequent subgraphs $\mathcal{M}$ which minimizes the residue set size. The mining problem can be decomposed into two subproblems of *maximal subgraph mining* and *orthogonal representative set generation*, which are discussed separately. Algorithm 21 shows the algorithm framework of ORIGAMI.

Algorithm 21 ORIGAMI(D, min_sup, α, β)

Input: Graph dataset D, minimum support min_sup, α, β
Output: α-orthogonal, β-representative set $\mathcal{R}$

1: EM=Edge-Map(D);
2: $\mathcal{F}_1$=Find-Frequent-Edges(D, min_sup);
3: $\widehat{\mathcal{M}} = \phi$;
4: **while** stopping_condition() $\neq$ **true do**
5:     M=Random-Maximal-Graph(D, $\mathcal{F}_1$, EM, min_sup);
6:     $\widehat{\mathcal{M}} = \widehat{\mathcal{M}} \cup M$;
7: $\mathcal{R}$=Orthogonal-Representative-Sets($\widehat{\mathcal{M}}$, α, β);
8: **return** $\mathcal{R}$;

4.2   Randomized Maximal Subgraph Mining

As the first step, ORIGAMI mines a set of maximal subgraphs, on which the α-orthogonal, β-representative graph pattern set is generated. This is based on the observation that the number of maximal frequent subgraphs is much fewer than that of frequent subgraphs, and the maximal subgraphs provide a synopsis of the frequent ones to some extent. Thus it is reasonable to mine the representative orthogonal pattern set based on the maximal subgraphs rather than the frequent ones. However, even mining all of maximal subgraphs could be infeasible in some real world applications. To avoid this problem, ORIGAMI first finds a sample $\widehat{\mathcal{M}}$ of the complete set of maximal frequent subgraphs $\mathcal{M}$.

The goal is to find a set of maximal subgraphs, $\widehat{\mathcal{M}}$, which is as diverse as possible. To achieve this goal, ORIGAMI avoids using combinatorial enumeration to mine maximal subgraph patterns. Instead, it adopts a random walk approach to enumerate a diverse set of maximal subgraphs from the positive border of such maximal patterns. The randomized mining algorithm starts with an empty pattern and iteratively adds a random edge during each extension, until a maximal subgraph M is generated and no more edges can be added. This process walks a random chain in the partial order of frequent subgraphs. To extend an intermediate pattern, $S_k \subseteq M$, it chooses a random vertex v from which the extension will be attempted. Then a random edge e incident on v is selected for extension. If no such edge is found, no extension is possible from the vertex. When no vertices can have any further extension in S_k, the random walk terminates and $S_k = M$ is the maximal graph. On the other hand, if a random edge e is found, the other endpoint v' of this edge is randomly selected. By adding the edge e and its endpoint v', a candidate subgraph pattern S_{k+1} is generated and its support is computed. This random walk process repeats until no further extension is possible on any vertex. Then the maximal subgraph M is returned.

Ideally, the random chain walks would cover different regions of the pattern space, thus would produce dissimilar maximal patterns. However, in practice, this may not be the case, since duplicate maximal subgraphs can be generated in the following ways: (1) multiple iterations following overlapping chains, or (2) multiple iterations following different chains but leading to the same maximal pattern. Let's consider a maximal subgraph M of size n. Let $e_1 e_2 ... e_n$ be a sequence of random edge extensions, corresponding to a random chain walk leading from an empty graph ϕ to the maximal graph M. The probability of a particular edge sequence leading from ϕ to M is given as

$$P[(e_1 e_2 ... e_n)] = P(e_1) \prod_{i=2}^{n} P(e_i | e_1 ... e_{i-1}). \qquad (4.1)$$

Let $ES(M)$ denote the set of all valid edge sequences for a graph M. The probability that a graph M is generated in a random walk is proportional to

$$\sum_{e_1 e_2 ... e_n \in ES(M)} P[(e_1 e_2 ... e_n)]. \qquad (4.2)$$

The probability of obtaining a specific maximal pattern depends on the number of chains or edge sequences leading to that pattern and the size of the pattern. According to Eq.(4.1), as a graph grows larger, the probability of the edge sequence becomes smaller. So this random walk approach in general favors a maximal subgraph of smaller size than one of larger size. To avoid generating duplicate maximal subgraphs, a termination condition is designed based on an estimate of the collision rate of the generated patterns. Intuitively the collision rate keeps track of the number of duplicate patterns seen within the same or across different random walks. As a random walk chain is traversed, ORIGAMI maintains the signature of the intermediate patterns in a bounded size hash table. As an intermediate or maximal subgraph is generated, its signature is added to the hash table and the collision rate is updated. If the collision rate exceeds a threshold ϵ, the method could (1) abort further extension along the current path and randomly choose another path; or (2) trigger the termination condition across different walks, since it implies that the same part of the search space is being revisited.

4.3   Orthogonal Representative Set Generation

Given a set of maximal subgraphs $\widehat{\mathcal{M}}$, the next step is to extract an α-orthogonal β-representative set from it. We can construct a meta-graph $\Gamma(\widehat{\mathcal{M}})$ to measure similarity between graph patterns in $\widehat{\mathcal{M}}$, in which each node represents a maximal subgraph pattern, and an edge exists between two nodes if their similarity is bounded by α. Then the problem of finding an α-orthogonal pattern set can be modeled as finding a maximal clique in the similarity graph $\Gamma(\widehat{\mathcal{M}})$.

For a given α, there could be multiple α-orthogonal pattern sets as feasible solutions. We could use the size of the residue set to measure the goodness of an α-orthogonal set. An optimal α-orthogonal β-representative set is the one which minimizes the size of the residue set. [10] proved that this problem is NP-hard.

Given the hardness result, ORIGAMI resorts to approximate algorithms to solve the problem which guarantees local optimality. The algorithm starts with a random maximal clique in the similarity graph $\Gamma(\widehat{\mathcal{M}})$ and tries to improve it. At each state transition, another maximal clique which is a local neighbor of the current maximal clique is chosen. If the new state has a better solution, the new state is accepted as the current state and the process continues. The process terminates when all neighbors of the current state have equal or larger residue sizes. Two maximal cliques of size m and n (assume $m \geq n$) are considered neighbors if they share exactly $n - 1$ vertices. The state transition procedure selectively removes one vertex from the maximal clique of the current state and then expands it to obtain another maximal clique which satisfies the neighborhood constraints.

5. Conclusions

Frequent subgraph mining is one of the fundamental tasks in graph data mining. The inherent complexity in graph data causes the combinatorial explosion problem. As a result, a mining algorithm may take a long time or even forever to complete the mining process on some real graph datasets.

In this chapter, we introduced several state-of-the-art methods that mine a compact set of significant or representative subgraphs without generating the complete set of graph patterns. The proposed mining and pruning techniques were discussed in details. These methods greatly reduce the computational cost, while at the same time, increase the applicability of the generated graph patterns. These research results have made significant progress on graph mining research with a set of new applications.

References

[1] T. Asai, K. Abe, S. Kawasoe, H. Arimura, H. Satamoto, and S. Arikawa. Efficient substructure discovery from large semi-structured data. In *Proc. 2002 SIAM Int. Conf. Data Mining (SDM'02)*, pages 158–174, 2002.

[2] C. Borgelt and M. R. Berthold. Mining molecular fragments: Finding relevant substructures of molecules. In *Proc. 2002 Int. Conf. Data Mining (ICDM'02)*, pages 211–218, 2002.

[3] B. Bringmann and S. Nijssen. What is frequent in a single graph? In *Proc. 2008 Pacific-Asia Conf. Knowledge Discovery and Data Mining (PAKDD'08)*, pages 858–863, 2008.

[4] H. Cheng, X. Yan, J. Han, and C.-W. Hsu. Discriminative frequent pattern analysis for effective classification. In *Proc. 2007 Int. Conf. Data Engineering (ICDE'07)*, pages 716–725, 2007.

[5] Y. Chi, Y. Xia, Y. Yang, and R. Muntz. Mining closed and maximal frequent subtrees from databases of labeled rooted trees. *IEEE Trans. Knowledge and Data Eng.*, 17:190–202, 2005.

[6] L. Dehaspe, H. Toivonen, and R. King. Finding frequent substructures in chemical compounds. In *Proc. 1998 Int. Conf. Knowledge Discovery and Data Mining (KDD'98)*, pages 30–36, 1998.

[7] M. Deshpande, M. Kuramochi, N. Wale, and G. Karypis. Frequent substructure-based approaches for classifying chemical compounds. *IEEE Trans. on Knowledge and Data Engineering*, 17:1036–1050, 2005.

[8] M. Fiedler and C. Borgelt. Support computation for mining frequent subgraphs in a single graph. In *Proc. 5th Int. Workshop on Mining and Learning with Graphs (MLG'07)*, 2007.

[9] Y. Freund and R. Schapire. A decision-theoretic generalization of on-line learning and an application to boosting. In *Proc. 2nd European Conf. Computational Learning Theory*, pages 23–27, 1995.

[10] M. Al Hasan, V. Chaoji, S. Salem, J. Besson, and M. J. Zaki. ORIGAMI: Mining representative orthogonal graph patterns. In *Proc. 2007 Int. Conf. Data Mining (ICDM'07)*, pages 153–162, 2007.

[11] H. He and A. K. Singh. Efficient algorithms for mining significant substructures in graphs with quality guarantees. In *Proc. 2007 Int. Conf. Data Mining (ICDM'07)*, pages 163–172, 2007.

[12] L. B. Holder, D. J. Cook, and S. Djoko. Substructure discovery in the subdue system. In *Proc. AAAI'94 Workshop Knowledge Discovery in Databases (KDD'94)*, pages 169–180, 1994.

[13] J. Huan, W. Wang, D. Bandyopadhyay, J. Snoeyink, J. Prins, and A. Tropsha. Mining spatial motifs from protein structure graphs. In *Proc. 8th Int. Conf. Research in Computational Molecular Biology (RECOMB)*, pages 308–315, 2004.

[14] J. Huan, W. Wang, and J. Prins. Efficient mining of frequent subgraph in the presence of isomorphism. In *Proc. 2003 Int. Conf. Data Mining (ICDM'03)*, pages 549–552, 2003.

[15] J. Huan, W. Wang, J. Prins, and J. Yang. SPIN: Mining maximal frequent subgraphs from graph databases. In *Proc. 2004 ACM SIGKDD Int. Conf. Knowledge Discovery in Databases (KDD'04)*, pages 581–586, 2004.

[16] A. Inokuchi, T. Washio, and H. Motoda. An apriori-based algorithm for mining frequent substructures from graph data. In *Proc. 2000 European Symp. Principle of Data Mining and Knowledge Discovery (PKDD'00)*, pages 13–23, 1998.

[17] R. Jin, C. Wang, D. Polshakov, S. Parthasarathy, and G. Agrawal. Discovering frequent topological structures from graph datasets. In *Proc. 2005 ACM SIGKDD Int. Conf. Knowledge Discovery in Databases (KDD'05)*, pages 606–611, 2005.

[18] M. Koyuturk, A. Grama, and W. Szpankowski. An efficient algorithm for detecting frequent subgraphs in biological networks. *Bioinformatics*, 20:I200–I207, 2004.

[19] T. Kudo, E. Maeda, and Y. Matsumoto. An application of boosting to graph classification. In *Advances in Neural Information Processing Systems 18 (NIPS'04)*, 2004.

[20] M. Kuramochi and G. Karypis. Frequent subgraph discovery. In *Proc. 2001 Int. Conf. Data Mining (ICDM'01)*, pages 313–320, 2001.

[21] M. Kuramochi and G. Karypis. Finding frequent patterns in a large sparse graph. *Data Mining and Knowledge Discovery*, 11:243–271, 2005.

[22] S. Nijssen and J. Kok. A quickstart in frequent structure mining can make a difference. In *Proc. 2004 ACM SIGKDD Int. Conf. Knowledge Discovery in Databases (KDD'04)*, pages 647–652, 2004.

[23] J. Pei, J. Han, B. Mortazavi-Asl, H. Pinto, Q. Chen, U. Dayal, and M.-C. Hsu. PrefixSpan: Mining sequential patterns efficiently by prefix-projected pattern growth. In *Proc. 2001 Int. Conf. Data Engineering (ICDE'01)*, pages 215–224, 2001.

[24] S. Ranu and A. K. Singh. GraphSig: A scalable approach to mining significant subgraphs in large graph databases. In *Proc. 2009 Int. Conf. Data Engineering (ICDE'09)*, pages 844–855, 2009.

[25] H. Saigo, N. Kramer, and K. Tsuda. Partial least squares regression for graph mining. In *Proc. 2008 ACM SIGKDD Int. Conf. Knowledge Discovery in Databases (KDD'08)*, pages 578–586, 2008.

[26] L. Thomas, S. Valluri, and K. Karlapalem. MARGIN: Maximal frequent subgraph mining. In *Proc. 2006 Int. Conf. on Data Mining (ICDM'06)*, pages 1097–1101, 2006.

[27] K. Tsuda. Entire regularization paths for graph data. In *Proc. 2007 Int. Conf. Machine Learning (ICML'07)*, pages 919–926, 2007.

[28] N. Vanetik, E. Gudes, and S. E. Shimony. Computing frequent graph patterns from semistructured data. In *Proc. 2002 Int. Conf. on Data Mining (ICDM'02)*, pages 458–465, 2002.

[29] C. Wang, W. Wang, J. Pei, Y. Zhu, and B. Shi. Scalable mining of large disk-base graph databases. In *Proc. 2004 ACM SIGKDD Int. Conf. Knowledge Discovery in Databases (KDD'04)*, pages 316–325, 2004.

[30] T. Washio and H. Motoda. State of the art of graph-based data mining. *SIGKDD Explorations*, 5:59–68, 2003.

[31] X. Yan, H. Cheng, J. Han, and P. S. Yu. Mining significant graph patterns by scalable leap search. In *Proc. 2008 ACM SIGMOD Int. Conf. on Management of Data (SIGMOD'08)*, pages 433–444, 2008.

[32] X. Yan and J. Han. gSpan: Graph-based substructure pattern mining. In *Proc. 2002 Int. Conf. Data Mining (ICDM'02)*, pages 721–724, 2002.

[33] X. Yan and J. Han. CloseGraph: Mining closed frequent graph patterns. In *Proc. 2003 ACM SIGKDD Int. Conf. Knowledge Discovery and Data Mining (KDD'03)*, pages 286–295, 2003.

[34] X. Yan and J. Han. Discovery of frequent substructures. In *D. Cook and L. Holder (eds.), Mining Graph Data*, pages 99–115, John Wiley Sons, 2007.

[35] X. Yan, P. S. Yu, and J. Han. Graph indexing: A frequent structure-based approach. In *Proc. 2004 ACM-SIGMOD Int. Conf. Management of Data (SIGMOD'04)*, pages 335–346, 2004.

[36] X. Yan, X. J. Zhou, and J. Han. Mining closed relational graphs with connectivity constraints. In *Proc. 2005 ACM SIGKDD Int. Conf. Knowledge Discovery in Databases (KDD'05)*, pages 324–333, 2005.

[37] M. J. Zaki. Efficiently mining frequent trees in a forest. In *Proc. 2002 ACM SIGKDD Int. Conf. Knowledge Discovery in Databases (KDD'02)*, pages 71–80, 2002.

Chapter 13

A SURVEY ON STREAMING ALGORITHMS FOR MASSIVE GRAPHS

Jian Zhang

Computer Science Department
Louisiana State University
zhang@csc.lsu.edu

Abstract   Streaming is an important paradigm for handling massive graphs that are too large to fit in the main memory. In the streaming computational model, algorithms are restricted to use much less space than they would need to store the input. Furthermore, the input is accessed in a sequential fashion, therefore, can be viewed as a stream of data elements. The restriction limits the model and yet, algorithms exist for many graph problems in the streaming model. We survey a set of algorithms that compute graph statistics, matching and distance in a graph, and random walks. These are basic graph problems and the algorithms that compute them may be used as building blocks in graph-data management and mining.

Keywords: Streaming algorithms, Massive graph, matching, graph distance, random walk on graph

1.   Introduction

In recent years, graphs of massive size have emerged in many applications. For example, in telecommunication networks, the phone numbers that call each other form a call graphs. On the Internet, the web pages and the links between them form the web graph. Also in applications such as structured data mining, the relationships among the data items in the data set are often modeled as graphs. These graphs are massive and often have a large number of nodes and connections (edges).

New challenges arise when computing with massive graphs. It is possible to store a massive graph on a large capacity storage device. However, large capacity comes with a price: random accesses in these devices are often quite

C.C. Aggarwal and H. Wang (eds.), *Managing and Mining Graph Data*,
Advances in Database Systems 40, DOI 10.1007/978-1-4419-6045-0_13,
© Springer Science+Business Media, LLC 2010

slow (comparing to random accesses in the main memory). In some cases, it is not necessary (or even not possible) to store the graph. Algorithms that deal with massive graphs have to consider these properties.

In traditional computational models, when calculating the complexity of an algorithm, all storage devices are treated indifferently and the union of them (main memory as well as disks) are abstracted as a single memory space. Basic operations in this memory space, such as access to a random location, take an equal (constant) amount of time. However, on a real computer, access to data in main memory takes much less time than access to data on the disk. Clearly, computational complexities derived using traditional model in these cases cannot reflect the complexity of the real computation.

To consider the difference between memory and storage types, several new computational models are proposed. In the external memory model [44] the memory is divide into two types: internal (main memory) and external (disks). Accesses to external memory are viewed as one measurement of the algorithm's complexity. Despite the fact that access to the external memory is slow and accounts heavily in the algorithm's complexity measure, an external algorithm can still store all the input data in the external memory and make random access to the data.

Compare to the external memory model, the streaming model of computation takes the difference between the main memory and the storage device to a new level. The streaming model completely eliminates random access to the input data. The main memory is viewed as a workspace in the streaming model where the computation sketches temporary results and performs frequent random accesses. The inputs to the computation, however, can only be accessed in a sequential fashion, i.e., as a stream of data items. (The input may be stored on some device and accessed sequentially or the input itself may be a stream of data items. For example, in the Internet routing system, the routers forward packets at such a high speed that there may not be enough time for them to store the packets on slow storage devices.) The size of the workspace is much smaller than the input data. In many cases, it is also expected that the process of each data item in the stream take small amount of time and therefore the computation be done in near-linear time (with respect to the size of the input).

Streaming computation model is different from sampling. In the sampling model, the computation is allowed to perform random accesses to the inputs. With these accesses, it takes a few samples from the input and then computes on these samples. In some cases, the computation may not see the whole input. In the streaming model, the computation is allowed to access the input only in a sequential fashion but the whole input data can be viewed in this way. A non-adaptive sampling algorithm can be trivially transformed into a streaming algorithm. However, an adaptive sampling algorithm that takes advantage of the random access allowed by the sampling model cannot be directly adapted into the streaming model.

The streaming model was formally proposed in [29]. However, before [29], there were studies [30, 25] that considered similar models, without using the term stream or streaming. Since [29], there has been a large body of work on algorithms and complexity in this model ([3, 24, 31, 27, 26, 37, 32] as a few examples). Some example problems considered in the streaming model are computing statistics, norms and histogram constructions. Muthu's book [37] gives an excellent elaboration on the topics of general streaming algorithms and applications.

In this survey, we consider streaming algorithms for graph problems. It is no surprise that the computation model is limited. Some graph problems therefore cannot be solved using the model. However, there are still many graph problems (e.g. graph connectivity [22, 45], spanning tree [22]) for which one can obtain solutions or approximation algorithms. We will survey approximate algorithms for computing graph statistics, matching and distance in a graph, and random walk on a graph in the streaming model. We remark that this is not a comprehensive survey covering all the graph problems that have been considered in the streaming model. Rather, we will focus on the set of aforementioned topics. Many papers cited here also give lower bounds for the problems as well as algorithms. We will focus on the algorithms and omit discussion on the lower bounds. Finally we remark that although these algorithms are not direct graph mining algorithms. They compute basic graph-theoretic problems and may be utilized in massive graph mining and management systems.

2.   Streaming Model for Massive Graphs

We give a detailed description of the streaming model in this section. Streaming is a general computation model and is not just for graphs. However, since this survey concerns only graph problems and algorithms, we will focus on graph streaming computation. We consider mainly undirected graphs. The graphs can be weighted—i.e., there is a weight function $w : E \rightarrow \mathbb{R}^+$ that assigns a non-negative weight to each edge. We denote by $G(V, E)$ a graph G with vertex (node) set $V = \{v_1, v_2, \ldots, v_n\}$ and edge set $E = \{e_1, e_2, \ldots, e_m\}$ where n is the number of vertices and m the number of edges.

Definition 13.1. *A graph stream is a sequence of edges $e_{i_1}, e_{i_2}, \ldots, e_{i_m}$, where $e_{i_j} \in E$ and $i_1, i_2, \ldots, i_m$ is an arbitrary permutation of $[m] = \{1, 2, \ldots, m\}$.*

While an algorithm goes through the stream, the graph is revealed one edge at a time. The edges may be presented in any order. There are variants of graph stream in which the adjacency matrix or the adjacency list of the graph is presented as a stream. In such cases, the edges incident to each vertex are grouped together in the stream. (It is sometimes called the *incident stream*.) Definition 13.1 is more general and accounts for graphs whose edges may be generated in an arbitrary order.

We now summarize the definitions of streaming computation in [29, 28, 16, 22] as follows: A *streaming algorithm* for massive graph is an algorithm that computes some function for the graph and has the following properties:

1 The input to the streaming algorithm is a graph stream.

2 The streaming algorithm accesses the data elements (edges) in the stream in a sequential order. The order of the data elements in the stream is not controlled by the algorithm.

3 The algorithm uses a workspace that is much smaller in size than the input. It can perform unrestricted random access in the workspace. The amount of workspace required by the streaming algorithm is an important complexity measure of the algorithm.

4 As the input data stream by, the algorithm needs to process each data element quickly. The time needed by the algorithm to process each data element in the stream is another important complexity measure of the algorithm.

5 The algorithms are restricted to access the input stream in a sequential fashion. However, they may go through the stream in multiple, but a small number, of passes. The third important complexity measure of the algorithm is the number of passes required.

These properties characterize the algorithm's behavior during the time when it goes through the input data stream. Before this time, the algorithm may perform certain pre-processing on the workspace (but not on the input stream). After going through the data stream, the algorithm may undertake some post-processing on the workspace. The pre and post-processing only concern the workspace and are essentially computations in the traditional model.

Considering the three complexity measures: the workspace size, the per-element processing time, and the number of passes that the algorithm needs to go through the stream, an efficient streaming computation means that the algorithm's complexity measures are small. For example, many streaming algorithms use polylog(N) space (polylog($\cdot$) means $O(\log^c(\cdot))$ where c is a constant) when the input size is N. The streaming-clustering algorithm in [28] uses $O(N^\epsilon)$ space for a small $0 < \epsilon < 1$. Many graph streaming algorithms uses $O(n \cdot \text{polylog}(n))$ space. Some streaming algorithms process a data element in $O(1)$ time. Others may need polylog(N) time. Many streaming algorithms access the input stream in one pass. There are also multiple-pass algorithms [36, 22, 40].

Other variants of the streaming model also exist. For example, the W-stream model [15] allows the algorithm to write to (annotate) the stream during each pass. These annotations can then be utilized by the algorithm in the successive passes. Another variant [1] augments the streaming model by adding a sorting primitive.

3. Statistics and Counting Triangles

In this section, we describe a set of problems that involve graphs but essentially can be reduced to problems whose input is an array presented as a stream of the array elements (or as a sequence of increments to the elements). For example, the array $a = [2, 1, 3]$ can be given as a stream $\{(a[1] + 1), (a[3] + 1), (a[3] + 1), (a[2] + 1), (a[1] + 1), (a[3] + 1)\}$. Assuming all the entries of the array take value 0 at the beginning of the stream, after the operations in the stream, we obtain the array a.

There are many streaming algorithms that computes, for this array, statistics such as frequency moments [3, 24, 31], heavy hitters [13, 10], and construct succinct data structures that support queries such as range queries [38]. These algorithms can be directly applied once the graph problem is reduced to the corresponding problem of an array.

We consider these problems involving the degree of the graph nodes. For an undirected graph, the degree of a node is the number of edges incident to the node. One may view that there is a virtual array D associated with each graph such that $D[i]$ is the degree of the i-th node. In the streaming setting, a stream of edges translates to updates to the array D. For example, the stream $\{(1, 2), (4, 8), (2, 7) \ldots\}$ means the operation sequence: $\{(D[1] + 1), (D[2] + 1), (D[4] + 1), (D[8] + 1), (D[2] + 1), (D[7] + 1), \ldots\}$. (The degree array can be extended to directed graph, where we may have one out-degree array and one in-degree array.)

The frequency moment problem is to compute the k-th moment $f_k = \sum_{i=1}^{n} (D[i])^k$ of the node degrees. The heavy hitter problem is to report, after seeing the graph stream, the nodes having the largest degrees. The range query requires to construct a succinct representation of the array (one that is much smaller in size than the array), from which $\sum_{i=j}^{k} D[i]$, given j and k as query input, can be calculated.

Cormode and Muthu show [14] that these problems can be solved using corresponding streaming algorithms that work for an array. They further provide algorithms for these problems when the graph is a multigraph, but the degree of a node is defined to count only the distinct edges. (e.g. if the stream for a multigraph has edges $(1, 2), (2, 5), (1, 2)$, the degree of the node 1 is 1, not 2 and the degree of the node 2 is 2, not 3.) The details of the algorithms are out of the scope of this survey. We refer readers to [14] and the aforementioned literatures for streaming algorithms that compute statistics and other queries for an array.

The node degree of a graph is also related to the entropy H of an unbiased random walk on the graph [9]. In particular, H is defined to be $H = \frac{1}{2|E|} \sum_{i=1}^{n} D[i] \log D[i]$. A streaming algorithm that computes the entropy for an array, of which the i-th entry represents the frequency of the i-th element in a set is given in [9]. The authors showed that the algorithm can be applied to compute the entropy when the array is the node-degree array D for a graph, and therefore the entropy of an unbiased random walk can be calculated for a graph stream. They also extended the algorithm to multigraphs where only distinct edges are counted for the degree.

Another problem that can be reduced to computing statistics of an array is the triangle counting problem, i.e., to find the number of triangles in an undirected graph. We describe here the reduction introduced by Bar-Yossef et. al [6]. Similar to the earlier problems, there is a virtual array P associated with the graph. Each entry in the array corresponds to an (unordered) triple of the graph nodes. e.g., if v_i, v_j, v_k are three nodes in the graph, there is an entry $P[(i, j, k)]$ in the array corresponds to the triple $\{v_i, v_j, v_k\}$. The value of the entry counts how many of the three pairs $\{v_i, v_j\}$, $\{v_i, v_k\}$, and $\{v_j, v_k\}$ are actual edges in the graph. There are 4 possible values for the entries. 0, 1, 2, and 3. Let T_0, T_1, T_2, and T_3 be the number of entries that take the corresponding value. Clearly, T_3 is exactly the number of triangles in the graph. (We will abuse the notation and also use T_i to denote the set of triples whose entry value is i.)

Different from the reduction described earlier, an edge in the graph stream here maps into updates of multiple entries in the array. If we see an edge (u, v), it means $(P[(u, v, s)] + 1)$ for all nodes $s \neq u, v$. Now consider the frequency moments of the array $f_k = \sum_t (P[t])^k$. It can be decomposed into $f_k = T_1 \cdot 1^k + T_2 \cdot 2^k + T_3 \cdot 3^k$ because each entry with value 1 contributes 1^k to f_k, with value 2, 2^k and with value 3, 3^k. We can have the following equations:

$$
\begin{pmatrix} f_0 \\ f_1 \\ f_2 \end{pmatrix} = \begin{pmatrix} 1 & 1 & 1 \\ 1 & 2 & 3 \\ 1 & 4 & 9 \end{pmatrix} \cdot \begin{pmatrix} T_1 \\ T_2 \\ T_3 \end{pmatrix}.
$$

Using streaming algorithms one can estimate f_0, f_2. f_1 can be easily obtained from the stream. Solving the above equation then gives us the estimate of T_3. (Although the size of the virtual array is larger than the size of the graph stream, e.g., a stream of m edges corresponds to an array with $m(n-2)$ entries, the estimate algorithms often use space logarithmic to the size of the array. Therefore, the memory space needed is not significantly affected by the reduction.)

In [6], Bar-Yossef *et al.*also proposed improved streaming frequency-moment estimate algorithms. Using the reduction and their frequency-moment estimation, they show that for $\epsilon, \delta > 0$, the number of triangles in a graph can be estimated within ϵ error (i.e., the estimate is bounded between $(1 - \epsilon)T_3$ and $(1 + \epsilon)T_3$) with at least $1 - \delta$ probability. The algorithm uses space

$$s = O\left(\frac{1}{\epsilon^3} \cdot \log \frac{1}{\delta} \cdot \left(\frac{T_1 + T_2 + T_3}{T_3}\right)^3 \cdot \log n\right)$$

and poly(s) process time for each edge. When the stream is an incident stream, they show that, the number of triangles can be (ϵ, δ)-estimated using space

$$O\left(\frac{1}{\epsilon^2} \cdot \log \frac{1}{\delta} \cdot \left(\frac{T_3 + T_2}{T_3}\right)^2 \cdot \log n + d_{max} \log n\right).$$

where d_{max} is the maximum degree.

In a follow-up work, Jowhari and Ghodsi [33] introduced several estimators for the number of triangles. One estimator uses sequences of random numbers in a way similar to [3]. Let R be an array of uniform, ± 1-valued random numbers, i.e., $P(R[i] = 1) = P(R[i] = -1) = 0.5$ and $E(R[i]) = 0$. The random numbers in the array are 12-wise independent. A family of such random arrays can be constructed using the BCH code [3] in log-space. While the edges stream by, one computes $Z = \sum_{(i,j) \in E} R[i]R[j]$. $X = Z^3/6$ is then an estimator for the number of triangles in the graph. This is so because $E(R^k[i]) = 0$ for odd k and the numbers in R are 12-wise independent. After the expansion of X, the expectations of the terms all evaluate to zero except those in form of $6R^2[i]R^2[j]R^2[k]$, which correspond to the triangles. Jowhari and Ghodsi showed that the variance of the estimator can be controlled such that only $O(\frac{1}{\epsilon^2} \cdot \log \frac{1}{\delta} \cdot (\frac{m^3 + mC_4 + C6}{T_3^2} + 1) \cdot \log n)$ space and per-edge processing time is needed for an (ϵ, δ)-estimation. (C_k is the number of cycles of length k in the graph.) Another two sample-based estimators are also proposed in [33].

Buriol *et al.*also proposed sample-based algorithms for counting triangles in [8]. We present one of their algorithms in Algorithm 13.1.

β is a $\{0, 1\}$-valued random variable whose expectation is $\frac{3T_3}{T_1 + 2T_2 + 3T_3}$. Because $T_1 + 2T_2 + 3T_3 = m(n - 2)$, (Consider the triples consist of two end nodes of an edge plus one node from the other $n - 2$. There are $m(n - 2)$ such combinations. On the other hand, this way of counting counts each triple in T_1 once, triples in T_2 twice and triples in T_3 three times. Hence the equality.) T_3 can be estimated using a set of samples of β. For making (ϵ, δ)-estimation, this algorithm uses $O((\frac{1}{\epsilon^2} \cdot \log \frac{1}{\delta} \cdot \frac{T_1 + T_2 + T_3}{T_3})$ memory space and constant expected per-edge process time.

Buriol *et al.*further showed that Algorithm 13.1 can be modified into a one-pass algorithm. The uniform sampling of the edges can be done in one pass by

---

Algorithm 13.1: Sample Triangle

---

1 **1st pass:** Count the number of edges in the graph.
2 **2nd pass:** Sample an edge (u, v) uniformly. Choose a node w uniformly
   from $V \setminus \{(u, v)\}$.
3 **3rd pass:**
4   **if** *Both (u, w) and (v, w) are actual edges in the stream* **then**
5       $\beta = 1$
6   **else**
7       $\beta = 0$
8   **end**
9 **return** β

---

reservoir sampling [43]. One difference here is that edges (u, w) and (v, w)
may arrive before (u, v) in the stream. When (u, v) gets selected as a sample,
we have missed (u, w) and (v, w) and would not detect u, v, w as an triangle.
This happens when (u, v) is not the first edge of the triangle in the stream and
it reduces the expectation of β by a factor of 3. Sample-based algorithms are
also proposed in [8] for incidence streams.

4.   Graph Matching

A matching in a graph is a set of edges without common nodes. For an unweighted graph, the maximum matching problem is to find a matching having
the largest cardinality (number of edges). For a weighted graph, the problem
is to find a matching whose edges give the largest weight sum. We survey unweighted and weighted matching algorithms for graph streams in the following
sections.

4.1   Unweighted Matching

An early algorithm for approximating unweighted bipartite matching in the
streaming model is given in [22]. We describe the algorithm here. It is easy to
see that a maximal matching (A matching no more edge can be added because
every edge outside the match share a vertex with some edge in the matching.)
can be constructed in one pass over the graph stream.

Given a matching M for a bipartite graph $G = (L \cup R, E)$, a length-3 augmenting path for an edge $e = (u, v) \in M$, $u \in L$ and $v \in R$, is a quadruple
(w_l, u, v, w_r) such that $(u, w_l), (w_r, v) \in E$, and w_l and w_r are free vertices.
We call w_l and w_r the *wing-tips* of the augmenting path, (u, w_l) the *left wing*
and (w_r, v) the *right wing*. A set of *simultaneously augmentable length-3 augmenting paths* is a set of length-3 augmenting paths that are vertex disjoint.

Algorithm 13.2: Find Augmenting Paths

Input: a graph $G = (L \cup R, E)$, a matching M for G and a parameter $0 < \delta < 1$.

1 **while** *true* **do**

2   In one pass, find a maximal set of disjoint left wings. If the number of left wings found is $\leq \delta M$, terminate.

3   In a second pass, for the edges in M with left wings, find a maximal set of disjoint right wings.

4   In a third pass we identify the set of vertices that are

- endpoints of a matched edge that got a left wing, or

- the wing tips of a matched edge that got *both* wings, or

- endpoints of a matched edge that is no longer 3 augmentable.

We remember these vertices and in subsequent passes, we ignore any edge incident on one of these vertices.

5 **end**

Given a bipartite graph and a matching in the graph, the subroutine in Algorithm 13.2 finds a set of simultaneously augmentable length-3 augmenting paths. It will be used in the main algorithm that computes the matching for a bipartite graph.

Let X be a maximum-sized set of simultaneously augmentable length-3 augmenting paths for the maximal matching M. Let $\alpha = \frac{|X|}{|M|}$. It is shown in [22] that Algorithm 13.2 finds at least $\frac{\alpha|M| - 2\delta|M|}{3}$ simultaneously augmentable length-3 augmenting paths in $3/\delta$ passes.

The main matching algorithm increases the size of a matching by repeatedly finding a set of simultaneously augmentable length-3 augmenting paths and augmenting the matching using these paths.

The for-loop in Algorithm 13.3 runs $\lceil \frac{\log 6\epsilon}{\log 8/9} \rceil$ times. During each run, the subroutine described in Algorithm 13.2 needs to go through the input graph stream $3/\delta$ passes. Therefore, Algorithm 13.3 in total goes through the stream $O\left(\frac{\log 1/\epsilon}{\epsilon}\right)$ passes. Each call to the subroutine will find a set of simultaneously augmentable length-3 augmenting paths which increases the size of the matching. The final matching size reaches at least $(2/3 - \epsilon)$ of the maximum matching. The algorithm processes each edge in $O(1)$ time in each pass except the first pass, in which the bipartition is found. The storage space required by the algorithm is $O(n \log n)$.

Algorithm 13.3: Unweighted Bipartite Matching

Input: a bipartite graph $G = (L \cup R, E)$ and a parameter $0 < \epsilon < 1/3$.

1 In one pass, find a maximal matching M and the bipartition of G.

2 **for** $k = 1, 2, \ldots, \lceil \frac{\log 6\epsilon}{\log 8/9} \rceil$ **do**

3   Run Algorithm 13.2 with G, M and $\delta = \frac{\epsilon}{2-3\epsilon}$.

4   **for** *each* $e = (u, v) \in M$ *for which an augmenting path* (w_l, u, v, w_r) *is found by algorithm 13.2* **do**

5   remove (u, v) from M and add (u, w_l) and (w_r, v) to M.

6   **end**

7 **end**

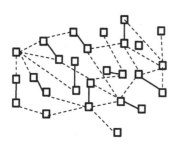 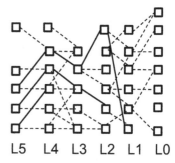

L5   L4   L3   L2   L1   L0

Figure 13.1. Layered Auxiliary Graph. Left, a graph with a matching (solid edges); Right, a layered auxiliary graph. (An illustration, not constructed from the graph on the left. The solid edges show potential augmenting paths.)

In [35], McGregor introduced an improved algorithm to find augmenting paths in an unweighted graph for which a maximal match has been constructed. Given the original input graph G and a matching M, McGregor constructed an auxiliary graph G_A to help searching for augment paths. Fig 13.1 gives an example of one auxiliary graph. The auxiliary graph is a layered graph with a small number, $k+2$, of layers. It is derived as follows: Let $L_0, L_1, \ldots, L_{k+1}$ be the layers in G_A. The free nodes in G, i.e. the nodes that haven't been covered by an edge in M, are randomly projected to be nodes in L_0 or L_{k+1}. The edges in M are projected to be a node in G_A and this node is randomly assigned to be in a layer of $L_1, L_2, \ldots, L_k$. There is an edge between a node $x \in L_i$ (that corresponding to $(v_1, v_2) \in M$) and a node $y \in L_{i-1}$ (that corresponding to $(v_3, v_4) \in M$) if $(v2, v3) \in G$. With this construction, an $(i + 1)$-length path in G_A can be mapped to a $(2i + 1)$-length augmenting path for M in G.

Identifying a set of augmenting paths for M in G now is transformed to find a set of node-disjoint paths in G_A. Because one doesn't have enough space to store the whole graph G in the streaming model, normally, the auxiliary graph G_A cannot be stored as a whole graph neither. However, the nodes in

G_A can be stored. While the algorithm passes through the input stream of G, the edges in G_A also gets revealed. Hence, the problem boils down to find a near-maximal set of node-disjoint paths in G_A.

A search algorithm was proposed in [35] for this purpose. The algorithm finds a maximal matching between layers L_{i-1} and L_i. Let $S_i \in L_i$ be the set of nodes involved in this matching. The algorithm then goes ahead to find a maximal matching between S_i and L_{i+1}. It continues in this fashion to grow a set of node-disjoint paths. Clearly, the size of S_i may decrease while i increases and may become empty before the last layer is reached. To avoid this, the path growth process may backtrack if the size of S_i becomes too small. The backtrack is done by marking the nodes in S_i as deadends, removing them from G_A and continuing path growth in the remaining of G_A.

For a particular G_A construction and path growth, the resulting set of paths may be small. However, the G_A construction is random because the nodes corresponding to the edges in M are randomly assigned to the layers. A matching algorithm is given in [35] that is similar to Algorithm 13.3 in structure but utilizes the G_A-based augmenting-path search. It is shown that, with high probability, this algorithm finds a matching in $O_\epsilon(1)$ (a function of ϵ and a constant is ϵ is constant) passes whose size is at least $\frac{1}{1+\epsilon}$ of the maximum matching.

4.2 Weighted Matching

The streaming version of the problem was first studied in [22] where a streaming algorithm (Algorithm 13.4) was proposed. The algorithm uses only one pass over the stream and manages to find a matching which is at least $\frac{1}{6}$ of the optimal size.

Algorithm 13.4: Weighted Matching

---

1 Maintain a matching M at all times.
2 **while** *there are edges in the stream* **do**
3   Let e be the next edge in the stream and $w(e)$ be the weight of e;
4   Let $w(C)$ be the sum of the weights of the edges in
    $C = \{e' | e' \in M$ and e' and e share an end point$\}$. ($w(C) = 0$ if C is empty.)
5   **if** $w(e) > 2w(C)$ **then**
6   update $M \leftarrow M \cup \{e\} \setminus C$.
7   **else**
8   ignore e
9   **end**
10 **end**

The following property of Algorithm 13.4 is shown in [22].

Theorem 13.2. *In 1 pass and $O(n \log n)$ storage, Algorithm 13.4 constructs a weighted matching that is at least $\frac{1}{6}$ of the optimal size.*

Proof: For any set of edges S, let $w(S) = \sum_{e \in S} w(e)$. We say that an edge is *selected* if it is ever part of M. We say that an edge is *dropped* if it was selected early but later replaced from M (step 6 in Algorithm 13.4) by a new heavier edge. This new edge *replaces* the dropped edge. We say an edge is a *survivor* if it is selected and never dropped. Let the set of survivors be S. The weight of the matching we find is therefore $w(S)$.

For each survivor e, let the *Trail of Drops* leading to this edge be $T(e) = C_1 \cup C_2 \cup \ldots$ where $C_0 = \{e\}$, $C_1 = \{$the edges replaced by $e\}$, and $C_i = \cup_{e' \in C_{i-1}}\{$the edges replaced by $e'\}$. We have $w(T(e)) \leq w(e)$. This is because for each replacing edge e, $w(e)$ is at least twice the cost of the replaced edges, and an edge has at most one replacing edge. Hence, for all i, $w(C_i) \geq 2w(C_{i+1})$ and

$$2w(T(e)) = \sum_{i \geq 1} 2w(C_i) \leq \sum_{i \geq 0} w(C_i) = w(T(e)) + w(e).$$

Now consider the optimal solution that includes edges opt $= \{o_1, o_2, \ldots\}$. We are going to charge the costs of the edges in opt to the survivors and their trail of drops, $\cup_{e \in S} T(e) \cup \{e\}$. We hold an edge e in this set *accountable* to $o \in$ opt if either $e = o$ or if o wasn't selected because e was in M when o arrived. Note that, in the second case, it is possible for two edges to be accountable to o. If only one edge is accountable for o then we charge $w(o)$ to e. If two edges e_1 and e_2 are accountable for o, then we charge $\frac{w(o)w(e_1)}{w(e_1)+w(e_2)}$ to e_1 and $\frac{w(o)w(e_2)}{w(e_1)+w(e_2)}$ to e_2. In either case, the amount charged by o to any edge e is at most $2w(e)$.

We now redistribute these charges as follows: (for distinct u_1, u_2, u_3) if $e = (u_1, v)$ gets charged by $o = (u_2, v)$, and e subsequently gets replaced by $e' = (u_3, v)$, we transfer the charge from e to e'. Note that we maintain the property that the amount charged by o to any edge e is at most $2w(e)$ because $w(e') \geq w(e)$. What this redistribution of charges achieves is that now every edge in a trail of drops is only charged by one edge in opt. Survivors can, however, be charged by two edges in opt. We charge $w(\text{opt})$ to the survivors and their trails of drops, and hence

$$w(\text{opt}) \leq \sum_{e \in S} (2w(T(e)) + 4w(e)).$$

Because $w(T(e)) \leq w(e)$,

$$\sum_{e \in S} (2w(T(e)) + 4w(e)) \leq 6w(S)$$

and the theorem follows. ☐

The condition on line 5 of Algorithm 13.4 can be generalized to be $w(e) > (1 + \gamma)w(C)$, $C = \{e'|e' \in M$ and e' and e share an end point$\}$. By setting γ appropriately and repeating Algorithm 13.4 until the improvement yielded falls below some threshold, a matching can be constructed [35] in $O_\epsilon(1)$ passes whose size is at least $\frac{1}{2+\epsilon}$ of the maximum matching.

Another improvement for weighted matching was made recently by Zelke [46]. Zelke's algorithm is also based on Algorithm 13.4, but incorporates some improvements. In particular, the algorithm stores a few edges that have been in M in the past but were replaced later, to potentially reinsert them into M in the future. Such edges are called in [46] the "shadow edges." With shadow edges, when a new edge arrives in the stream, besides the (two) edges that sharing the endpoints with the new edge, a few other edges (edges in M as well as the shadow edges) in the vincinity of the new edge can be examined to find potential augmenting path. This improves the approximation from 1/5.82 (by an algorithm in [35]) to 1/5.58.

5. Graph Distance

We consider the shortest-path distance in a graph. The shortest path between two vertices in a graph is the path that has the smallest number of edges (for an unweighted graph) or the smallest sum of the weights of the path edges (for a weighted graph). There may be more than one such shortest path.

A structure often used in approximating graph distance is the *graph spanner* [39, 11, 18]. An undirected graph $G = (V, E)$ induces a *metric space* $\mathcal{U}$ in which the vertex set V serves as the set of points, and the shortest-path distances serve as the *distances* between the points. The graph spanner $G' = (V, H), H \subseteq E$, is a *sparse skeleton* of the graph G whose induced metric space $\mathcal{U}'$ is a close approximation of the metric space $\mathcal{U}$ of the graph G. That is, the distance between two vertices in G' is not far from the distance between the same two vertices in G. For example, a subgraph $G' = (V, H)$, $H \subseteq E$ is a (multiplicative) t-spanner of the graph G, if for every pair of vertices $u, v \in V$, $dist_{G'}(u, v) \leq t \cdot dist_G(u, v)$ (where $dist_G(u, v)$ stands for the distance between the vertices u and v in the graph G). The *stretch factor* of a spanner is the parameter(s) that determines how close the spanner approximates the distances in the original graph, e.g., in the case of a t-spanner, the parameter t.

Clearly, if a spanner can be constructed for a massive graph, one can approximate the node distance in the graph using the spanner. Because the spanner is much smaller than the original graph, it can often be stored in the main memory. In fact, an early application of spanners is to maintain a succinct representation of the routing information [39, 11]. Instead of the original network graph, spanners are passed and stored by the routers for calculating the routing paths. Besides distances, the diameter of a graph can be approximated using the spanner diameter.

In [22], Feigenbaum *et al.* gave a simple streaming algorithm for spanner-construction by adapting the technique of [4]. It displays a certain connection between the girth of a graph and the spanner. (The *girth* of a graph is the length of the shortest cycle in the graph.) However, in the worst case, the algorithm needs more than $O(n)$ time to process an edge. Such a processing time is prohibitively high for the streaming model.

For an unweighted graph, the algorithm of [22] in one pass constructs a $(\log n / \log \log n)$-spanner S: Because a graph whose girth is larger than k have at most $\lceil n^{1+2/(k-2)} \rceil$ edges [7, 17, 2], the algorithm constructs S by adding an edge in the stream to S if the edge does not cause a cycle of length less than $\log n / \log \log n$ in the S constructed so far. Otherwise, the edge is ignored. Note that for each ignored edge, there is a path P of length at most $\log n / \log \log n$ in S that connects the two endpoints of this edge. Any shortest path in the original graph that uses this edge can be replaced by a path in S that uses P. Therefore, S is a $\log n / \log \log n$ spanner of the original graph.

For a weighted graph, however, the construction in [4] requires sorting the edges according to their weights, which is difficult in the streaming model. Instead of sorting, a geometric grouping technique is used in [22] to extend the spanner construction for unweighted graphs to a construction for weighted graphs. This technique is similar to the one used in [12]. Let ω_{min} be the minimum weight and ω_{max} be the maximum weight. We divide the range $[\omega_{min}, \omega_{max}]$ into intervals of the form $[(1 + \epsilon)^i \omega_{min}, (1 + \epsilon)^{i+1} \omega_{min})$ and round all the weights in the interval $[(1+\epsilon)^i \omega_{min}, (1+\epsilon)^{i+1} \omega_{min})$ down to $(1+\epsilon)^i \omega_{min}$. For each induced graph $G^i = (V, E^i)$, where E^i is the set of edges in E whose weight is in the interval $[(1+\epsilon)^i \omega_{min}, (1+\epsilon)^{i+1} \omega_{min})$, a spanner can be constructed in parallel using the above construction for unweighted graphs. The union of the spanners for all the G^i, $i \in \{0, 1, \ldots, \log_{(1+\epsilon)} \frac{\omega_{max}}{\omega_{min}} - 1\}$, forms a spanner for the graph G. Note that this can be done without prior knowledge of ω_{min} and ω_{max}. The goal is to break the range $[\omega_{min}, \omega_{max}]$ into a small number of intervals. Given any value $\omega \in [\omega_{min}, \omega_{max}]$, we can use the set of intervals of the form $[(1 + \epsilon)^i \omega, (1 + \epsilon)^{i+1} \omega)$ and $[\frac{\omega}{(1+\epsilon)^{i+1}}, \frac{\omega}{(1+\epsilon)^i})$. Therefore, we can determine the intervals without the prior knowledge of ω_{min} and ω_{max}.

5.1   Distance Approximation using Multiple Passes

Elkin and Zhang gave a multiple-pass streaming spanner construction in [21]. This algorithm builds an additive spanner. A subgraph $G' = (V, H)$

of the graph $G = (V, E)$ is an (additive) (α, β)-spanner of G if for every pair of vertices $u, v \in V$, $dist_{G'}(u, v) \leq \alpha \cdot dist_G(u, v) + \beta$.

We describe the algorithm of [21] and its subroutine in the following fashion. We describe first the distributed version of the algorithm and then its adaptation to the streaming model. As observed in [21], leaving space complexity aside, it is easy to see that many distributed algorithms with time complexity T translate directly into streaming algorithms that use T passes. For example, a straightforward streaming adaptation of a synchronous distributed algorithm for constructing a BFS tree would be the following: in each pass over the input stream, the BFS tree grows one more level. An exploration of d levels would result in d passes over the input stream. On the other hand, there are cases in which the running time of a synchronous algorithm may not translate directly to the number of passes of the streaming adaptation. In the example of the BFS tree, if two BFS trees are being constructed in parallel, some edges may be explored by both constructions, resulting in congestion that may increase the running time of the distributed algorithm. But for a streaming algorithm, both explorations of the same edge can be done using only one pass over the stream.

We follow the notations used in [21]. Let $diam(G)$ denote the *diameter* of the graph G, i.e., $diam(G) = max_{u,v \in V} dist_G(u, v)$. Given a subset $V' \subseteq V$, denote by $E_G(V')$ the set of edges in G *induced* by V', i.e., $E_G(V') = \{(u, w) \mid (u, w) \in E \text{ and } u, w \in V'\}$. Let $G(V') = (V', E_G(V'))$. Denote by $\Gamma_k(v, V')$ the k-*neighborhood* of vertex v in the graph $G(V')$, i.e., $\Gamma_k(v, V') = \{u \mid u \in V' \text{ and } dist_{(V', E_G(V'))}(u, v) \leq k\}$. The diameter of a subset $V' \subseteq V$, denoted by $diam(V')$, is the maximum pairwise distance in G between a pair of vertices from V'. For a collection $\mathcal{F}$ of subsets $V' \subseteq V$, let $diam(\mathcal{F}) = max_{V' \in \mathcal{F}}\{diam(V')\}$.

The spanner construction utilizes graph covers. For a graph $G = (V, E)$ and two integers $\kappa, W > 0$, a (κ, W)-*cover* [5, 11, 18] $\mathcal{C}$ is a collection of not necessarily disjoint subsets (or clusters) $C \subseteq V$ that satisfy the following conditions. (1) $\bigcup_{C \in \mathcal{C}} C = V$. (2) $diam(\mathcal{C}) = O(\kappa W)$. (3) The *size* of the cover $s(\mathcal{C}) = \sum_{C \in \mathcal{C}} |C|$ is $O(n^{1+1/\kappa})$, and furthermore, every vertex belongs to $\text{polylog}(n) \cdot n^{1/\kappa}$ clusters. (4) For every pair of vertices $u, v \in V$ that are at distance at most W from one another, there exists a cluster $C \in \mathcal{C}$ that contains both vertices, along with the shortest path between them. Note that many constructions of (κ, W)-cover will also build one BFS tree for each cluster in the cover as a by-product. The BFS tree spans the whole cluster and is rooted at one vertex in the cluster.

Algorithm 13.5 shows the construction [11, 19, 21] of (κ, W)-covers. It will be used as a subroutine in the spanner construction. Algorithm 13.5 builds a (κ, W)-cover in κ phases. A vertex v in graph G is called *covered* if there is a cluster $C \in \mathcal{C}$ such that $\Gamma_W(v, V) \subseteq C$. Let U_i be the set of uncovered

---

Algorithm 13.5: Cover

Input: a graph $G = (V, E)$ and two positive integer parameters κ and W.

1    $U_1 \leftarrow V$.

2   **for** $i = 1, 2, \ldots, \kappa$ **do**

3   Include each vertex $v \in U_i$ independently at random with probability $p_i = min\{1, \frac{n^{i/\kappa}}{n} \cdot \log n\}$ in the set S_i of phase i.

4   Each vertex $s \in S_i$ constructs a cluster by growing a BFS tree of depth $d_{i-1} = 2((\kappa - i) + 1)W$ in the graph $(U_i, E(U_i))$. We call s the *center* of the cluster and the set $\Gamma_{2(\kappa-i)W}(s, U_i)$ the *core set* of the cluster $\Gamma_{2((\kappa-i)+1)W}(s, U_i)$.

5   Let R_i be the union of the core sets of the clusters constructed in step 4. Set $U_{i+1} \leftarrow U_i \setminus R_i$.

6   **end**

---

vertices at phase i. At the beginning, $U_1 = V$. At each phase i, a subset of vertices is covered and removed from U_i.

A streaming version of Algorithm 13.5 is also given in [21]. The streaming version proceeds in κ phases. In each phase i, the algorithm passes through the input stream d_{i-1} times to build the BFS trees $\tau(v)$ of depth d_{i-1} for each selected vertex $v \in S_i$. The cluster and its core set can be computed during the construction of these BFS trees. Note that for any i, $d_{i-1} \leq 2\kappa W$. Therefore, with high probability, the streaming version of Algorithm 13.5 constructs a (κ, W)-cover using at most $2\kappa^2 W$ passes over the input stream.

We now describe the distributed algorithm in [21] that constructs the spanner. Given a cluster C, let $\mathscr{C}(C)$ be the cover constructed for the graph $(C, E_G(C))$. For a cluster $C' \in \mathscr{C}(C)$, we define $Parent(C') = C$. An execution of the algorithm can be divided into ℓ stages (levels). The original graph is viewed as a cluster on level 0. The algorithm starts level 1 by constructing a cover for this cluster. Recall that a cover is also a collection of clusters. The clusters of $\cup \mathscr{C}(C)$, where the union is over all the clusters C on level 0, are called *clusters on level 1*, and we denote the set of those clusters by $\mathscr{C}_1$. If a cluster $C \in \mathscr{C}_1$ satisfies $|C| \geq |Parent(C)|^{1-\nu}$, we say that C is a *large cluster* on level 1. Otherwise, we say that C is a *small cluster* on level 1. We denote by $\mathscr{C}_1^H$ the set of large clusters on level 1 and $\mathscr{C}_1^L$ the set of small clusters on level 1. Note that the cover-construction subroutine (Algorithm 13.5) builds a BFS-spanning tree for each cluster in the cover. The algorithm includes all the BFS-spanning trees in the spanner and then goes on to make interconnections between all pairs of clusters in $\mathscr{C}_1^H$ that are close to each other.

Algorithm 13.6: Additive Spanner Construction

Input: a graph $G = (V, E)$ on n vertices and four parameters κ, ν, D,
and Δ, where κ, D, and Δ are positive integers and $0 < \nu < 1$.

1 $\mathscr{C}_0^L \leftarrow \{V\}, \mathscr{C}_0^H = \phi$.

2 **for** *level* $i = 1, 2, \ldots, \ell = \lceil \log_{1/(1-\nu)} \log_\Delta n \rceil$ **do**

3   **Cover Construction**: For all clusters $C \in \mathscr{C}_{i-1}^L$, in parallel, construct
       (κ, D^ℓ)-covers using Algorithm 13.5. (Invoking Algorithm 13.5 with
       parameters κ and $W = D^\ell$.)

4   Include the edges of the BFS-spanning trees of all the clusters in the
       spanner. Set $\mathscr{C}_i \leftarrow \bigcup_{C \in \mathscr{C}_{i-1}^L} \mathscr{C}(C)$,
       $\mathscr{C}_i^H \leftarrow \{C \in \mathscr{C}_i \mid |C| \geq |Parent(C)|^{(1-\nu)}\}, \mathscr{C}_i^L \leftarrow \mathscr{C}_i \setminus \mathscr{C}_i^H$.

5   **Interconnection**: For all clusters $C' \in \mathscr{C}_i^H$, in parallel, construct BFS
       trees in $G(C)$, where $C = Parent(C')$. For each cluster C', the BFS
       tree is rooted at the center of the cluster, and the depth of the BFS tree
       is $2D^h + D^{h+1}$, where $h = \lceil \log_{1/(1-\nu)} \log_\Delta |Parent(C')| \rceil$.

6   For all the clusters C'' whose center vertex is in the BFS tree, if
       $C'' \in \mathscr{C}_i^H$ and $Parent(C'') = Parent(C')$, add to the spanner the
       shortest path between the center of C' and the center of C''.

7 **end**

8 Add to the spanner all the edges of the set $\bigcup_{C \in \mathscr{C}_{\ell+1}} E_G(C)$.

After these interconnections are completed, the algorithm enters level 2. For
each cluster in $\mathscr{C}_1^L$, it constructs a cover. We call the clusters in each of these
covers the *clusters on level 2*. The union of all the level-2 clusters is denoted
by $\mathscr{C}_2$. If a cluster $C \in \mathscr{C}_2$ satisfies $|C| \geq |Parent(C)|^{1-\nu}$, we say that C
is a *large cluster* on level 2. Otherwise, we say that C is a *small cluster* on
level 2. Again, we denote by $\mathscr{C}_2^H$ the set of large clusters on level 2 and $\mathscr{C}_2^L$
the set of small clusters on level 2. The BFS-spanning trees of all the clusters
in $\mathscr{C}_2$ are included into the spanner and all the close pairs of clusters in $\mathscr{C}_2^H$ get
interconnected by the algorithm.

The algorithm proceeds in a similar fashion at levels 3 and above. That is,
at level i, the algorithm constructs covers for each small cluster in $\mathscr{C}_{i-1}^L$, and
interconnects all the close pairs of large clusters in $\mathscr{C}_i^H$. Similarly, we denote
by $\mathscr{C}_i$ the collection of all the clusters in the covers constructed at level i, by
$\mathscr{C}_i^H$ the set of large clusters of $\mathscr{C}_i$, and $\mathscr{C}_i^L$ the set of small clusters of $\mathscr{C}_i$. After
level ℓ, each of the small clusters of level ℓ contains very few vertices and the
algorithm can include in the spanner all the edges induced by these clusters. A
description of the detailed algorithm is given in Algorithm 13.6.

See Figure 13.2 for an example of covers and clusters constructed by the
algorithm. The circles in the figure represent the clusters. $\mathscr{C}_1 = \{C_1, C_2, C_3\}$,

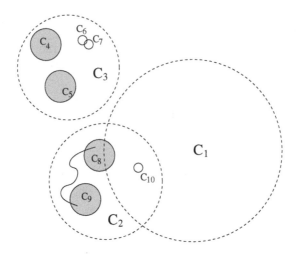

Figure 13.2. Example of clusters in covers.

$\mathscr{C}_1^H = \{C_1\}$ and $\mathscr{C}_1^L = \{C_2, C_3\}$. Note that for each cluster in $\mathscr{C}_1^L$, a cover is constructed. The union of the clusters in these covers forms $\mathscr{C}_2$, *i.e.*, $\mathscr{C}_2 = \{C_4, C_5, C_6, C_7, C_8, C_9, C_{10}\}$. The large clusters in $\mathscr{C}_2$ form $\mathscr{C}_2^H = \{C_4, C_5, C_8, C_9\}$ and the small clusters in $\mathscr{C}_2$ form $\mathscr{C}_2^L = \{C_6, C_7, C_{10}\}$. Also note that a pair of close, large clusters C_8 and C_9 is interconnected by a shortest path between them.

The streaming version [21] of Algorithm 13.6 is recursive, and the recursion has ℓ levels. At level i, a cover is constructed for each of the small clusters in $\mathscr{C}_{i-1}^L$ using the streaming algorithm for constructing covers described above. Because the processes of building BFS trees for constructing covers are independent, they can be carried out in parallel. That is, when the algorithm encounters an edge in the input stream, it examines its two endpoints. For each of the clusters in $\mathscr{C}_{i-1}^L$ that contains both endpoints, for each of the BFS-tree constructions in those clusters that has reached one of the endpoints, the algorithm checks whether the edge would help to extend the BFS tree. If so, the edge would be added to that BFS tree. After the construction of the covers is completed, the algorithm makes interconnections between close, large clusters of each cover. Again, the constructions of the BFS trees that are invoked by different interconnection subroutines are independent and can be performed in parallel. It is shown [21] that:

Theorem 13.3. *Given an unweighted, undirected graph on n vertices, presented as a stream of edges, and constants $0 < \rho, \delta, \epsilon < 1$, such that $\delta/2 + 1/3 > \rho > \delta/2$, the streaming adaptation of Algorithm 13.6, with high probability, constructs a $(1 + \epsilon, \beta)$-spanner of size $O(n^{1+\delta})$. The adaptation* *accesses the stream sequentially in $O(1)$ passes, uses $O(n^{1+\delta} \cdot \log n)$ bits of space, and processes each edge of the stream in $O(n^\rho)$ time.*

The parameters for Algorithm 13.6 are determined as follows: Set $\Delta = n^{\delta/2}$, $\frac{1}{\kappa\nu} = \delta/2$, $\frac{1}{\kappa\nu} + \nu = \rho$. This gives $\nu = \rho - \frac{\delta}{2} > 0$, $\nu = O(1)$, $\kappa = \frac{2}{(\rho-\delta/2)\delta} = O(1)$, and $\ell = \log_{1/(1-\nu)} \log_\Delta n = \log_{1/(1-\nu)} \frac{2}{\delta} = O(1)$, satisfying the requirement that κ, ν, and ℓ are all constants. Also set $D = O(\frac{\kappa\ell}{\epsilon})$. Then $\beta = O(\kappa D^\ell) = O(1)$.

Note that once the spanner is computed, the algorithm is able to compute all-pairs, almost-shortest paths and distances in the graph by computing the *exactly* shortest paths and distances in the spanner using the same space. This computation of the shortest paths in the spanner requires no additional passes through the input, and also, no additional space if one does not need to store the paths found.

5.2   Distance Approximation in One Pass

A one-pass spanner construction is given by Feigenbaum *et al.*in [23]. The algorithm is randomized and constructs a multiplicative $(2t + 1)$-spanner for an unweighted, undirected graph in one pass. With high probability, it uses $O(t \cdot n^{1+1/t} \log^2 n)$ bits of space and processes each edge in the stream in $O(t^2 \cdot n^{1/t} \log n)$ time. It is also shown in [23] that, with $O(n^{1+1/t})$ space, we cannot approximate the distance between two vertices better than by a factor of t. Therefore, this algorithm is close to the optimal.

The algorithm labels the vertices of the graph while going through the stream of edges. A label l is a positive integer. Given two parameters n and t, the set of labels L used by the algorithm is generated in the following way. Initially, we have the labels $1, 2, \ldots, n$. We denote by L^0 this set of labels and call them the *level* 0 labels. Independently, and with probability $\frac{1}{n^{1/t}}$, each label $l \in L^0$ will be selected for membership in the set S^0 and l will be marked as *selected*. From each label l in S^0, we generate a new label $l' = l + n$. We denote by L^1 the set of newly generated labels and call them level 1 labels. We then apply the above selection and new-label-generation procedure to L^1 to get the set of level 2 labels L^2. We continue this until the level $\lfloor \frac{t}{2} \rfloor$ labels $L^{\lfloor \frac{t}{2} \rfloor}$ are generated. If a level $i + 1$ label l is generated from a level i label l', we call l the *successor* of l' and denote this by $Succ(l') = l$. The set of labels used in the algorithm is the union of labels of level $1, 2, \ldots, \lfloor \frac{t}{2} \rfloor$, i.e., $L = \cup L^i$. Note that L can be generated before the algorithm sees the edges in the stream. But, in order to generate the labels, except in the case $t = O(\log n)$, the algorithm needs to know n, the number of vertices in the graph, before seeing the edges in the input stream. For $t = O(\log n)$, a simple modification of the above method can be used to generate L without knowing n, because the probability of a label's being selected can be any constant smaller than $\frac{1}{2}$.

While going through the stream, the algorithm labels each vertex with labels chosen from L. Let $C(l)$ be the collection of vertices that are labeled with l. We call the subgraph induced by the vertices in $C(l)$ a *cluster*, and we say that the label of the cluster is l. Each label thus defines a cluster.

The algorithm may label a vertex v with multiple labels; however, v will be labeled by at most one label from L^i, for $i = 1, 2, \ldots, \lfloor \frac{t}{2} \rfloor$. Moreover, if v is labeled by a label l, and l is selected, the algorithm also labels v with the label $Succ(l)$.

Denote by l^i a label of level i, i.e., $l^i \in L^i$. Let $L(v) = \{l^0, l^{k_1}, l^{k_2}, \ldots, l^{k_j}\}$, $0 < k_1 < k_2 < \ldots < k_j < t/2$ be the collection of labels that has been assigned to the vertex v. Let $Height(v) = max\{j | l^j \in L(v)\}$ and $Top(v) = l^k \in L(v)$ s.t. $k = Height(v)$.

At the beginning of the algorithm, the set $L(v_i)$ contains only the label $i \in L^0$. The set $C(l) = \{v_l\}$ for $l = 1, 2, \ldots, n$ and is empty for other labels. $L(v)$ and $C(l)$ grow while the algorithm goes through the stream and labels the vertices. For each $C(l)$, the algorithm stores a rooted spanning tree $Tree(l)$, on the vertices of $C(l)$. For $l \in L^i$, the depth of the spanning tree is at most i, i.e., the deepest leaf is at distance i from the root.

We say an edge (u, v) connects $C(l)$ and $C(l')$ if u is labeled with l and v is labeled with l'. If there are edges connecting two clusters at level $\lfloor \frac{t}{2} \rfloor$, the algorithm stores one such edge for this pair of clusters. We denote by H the set of these edges stored by the algorithm. Another small set of edges is also stored for each vertex. We denote by $M(v)$ the edges in the set for the vertex v. The spanner constructed by the algorithm is the union of the spanning trees for all the clusters, $M(v)$ for all the vertices, and the set H. The detailed algorithm is given in Algorithm 13.7.

In a later work [20], Elkin gave an improved algorithm that constructs a $(2t - 1)$-spanner in one pass over the stream. The size of the spanner is $O(t \cdot (\log n)^{1-1/t} \cdot n^{1+1/t})$ with high probability. The algorithm processes each edge in the stream with $O(1)$ time.

6.   Random Walks on Graphs

The Construction of actual random walk on a graph in the streaming model is considered by Sarma *et al.* in [40]. The algorithm of [40] that constructs a random walk from a single starting node is presented in Algorithm 13.8. The algorithm begins by randomly sampling a set of nodes, each independently with probability α. Using each sampled node as a starting point, it performs a short random walk of length w. (w is a parameter that will be set later.) This can be done in w passes over the stream. It then tries to stitch together the short random walks one by one to form a long walk and eventually produce a walk of the required length.

Algorithm 13.7: One-Pass Spanner Construction

Input: an unweighted, undirected graph $G = (V, E)$, presented as a
   stream of edges, and two positive integer parameters n and t.

1 Generate the set L of labels as described. $\forall\, v_i \in V$, label vertex v_i with
   label $i \in L^0$. If i is selected, label v_i with $Succ(i)$. Continue until we see
   a label that is not selected. Set $H \leftarrow \phi$ and $M(v_i) \leftarrow \phi$;

2 **for** *each edge (u, v) in the stream* **do**

3   **if** $L(v) \cap L(u) = \emptyset$ **then**

4   **if** $Height(v) = Height(u) = \lfloor \frac{t}{2} \rfloor$, *and there is no edge in H that*
   connects $C(Top(v))$ and $C(Top(u))$ **then**

5   set $H \leftarrow H \cup \{(u, v)\}$;

6   **else**

7   Assume, *without loss of generality*,
   $\lfloor \frac{t}{2} \rfloor \geq Height(u) \geq Height(v)$. Consider the collection of
   labels $L_v(u) = \{l^{k_1}, l^{k_2}, \ldots, l^{Height(u)}\} \subseteq L(u)$, where
   $k_1 \geq Height(v)$ and $k_1 < k_2 < \ldots < Height(u)$. Let
   $l = l^i \in L_v(u)$ such that l^i is marked as selected and there is
   no $l^j \in L_v(u)$ with $j < i$ that is marked as selected.

8   **if** *such a label l exists* **then**

9   label the vertex v with the successor $l' = Succ(l)$ of l, i.e.,
   $L(v) \leftarrow L(v) \cup \{l'\}$. Incorporate the edge in the spanning
   tree $Tree(l')$. If l' is selected, label v with $l'' = Succ(l')$
   and incorporate the edge in the tree $Tree(l'')$. Continue
   this until we see a label that is not marked as selected;

10   **else**

11   **if** *There is no edge (u', v) in $M(v)$ such that u, u' are*
   labeled with the same label $l \in L_v(u)$ **then**

12   add (u, v) to $M(v)$, i.e., set
   $M(v) \leftarrow M(v) \cup \{(u, v)\}$;

13   **end**

14   **end**

15   **end**

16   **end**

17 **end**

18 After seeing all the edges in the stream, output the union of the spanning
   trees for all the clusters, $M(v)$ for all the vertices, and the set H as the
   spanner.

The stitch works if the w-length random walk from a node u ends on a node v such that u and v are both in the set T of sampled nodes and the w-length random walk from v has not been used previously in the stitch process. If the random walk from u ends on a node outside of T or on a node in T but whose random-walk path has already been used, the stitch process gets stuck. This situation is dealt by the subroutine described in Algorithm 13.9.

Algorithm 13.8: Random Walk

Input: starting node u, walk length l, control parameter $0 < \alpha \leq 1$.

1 $T \leftarrow$ sample each node independently with probaility α.
2 In w passes, perform walks of length w from every node in T. Let $W[t]$ be the end point of the w-length walk from $t \in T$.
3 $S \leftarrow \{\}$.
4 Let $\mathscr{L}_u$ be the random walk from u to be constructed. Initialize $\mathscr{L}_u$ to be u. Let $x \leftarrow u$.
5 **while** $|\mathscr{L}_u| < l$ **do**
6   **if** $x \in T$ and $x \notin S$ **then**
7   Extend $\mathscr{L}_u$ by appending the walk $W[x]$. $S \leftarrow S \cup \{x\}$. $x \leftarrow W[x]$.
8   **else**
9   HanddleStuckNode$(x, T, S, \mathscr{L}_u, l)$.
10   **end**
11 **end**

Algorithm 13.9 first tries to extend the random walk by a length s. (s is another parameter whose value will be determined later.) It does so by randomly sample (with repetition) s edges for the node on which the stitch process is currently stuck plus each node in T whose w-length path has been used in the stitch process up to now. Let O be the set of the nodes for which we sample edges. ($O = S \cup R$ where S and R are the notations used in Algorithm 13.9.) The random walk can be extended (as far as possible) using these edges. Let x be the end node of this extension. If x is one of the nodes in O, we repeat the sampling and the extension. If x is outside O but in T, and the w-length random-walk path from x has not been used, we go back to Algorithm 13.8 and continue the stitch process. Finally, if x falls on a new node that is neither in T nor in O, we add x to O and perform the sampling and the extension again.

Each stitch extends the random walk by length w. When handling the stuck situation, either an s-progress is made or the algorithm encounters a node outside of O. With probability α, this node is in T (because T is the set of nodes sampled with probability α) and the algorithm can make a w-progress. Therefore, after a pass over the stream to sample the edges for the nodes in O, Algo-

---

Algorithm 13.9: HandleStuckNode

---

1   $R \leftarrow x$.

2   **while** $|\mathscr{L}_u| < l$ **do**

3      $E \leftarrow$ sample s edges (with repetition) out of each node in $S \cup R$.

4   Extend $\mathscr{L}_u$ as far as possible by walking along the edges in E.

5      $x \leftarrow$ new end point of $\mathscr{L}_u$. One of the following arise:

        1   if $(x \in S \cup R)$ continue;

        2   if $(x \in T$ and $x \notin S \cup R)$ **return**;

        3   if $(x \notin T$ and $x \notin S \cup R)$ $R \leftarrow R \cup \{x\}$.

6   **end**

---

rithm 13.9 can make a progress whose length is at least $\min(s, \alpha w)$ on average. Sarma *et al.*showed in [40] that, by setting $w = \sqrt{l/\alpha}$ and $s = \sqrt{l\alpha}$, the l-length random walk from a single start node can be performed in $O(\sqrt{l/\alpha})$ passes and $O(n\alpha + \sqrt{l/\alpha})$ space for any $0 < \alpha \leq 1$.

This single starting-point random walk is then extended to perform a large number K of random walks. A naive extension would simply run K copies of the single random walk in parallel. Sarma *et al.*introduced an extension that uses much less space than the naive one. They estimate the probability that the w-length walk would be used for each node. Based on this probability they store an appropriate number of w-length walks for each sampled node for K execution of Algorithm 13.8. In this way, instead of $O(K(n\alpha + \sqrt{l/\alpha}))$ space, one needs only $\tilde{O}(n\alpha + K\sqrt{l/\alpha} + Kl\alpha)$ space. (An alternative algorithm for running multiple random walks is also given in [40] that uses $\tilde{O}(n\alpha\sqrt{l\alpha} + K\sqrt{l/\alpha} + l)$ space. Combining the two, the space requirement for performing a large number of walks is $\tilde{O}(\min\{n\alpha + K\sqrt{l/\alpha} + Kl\alpha, n\alpha\sqrt{l\alpha} + K\sqrt{l/\alpha} + l\})$.) Sarma *et al.*further shows that the algorithms can be used to estimate probability distributions and to approximate mixing time.

In a later work [41], Sarma *et al.*modify and apply the above random-walk algorithms to compute sparse graph cut.

Definition 13.4. *The conductance of a cut S is defined as* $\Phi(S) = \frac{E(S,V \setminus S)}{\min\{E(S),E(V \setminus S)\}}$ *where $E(S, V \setminus S)$ is the number of edges crossing the cut $(S, V \setminus S)$ and $E(S)$ is the number of edges with at least one endpoint in S. The conductance of a graph $G = (V, E)$ is defined as $\Phi = \min_{S:E(S) \leq E(V)/2} \frac{E(S,V \setminus S)}{E(S)}$. For d-regular graphs, $\Phi =$*

$\min_{S: |S| \le |V|/2} \frac{E(S, V \setminus S)}{d|S|}$. *The sparsity of a d-regular graph is related to the conductance by a factor d.*

It is well known that a sparse cut of a graph can be obtained by performing random walks [34, 42]. In particular, one can start from a random source and perform a random walk of length about $1/\Phi$. The random walk defines a probability p_i for each node i that is the probability of the random walk landing on node i. One can sort the nodes in decreasing order of p_i. Each prefix of this ordered sequence of nodes gives a cut. Lovasz and Simonovits [34] showed that one of the n cuts can be sparse. Sarma *et al.*extended the result to the case where an estimate $\tilde{p}_i$ of p_i is available. Let $\rho_p(i) = p_i/d_i$ (where d_i is the degree of the i-th node). They show in [41] that:

Theorem 13.5. *Let $\tilde{p}_i$ be an estimate for p_i where the error $|\tilde{p}_i - p_i| \le \epsilon(p + \sqrt{p/n} + 1/n)$ for a source s from U, where there is a cut $(U, V \setminus U)$ of conductance at most Φ (with $|U| \le |V|/2$), and a random walk of length l. Order the nodes in decreasing order of $\rho_{\tilde{p}}(i)$. Each prefix of this ordered sequence gives a cut. If the source node s is chosen randomly and l is chosen randomly in the range $\{1, 2, \ldots, O(1/\Phi)\}$, then one of the n cuts S gives $\Phi(S) \le \tilde{O}(\sqrt{\Phi})$ if $\epsilon \le o(\Phi)$, with constant probability.*

Following Theorem 13.5 and using a modified version of the random walk algorithm of [40], Sarma *et al.*provided an algorithm that finds, with high probability, a cut of conductance at most $\tilde{O}(\sqrt{\Phi})$ for any d-regular graph that has a cut of conductance at most Φ and balance b. The algorithm goes through the graph stream $\tilde{O}(\sqrt{\frac{1}{\Phi \alpha}})$ passes and uses space $\tilde{O}(\min\{n\alpha + \frac{1}{b}(\frac{n\alpha}{d\Phi^3} + \frac{n}{d\sqrt{\alpha}\Phi^{2.5}}), (n\alpha + \frac{1}{b}\frac{n}{d\alpha\Phi^2})\sqrt{\frac{1}{\Phi\alpha}} + \frac{1}{\Phi}\})$. In [41], they also give algorithms that computes sparse projected cuts.

7.   Conclusions

Massive graphs emerged in recent years that may be too large to fit into main memory. Streaming is considered as a computation model to handle massive data sets (including massive graphs). Despite the restriction imposed by the model, there are algorithms for many graph problems. We surveyed recent algorithms for computing graph statistics, matching and distance in a graph, and random walks on a graph. Due to the limitation of the model, many algorithms output an approximate result. Streaming algorithms are a topic of considerable research interest. Efforts are being made to improve the approximation and to design more algorithms for problems arising from applications.

References

[1] G. Aggarwal, M. Datar, S. Rajagopalan, and M. Ruhl. On the streaming model augmented with a sorting primitive. In *IEEE Symposium on Foundations of Computer Science*, pages 540–549, 2004.

[2] N. Alon, S. Hoory, and N. Linial. The moore bound for irregular graphs. *Graphs and Combinatorics*, 18(1):53–57, 2002.

[3] N. Alon, Y. Matias, and M. Szegedy. The space complexity of approximating the frequency moments. *Journal of Computer and System Sciences*, 58(1):137–147, 1999.

[4] I. Althefer, G. Das, D. Dobkin, and D. Joseph. Generating sparse spanners for weighted graphs. In *Proc. 2nd Scandinavian Workshop on Algorithm Theory, LNCS 447*, pages 26–37, 1990.

[5] B. Awerbuch, B. Berger, L. Cowen, and D. Peleg. Near-linear time construction of sparse neighborhood covers. *SIAM Journal on Computing*, 28(1):263–277, 1998.

[6] Z. Bar-Yossef, R. Kumar, and D. Sivakumar. Reductions in streaming algorithms, with an application to counting triangles in graphs. In *Proc. 13th ACM-SIAM Symposium on Discrete Algorithms*, pages 623–632, 2002.

[7] B. Bollobas. *Extremal Graph Theory*. Academic Press, New York, 1978.

[8] L. S. Buriol, G. Frahling, S. Leonardi, A. Marchetti-Spaccamela, and C. Sohler. Counting triangles in data streams. In *Proceedings of ACM Symposium on Principles of Database Systems*, pages 253–262, 2006.

[9] A. Chakrabarti, G. Cormode, and A. McGregor. A near-optimal algorithm for computing the entropy of a stream. In *ACM-SIAM Symposium on Discrete Algorithms*, pages 328–335, 2007.

[10] M. Charikar, K. Chen, and M. Farach-Colton. Finding frequent items in data streams. *Theoretical Computer Science*, 312, 2004.

[11] E. Cohen. Fast algorithms for t-spanners and stretch-t paths. In *Proc. 34th IEEE Symposium on Foundation of Computer Science*, pages 648–658, 1993.

[12] E. Cohen. Fast algorithms for constructing t-spanners and paths with stretch t. *SIAM Journal on Computing*, 28:210–236, 1998.

[13] Cormode and Muthukrishnan. What's hot and what's not: Tracking most frequent items dynamically. *ACM Transactions on Database Systems*, 30, 2005.

[14] G. Cormode and S. Muthukrishnan. Space efficient mining of multigraph streams. In *Proceedings of ACM Symposium on Principles of Database Systems*, pages 271–282, 2005.

[15] C. Demetrescu, I. Finocchi, and A. Ribichini. Trading of space for passes in graph streaming problems. In *ACM-SIAM Symposium on Discrete Algorithms*, pages 714–723, 2006.

[16] P. Drineas and R. Kannan. Pass efficient algorithms for approximating large matrices. In *Proc. 14th ACM-SIAM Symposium on Discrete Algorithms*, pages 223–232, 2003.

[17] R. D. Dutton and R. C. Brigham. Edges in graphs with large girth. *Graphs and Combinatorics*, 7(4):315–321, 1991.

[18] M. Elkin. Computing almost shortest paths. In *Proc. 20th ACM Symposium on Principles of Distributed Computing*, pages 53–62, 2001.

[19] M. Elkin. A fast distributed protocol for constructing the minimum spanning tree. In *Proc. 15th ACM-SIAM Symposium on Discrete Algorithms*, pages 352–361, 2004.

[20] M. Elkin. Streaming and fully dynamic centralized algorithms for constructing and maintaining sparse spanners. In *International Col loquium on Automata, Languages and Programming*, pages 716–727, 2007.

[21] M. Elkin and J. Zhang. Efficient algorithms for constructing $(1 + \epsilon, \beta)$-spanners in the distributed and streaming models. In *Proc. 23rd ACM Symposium on Principles of Distributed Computing*, pages 160–168, 2004.

[22] J. Feigenbaum, S. Kannan, A. McGregor, S. Suri, and J. Zhang. On graph problems in a semi-streaming model. In *Proc. 31st International Colloquium on Automata, Languages and Programming, LNCS 3142*, pages 531–543, 2004.

[23] J. Feigenbaum, S. Kannan, A. McGregor, S. Suri, and J. Zhang. Graph distances in the streaming model: The value of space. In *Proc. 16th ACM-SIAM Symposium on Discrete Algorithms*, pages 745–754, 2005.

[24] J. Feigenbaum, S. Kannan, M. Strauss, and M. Viswanathan. An approximate L^1 difference algorithm for massive data streams. *SIAM Journal on Computing*, 32(1):131–151, 2002.

[25] P. Flajolet and G. Martin. Probabilistic counting. In *Proc. 24th IEEE Symposium on Foundation of Computer Science*, pages 76–82, 1983.

[26] A. C. Gilbert, S. Guha, P. Indyk, Y. Kotidis, S. Muthukrishnan, and M. Strauss. Fast, small-space algorithms for approximate histogram maintenance. In *Proc. 34th ACM Symposium on Theory of Computing*, pages 389–398, 2002.

[27] S. Guha, N. Koudas, and K. Shim. Data-streams and histograms. In *Proc. 33rd ACM Symposium on Theory of Computing*, pages 471–475, 2001.

[28] S. Guha, N. Mishra, R. Motwani, and L. O'Callaghan. Clustering data streams. In *Proc. 41st IEEE Symposium on Foundations of Computer Science*, pages 359–366, 2000.

[29] M. R. Henzinger, P. Raghavan, and S. Rajagopalan. Computing on data streams. *Technical Report 1998-001, DEC Systems Research Center*, 1998.

[30] J. Hopcroft and J. Ullman. Some results on tape-bounded turing machines. *Journal of the ACM*, 16:160–177, 1969.

[31] P. Indyk. Stable distributions, pseudorandom generators, embeddings and data stream computation. In *Proc. 41st IEEE Symposium on Foundations of Computer Science*, pages 189–197, 2000.

[32] P. Indyk. Algorithms for dynamic geometric problems over data streams. In *Proc. 36th ACM Symposium on Theory of Computing*, pages 373–380, 2004.

[33] Jowhari and Ghodsi. New streaming algorithms for counting triangles in graphs. In *Annual International Conference on Computing and Combinatorics*, pages 710–716, 2005.

[34] L. Lovasz and M. Simonovits. The mixing rate of markov chains, an isoperimetric inequality, and computing the volume. In *IEEE Symposium on Foundations of Computer Science*, pages 346–354, 1990.

[35] A. McGregor. Finding graph matchings in data streams. In *APPROX-RANDOM*, pages 170–181, 2005.

[36] J. Munro and M. Paterson. Selection and sorting with limited storage. *Theoretical Computer Science*, 12:315–323, 1980.

[37] S. Muthukrishnan. *Data Streams: Algorithms and Applications*. Now Publishers, 2006.

[38] S. Muthukrishnan and M. Strauss. Rangesum histograms. In *ACM-SIAM Symposium on Discrete Algorithms*, pages 233–242, 2003.

[39] D. Peleg and J. Ullman. An optimal synchronizer for the hypercube. *SIAM Journal on Computing*, 18:740–747, 1989.

[40] A. D. Sarma, S. Gollapudi, and R. Panigrahy. Estimating pagerank on graph streams. In *ACM Symposium on Principles of Database Systems*, pages 69–78, 2008.

[41] A. D. Sarma, S. Gollapudi, and R. Panigrahy. Sparse cut projections in graph streams. In *European Symposium on Algorithms*, 2009.

[42] D. Spielman and S.-H. Teng. Nearly-linear time algorithms for graph partitioning, graph sparsification, and solving linear systems. In *ACM Symposium on Theory of Computing*, pages 81–90, 2004.

[43] J. Vitter. Random sampling with a reservoir. *ACM Trans. Math. Softw*, 11(1):37–57, 1985.

[44] J. S. Vitter. External memory algorithms and data structures: Dealing with massive data. *ACM Computing Surveys*, 33(2):209–271, 2001.

[45] M. Zelke. k-connectivity in the semi-streaming model. *CoRR*, cs/0608066, 2006.

[46] M. Zelke. Weighted matching in the semi-streaming model. In *Symposium on Theoretical Aspects of Computer Science*, pages 669–680, 2008.

Chapter 14

A SURVEY OF PRIVACY-PRESERVATION OF GRAPHS AND SOCIAL NETWORKS

Xintao Wu
University of North Carolina at Charlotte
xwu@uncc.edu

Xiaowei Ying
University of North Carolina at Charlotte
xying@uncc.edu

Kun Liu
Yahoo! Labs
kun@yahoo-inc.com

Lei Chen
Hong Kong University of Science and Technology
leichen@cs.ust.hk

Abstract Social networks have received dramatic interest in research and development. In this chapter, we survey the very recent research development on privacy-preserving publishing of graphs and social network data. We categorize the state-of-the-art anonymization methods on simple graphs in three main categories: K-anonymity based privacy preservation via edge modification, probabilistic privacy preservation via edge randomization, and privacy preservation via generalization. We then review anonymization methods on rich graphs. We finally discuss challenges and propose new research directions in this area.

Keywords: Anonymization, Randomization, Generalization, Privacy Disclosure, Social Networks

C.C. Aggarwal and H. Wang (eds.), *Managing and Mining Graph Data*,
Advances in Database Systems 40, DOI 10.1007/978-1-4419-6045-0_14,
© Springer Science+Business Media, LLC 2010

1.   Introduction

Graphs and social networks are of significant importance in various application domains such as marketing, psychology, epidemiology and homeland security. The management and analysis of these networks have attracted increasing interests in the sociology, database, data mining and theory communities. Most previous studies are focused on revealing interesting properties of networks and discovering efficient and effective analysis methods [24, 37, 39, 5, 25, 7, 27, 14, 38, 6, 15, 23, 40, 36]. This chapter will provide a survey of methods for privacy-preservation of graphs, with a special emphasis towards social networks.

Social networks often contain some private attribute information about individuals as well as their sensitive relationships. Many applications of social networks such as anonymous Web browsing require identity and/or relationship anonymity due to the sensitive, stigmatizing, or confidential nature of user identities and their behaviors. The privacy concerns associated with data analysis over social networks have incurred the recent research. In particular, privacy disclosure risks arise when the data owner wants to publish or share the social network data with another party for research or business-related applications. Privacy-preserving social network publishing techniques are usually adopted to protect privacy through masking, modifying and/or generalizing the original data while without sacrificing much data utility. In this chapter, we provide a detailed survey of the *very recent* work on this topic in an effort to allow readers to observe common themes and future directions.

1.1   Privacy in Publishing Social Networks

In a social network, nodes usually correspond to individuals or other social entities, and an edge corresponds to the relationship between two entities. Each entity can have a number of attributes, such as age, gender, income, and a unique identifier. One common practice to protect privacy is to publish a naive node-anonymized version of the network, e.g., by replacing the identifying information of the nodes with random IDs. While the naive node-anonymized network permits useful analysis, as first pointed out in [4, 20], this simple technique does not guarantee privacy since adversaries may re-identify a target individual from the anonymized graph by exploiting some known structural information of his neighborhood.

The privacy breaches in social networks can be grouped to three categories: *identity disclosure, link disclosure,* and *attribute disclosure.* The identity disclosure corresponds to the scenario where the identity of an individual who is associated with a node is revealed. The link disclosure corresponds to the scenario where the sensitive relationship between two individuals is disclosed.

The attribute disclosure denotes the sensitive data associated with each node is compromised. Compared with existing anonymization and perturbation techniques of tabular data, it is more challenging to design effective anonymization techniques for social network data because of difficulties in modeling background knowledge and quantifying information loss.

1.2   Background Knowledge

Adversaries usually rely on background knowledge to de-anonymize nodes and learn the link relations between de-anonymized individuals from the released anonymized graph. The assumptions of the adversary's background knowledge play a critical role in modeling privacy attacks and developing methods to protect privacy in social network data. In [51], Zhou et al. listed several types of background knowledge: attributes of vertices, specific link relationships between some target individuals, vertex degrees, neighborhoods of some target individuals, embedded subgraphs, and graph metrics (e.g., betweenness, closeness, centrality).

For simple graphs in which nodes are not associated with attributes and links are unlabeled, adversaries only have structural background knowledge in their attacks (e.g., vertex degrees, neighborhoods, embedded subgraphs, graph metrics). For example, Liu and Terzi [31] considered vertex degrees as background knowledge of the adversaries to breach the privacy of target individuals, the authors of [20, 50, 19] used neighborhood structural information of some target individuals, the authors of [4, 52] proposed the use of embedded subgraphs, and Ying and Wu [47] exploited the topological similarity/distance to breach the link privacy.

For rich graphs in which nodes are associated with various attributes and links may have different types of relationships, it is imperative to study the impact on privacy disclosures when adversaries combine attributes and structural information together in their attacks. Re-identification with attribute knowledge of individuals has been well-studied and resiting techniques have been developed for tabular data (see, e.g., the survey book [1]). However, applying those techniques directly on network data erases inherent graph structural properties. The authors, in [11, 8, 9, 49], investigated anonymization techniques for different types of rich graphs against complex background knowledge.

As pointed out in two earlier surveys [30, 51], it is very challenging to model all types of background knowledge of adversaries and quantify their impacts on privacy breaches in the scenario of publishing social networks with privacy preservation.

1.3   Utility Preservation

An important goal of publishing social network data is to permit useful analysis tasks. Different analysis tasks may expect different utility properties to be preserved. So far, three types of utility have been considered.

- Graph topological properties. One of the most important applications of social network data is for analyzing graph properties. To understand and utilize the information in a network, researches have developed various measures to indicate the structure and characteristics of the network from different perspectives [12]. Properties including degree sequences, shortest connecting paths, and clustering coefficients are addressed in [20, 45, 31, 19, 50, 46].

- Graph spectral properties. The spectrum of a graph is usually defined as the set of eigenvalues of the graph's adjacency matrix or other derived matrices. The graph spectrum has close relations with many graph characteristics and can provide global measures for some network properties [36]. Spectral properties are adopted to preserve utility of randomized graphs in [45, 46].

- Aggregate network queries. An aggregate network query calculates the aggregate on some paths or subgraphs satisfying some query conditions. One example is that the average distance from a medical doctor vertex to a teacher vertex in a network. In [52, 50, 8, 11], the authors considered the accuracy of answering aggregate network queries as the measure of utility preservation.

In general, it is very challenging to quantify the information loss in anonymizing social networks. For tabular data, since each tuple is usually assumed to be independent, we can measure the information loss of the anonymized table using the sum of the information loss of each individual tuple. However, for social network data, the information loss due to the graph structure change should also be taken into account in addition to the information loss associated with node attribute changes. In [52], Zou et al. used the number of modified edges between the original graph and the released one to quantify information loss due to structure change. The rationale of using anonymization cost to measure the information loss is that a lower anonymization cost indicates that fewer changes have been made to the original graph.

1.4   Anonymization Approaches

Similar to the design of anonymization methods for tabular data, the design of anonymization methods also need take into account the attacking models and the utility of the data. We categorize the state-of-the-art anonymization methods on simple network data into three categories as follows.

- K-anonymity privacy preservation via edge modification. This approach modifies graph structure via a sequence of edge deletions and additions such that each node in the modified graph is indistinguishable with at least $K - 1$ other nodes in terms of some types of structural patterns.

- Edge randomization. This approach modifies graph structure by randomly adding/deleting edges or switching edges. It protects against re-identification in a probabilistic manner.

- Clustering-based generalization. This approach clusters nodes and edges into groups and anonymizes a subgraph into a super-node. The details about individuals are hidden.

The above anonymization approaches have been shown as a necessity in addition to naive anonymization to preserve privacy in publishing social network data.

In the following, we first focus on *simple graphs* in Section 2 to 5. Specifically, we revisit existing attacks on naive anonymized graphs in Section 2, K-anonymity approaches via edge modification in Section 3, edge randomization approaches in Section 4, and clustering-based generalization approaches in Section 5 respectively. We then survey the recent development of anonymization techniques for *rich graphs* in Section 6. Section 7 is dedicated to other privacy issues in online social networks in addition to those on publishing social network data. We give conclusions and point out future directions in Section 8.

1.5   Notations

A network $G(V, E)$ is a set of n nodes connected by a set of m links, where V denotes the set of nodes and $E \subseteq V \times V$ is the set of links. The network considered here is binary, symmetric, and without self-loops. $A = (a_{ij})_{n \times n}$ is the adjacency matrix of G: $a_{ij} = 1$ if node i and j are connected and $a_{ij} = 0$ otherwise. The degree of node i, d_i, is the number of the nodes connected to node i, i.e., $d_i = \sum_j a_{ij}$, and $d = \{d_1, \ldots, d_n\}$ denotes the degree sequence. The released graph after perturbation is denoted by $\widetilde{G}(\widetilde{V}, \widetilde{E})$. $\widetilde{A} = (\tilde{a}_{ij})_{n \times n}$ is the adjacency matrix of $\widetilde{G}$, and $\tilde{d}_i$ and $\tilde{d}$ are the degree and degree sequence of $\widetilde{G}$ respectively.

Note that, for ease of presentation, we use the following pairs of terms interchangeably: "graph" and "network", "node" and "vertex", "edge" and "link", "entity" and "individual", "attacker" and "adversary".

2.   Privacy Attacks on Naive Anonymized Networks

The practice of naive anonymization replaces the personally identifying information associated with each node with a random ID. However, an adversary can potentially combine external knowledge with the observed graph structure to compromise privacy, de-anonymize nodes, and learn the existence of sensitive relationships between explicitly de-anonymized individuals.

2.1   Active Attacks and Passive Attacks

In [24], Backstrom et al. presented two different types of attacks on anonymized social networks.

- **Active attacks.** An adversary chooses an arbitrary set of target individuals, creates a small number of new user accounts with edges to these target individuals, and establishes a highly distinguishable pattern of links among the new accounts. The adversary can then efficiently find these new accounts together with the target individuals in the released anonymized network.

- **Passive attacks.** An adversary does not create any new nodes or edges. Instead, he simply constructs a coalition, tries to identify the subgraph of this coalition in the released network, and compromises the privacy of neighboring nodes as well as edges among them.

The *active attack* is based on the uniqueness of small subgraphs embedded in the network. The constructed subgraph H by the adversary needs to satisfy the following three properties in order to make the *active attack* succeed:

- There is no other subgraph S in G such that S and H are isomorphic.

- H is uniquely and efficiently identifiable regardless of G.

- The subgraph H has no non-trivial automorphisms.

It has been shown theoretically that a randomly generated subgraph H formed by $O(\sqrt{\log n})$ nodes can compromise the privacy of arbitrarily target nodes with high probability for any network. The *passive attack* is based on the observation that most nodes in real social network data already belong to a small uniquely identifiable subgraph. A coalition X of size k is initiated by one adversary who recruits $k-1$ of his neighbors to join the coalition. It assumes that the users in the coalition know both the edges amongst themselves (i.e., the internal structure of H) and the names of their neighbors outside X. Since the structure of H is not randomly generated, there is no guarantee that it can be uniquely identified. The primary disadvantage of the *passive attack* in practice, compared to the *active attack*, is that it does not allow one to compromise the privacy of arbitrary users. The adversaries can adopt a hybrid *semi-passive* attack: they create no new accounts, but simply create a few additional out-links to target users before the anonymized network is released. We refer readers to [24] for more details on theoretical results and empirical evaluations on a real social network with 4.4 million nodes and 77 million edges extracted from LiveJournal.com.

2.2   Structural Queries

In [19], Hay et al. studied three types of background knowledge to be used by adversaries to attack naively-anonymized networks. They modeled adversaries' external information as the access to a source that provides answers to a *restricted knowledge query Q* about a single target node in the original graph. Specifically, background knowledge of adversaries is modeled using the following three types of queries.

- **Vertex refinement queries.** These queries describe the local structure of the graph around a node in an iterative refinement way. The weakest knowledge query, $\mathcal{H}_0(x)$, simply returns the label of the node x; $\mathcal{H}_1(x)$ returns the degree of x; $\mathcal{H}_2(x)$ returns the multiset of each neighbors' degree, and $\mathcal{H}_i(x)$ can be recursively defined as:

$$\mathcal{H}_i(x) = \{\mathcal{H}_{i-1}(z_1), \mathcal{H}_{i-1}(z_2), \cdots, \mathcal{H}_{i-1}(z_{d_x})\}$$

  where $z_1, \cdots, z_{d_x}$ are the nodes adjacent to x.

- **Subgraph queries.** These queries can assert the existence of a subgraph around the target node. The descriptive power of a query is measured by counting the number of edges in the described subgraph. The adversary is capable of gathering some fixed number of edges focused around the target x. By exploring the neighborhood of x, the adversary learns the existence of a subgraph around x representing partial information about the structure around x.

- **Hub fingerprint queries.** A hub is a node in a network with high degree and high betweenness centrality. A hub fingerprint for a target node x, $\mathcal{F}_i(x)$, is a description of the node's connections to a set of designated hubs in the network where the subscript i places a limit on the maximum distance of observable hub connections.

The above queries represent a range of structural information that may be available to adversaries, including complete and partial descriptions of node's local neighborhoods, and node's connections to hubs in the network.

Vertex refinement queries provide complete information about node degree while a subgraph query can never express $\mathcal{H}_i$ knowledge because subgraph queries are existential and cannot assert exact degree constraints or the absence of edges in a graph. The semantics of subgraph queries seem to model realistic adversary capabilities more accurately. It is usually difficult for an adversary to acquire the complete detailed structural description of higher-order vertex refinement queries.

2.3   Other Attacks

In [34], Narayanan and Shmatikov assumed that the adversary has two types of background knowledge: aggregate auxiliary information and individual auxiliary information. The aggregate auxiliary information includes an auxiliary graph $G_{aux}(V_{aux}, E_{aux})$ whose members overlap with the anonymized target graph and a set of probability distributions defined on attributes of nodes and edges. These distributions represent the adversary's (imperfect) knowledge of the corresponding attribute values. The individual auxiliary information is the detailed information about a very small number of individuals (called *seeds*) in both the auxiliary graph and the target graph.

After re-identifying the seeds in target graph, the adversaries immediately get a set of de-anonymized nodes. Then, by comparing the neighborhoods of the de-anonymized nodes in the target graph with the auxiliary graph, the adversary can gradually enlarge the set of de-anonymized nodes. During this *propagation* process, known information such as probability distributions and mappings are updated repeatedly to reduce the error. The authors showed that even some edge addition and deletion are applied independently to the released graph and the auxiliary graph, their de-anonymizing algorithm can correctly re-identify a large number of nodes in the released graph.

To protect against these attacks, researchers have developed many different privacy models and graph anonymization methods. Next, we will provide a detailed survey on these techniques.

3.   *K*-Anonymity Privacy Preservation via Edge Modification

The adversary aims to locate the vertex in the network that corresponds to the target individual by analyzing topological features of the vertex based on his background knowledge about the individual. Whether individuals can be re-identified depends on the descriptive power of the adversary's background knowledge and the structural similarity of nodes. To quantify the privacy breach, Hey et al. [19] proposed a general model for social networks as follows:

Definition 14.1. *K-candidate anonymity. A node x is K-candidate anonymous with respect to a structure query Q if there exist at least $K - 1$ other nodes in the graph that match query Q. In other words, $|cand_Q(x)| \geq K$*

where $cand_Q(x) = \{y \in V | Q(y) = Q(x)\}$. *A graph satisfies K-candidate anonymity with respect to Q if all the nodes are K-candidate anonymous with respect to Q.*

Three types of queries (vertex refinement queries, subgraph queries, and hub fingerprint queries) were presented and evaluated on the naive anonymized graphs. In [20], Hay et al. studied an edge randomization technique that modifies the graph via a sequence of random edge deletions followed by edge additions. In [19] Hay et al. presented a generalization technique that groups nodes into super-nodes and edges into super-edges to satisfy the K-anonymity. We will introduce their techniques in Section 4.1 and 5 in details respectively.

Several methods have been investigated to prevent node re-identification based on the K-anonymity concept. These methods differ in the types of the structural background knowledge that an adversary may use. In [31], Liu and Terzi assumed that the adversary knows only the degree of the node of a target individual. In [50], Zhou and Pei assumed one specific subgraph constructed by the immediate neighbors of a target node is known. In [52], Zou et al. considered all possible structural information around the target and proposed K-automorphism to guarantee privacy under any structural attack.

3.1 K-Degree Generalization

In [31], Liu and Terzi pointed out that the degree sequences of real-world graphs are highly skewed, and it is usually easy for adversaries to collect the degree information of a target individual. They investigated how to modify a graph via a set of edge addition (and/or deletion) operations in order to construct a new K-degree anonymous graph, in which every node has the same degree with at least $K - 1$ other nodes. The authors imposed a requirement that the minimum number of edge-modifications is made in order to preserve the utility. The K-degree anonymity property prevents the re-identification of individuals by the adversaries with prior knowledge on the number of social relationships of certain people (i.e., vertex background knowledge).

Definition 14.2. *K-degree anonymity.* *A graph $G(V, E)$ is K-degree anonymous if every node $u \in V$ has the same degree with at least $K - 1$ other nodes.*

Problem 1. *Given a graph $G(V, E)$, construct a new graph $\widetilde{G}(\widetilde{V}, \widetilde{E})$ via a set of edge-addition operations such that 1) $\widetilde{G}$ is K-degree anonymous; 2)$V = \widetilde{V}$; and 3) $\widetilde{E} \cap E = E$.*

The proposed algorithm is outlined below.

1 Starting from the degree sequence d of the original graph $G(V, E)$, construct a new degree sequence $\tilde{d}$ that is K-anonymous and the L_1 distance, $\|\tilde{d} - d\|_1$ is minimized.

2 Construct a new graph $\widetilde{G}(\widetilde{V}, \widetilde{E})$ such that $d_{\widetilde{G}} = \tilde{d}$, $\widetilde{V} = V$, and $\widetilde{E} = E$ (or $\widetilde{E} \cap E \approx E$ in the relaxed version).

The first step is solved by a linear-time dynamic programming algorithm while the second step is based on a set of graph-construction algorithms given a degree sequence. The authors also extended their algorithms to allow for simultaneous edge additions and deletions. Their empirical evaluations showed that the proposed algorithms can effectively preserve the graph utility (in terms of topological features) while satisfying the K-degree anonymity.

3.2 K-Neighborhood Anonymity

In [50], Zhou and Pei assumed that the adversary knows subgraph constructed by the immediate neighbors of a target node. The proposed greedy graph-modification algorithm generalizes node labels and inserts edges until each neighborhood is indistinguishable to at least $K - 1$ others.

Definition 14.3. *K-neighborhood anonymity.* *A node u is K-neighborhood anonymous if there exist at least $K - 1$ other nodes $v_1, \ldots, v_{K-1} \in V$ such that the subgraph constructed by the immediate neighbors of each node $v_1, \cdots, v_{K-1}$ is isomorphic to the subgraph constructed by the immediate neighbors of u. A graph satisfies K-neighborhood anonymity if all the nodes are K-neighborhood anonymous.*

The definition can be extended from the immediate neighbor to the d-neighbors ($d > 1$) of the target vertex, i.e., the vertices within distance d to the target vertex in the network.

Problem 2. *Given a graph $G(V, E)$, construct a new graph $\widetilde{G}(\widetilde{V}, \widetilde{E})$ satisfying the following conditions: 1) $\widetilde{G}$ is K-neighborhood anonymous; 2)$V = \widetilde{V}$; 3) $\widetilde{E} \cap E = E$; and 4) $\widetilde{G}$ can be used to answer aggregate network queries as accurately as possible.*

The simple case of constructing a K-neighborhood anonymous graph satisfying condition 1-3) was shown as *NP*-hard [50]. The proposed algorithm is outlined below.

1 Extract the neighborhoods of all vertices in the network. A *neighborhood component coding* technique, which can represent the neighborhoods in a concise way, is used to facilitate the comparisons among neighborhoods of different vertices including the isomorphism tests.

2 Organize vertices into groups and anonymize the neighborhoods of vertices in the same group until the graph satisfies K-anonymity. A heuristic of starting with vertices with high degrees is adopted since these vertices are more likely to be vulnerable to structural attacks.

In [50], Zhou and Pei studied social networks with vertex attributes information in addition to the unlabeled network topology. The vertex attributes form a hierarchy. Hence, there are two ways to anonymize the neighborhoods of vertices: generalizing vertex labels and adding edges. In terms of utility, it focuses on using anonymized social networks to answer aggregate network queries.

3.3 K-Automorphism Anonymity

Zou et al. in [52] adopted a more general assumption: the adversary can know any subgraph around a certain individual α. If such a subgraph can be identified in the anonymized graph with high probability, user α has a high identity disclosure risk. The authors aimed to construct a graph $\widetilde{G}$ so that for any subgraph $X \subset G$, $\widetilde{G}$ contains at least K subgraphs isomorphic to X. We first give some definitions introduced in [52]:

Definition 14.4. *Graph isomorphism and automorphism.* *Given two graphs $G_1(V_1, E_1)$ and $G_2(V_2, E_2)$, G_1 is isomorphic to G_2 if there exists a bijective function $f : V_1 \rightarrow V_2$ such that for any two nodes $u, v \in V_1$, $(u, v) \in E_1$ if and only if $(f(u), f(v)) \in E_2$. If G_1 is isomorphic to itself under function f, G_1 is an automorphic graph, and f is called an automorphic function of G_1.*

Definition 14.5. *K-automorphic graph.* *Graph G is a K-automorphic graph if 1) there exist $K - 1$ non-trivial automorphic functions of G, $f_1, \ldots, f_{K-1}$; and 2) for any node u, $f_i(u) \neq f_j(u)$ $(i \neq j)$.*

If the released graph $\widetilde{G}$ is a K-automorphic graph, when the adversary tries to re-identify node u through a subgraph, he will always get at least K different subgraphs in $\widetilde{G}$ that match his subgraph query. With the second condition in Definition 14.5, it is guaranteed that the probability of a successful re-identification is no more than $\frac{1}{K}$. The second condition in Definition 14.5 is necessary to guarantee the privacy safety. If it is violated, the worst case is that for a certain node u and any $i = 1, 2, \ldots, K - 1$, $f_i(u) \equiv u$, and the adversary can then successfully re-identify node u in $\widetilde{G}$. For example, consider a l-asteroid graph in which a central node is connected by l satellite nodes and the l satellite nodes are not connected to each other. This l-asteroid graph has at least l automorphic functions. However the central node is always mapped to itself by any automorphic function. Condition 2 prevents such cases from happening in the released graph $\widetilde{G}$. The authors then considered the following problem:

Problem 3. *Given the original graph G, construct graph $\widetilde{G}$ such that $E \subseteq \widetilde{E}$ and $\widetilde{G}$ is a K-automorphic graph.*

The following steps briefly show the framework of their algorithm:

1. Partition graph G into several groups of subgraphs $\{U_i\}$, and each group U_i contains $K_i \geq K$ subgraphs $\{P_{i1}, P_{i2}, \ldots, P_{iK_i}\}$ where any two subgraphs do not share a node or edge.

2. For each U_i, make $P_{ij} \in U_i$ isomorphic to each other by adding edges. Then, there exists function $f_{s,t}^{(i)}(\cdot)$ under which P_{is} is isomorphic to P_{it}.

3. For each edge (u, v) across two subgraphs, i.e. $u \in P_{ij}$ and $v \in P_{st}$ $(P_{ij} \neq P_{st})$, add edge $\left(f_{j,\pi_j(r)}^{(i)}(u), f_{t,\pi_t(r)}^{(s)}(v) \right)$, where $\pi_j(r) = (j + r)$ mod K, $r = 1, 2, \ldots, K - 1$.

After the modification, for any node u, suppose $u \in P_{ij}$, define $f_r(\cdot)$ as $f_r(u) = f_{j,\pi_j(r)}^{(i)}(u)$, $r = 1, \ldots, K - 1$. Then, $f_r(u)$, $r = 1, \ldots, K - 1$, are $K - 1$ non-trivial automorphic functions of $\widetilde{G}$, and for any $s \neq t$, $f_s(u) \neq f_t(u)$, which guarantees the K-automorphism.

To better preserve the utility, the authors expected that the above algorithm introduces the minimal number of fake edges, which implies that subgraphs within one group U_i should be very similar to each other (so that Step 2 only introduces a small number of edges), and there are few edges across different subgraphs (so that Step 3 will not add many edges). This depends on how the graph is partitioned. If G is partitioned into fewer subgraphs, there are fewer crossing edges to be added. However, fewer subgraphs imply that the size of each subgraph is large, and more edges within each subgraph need to be added in Step 2. The authors proved that to find the optimal solution is *NP*-complete, and they proposed a greedy algorithm to achieve the goal.

In addition to proposing the K-automorphism idea to protect the graph under any structural attack, the authors also studied an interesting problem with respect to privacy protection over dynamic releases of graphs. Specially, the requirements of social network analysis and mining demand releasing the network data from time to time in order to capture the evolution trends of these data. The existing privacy-preserving methods only consider the privacy protection in "one-time" release. The adversary can easily collect the multiple releases and identify the target through comparing the difference among these releases. Zou et al. [52] extended the solution of K-automorphism by publishing the vertex ID set instead of single vertex ID for the high risk nodes.

4.   Privacy Preservation via Randomization

Besides K-anonymity approaches, randomization is another widely adopted strategy for privacy-preserving data analysis. Additive noise based randomization approaches have been well investigated in privacy-preserving data mining for numerical data (e.g., [3, 2]). For social networks, two edge-based randomization strategies have been commonly adopted.

- *Rand Add/Del*: randomly add k false edges followed by deleting k true edges. This strategy preserves the total number of edges in the original graph.

- *Rand Switch*: randomly switch a pair of existing edges (t, w) and (u, v) (satisfying edge (t, v) and edge (u, w) do not exist in G) to (t, v) and (u, w), and repeat this process for k times. This strategy preserves the degree of each vertex.

The process of randomization and the randomization parameter k are assumed to be published along with the released graph. By using adjacency matrix, the edge randomization process can be expressed in the matrix form $\widetilde{A} = A + E$, where E is the perturbation matrix: $E(i, j) = E(j, i) = 1$ if edge (i, j) is added, $E(i, j) = E(j, i) = -1$ if edge (i, j) is deleted, and 0 otherwise. Naturally, edge randomization can also be considered as an additive-noise perturbation. After the randomization, the randomized graph is expected to be different from the original one. As a result, the node identities as well as the true sensitive or confidential relationship between two nodes are protected.

In this section, we first discuss why randomized graphs are resilient to structural attacks and how well randomization approaches can protect node identity in Section 4.1. Notice that the randomization approaches protect against re-identification in a probabilistic manner, and hence they cannot guarantee that the randomized graphs satisfy K-anonymity strictly.

There exist some scenarios that node identities (and even entity attributes) are not confidential but sensitive links between target individuals are confidential and should be protected. For example, in a transaction network, an edge denoting a financial transaction between two individuals is considered confidential while nodes corresponding to individual accounts is non-confidential. In these cases, data owners can release the edge randomized graph without removing node annotations. We study how well the randomization approaches protect sensitive links in Section 4.2.

An advantage of randomization is that many features could be accurately reconstructed from the released randomized graph. However, distribution reconstruction methods (e.g., [3, 2]) designed for numerical data could not be applied on network data directly since the randomization mechanism in social networks (based on the positions of randomly chosen edges) is much different from the additive noise randomization (based on random values for all entries). We give an overview of low rank approximation based reconstruction methods in Section 4.3.

Edge randomization may significantly affect the utility of the released randomized graph. We survey some randomization strategies that can preserve structural properties in Section 4.4.

4.1   Resilience to Structural Attacks

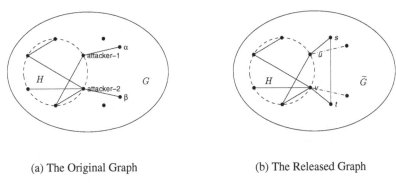

(a) The Original Graph   (b) The Released Graph

Figure 14.1. Resilient to subgraph attacks

Recall that in both *active attacks* and *passive attacks* [4], the adversary needs to construct a highly distinguishable subgraph H with edges to a set of target nodes, and then to re-identify the subgraph and consequently the targets in the released anonymized network. As shown in Figure 14.1(a), attackers form an subgraph H in the original graph G, and attacker 1 and 2 send links to the target individuals α and β. After randomization using either *Rand Add/Del* or *Rand Switch*, the structure of subgraph H as well G is changed. The re-identifiability of the subgraph H from the randomized released graph $\widetilde{G}$ may significantly decrease when the magnitude of perturbation is medium or large. Even if the subgraph H can still be distinguished, as shown in Figure 14.1(b), link (u, s) and (v, t) in $\widetilde{G}$ can be false links. Hence node s and t do not correspond to target individuals α and β. Furthermore, even individuals α and β have been identified, the observed link between α and β can still be a false link. Hence, the link privacy can still be protected. In summary, it is more difficult for the adversary to breach the identity privacy and link privacy.

Similarly for structural queries [20], because of randomization, the adversary cannot simply exclude from those nodes that do not match the structural properties of the target. Instead, the adversary needs to consider the set of all possible graphs implied by $\widetilde{G}$ and k. Informally, this set contains any graph G_p that could result in $\widetilde{G}$ under k perturbations from G_p, and the size of the set is

$\binom{m}{k}\binom{\binom{n}{2}-m}{k}$. The candidate set of a target node includes every node y if it is a candidate in some possible graph. The probability associated with a candidate y is the probability of choosing a possible graph in which y is a candidate. The computation is equivalent to compute a query answer over a probabilistic database and is likely to be intractable.

We would emphasize that it is very challenging to formally quantify identity disclosure in the presence of complex background knowledge of adversaries (such as embedded subgraphs or graph metrics). Ying et al. [44] quantified the risk of identity disclosure (and link disclosure) when adversaries adopt one specific type of background knowledge (i.e., knowing the degree of target individuals). The node identification problem is that given the true degree d_α of a target individual α, the adversary aims to discover which node in the randomized graph $\tilde{G}$ corresponds to individual α. However, it is unclear whether the quantification of disclosure risk can be derived for complex background knowledge based attacks.

4.2 Link Disclosure Analysis

Note that link disclosure can occur even if each vertex is K-anonymous. For example, in a K-degree anonymous graph, nodes with the same degree can form an equivalent class (EC). For two target individuals α and β, if every node in the EC of individual α has an edge with every node in the EC of β, the adversary can infer with probability 100% that an edge exists between the two target individuals, even if the adversary may not be able to identify the two individuals within their respective ECs. In [48], L. Zhang and W. Zhang described an attacking method in which the adversary estimates the probability of existing link (i, j) through the link density between the two equivalence classes. The authors then proposed a greedy algorithm aiming to reduce the probabilities of link disclosure to a tolerance threshold τ via a minimum series of edge deletions or switches.

In [45–47], the authors investigated link disclosure of edge-randomized graphs. They focused on networks where node identities (and even entity attributes) are not confidential but sensitive links between target individuals are confidential. The problem can be regarded as, compared to not releasing the graph, to what extent releasing a randomized graph $\tilde{G}$ jeopardizes the link privacy. They assumed that adversaries are capable of calculating posterior probabilities.

In [45], Ying and Wu investigated the link privacy under randomization strategies (*Rand Add/Del* and *Rand Switch*). The adversary's prior belief about the existence of edge (i, j) (without exploiting the released graph) can be calculated as $P(a_{ij} = 1) = \frac{2m}{n(n-1)}$, where n is the number of nodes and m is the number of edges. For *Rand Add/Del*, with the released graph and

perturbation parameter k, the posterior belief when observing $\tilde{a}_{ij} = 1$ is
$P(a_{ij} = 1|\tilde{a}_{ij} = 1) = \frac{m-k}{m}$.

An attacking model, which exploits the relationship between the probability of existence of a link and the similarity measure values of node pairs in the released randomized graph, was presented in [47]. Proximity measures have been shown to be effective in the classic link prediction problem [28] (i.e., predicting the future existence of links among nodes given a snapshot of a current graph). The authors investigated four proximity measures (common neighbors, Katz measure, Adamic/Adar measure, and commute time) and quantified how much the posterior belief on the existence of a link can be enhanced by exploiting those similarity values derived from the released graph which is randomized by the *Rand Add/Del* strategy. The enhanced posterior belief is given by

$$P(a_{ij} = 1|\tilde{a}_{ij} = 1, \tilde{m}_{ij} = x) = \frac{(1 - p_1)\rho_x}{(1 - p_1)\rho_x + p_2(1 - \rho_x)}$$

where $p_1 = \frac{k}{m}$ denotes the probability of deleting a true edge, $p_2 = \frac{k}{\binom{n}{2}-m}$ denotes the probability of adding a false edge, $\tilde{m}_{ij}$ denotes the similarity measure between node i and j in $\widetilde{G}$, and $\rho_x = P(a_{ij} = 1|\tilde{m}_{ij} = x)$ denotes the proportion of true edges in the node pairs with $\tilde{m}_{ij} = x$. The maximum likelihood estimator (MLE) of ρ_x can be calculated from the randomized graph.

The authors further theoretically studied the relationship among the prior beliefs, posterior beliefs without exploiting similarity measures, and the enhanced posterior beliefs with exploiting similarity measures. One result is that, for those observed links with high similarity values, the enhanced posterior belief $P(a_{ij} = 1|\tilde{a}_{ij} = 1, \tilde{m}_{ij} = x)$ is significantly greater than $P(a_{ij} = 1|\tilde{a}_{ij} = 1)$ (the posterior belief without exploiting similarity measures). Another result is that the sum of the enhanced posterior belief (with exploiting similarity measures) approaches to m, i.e.,

$$\sum_{i<j} P(a_{ij} = 1|\tilde{a}_{ij}, \tilde{m}_{ij}) \rightarrow m \quad \text{as} \quad n \rightarrow \infty,$$

while the sum of the prior beliefs and the sum of posterior beliefs (without exploiting similarity measures) over all node pairs equal to m. Notice that it is more desirable to quantify the probability of existing true link (i, j) via comprehensive information of $\widetilde{G}$, i.e., $P(a_{ij} = 1|\widetilde{G})$. However, this is very challenging.

A different attacking model was presented in [46]. It is based on the distribution of the probability of existence of a link across all possible graphs in the graph space $\mathcal{G}$ implied by G and k. If many graphs in $\mathcal{G}$ have an edge (i, j), the original graph is also very likely to have the edge (i, j). Hence the proportion of graphs with edge (i, j) can be used to denote the posterior probability of existence of edge (i, j) in the original graph. More details will be provided in Section 4.4.0.

4.3   Reconstruction

Recall that the edge randomization process can be written in the matrix form $\widetilde{A} = A + E$, where A ($\widetilde{A}$) is the adjacency matrix of the original (randomized) graph and E is the perturbation matrix. In the setting of randomizing numerical data, a data set U with m records of n attributes is perturbed to $\widetilde{U}$ by an additive noise data set V with the same dimensions as U. In other words, $\widetilde{U} = U + V$. Distributions of U can be approximately reconstructed from the perturbed data $\widetilde{U}$ using distribution reconstruction approaches (e.g., [3, 2]) when some a-priori knowledge (e.g., distribution, statistics etc.) about the noise V is available. Specifically, Agrawal and Aggawal [2] provided an expectation-maximization (EM) algorithm for reconstructing the distribution of the original data from perturbed observations. However, it is unclear whether similar distribution reconstruction methods can be derived for network data. This is because 1) it is hard to define distribution for network data; and 2) the randomization mechanism for network data is based on the positions of randomly chosen edges rather than the independent random additive values for all entries for numerical data.

In [41], Wu et al. investigated the use of low rank approximation methods to reconstruct structural features from the graph randomized via *Rand Add/Del*. Let λ_i ($\widetilde{\lambda}_i$) be A's ($\widetilde{A}$'s) i-th largest eigenvalue in magnitude whose eigenvector is x_i ($\widetilde{x}_i$). Then, the rank l approximations of A and $\widetilde{A}$ are respectively given by:

$$A_l = \sum_{i=1}^{l} \lambda_i x_i x_i^T \quad \text{and} \quad \widetilde{A}_l = \sum_{i=1}^{l} \widetilde{\lambda}_i \widetilde{x}_i \widetilde{x}_i^T.$$

By choosing a proper l, Wu et al. [41] showed that $\widetilde{A}_l$ can preserve the major information of the original graph and filter out noises added in the rest dimensions. This is because real-world data is usually highly correlated in a low dimensional space while the randomly added noise is distributed (approximately) equally over all dimensions. In $\widetilde{A}_l$, those entries close to 1 are more likely to have true edges while those entries close to 0 are less likely to have edges. They simply derived the reconstructed graph $\hat{A}$ by setting the $2m$ largest off-diagonal entries in $\widetilde{A}_l$ as 1, and 0 otherwise. Empirical evaluations showed that more accurate features can be reconstructed via the low rank approximation even when the magnitude of additive noise k equals to $0.8m$.

Note that the low rank approximation has been well investigated as a pointwise reconstruction method in the numerical setting. A spectral filtering based reconstruction method was first proposed in [22] to reconstruct original data values from the perturbed data. Similar methods (e.g., PCA based reconstruction method [21], SVD based reconstruction method [17]) were also investigated. All methods exploited spectral properties of the correlated data to remove the noise from the perturbed one. Preliminary results [41] showed that the accuracy of the reconstructed individual data (i.e., edge entries of the adjacency matrix) using the low rank approximation is not as good as that of the reconstructed numerical data.

We would emphasize that reconstruction methods on purely randomized graphs need further investigations so that more accurate analysis can be conducted on reconstructed graphs while individual privacy can be preserved. It is our conjecture that it is very hard, if not impossible, to figure out reconstruction methods on the released data randomized using K-anonymity schemes. This is because in K-anonymity based modification schemes, modified edge entries are not randomly chosen. For example, the K-degree scheme examines the degree sequence of nodes and chooses a subset of nodes (that violates the K-degree anonymity property) for edge modification.

4.4   Feature Preserving Randomization

Edge randomization may significantly affect the utility of the released randomized graph. To preserve utility, certain aggregate characteristics (a.k.a., feature) of the original graph should remain basically unchanged or at least some properties can be reconstructed from the randomized graph. However, as shown in [45], many topological features are lost due to randomization. In this section, we summarize randomization strategies that can preserve structural properties. We would emphasize that it is very challenging to quantify disclosures since the process of feature preserving strategies or generalization strategies is more complicated than that of randomization strategies.

Instead of randomizing the original graph via *Add/Del* or *Switch*, researchers also considered the problem of directly generating synthetic graphs given a set of features. We refer interested readers to a recent survey [10] and the references wherein for more details.

Spectrum Preserving Randomization.   In [45], Ying and Wu presented a randomization strategy that can preserve the spectral properties of the graph. The spectra of graph matrices have close relations with many important topological properties such as diameter, presence of cohesive clusters, long paths and bottlenecks, and randomness of the graph [36]. The authors aimed to preserve the data utility by preserving two important eigenvalues during the randomization: the largest eigenvalue of the adjacency matrix and the second smallest eigenvalue of the Laplacian matrix.

The authors showed that pure randomization tends to move the eigenvalues toward one direction, and the randomized eigenvalues can be significantly different from the original values. The two proposed algorithms, *Spctr Add/Del* and *Spctr Switch*, selectively pick up those edges that can increase (or decrease) the target eigenvalue by examining the eigenvector values of the nodes involved in the randomization, and apply the randomizing operation, which guarantees the randomized eigenvalues do not move far from the original value. Their empirical evaluations showed that the proposed algorithms can keep the spectral features as well as many topological features close to the original ones even when the magnitude of randomization is large.

Although they empirically showed that the spectrum preserving approach can achieve similar privacy protection as the random perturbation approach, however, they did not derive the formula of the protection measure for either *Spctr Add/Del* or *Spctr Switch* since the number of false edges in the randomization cannot be explicitly expressed.

Markov Chain based Feature Preserving Randomization.   The degree sequence and topological features are of great importance to the graph structure. One natural idea is that it can better preserve the data utility if the released graph $\widetilde{G}$ preserves the original degree sequence and a certain topological feature, such as transitivity or average shortest distance. In [46, 18], the authors investigated switch based randomization algorithms that can preserve various properties of a real social network in addition to a given degree sequence.

To preserve data utility, data owners may want to preserve some particular feature S within a precise range in the released graph. All the graphs that satisfy the degree sequence d and the feature constraint S form a graph space $\mathcal{G}_{d,S}$ (or $\mathcal{G}_d$ if no feature constraint). Starting with the original graph, series of switches form a Markov chain that can explore the graph space $\mathcal{G}_{d,S}$. Ying and Wu [46] proposed an algorithm that can generate any graph in $\mathcal{G}_{d,S}$ with equal probability, and Hanhijarvi et al. [18] proposed an algorithm that generates a graph whose feature is close to the original value with high probability.

One concern on the privacy is that the feature constraint may reduce the graph space and increase the risk of privacy disclosure. In [46], Ying and Wu also studied how adversaries exploit the released graph as well as feature constraints to breach link privacy. The adversary can calculate the posterior probability of existence of a certain link by exploiting the graph space $\mathcal{G}_{d,S}$. If many graphs in the graph space have link (i, j), the original graph is also very likely to have link (i, j), and hence the adversary's posterior belief about link (i, j) is given by

$$P[G(i,j) = 1|\mathcal{G}_{d,S}] = \frac{1}{|\mathcal{G}_{d,S}|} \sum_{G_t \in \mathcal{G}_{d,S}} G_t(i,j).$$

The attacking model works as follows: knowing the degree sequence d and the feature constraint S, the adversary generates N samples $G_t \in \mathcal{G}_{d,S}$

$(t = 1, 2, \ldots, N)$ via the Markov chain that starts with the released graph $\widetilde{G}$ and converges to the uniform stationary distribution over the graph space. Then, $P[G(i, j) = 1 | \mathcal{G}_{d,S}]$ can be simply estimated by $\frac{1}{N} \sum_{t=1}^{N} G_t(i, j)$. The adversary can take the node pairs with highest posterior beliefs as candidate links. This attacking model works because the convergence of the Markov chain does not depend on the initial point. Their evaluations showed that some feature constraints can significantly enhance the adversary's attacking accuracy and the extent to which a feature constraint jeopardizes link privacy varies for different graphs.

5.   Privacy Preservation via Generalization

To preserve privacy, both K-anonymity and randomization approaches modify the graph structure by adding/deleting edges and then release the detailed graph. Different from the above two approaches, generalization approaches can be essentially regarded as grouping nodes and edges into partitions called *super-nodes* and *super-edges*. The idea of generalization has been well adopted in anonymizing tabular data. For social network data, the generalized graph, which contains the link structures among partitions as well as the aggregate description of each partition, can still be used to study macro-properties of the original graph.

In [19], Hay et al. applied structural generalization approaches that groups nodes into clusters, by which privacy details about individuals can be hidden properly. To ensure node anonymity, they proposed to use the size of a partition as a basic guarantee against re-identification attacks. Their method obtains a vertex K-anonymous super-graph by aggregating nodes into super-nodes and edges into super-edges, such that, each super-node represents at least K nodes and each super-edge represents all the edges between nodes in two super-nodes. Because only the edge density is published for each partition, it is impossible for the adversary to distinguish between individuals in partition. Note that more than one partition may be consistent with a knowledge query about target individual x. Hence, the size of a partition is used to provide a conservative guarantee against re-identification and there exists an improved bound on the size of candidate sets.

To retain utility, the partitions should fit the original network as closely as possible given the anonymity condition. The proposed method estimates fitness via a maximum likelihood approach. The likelihood is defined as one over the size of possible worlds implied by the partition. For any generalization $\mathcal{G}$, the number of edges in the super-node X is denoted as $c(X, X)$, the number of edges between X and Y is denoted as $c(X, Y)$, the set of possible worlds that are consistent with $\mathcal{G}$ is denoted by $\mathcal{W}(\mathcal{G})$ whose size is given by:

$$|\mathcal{W}(\mathcal{G})| = \prod_{X \in \mathcal{V}} \left(\frac{\frac{1}{2}|X|(|X|-1)}{c(X,X)} \right) \prod_{X,Y \in \mathcal{V}} \left(\frac{|X||Y|}{c(X,Y)} \right)$$

The likelihood for a graph $g \in \mathcal{W}(\mathcal{G})$ is then $1/|\mathcal{W}(\mathcal{G})|$. The partitioning of nodes is chosen so that the generalized graph satisfies privacy constraints and maximizes the utility $(1/|\mathcal{W}(\mathcal{G})|)$.

Their algorithm searches the approximate optimal partitioning, using simulated annealing [35]. Starting with a single partition containing all nodes, the algorithm proposes a change of state by splitting a partition, merging two partitions, or moving a node to a different partition. The movement from one partition to next valid partition is always accepted if it increases the likelihood and accepted with some probability if it decreases the likelihood. Search terminates when it reaches a local maximum.

The authors evaluated the effectiveness of structural queries on real networks from various domains and random graphs. Their results showed that networks are diverse in their resistance to attacks: social and communication networks tend to be more resistant than some random graph models (Erdos-Renyi and power-law graphs) would suggest, and hubs cannot be used to re-identify many of their neighbors.

One problem of this generalization approach is that since the released network only contains a summary of structural information about the original network (e.g., degree distribution, path lengths, and transitivity), users have to generate some random sample instances of the released network. As a result, uncertainty may arise in the later analysis since the samples come from a large number of possible worlds.

6.   Anonymizing Rich Graphs

Real social network sources usually contain much richer information in addition to the simple graph structure. For example, in an online social network, the main entities in the data are individuals whose profiles can list lots of demographic information, such as age, gender and location, as well as other sensitive personal data, such as political and religious preferences, relationship status, etc. Between users, there are many different kinds of interactions such as friendship and email communication. Interactions can also involve more than two participants, e.g., many users can play a game together. Bhagat et al. [8] referred to the connections formed in the social networks as *rich interaction graphs*. Various queries on the network data are not simply about properties of the entities in the data, or simply about the pattern of the link structure in the graph, but rather on their combination. Thus it is important for the anonymization to mask the associations between entities and their interactions.

Notice that for rich social networks, a K-anonymous social network may still leak privacy. For example, if all nodes in a K-anonymous group are associated with some sensitive information, the adversary can derive that sensitive attribute of target individuals. Mechanism analogous to l-diversity [33] can be applied here. Several rich graph data models, which may contain labeled vertices/edges in addition to the structural information associated with the network, have been investigated in the privacy-preserving network analysis.

6.1   Link Protection in Rich Graphs

In [49], Zheleva et al. considered a graph model, in which there are multiple types of edges but only one type of nodes. Edges are classified as either sensitive or non-sensitive. The problem of link re-identification is defined as inferring sensitive relationships from non-sensitive ones. The goal is to attain privacy preservation of the sensitive relationships, while still producing useful anonymized graph data. They proposed to use the number of removed non-sensitive edges to measure the utility loss. Several graph anonymization strategies were proposed, including the removal of all sensitive edges and/or some non-sensitive edges, and the cluster-edge anonymization. In the cluster-edge anonymization approach, all the anonymized nodes in an equivalence class are collapsed into a single super-node and a decision is made on which edges to be included the collapsed graph. One feasible way is to separately publish the number of edges of each type between two equivalence classes.

The difference between the cluster-edge anonymization approach and the generalization approach in [19] is that the former aggregates edges by type to protect link privacy while the latter clusters vertices to protect node identities.

In [9], Campan and Truta considered an undirected graph model, in which edges are not labeled but vertices are associated with some attributes including identifier, quasi-identifier, and sensitive attributes. Those identifier attributes such as name and SSN are removed while the quasi-identifier and the sensitive attributes as well as the graph structure are released. To protect privacy in network data, they adopted the K-anonymity model for both the quasi-identifier attributes and the quasi-identifier relationship homogeneity. The goal is that any two nodes from any cluster are indistinguishable based on either their relationships or their attributes.

For structural anonymization, they proposed an edge generalization based method that does not insert or remove edges from the network data. They perform social network data clustering followed by anonymization through cluster collapsing. Specifically, the method first partitions vertices into clusters and attaches the structural description (i.e., the number of nodes and the number of edges) to each cluster. From the privacy standpoint, an original node within such a cluster is indistinguishable from other nodes. Then all vertices in the same cluster are made uniform with respect to the quasi-identifier attributes and the quasi-identifier relationship. This homogenization is achieved by using generalization, for both the quasi-identifier attributes and the quasi-identifier relationship. All vertices in the same cluster are collapsed into one single vertex (labeled by the number of vertices and edges in the cluster) and edges between two clusters are collapsed into a single edge (labeled with the number of edges between them). The method takes into account the information loss due to both the attribute generalization and the changes of structural properties. Users can tune the process to balance the tradeoff between preserving more structural information and preserving more vertex attribute information.

6.2   Anonymizing Bipartite Graphs

Cormode et al. [11] studied a particular type of network data that can be modeled as bipartite graphs – there are two types of entities, and an association only exists between two entities of different types. One example is the pharmacy (customers buy products). The association between two nodes (e.g., who bought what products) is considered to be private and needs to be protected while properties of some entities (e.g., product information or customer information) are public.

Their anonymization method can preserve the graph structure exactly by masking the mapping from entities to nodes rather than masking or altering the graph structure. As a result, analysis principally based on the graph structure is correct. Privacy is ensured in this approach because given a group of nodes, there is a secret mapping from these nodes to the corresponding group of entities. There is no information published that would allow an adversary to learn, within a group, which node corresponds to which entity.

They evaluated the utility using three types of aggregate queries with increasing complexity for the bipartite graphs:

- Type 0 - Graph structure only: compute an aggregate over all neighbors of nodes in V that satisfy some P_n (i.e., predicates over solely graph properties of nodes), such as the average number of products bought by each customer.

- Type 1 - Attribute predicate on one side only: compute an aggregate for nodes in V satisfying P_a (i.e., predicates over attributes of the entities), such as the average number of products for NJ customers.

- Type 2 - Attribute predicate on both sides: compute an aggregate for nodes in V satisfying P_a and nodes in W satisfying P'_a, such as the total number of OTC products bought by NJ customers.

6.3   Anonymizing Rich Interaction Graphs

In [8], Bhagat et al. adopted a flexible representation of rich interaction graphs which is capable of encoding multiple types of interactions between entities. Interactions involving large number of participants are represented by a hypergraph, denoted by $G(V, I, E)$. V is the node set. Each entity $v \in V$ has a hidden identifier u and a set of properties. Each entity in I is an interaction between/among a subset of entities in V. E is the set of hyperedges: for $v \in V$ and $i \in I$, an edge $(v, i) \in E$ represents node v participates in interaction i. One simple example of a hypergraph is shown in Figure 14.2(a).

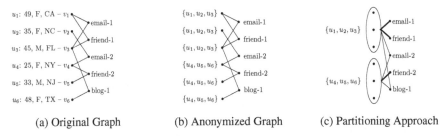

(a) Original Graph   (b) Anonymized Graph   (c) Partitioning Approach

Figure 14.2. The interaction graph example and its generalization results

The authors assumed that adversaries know part of the links and nodes in the graph. They presented two types of anonymization techniques based on the idea of grouping nodes in V into several classes. The authors pointed out that merely grouping nodes into several classes can not guarantee the privacy. For example, consider the case where the nodes within one class form a complete graph via a certain interaction. Then, once the adversary knows the target is in the class, he can be sure that the target must participate in the interaction. The authors provided a safety condition, called *class safety* to ensure that the pattern of links between classes does not leak information: each node cannot have interactions with two (or more) nodes from the same group.

Their algorithm is briefly summarized as follows:

1 Sort the nodes according to attribute values.

2 Group the nodes in V into groups $\{C_i\}$ that satisfy the *class safety* property and $|C_i| \geq s$.

3 For node $v \in C_j$, replace the true identifier of v by a *label list* $l(v)$ containing $t \leq s$ identifiers, $l(v) = \{u_1, u_2, \ldots, u_t\}$. $l(v)$ contains the true identifier of v, and $\forall u_i \in l(v) \Rightarrow u_i \in C_j$.

After modification, graph G and the *label lists* are released. Figure 14.2(b) shows a special case where $s = t$ for the *label list*. In Figure 14.2(b), node v_1 has interactions with v_3 through an email and the friendship. This is allowed in the *class safety* property, as it allows two nodes to share multiple interactions, but prohibits a node having multiple friends in the same class. The authors also showed that the *label lists* are structured to ensure that the true identity cannot be inferred. Hence, the above procedures can greatly reduce the probability that an adversary can learn about other nodes and interactions through known nodes and interactions.

Note that the released graph contains the full topological structure of the original graph, some structural attacks such as the *active attack* and *passive attack* [4] can be applied here to de-anonymize the nodes in V. However, the adversary cannot further obtain the attributes of the target, for the attributes of those nodes within the same class are mixed together, which is similar to the anatomy approach [42] for the tabular database.

To prevent identity disclosure, the authors further proposed a solution, called *partitioning approach*, which groups edges in the anonymized graph and only releases the number of interactions between two groups, as illustrated in Figure 14.2(c). This method describes the number of interactions at the level of classes rather than nodes. The authors proved that this procedure guarantees that the adversary can correctly guess which nodes participate in the unknown links with probability at most $\frac{1}{s}$.

In term of the utility, the authors focused on the accuracy of aggregate queries on the graph data. They observed that if the nodes within one class have the same attribute values, the results of some queries can still be accurate, for the nodes of the class are either all included or all excluded in the result. Based on this idea, the proposed algorithms first sort all the nodes according to their attribute values, and then partition the nodes into classes that satisfy the *class safety* property. After partition, nodes within one class may not have exactly the same attribute values due to the *class safety* restriction, but they still have similar attribute values. The authors empirically showed that when the sorting order is appropriate, the query results based on the modified graph are not much different from the results based on the original graph.

6.4   Anonymizing Edge-Weighted Graphs

Beyond the ongoing privacy-preserving social network analysis which mainly focuses on un-weighted social networks, in [32, 13], the authors studied the situations in which the network edges as well as the corresponding weights are considered to be private.

In [13], Das et al. considered the problem of anonymizing the weights of edges in the social network. The authors proposed a framework to re-assign weights to edges so that a certain *linear property* of the original graph can be preserved in the anonymized graph. A *linear property* is the property that can be expressed by a specific set of linear inequalities of edge weights. If the newly assigned edge weights also satisfy the set of linear inequalities, the corresponding *linear property* is also preserved. Then, finding new weight for each edge is a linear programming problem. The authors discussed two linear properties in details, single source shortest paths and all pairs shortest paths, and proposed the algorithms that can efficiently construct the corresponding linear inequality sets. Their empirical evaluations showed that the proposed algorithms can considerably improve the edge k-anonymity of the modified graph, which prevents the adversary to identify an edge by its weight.

In [32], Liu et al. also proposed two randomization strategies aiming to preserve the shortest paths in the weighted social network. The first one, which is easier to implement, is the Gaussian randomization multiplication strategy. The algorithm multiplies the original weight of each edge by an i.i.d. Gaussian random variable with mean 1 and variance σ^2. In the original graph, if the total weight of the shortest path between two nodes is much smaller than that of the second shortest path, the strategy can preserve the original shortest path with high probability. The authors further proposed the second strategy which can preserve a set of the target shortest paths or even all the shortest paths in the graph. The authors pointed out that all edges can be divided into three categories: the *all-visited edge* which belongs to all shortest paths, the *non-visited edge* which belongs to no shortest path, and the *partially-visited edge* which belongs to some but not all shortest paths. In order to preserve the target shortest paths, one can then reduce the weight of *all-visited edges*, increase the weight of *non-visited edges*, and perturb the weight of *partially-visited edges* within a certain range. The weight sum of a target shortest path is changed and is probably not the same as the original one, but the difference is minimized by the proposed greedy perturbation algorithm.

In both works of [13] and [32], the authors did not apply addition, deletion or generalization process to links or nodes. They only adjusted the weights of each links. However, their algorithms can be incorporated with some other graph modification algorithms.

7.   Other Privacy Issues in Online Social Networks

We have restricted our discussion to the problem of privacy-preserving social network publishing so far. In this section, we give an overview about recent studies on other privacy issues in the real online social networks such as Facebook and MySpace.

7.1   Deriving Link Structure of the Entire Network

In [26], Korolova et al. considered a particular threat in which an adversary subverts user accounts to gain information about local neighborhoods in the network and pieces them together to build a global information about the social graph. It considered the case where no underlying graph is released, and, in fact, the owner of the network would like to keep the entire structure of the graph hidden from any one. The goal of the adversary is, rather than to de-anonymize particular individuals from that graph, to compromise the link privacy of as many individuals as possible. Specifically, the adversary determines the link structure of the graph based on the local neighborhood views of the graph from the perspective of several non-anonymous users.

Analysis showed that the number of users that need to be compromised in order to cover a constant fraction of the entire network drops exponentially with increase in the lookahead parameter l provided by the network data owner. Here a network has a lookahead l if a registered user can see all the links and nodes incident to him within distance l from him. For example, $l = 0$ if a user can see exactly who he links to; $l = 1$ if a user can see exactly the friends that he links to as well as the friends that his friends link to.

Each time the adversary gains access to a user account, he immediately covers all nodes that are at distance no more than the lookahead distance l enabled by the social network. In other words, he learns about all the edges incident to these nodes. Thus by gaining access to the account of user u, an adversary immediately covers all nodes that are within distance l of u. Additionally, he learns about the existence of all nodes within distance $l+1$ from u. The authors studied several attacking strategies shown as below.

- Benchmark-Greedy: Among all users in the social network, pick the next user to bribe as the one whose perspective on the network gives the largest possible amount of new information. Formally, at each step the adversary picks the node covering the maximum number of nodes not yet covered.

- Heuristically Greedy: Pick the next user to bribe as the one who can offer the largest possible amount of new information, according to some heuristic measure. For example, Degree-Greedy picks the next user to bribe as the one with the maximum unseen degree, i.e., its degree minus the number of edges incident to it already known by the adversary.

- Highest-Degree: Bribe users in the descending order of their degrees.

- Random: Pick the users to bribe at random.

- Crawler: Similar to the Heuristically Greedy strategy, but choose the next node to bribe only from the nodes already seen (within distance $l + 1$ of some bribed node). One example is Degree-Greedy-Crawler that picks, from all users already seen, the next user to bribe as the one with the maximum unseen degree.

Experiments on a $572,949$-node friendship graph extracted from Live-Joural.com indicated that 1) Highest-Degree yields the best performance while Random performs the worst; 2) in order to obtain 80% coverage of the graph using lookahead 2, Highest-Degree needs to bribe $6,308$ users while it only needs to bribe 36 users to obtain the same coverage using lookahead 3. The authors suggested that as a general rule, the social network owners should refrain from permitting a lookahead higher than 2. Data owner may also want to decrease their vulnerability of the social network by not showing the exact number of connections that each user has, or by varying the lookahead available to users based on their trustworthiness.

7.2   Deriving Personal Identifying Information from Social Networking Sites

Online network users often publish their profiles as well as their connections that contain vast amounts of personal and sometimes sensitive information (e.g., photo, birth date, phone number, current residence, various interests, and their friends). Acquisti and Gross in [16] studied the privacy risk associated with these networks. The user's profile information can be used to estimate a person's social security number and exposes his/her to identity theft. Their studies showed that only a small number of Facebook members change the default privacy preferences. As a result, users expose themselves to various physical and cyber risks, and make it extremely easy for third parties to create digital dossiers of their behavior. Their study quantified patterns of information revelation and inferred usage of privacy settings from actual field data.

8.   Conclusion and Future Work

We surveyed recent studies on anonymization techniques for privacy-preserving publishing of social network data. The research and development of privacy-preserving social network analysis is still in its early stage compared with much better studied privacy-preserving data analysis for tabular data. We revisited the naive anonymization approach and several structural attacks which can be exploited on the naive anonymized graphs. We categorized the state-of-the-art anonymization methods on simple graphs in three main categories: K-anonymity based privacy preservation via edge modification, probabilistic privacy preservation via edge randomization, and privacy preservation via generalization. We then review anonymization methods on rich graphs. Since social network data is more complicated than tabular data, privacy preservation in social networks is much more challenging than privacy preservation in tabular data. While ideas and methods can be borrowed from the well studied privacy preservation in tabular data, many serious efforts are greatly needed due to new challenges (see Section 1.2 and 1.3) associated with the network data. We present a set of recommendations for future research in this emerging area.

- Develop privacy models for graphs and networks. Investigate how well different strategies protect privacy (identity, link privacy, and attribute privacy) when adversaries exploit various complex background knowledge in their attacks. How to model various background knowledge and quantify disclosures when complex attacks are used needs to be investigated.

- Since how to preserve utility in the released graph is an important issue in privacy-preserving social network analysis, measures and methodologies need to be developed to quantify utility and information loss. It is important to develop workload-aware metrics that adequately quantify levels of information loss of graph data. Furthermore, various anonymization strategies need to be evaluated in terms of the tradeoff between privacy and utility.

- Existing studies except [52] do not consider dynamic releases. Many applications of evolutionary networks and dynamic social network analysis require publishing data periodically to support dynamic analysis. The "one-time" released network data from existing annonymization methods cannot guarantee privacy when adversaries collect historical information from multiple releases.

- Distributed privacy-preserving social network analysis based on secure multi-party computation [43]. Distributed privacy-preserving data analysis on tabular data has been well studied (e.g., [29]; refer to the book [1] for surveys). However, distributed privacy-preserving social network analysis has not been well reported in literature.

- Create a benchmark graph data repository. Researchers can compare and learn how different approaches work in terms of the privacy-utility tradeoff. The scalability issue needs to be studied and empirical evaluations need to be conducted on large social networks.

Acknowledgments

Authors Wu and Ying were supported in part by U.S. National Science Foundation IIS-0546027 and CNS-0831204.

References

[1] C. C. Aggarwal and P. S. Yu. *Privacy-Preserving Data Mining: Models and Algorithms.* Springer, 2008.

[2] D. Agrawal and C. Agrawal. On the design and quantification of privacy preserving data mining algorithms. In *Proceedings of the 20th Symposium on Principles of Database Systems*, 2001.

[3] R. Agrawal and R. Srikant. Privacy-preserving data mining. In *Proceedings of the ACM SIGMOD International Conference on Management of Data*, pages 439–450. Dallas, Texas, May 2000.

[4] L. Backstrom, C. Dwork, and J. Kleinberg. Wherefore art thou r3579x?: anonymized social networks, hidden patterns, and structural steganography. In *WWW '07: Proceedings of the 16th international conference on World Wide Web*, pages 181–190, New York, NY, USA, 2007. ACM Press.

[5] L. Backstrom, D. Huttenlocher, J. Kleinberg, and X. Lan. Group formation in large social networks: membership, growth, and evolution. In *KDD '06: Proceedings of the 12th ACM SIGKDD international conference on Knowledge discovery and data mining*, pages 44–54, New York, NY, USA, 2006. ACM.

[6] J. Baumes, M. K. Goldberg, M. Magdon-Ismail, and W. A. Wallace. Discovering hidden groups in communication networks. In *ISI*, pages 378–389, 2004.

[7] T. Y. Berger-Wolf and J. Saia. A framework for analysis of dynamic social networks. In *KDD*, pages 523–528, 2006.

[8] S. Bhagat, G. Cormode, B. Krishnamurthy, and D. Srivastava. Class-based graph anaonymization for social network data. In *Proc. of 35th International Conference on Very Large Data Base*, 2009.

[9] A. Campan and T. M. Truta. A clustering approach for data and structural anonymity in social networks. In *PinKDD*, 2008.

[10] D. Chakrabarti, C. Faloutsos, and M. McGlohon. *Graph Mining: Laws and Generators.* Springer, 2010.

[11] G. Cormode, D. Srivastava, T. Yu, and Q. Zhang. Anonymizing bipartite graph data using safe groupings. In *Proc. of VLDB08*, pages 833–844, 2008.

[12] L. da F. Costa, F. A. Rodrigues, G. Travieso, and P. R. V. Boas. Characterization of complex networks: A survey of measurements. *Advances In Physics*, 56:167, 2007.

[13] S. Das, Ömer Egecioglu, and A. E. Abbadi. Anonymizing edge-weighted social network graphs. Technical report, UCSB CS, March 2009.

[14] A. Fast, D. Jensen, and B. N. Levine. Creating social networks to improve peer-to-peer networking. In *KDD*, pages 568–573, 2005.

[15] M. Girvan and M. E. Newman. Community structure in social and biological networks. *Proc. Natl. Acad. Sci. USA*, 99(12):7821–7826, June 2002.

[16] R. Gross and A. Acquisti. Information revelation and privacy in online social networks (the Facebook case). *Proceedings of the Workshop on Privacy in the Electronic Society*, 2005.

[17] S. Guo, X. Wu, and Y. Li. Determining error bounds for spectral filtering based reconstruction methods in privacy preserving data mining. *Knowl. Inf. Syst.*, 17(2):217–240, 2008.

[18] S. Hanhijarvi, G. C. Garriga, and K. Puolamaki. Randomization techniques for graphs. In *Proc. of the 9th SIAM Conference on Data Mining*, 2009.

[19] M. Hay, G. Miklau, D. Jensen, D. Towsely, and P. Weis. Resisting structural re-identification in anonymized social networks. In *VLDB*, 2008.

[20] M. Hay, G. Miklau, D. Jensen, P. Weis, and S. Srivastava. Anonymizing social networks. *University of Massachusetts Technical Report*, 07-19, 2007.

[21] Z. Huang, W. Du, and B. Chen. Deriving private information from randomized data. In *Proceedings of the ACM SIGMOD Conference on Management of Data*. Baltimore, MA, 2005.

[22] H. Kargupta, S. Datta, Q. Wang, and K. Sivakumar. On the privacy preserving properties of random data perturbation techniques. In *Proc. of the 3rd Int'l Conf. on Data Mining*, pages 99–106, 2003.

[23] D. Kempe, J. M. Kleinberg, and E. Tardos. Maximizing the spread of influence through a social network. In *KDD*, pages 137–146, 2003.

[24] J. M. Kleinberg. Challenges in mining social network data: processes, privacy, and paradoxes. In *KDD*, pages 4–5, 2007.

[25] Y. Koren, S. C. North, and C. Volinsky. Measuring and extracting proximity in networks. In *KDD*, pages 245–255, 2006.

[26] A. Korolova, R. Motwani, S. Nabar, and Y. Xu. Link privacy in social networks. In *Proceedings of the 24th International Conference on Data Engineering*, Cancun, Mexico, 2008.

[27] R. Kumar, J. Novak, and A. Tomkins. Structure and evolution of online social networks. In *KDD*, pages 611–617, 2006.

[28] D. Liben-Nowell and J. Kleinberg. The link prediction problem for social networks. In *CIKM '03: Proceedings of the twelfth international conference on Information and knowledge management*, pages 556–559, New York, NY, USA, 2003. ACM.

[29] Y. Lindell and B. Pinkas. Privacy preserving data mining. In *Advances in Cryptology (CRYPTO'00)*, pages 36–53. Springer-Verlag, 2000.

[30] K. Liu, K. Das, T. Grandison, and H. Kargupta. Privacy-preserving data analysis on graphs and social networks, 2008.

[31] K. Liu and E. Terzi. Towards identity anonymization on graphs. In *Proceedings of the ACM SIGMOD Conference*, Vancouver, Canada, 2008. ACM Press.

[32] L. Liu, J. Wang, J. Liu, and J. Zhang. Privacy preservation in social networks with sensitive edge weights. In *SDM*, pages 954–965, 2009.

[33] A. Machanavajjhala, J. Gehrke, D. Kifer, and M. Venkitasubramaniam. l-diversity: privacy beyond k-anonymity. In *Proceedings of the IEEE ICDE Conference*, 2006.

[34] A. Narayanan and V. Shmatikov. De-anonymizing social networks. In *IEEE Security & Privacy '09*, 2009.

[35] S. Russell and P. Norvig. *Artifical Intelligence: A Modern Approach.* Pearson Education, 2003.

[36] A. Seary and W. Richards. Spectral methods for analyzing and visualizing networks: an introduction. *National Research Council, Dynamic Social Network Modelling and Analysis: Workshop Summary and Papers*, pages 209–228, 2003.

[37] M. Shiga, I. Takigawa, and H. Mamitsuka. A spectral clustering approach to optimally combining numericalvectors with a modular network. In *KDD*, pages 647–656, 2007.

[38] E. Spertus, M. Sahami, and O. Buyukkokten. Evaluating similarity measures: a large-scale study in the orkut social network. In *KDD*, pages 678–684, 2005.

[39] C. Tantipathananandh, T. Y. Berger-Wolf, and D. Kempe. A framework for community identification in dynamic social networks. In *KDD*, pages 717–726, 2007.

[40] S. White and P. Smyth. Algorithms for estimating relative importance in networks. In *KDD*, pages 266–275, 2003.

[41] L. Wu, X. Ying, and X. Wu. Reconstruction of randomized graph via low rank approximation. Technical report, UNC-Charlotte, SIS, 2009.

[42] X. Xiao and Y. Tao. Anatomy: Simple and effective privacy preservation. In *Proceedings of the 32nd International Conference on Very Large Data Bases*, pages 139–150, September 2006.

[43] A. C. Yao. How to generate and exchange secrets. In *SFCS '86: Proceedings of the 27th Annual Symposium on Foundations of Computer Science*, pages 162–167. IEEE Computer Society, 1986.

[44] X. Ying, K. Pan, X. Wu, and L. Guo. Comparisons of randomization and k-degree anonymization schemes for privacy preserving social network publishing. In *SNA-KDD '09: Proceedings of the 3rd SIGKDD Workshop on Social Network Mining and Analysis (SNA-KDD)*, 2009.

[45] X. Ying and X. Wu. Randomizing social networks: a spectrum preserving approach. In *Proc. of the 8th SIAM Conference on Data Mining*, April 2008.

[46] X. Ying and X. Wu. Graph generation with prescribed feature constraints. In *Proc. of the 9th SIAM Conference on Data Mining*, 2009.

[47] X. Ying and X. Wu. On link privacy in randomizing social networks. In *PAKDD*, 2009.

[48] L. Zhang and W. Zhang. Edge anonymity in social graphs. In *Proceedings of the 2009 International Conference on Social Computing*, 2009.

[49] E. Zheleva and L. Getoor. Preserving the privacy of sensitive relationships in graph data. In *PinKDD*, pages 153–171, 2007.

[50] B. Zhou and J. Pei. Preserving privacy in social networks against neighborhood attacks. *IEEE 24th International Conference on Data Engineering*, pages 506–515, 2008.

[51] B. Zhou, J. Pei, and W.-S. Luk. A brief survey on anonymization techniques for privacy preserving publishing of social network data. *SIGKDD Explorations*, 10(2), 2009.

[52] L. Zou, L. Chen, and M. T. Özsu. K-automorphism: A general framework for privacy preserving network publication. In *Proc. of 35th International Conference on Very Large Data Base*, 2009.

Chapter 15

A SURVEY OF GRAPH MINING FOR WEB APPLICATIONS

Debora Donato
Yahoo! Research
Avd Diagonal 177, Barcelona, Spain
debora@yahoo-inc.com

Aristides Gionis
Yahoo! Research
Avd Diagonal 177, Barcelona, Spain
gionis@yahoo-inc.com

Abstract   Graph structures provide a general framework for modeling entities and their relationships, and they are routinely used to describe a wide variety of data such as the Internet, the web, social networks, metabolic networks, protein-interaction networks, food webs, citation networks, and many more. In recent years, there has been an increasing amount of literature on studying properties, models, and algorithms for graph data. In this chapter we provide a brief overview of graph-mining algorithms for web and social-media applications. We review a wide range of algorithms, such as those for estimating reputation and popularity of items in a network, mining query logs and performing query recommendations. The main goal of the chapter is to provide the reader with an understanding of how graph structural mining algorithms can be exploited in the context of web applications. This highlights the challenges of, and provides an understanding of the power of graph mining in the context of web and social-media applications.

Keywords: Graph Mining, Link Mining, Web Mining, Social Network Analysis, World Wide Web, Query-Log Mining, Query Recommendation

C.C. Aggarwal and H. Wang (eds.), *Managing and Mining Graph Data*,
Advances in Database Systems 40, DOI 10.1007/978-1-4419-6045-0_15,
© Springer Science+Business Media, LLC 2010

1.   Introduction

Graph mining has been widely used to study relationships among various types of entities. Real-world graphs are also referred to as *networks*, and the interactions between the entities represented in the networks are modeled as *links*. The problems of studying the properties of real-world networks, designing algorithms for mining such networks, and developing applications on top of network data has been of increasing interest in the past few years. This has led to the birth of a very active area of scientific research, which is known as *analysis of complex networks* [7, 16, 55].

One of the most pervasive properties of real-world networks is the emergence of *power-law* distributions that tend to characterize many of networks statistical properties [6, 26]. Power laws have intrigued the interest of researchers, who have proposed various models that attempt to explain the presence of power-law distributions in real graphs. For examples of such models, see [6, 25, 40].

In this chapter, we deviate from the classical exposition of properties and generative models for complex networks, and we focus on graph-mining applications that appear in the context of the web and social-media. Such graphs include data that model the interaction of users in a social network. For example, this may correspond to comments of users in a blog, user activity in a question-answering portal, or query-log data that summarize the interaction of users with a search engine. Understanding the structure of such graphs, modeling the complex interactions between entities, and designing algorithms for leveraging the latent knowledge (also known as *the wisdom of the crowds*) in those graphs introduces new challenges in the field of graph mining. One important difference with networks that have been previously studied, is that in social-media and web-usage graphs the links represent many different types of interactions and activities among nodes. For instance in a question-answering portal, users ask questions, answer questions for other users, vote for favorite answers, interesting questions, assign answers to categories of a hierarchy, and much more. Hence graphs from such applications are characterized by having different types of nodes and high degree of heterogeneity in the types of interactions among nodes. Consequently, algorithms and methodologies widely applied in the web and other complex networks have to be adapted to this new multifaceted scenario, which allows for the different meanings that are implicitly or explicitly captured by each link.

This chapter is organized as follows. In Section 2 we briefly introduce measures and algorithms that have been extensively used as basic tools for graph mining. Then we focus on two different areas of graph mining in the context of social-media and web applications. In Section 3, we review techniques for identifying items of high quality in social-media networks. We discuss two concrete examples: (1) predicting the number of citations of authors in a bibliographic data set, and (2) finding high-quality items in a question answering system. In both cases, the examples rely on adapting link-mining algorithms for computing authoritativeness scores in linked environments. In Section 4 we discuss algorithms for mining graph structures that represented information collected in the query logs of search engines. We first discuss various graph representations of query logs, and then discuss how to use these representations in order to perform the task of query recommendation. The conclusions are presented in Section 5.

2.   Preliminaries

An undirected graph $\mathcal{G} = (V, E)$ consists of a set of nodes V, also called vertices, and a set E of pairs of distinct nodes, which are called edges or arcs. A directed graph, or digraph, is distinguished from the undirected version by the fact that its edges are *ordered pairs* of nodes. In an undirected graph, the *degree* of a node is the number of edges incident to it. For a directed graph, we define the in-degree and the out-degree of a node to be the number of incoming and out-going edges, respectively.

In an undirected graph $\mathcal{G}$, a set of nodes S forms a *connected component* (CC), if for every pair of nodes $u, v \in S$ there exists a path from u to v (which is also a path from v to u). In a directed graph $\mathcal{G}$, a set of nodes S forms a *strongly connected component* (SCC), if for every pair of nodes $u, v \in S$, there exists a (directed) path from u to v, and a path from v to u. A set of nodes S forms a *weakly connected component* (WCC), if and only if the set S is a connected component in the undirected graph $\mathcal{G}_u$ that is obtained by ignoring the directionality of the edges in $\mathcal{G}$.

Power laws and scale-free networks.   Power-law distributions ubiquitously characterize real-world networks. We say that a discrete random variable X follows a power-law distribution if the probability distribution is defined for each discrete value k as follows:

$$\Pr[X = k] \propto k^{-\gamma}$$

The value γ is called the exponent of the power-law. We assume that $\gamma \geq 0$. Detailed surveys on power laws may be found in [45] and [46].

If a random variable X follows a power-law distribution, then we know that the conditional probability $\Pr[X \geq k \mid X \geq m]$ is the same as $\Pr[X \geq k]$. In other words, conditioning on the size does not yield any additional information. For this reason, networks that have attributes that follow a power-law distribution are also called *scale-free* networks.

Degree and Assortativeness.   The degree of the nodes a graph can be of great interest in social-media applications. The out-degree of a node might provide an indication of its capacity to influence his neighbors. This property is called *expansiveness* [58]. On the other hand, the in-degree is the most straightforward measure for the *popularity* of each node in the network. Complex networks exhibit large variance in the values of their degrees: very few nodes have the capacity of attracting a large fraction of links while the largest majority of nodes are connected to the network by few in-coming and out-going links.

Significant insight on the nature of the graph can be obtained by measuring the correlation between the degrees of adjacent vertexes [47]. This is also referred to as *assortative mixing*. Complex networks can be divided into three types based on the value of their mixing coefficient r: (*i*) *disassortative* if $r < 0$; (*ii*) *neutral* if $r \approx 0$; and (*iii*) *assortative* if $r > 0$. An alternative way to identify assortative or disassortative network is by using the average degree $E[k_{nn}(k)]$ of a neighboring vertex of a vertex with degree k [47]. As k increases, the expectation $E[k_{nn}(k)]$ increases for an assortative network and decreases for a disassortative one. In particular, a power-law equation $E[k_{nn}(k)] \approx k^{-\gamma}$ is satisfied, where γ is negative for an assortative network and positive for a disassortative one [49]. Social networks such as friendship networks are mostly assortative mixed, but technological and biological networks tend to be disassortative [62]. "Assortative mating" is a well-known social phenomenon that captures the likelihood that marriage partners will share common background characteristics, whether it is income, education, or social status. In online activity networks such as question-answering portals and newsgroups, the degree correlation provides information about user tendency to provide help. Such kind of networks are neutral or slightly disassortative: active users are prone to contribute without considering the expertise or the involvements of the users searching for help [63, 20].

Centrality and prestige. A key issue in social network analysis is the identification of the most important or prominent nodes. The measure of *centrality* captures whether a node is involved in a high number of ties regardless the directionality of the edges. Various definitions of centrality have been suggested. For instance, the *closeness centrality* is just the degree of a node eventually normalized by the number of all nodes V in the network. Two alternative measures of centrality are the *distance centrality* and the *betweenness centrality*. The closeness centrality $\mathcal{D}_c$ of a node u is the average distance of u to the rest of the nodes in the graph:

$$\mathcal{D}_c(u) = \frac{1}{|V| - 1} \sum_{v \neq u} d(u, v),$$

where $d(u, v)$ is the shortest-path distance between u and v. Similarly, the betweenness centrality $\mathcal{B}_c$ of a node u is the average number of shortest paths that pass through u:

$$\mathcal{B}_c(u) = \sum_{s \neq u \neq t} \frac{\sigma_{st}(u)}{\sigma_{st}},$$

where $\sigma_{st}(u)$ is the number of shortest paths from the node s to the node t that pass through node u, and σ_{st} is the total number of shortest paths from s to t.

A different concept for identifying important nodes is the measure of *prestige*, which exclusively considers the capacity of the node to attract incoming links, and ignores the capacity of initiating any outgoing ties. The basic intuition behind the prestige definition is the idea that a link from node u to node v denotes endorsement. In its simplest form, the prestige of a node is defined to be its in-degree, but there are other alternative definitions of prestige [58]. This concept is also at the core of a number of link analysis algorithms, an issue which we will explore in the next section.

2.1   Link Analysis Ranking Algorithms

PageRank.   Although we can view the existence of a link between two pages as an endorsement of authority from the former to the latter, the in-degree measure is a rather superficial way to examine page authoritativeness. This is because such a measure can easily be manipulated by creating spam pages which point to a particular target page in order to improve its authority. A smarter method of assigning authority score to a node is by using the *PageRank* algorithm [48], which uses the authoritative information of both the source and target page in an iterative way in order to determine the rank. The PageRank algorithm models the behavior of a "random surfer" on the Web graph. The surfer essentially browses the documents by following hyperlinks randomly. More specifically, the surfer starts from some node arbitrarily. At each step the surfer proceeds as follows:

- With probability α an outgoing hyperlink is selected randomly from the current document, and the surfer moves to the document pointed by the hyperlink.

- With probability $1 - \alpha$ the surfer jumps to a random page chosen according to some distribution. This distribution is typically chosen to be the uniform distribution.

The value $\text{Rank}(i)$ of a node i (called the PageRank value of node i) is the fraction of time that the surfer spends at node i. Intuitively, $\text{Rank}(i)$ is a measure of the importance of node i.

PageRank is expressed in matrix notation as follows. Let N be the number of nodes of the graph and let $n(j)$ be the out-degree of node j. We define the square matrix M as one in which the entry $M_{ij} = \frac{1}{n(j)}$ if there is a link from node j to node i. We define the square matrix $\left[\frac{1}{N}\right]$ of size $N \times N$ that has all entries equal to $\frac{1}{N}$. This matrix models the uniform distribution of jumping to a random node in the graph. The vector Rank stores the PageRank values that are computed for each node in the graph. A matrix M' is then derived by adding transition edges of probability $\frac{1-\alpha}{N}$ between every pair of nodes to include the case of jumping to a random node of the graph.

$$M' = \alpha M + (1 - \alpha) \left[\frac{1}{N}\right]$$

Since the PageRank algorithm computes the stationary distribution of the random surfer, we have M'Rank = Rank. In other words, Rank is the principal eigenvector of the matrix M', and thus it can be computed by the power-iteration method [15].

The notion of PageRank has inspired a large body of research on designing improved algorithms for more efficient computation of PageRank [24, 54, 36, 42], and for providing alternative definitions that can be used to address specific issues in search, such as personalization [27], topic-specific search [12, 32], and spam detection [8, 31].

One disadvantage of the PageRank algorithm is that while it is superior to a simple indegree measure, it continues to be prone to adversarial manipulation. For instance, one of the methods that owners of spam pages use to boost the ranking of their pages is to create a large number of auxiliary pages and hyperlinks among them, called *link-farms*, which result in boosting the PageRank score of certain target spam pages [8].

HITS. The main intuition behind PageRank is that authoritative nodes are linked to by other authoritative nodes. The HITS algorithm, proposed by Jon Kleinberg [38], introduced a double-tier paradigm for measuring authority. In the HITS framework, every page can be thought of as having a hub and an authority identity. There is a mutually reinforcing relationship between the two: a good hub is a page that points to many good authorities, while a good authority is a page that is pointed to by many good hubs.

In order to quantify the quality of a page as a hub and as an authority, Kleinberg associated every page with a hub and an authority *score*, and he proposed the following iterative algorithm: Assuming n pages with hyperlinks among them, let h and a denote n-dimensional hub and authority score vectors. Let also W be an $n \times n$ matrix, whose (i, j)-th entry is 1 if page i points to page j and 0 otherwise. Initially, all scores are set to 1. The algorithm iteratively updates the hub and authority scores sequentially one after the other and vice-versa. For a node i, the authority score of node i is set to be the sum of the hub scores of the nodes that point to i, while the hub score of node i is the authority score of the nodes pointed by i. In matrix-vector terms this is equivalent to setting $h = Wa$ and $a = W^T h$. A normalization step is then applied, so that the vectors h and a become unit vectors. The vectors a and h converge to the principal eigenvectors of the matrices $W^T W$ and WW^T, respectively. The vectors a and h correspond to the right and left *singular vectors* of the matrix W.

Given a user query, the HITS algorithm determines a set of relevant pages for which it computes the hub and authorities scores. Kleinberg's approach obtains such an initial set of pages by submitting the query to a text-based search engine. The pages returned by the search engine are considered as a root set, which is consequently expanded by adding other pages that either point to a page in the root set or are pointed by a page in the root set.

Kleinberg showed that additional information can be obtained by using more eigenvectors, in addition to the principal ones. Those additional eigenvectors correspond to clusters or distinct topics associated with the user query. One important characteristic of the HITS algorithm is that it computes page scores that depend on the user query: one particular page might be highly authoritative with respect to one query, but not such an important source of information with respect to another query. On the other hand, it is computationally expensive to compute eigenvectors for each query. This makes the algorithm computationally demanding. In contrast, the authority scores computed by the PageRank algorithm are not query-sensitive, and thus, they can be computed in a preprocessing stage.

3. Mining High-Quality Items

Online expertise-sharing communities have recently become extremely popular. The online media that allow the spread of this enormous amount of knowledge can take many different forms: users are sharing their knowledge in blogs, newsgroups, newsletters, forums, wikis, and question/answering portals. Those social-media environments can be represented as graphs with nodes of different types and with various types of relations among nodes. In the rest of the section we describe particular characteristics of the graphs arising in social-media environments, and their importance in driving the graph-mining process.

There are two main factors that differentiate social media from the traditional Web: (i) content-quality variance and (ii) interaction multiplicity. Differently from the traditional Web, in which the content is mediated by professional publishers, in social-media environments the content is provided by users. The massive contribution of users to the system leads to a high variance in the distribution of the quality of available content. With everyone able to create content and share any single opinion and thought, Thus the problem of determining items of high quality in an environment of excessive content is

 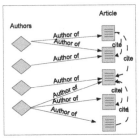 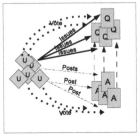

(a) Single Item:	(b) Double Item:	(c) Multiple Items:
Single Relation Model	Double Relation Model	Multiple Relation Model

Figure 15.1. Relation Models for Single Item, Double Item and Multiple Items

one of the most important issues to be solved. Furthermore, filtering out and ranking relevant items is more complex than in other domains.

The second aspect that must be considered is the wide variety of types of nodes, of relations among such nodes, and of interactions among users. For instance, the PageRank and HITS algorithms considers a simple graph model with one type of nodes (documents) and one type of edges (hyperlinks), see Figure 15.1(a).

On the other hand, social media are characterized by much more heterogeneous and rich structure, with a wide variety of user-to-document relation types and user-to-user interactions. In Figure 15.1(b) is shown the structure of a citation network as CiteSeer [21]. In this case, nodes can be of two types: `author` and `article`. Edges can also be of two types, `is-an-author-of` between a node of type `author` and a node of type `article`, and `cites` between two nodes of type `article`.

A more complex structure can be found in a question-answering portal, such as Yahoo! Answers [61], a graphical representation of which is shown in Figure 15.1(c). The main types of nodes are the following:

- `user`, representing the users registered with the system; they can act as askers or answerers, and can vote or comment questions and answers provided by other users,
- `question`, representing the questions asked by the users,
- `answer`, prepresenting the answers provided by the users.

Potential interesting research questions to ask for this type of application are the following: (i) find items of high-quality, (ii) predict which items will become successful in the future (assuming a dynamic environment), (iii) identify experts on a particular topic.

As in the case of other social-media applications, the variance of content quality in Yahoo! Answers is very high. According to Su et al. [56], the number of correct answers to specific questions varies from 17% to 45%, meanwhile the number of questions with at least one good answer is between 65% and 90%.

When a higher number of nodes and relations are involved, the features that can be exploited for developing successful ranking algorithms become notably more complex. Algorithms based on *single-item* models may still be profitably used, provided that the underlying multi-graphs can be projected on a single dimension. The results obtained at each projection provide a multifaceted set of features that can be profitably used for tuning automatic classifiers able to discern high-quality items, or to identify experts.

In the rest of this chapter we detail a methodology for mining multi-item multi-relation graphs for two particular study cases. In the first case we describe the methodology presented in [18] for predicting successful items in a co-citation network, while in the second case we report the work of Agichtein et al. [2] for determining high-quality items in a question-answering portal.

3.1 Prediction of Successful Items in a Co-citation Network

Predicting the impact that a book or an article might have on readers is of great interest for publishers and editors for the purpose of planning marketing campaigns or deciding the number of copies to print. This problem was addressed in [18], where the authors present a methodology to estimate the number of citations that an article will receive, which is one measure of impact in a scientific community. The data was extracted by the large collection of academic articles made publicly available by CiteSeer [21] through an Open Archives Initiative (OAI) interface.

The two main objects in bibliometric networks are authors and papers. A bibliographic network can be modeled by a graph $\mathcal{G} = (V_a \cup V_p, E_a \cup E_c)$, where (i) V_a represents the set of authors, (ii) V_p represents the set of the papers, (iii) $E_a \subseteq V_a \times V_p$ represents the edges that express which author has written which paper, and (iv) $E_c \subseteq V_p \times V_p$ represents the edges that express which paper cites which. To model the dynamics of the citation network, different snapshots can be considered, with $\mathcal{G}_t = (V_{t,a} \cup V_{t,p}, E_{a,t} \cup E_{t,c})$ representing the snapshot at time t. The set of edges $E_{a,t}$ and $E_{c,t}$ can also be represented by matrices $P_{a,t}$ and $P_{c,t}$ respectively.

One way to model the network is by assigning a dual role to each author: in one role, an author produces original content (i.e., as authorities in the Kleinberg model. In the other role, an author provides an implicit evaluation of other authors (i.e., as a hub) with the use of citations. Fujimura and Tanimoto [29] present an algorithm, called *EigenRumor*, for ranking object and users when they act in this dual role. In their framework, the authorship relation $P_{a,t}$ is called *information provisioning*, while the citation relation $P_{c,t}$ is called *infor-*

mation evaluation. One of the main advantages of the EigenRumor algorithm is that the relations implied by both information provisioning and information evaluation are used to address the problem of correctly ranking items produced by sources that have been proven to be authoritative, even if the items themselves have not still collected a high number of in-links. The EigenRumor algorithm has been proposed in order to overcome the problem of algorithms like PageRank, which tend to favor items that have been present in the network for a period of time long enough to accumulate many links.

For the task of predicting the number of citations of a paper, Castillo et al. [18] use supervised learning methods that rely on features extracted from the co-citation network. In particular, they propose to exploit features that determine popularity, and then to train a classifier. Three different types of features are extracted: (1) *a priori* author-based features, (2) *a priori* link-based features, and (3) *a posteriori* features.

- **A priori author-based features.** These features capture the popularity of previous papers of the same authors. At time t, the past publication history of a given author a can be expressed in terms of:

  (i) Total number of citations $C_t(a)$ received by the author i from all the papers published before time t.

  (ii) Total number of papers $M_t(a)$ published by the author a before time t

  $$M_t(a) = |\{p|(a,p) \in E_a \wedge \text{time}(p) < t\}|.$$

  (iii) Total number of coauthors $A_t(a)$ for papers published before time t

  $$A_t(a) = |\{a'|(a',p) \in E_a \wedge (a,p) \in E_a \wedge \text{time}(p) < t \wedge a' \neq a\}|$$

  Given that one paper can have multiple authors, the previous three kinds of features are aggregated. For each, we consider the maximum, the average and the sum over all the co-authors of each paper.

- **A priori link-based features.** These features are based on the intuition that mutual reinforcement characterizes the relation between citing and cited authors: good authors are probably aware of the best previous articles written in a certain field, and hence they tend to cite the most relevant of them. As mentioned previously, the EigenRumor algorithm [29] can be used for ranking objects and users.

  The reputation score of a paper p is denoted by $r(p)$. The authority and the hub values of the author a are denoted by $a_t(a)$ and $h_t(a)$ respectively. The EigenRumor algorithm is formalized as follows:

- $\mathbf{r} = P_{a,t}^{T}\mathbf{a}_t$ expresses the fact that good papers are likely to be written by good authors,
- $\mathbf{r} = P_{c,t}^{T}\mathbf{h}_t$ expresses the fact that good papers are likely to be cited by good authors,
- $\mathbf{a}_t = P_{a,t}\mathbf{r}$ expresses the fact that good authors usually write good papers,
- $\mathbf{h}_t = P_{c,t}\mathbf{r}$ expresses the fact that good authors usually cite good papers.

Combining the previous equations with a mixing parameter α, gives the following formula for the score vector:

$$\mathbf{r} = \alpha P_{a,t}^{T}\mathbf{a}_t + (1 - \alpha)P_{c,t}^{T}\mathbf{h}_t.$$

- **A posteriori features.** These features are simply used to count the number of citations of a paper at the end of a few time intervals that are much shorter than the target time for the prediction that has to be made.

With respect to the case in which only *a posteriori* citations are used, *a priori* information about the authors helps in predicting the number of citations it will receive in the future. It is worth noting that a priori information about authors degrades quickly. When the features describing the reputation of an author are calculated at a certain time, and re-used without taking into account the last papers the author has published, the predictions tend to be much less accurate. These results are even more interesting if the reader considers that many other factors can be taken into consideration. For instance, the venue where the paper was published is related to the content of the paper itself.

3.2 Finding High-Quality Content in Question-Answering Portals

Yahoo! Answer is one of the largest question-answering portals, where users can issue question and find answers. Questions are the central elements. Each question has a life cycle. After it is "opened", it can receive answers. When the person who has asked the question is satisfied by an answer or after the expiration of an automatic timer, the question is considered "closed", and can not receive any other answers. However, the question and the answers can be voted on by other users. The question is "resolved" once a best answer is chosen. Because of its extremely rich set of user-document relations, Yahoo! Answers has recently been the subject of much research [1, 2, 11]. In [2], the authors focus on the task of finding high quality items in social networks and they use Yahoo! Answers as cases of study. The general approach is similar to the one used in the previous case for predicting successful items in co-citation networks, i.e., exploiting features that are correlated with quality in social media and then training a classifier to select and weight features for this task. In

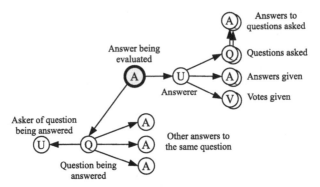

(a) Features for Inferring Answer Quality

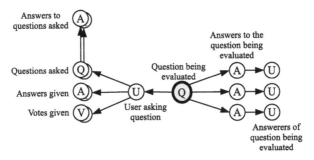

(b) Features for Inferring Question Quality

Figure 15.2. Types of Features Available for Inferring the Quality of Questions and Answers

the remainder of this section, the features for quality classification are considered. As in the previous case, three different types of features are used: (1) intrinsic content quality features, (2) link-based (or relation-based) features, and (3) content usage statistics.

- **Intrinsic content quality features.** For text-based social media the intrinsic content quality is mainly related with the text quality. This can be measured using *lexical, syntactic* and *semantic* features.

  Lexical features include word length, word and phrase frequencies, and the average number of syllables in the words.

  All the word n-grams up to length 5 that appear in the documents more than 3 times are used as syntactic features.

  Semantic features try to capture (1) the *visual quality* of the text (i.e., ignored capitalization rules, excessive punctuation, spacing density,etc.), (2)*semantic complexity* (i.e., entropy of word length, readability measures [30, 43, 37], etc.) and (3) *grammaticality* (i.e., features that try to capture the correctness of grammatical forms, etc).

In the QA domain, additional features are required to explicitly model the relationship between the question and the answer. In [2] such a relation was modeled using the KL-divergence between the language models of the two texts, their non-stopword overlap, the ratio between their lengths, and other similar features.

- **Link-based features.** As mentioned earlier, Yahoo! Answers is characterized by nodes of multiple types (e.g., questions, answers and users) and interactions with different semantics (e.g., "answers", "votes for", "gives a star to", "gives a best answer"), that are modeled using a complex multiple-node multiple-relations graph. Traditional link-analysis algorithms, including HITS and PageRank, are proven to still be useful for quality classification whether applied to the projections obtained from the graph G considering one type of relation at the time.

Answer features. In Figure 15.2(a), the relationship data related to a particular answer are shown. These relationships form a *tree*, in which the type "Answer" is the root. Two main subtrees start from the answer being evaluated: one related to the question Q being answered, and the other related to the user U contributing the answer.

By following paths through the question subtree, it is also possible to derive features QU about the questioner, or features QA concerning the other answers to the same question. By following paths through the user subtree, we can derive features UA from the answers of the user, features UQ from questions of the user, features UV from the votes of the user, and features UQA from answers received to the user's questions.

Question features. Figure 15.2(b) represents user relationships around a question. Again, there are two subtrees: one related to the asker of the question, and the other related to the answers received. The types of features on the answers subtree are: features A directly from the answers received and features AU from the answerers of the question being answered. The types of features on the user subtree are the same as the ones above for evaluating answers.

Implicit user-user relations To apply link-analysis algorithms, it is necessary to consider the user-user graph. This is the graph $G = (V, E)$ in which the set of vertices V is composed of the set of users and the set $E = E_a \cup E_b \cup E_v \cup E_s \cup E_+ \cup E_-$ represents the relationships between users as follows:

- E_a represents the answers: $(u, v) \in E_a$ if user u has answered at least one question asked by user v.

- E_b represents the best answers: $(u, v) \in E_b$ if user u has provided at least one best answer to a question asked by user v.
- E_v represents the votes for best answer: $(u, v) \in E_v$ if user u has voted for best answer at least one answer given by user v.
- E_s represents the stars given to questions: $(u, v) \in E_v$ if user u has given a star to at least one question asked by user v.
- E_+/E_- represents the thumbs up/down: $(u, v) \in E_+/E_-$ if user u has given a "thumbs up/down" to an answer by user v.

For each graph $G_x = (V, E_x)$, h_x is the vector of hub scores on the vertices V, a_x the vector of authority scores, and p_x the vector of PageRank scores. Moreover p'_x is the vector of PageRank scores in the transposed graph.

To classify these features in our framework, PageRank and authority scores are assumed to be related mostly to in-links, while the hub score deals mostly with out-links. For instance, let us consider h_b. It is the hub score in the "best answer" graph, in which an out-link from u to v means that u gave a best answer to user v. Then, h_b represents the answers of users, and is assigned to the record (UA) of the person answering the question.

- **Content usage statistics.** Usage statistics such as the number of clicks on the item and time spent on the item have been shown useful in the context of identifying high quality web search results. These are complementary to link-analysis based methods. Intuitively, usage statistics measures are useful for social media content, but require different interpretation from the previously studied settings.

  In the QA settings, it is possible to exploit the rich set of metadata available for each question. This includes temporal statistics, e.g., how long ago the question was posted, which allows us to give a better interpretation to the number of views of a question. Also, given that clickthrough counts on a question are heavily influenced by the topical and genre category, we also use derived statistics. These statistics include the expected number of views for a given category, the deviation from the expected number of views, and other second-order statistics designed to normalize the values for each item type. For example, one of the features is computed as the click frequency normalized by subtracting the expected click frequency for that category, divided by the standard deviation of click frequency for the category.

The conclusion of Agichtein et al. [2] from analyzing the above features, is that many of the features are complementary and their combination enhances the robustness of the classifier. Even though the analysis was based on a particular question-answering system, the ideas and the insights are applicable to other social media settings, and to other emerging domains centered around user contributed-content.

4.   Mining Query Logs

A query log contains information about the interaction of users with search engines. This information can be characterized in terms of the queries that users make, the results returned by the search engines, and the documents that users click in the search results. The wealth of explicit and implicit information contained in the query logs can be a valuable source of knowledge for a large number of applications. Examples of such applications include the following:

- (*i*) analyzing the interests of users and their searching behavior,

- (*ii*) finding semantic relations between queries (which terms are similar to each other or which one is a specialization of another) allowing to build taxonomies that are much richer than any human-built taxonomy,

- (*iii*) improving the results provided by search engines by analysis of the documents clicked by users and understanding the user information needs,

- (*iv*) fixing spelling errors and suggesting related queries,

- (*v*) improving advertising algorithms and helping advertisers select bidding keywords.

As a result of the wide range of applications which work with query-logs, considerable research has recently been performed in this area. Many of these papers discuss related problems such as analyzing query logs and on addressing various data-mining problems which work off the properties of the query-logs. On the other hand, query logs contain sensitive information about users and search-engine companies are not willing to release such data in order to protect the privacy of their users. Many papers have demonstrated the security breaches that may occur as a result of the release of query-log data even after anonymization operations have been applied and the data appears to be secure [34, 35, 41]. Nevertheless, some query log data that have been carefully anonymized have been released to the research community [22], and researchers are working actively on the problem of anonymizing query logs without destroying the utility of the released data. Recent advances on the anonymization problem are discussed in Korolova et al. [39]. Because of the wide range of knowledge embedded in query logs, this area is a central problem for the entire research community, and is not restricted to researchers working on problems related to search engines. Because of the natural ability to construct graph representations of query-log data, the graph mining area is particularly related to problems associated with query-log mining. In the next sections, we discuss graph representations of query log data, and consequently we present techniques for mining and analyzing the resulting graph structures.

4.1   Description of Query Logs

Query log. A typical query log $\mathcal{L}$ is a set of records $\langle q_i, u_i, t_i, V_i, C_i \rangle$, where q_i is the submitted query, u_i is an anonymized identifier for the user who submitted the query, t_i is a timestamp, V_i is the set of documents returned as results to the query, and C_i is the set of documents clicked by the user. We denote by Q, U, and D the set of queries, users, and documents, respectively. Thus, we have $q_i \in Q$, $u_i \in U$, and $C_i \subseteq V_i \subseteq D$.

Sessions. A *user query session*, or just *session*, is defined as the sequence of queries of one particular user within a specific time limit. More formally, if t_θ is a timeout threshold, a user query session S is a *maximal* ordered sequence

$$ S = \big\langle\, \langle q_{i_1}, u_{i_1}, t_{i_1} \rangle, \ldots, \langle q_{i_k}, u_{i_k}, t_{i_k} \rangle \,\big\rangle, $$

where $u_{i_1} = \cdots = u_{i_k} = u \in U$, $t_{i_1} \leq \cdots \leq t_{i_k}$, and $t_{i_{j+1}} - t_{i_j} \leq t_\theta$, for all $j = 1, 2, \ldots, k - 1$. The typical timeout threshold used for splitting sessions in query log analysis is $t_\theta = 30$ minutes [13, 19, 50, 57].

Supersessions. The temporally ordered sequence of all the queries of a user in the query log is called a *supersession*. Thus, a supersession is a sequence of sessions in which consecutive sessions are separated by time periods larger than t_θ.

Chains. A chain is a topically coherent sequence of queries of one user. Radlinski and Joachims [53] defined a chain as *"a sequence of queries with a similar information need"*. For instance, a query chain may contain the following sequence of queries [33]: "brake pads"; "auto repair"; "auto body shop"; "batteries"; "car batteries"; "buy car battery online". Clearly, all of these queries are closely related to the concept of car-repair. The concept of chain is also referred to in the literature with the terms *mission* [33] and *logical session* [3]. Unlike the straightforward definition of a session, chains involve relating queries based on an *analysis* of the user information need. This is a very complex problem, since it is based on an analysis of the information need, rather than in a crisp way, as in the case of a session. We do not try to give a formal definition of chains here, since this is beyond the scope of the chapter.

4.2   Query Log Graphs

Query graphs. In a recent paper about extracting semantic relations from query logs, Baeza-Yates and Tiberi define a graph structure derived from the query log. This takes into account not only the queries of the users, but also the actions of the users (clicked documents) after submitting their queries [4]. The analysis of the resulting graph captures different aspects of user behavior and topic distributions of what people search in the web. The graph representation introduced in [4] allows us to infer interesting semantic relationships among queries. This can be used in many applications.

The basic idea in [4] is to start from a weighted query-click bipartite graph, which is defined as the graph that has all distinct queries and all distinct documents as two partitions. We define an edge (q, u) between query q and document d, if a user who has submitted query q has clicked on document d. Obviously, d has to be in the result set of query q. The bipartite graph that has queries and documents as two partitions is also called the *click graph* [23]. Baeza-Yates and Tiberi define the *url cover* $\text{uc}(q)$ of a query q to be the set of neighbor documents of q in the click graph. The weight $w(q, d)$ of the edge (q, d) is defined to be the fraction of the clicks from q to d. Therefore, we have $\sum_{d \in \text{uc}(q)} w(q, d) = 1$. The url cover $\text{uc}(q)$ can be viewed as a vector representation for the query q, and we can then define the similarity between two queries q_1 and q_2 to be the *cosine similarity* of their corresponding url-cover vectors. This is denoted by $\cos(\text{uc}(q_1), \text{uc}(q_2))$. The next step in [4] is to define a graph G_q among queries, where the weight between two queries q_1 and q_2 is defined by their similarity value $\cos(\text{uc}(q_1), \text{uc}(q_2))$.

Using the url cover of the queries, Baeza-Yates and Tiberi define the following semantic relationship among queries:

- **Identical cover:** $\text{uc}(q_1) = \text{uc}(q_2)$. Those are undirected edges in the graph G_q, which are denoted as *red* edges or edges of type I. These imply that the two queries q_1 and q_2 are equivalent in practice.

- **Strict complete cover:** $\text{uc}(q_1) \subset \text{uc}(q_2)$. Those are directed edges, which are denoted as *green* edges or edges of type II. These imply that q_1 is more specific than q_2.

- **Partial complete cover:** $\text{uc}(q_1) \cap \text{uc}(q_2) \neq \emptyset$ and none of the previous two conditions are fulfilled. These are denoted as *black* edges or edges of type III. They are the most common edges and exist due to multi-topic documents or related queries, among other reasons.

The authors of [4] also define relaxed versions of the above concepts. In particular, they define α-red edges and α-green edges, when equality and inclusion hold with a slackness factor of α.

The resulting graph is very rich and may lead to many interesting applications. The mining tasks can be guided both by the semantic relationships of the edges as well as the graph structure. Baeza-Yates and Tiberi demonstrate an application of finding multi-topic documents. The idea is that edges with low weight are most likely caused by multi-topic documents e.g., e-commerce sites to which many different queries may lead. Thus, low-weight edges are considered as voters for the documents shared by the two corresponding queries. Documents are sorted according to the number of votes they received: the more votes a document gets, the more multitopical it is. Then the multi-topic documents may be removed from the graph (on a basis of a threshold value) and a new graph of better quality can be computed.

As Baeza-Yates and Tiberi point out, the analysis described in their paper is only the tip of the iceberg, and the potential number of applications of query graphs is huge. For instance, in addition to the graph defined in [4], Baeza-Yates [3] identifies five different types of graphs whose nodes are queries, and an edge between two queries implies that: (*i*) the queries contain the same word(s) (*word graph*), (*ii*) the queries belong to the same session (*session graph*), (*iii*) users clicked on the same urls in the list of their results (*url cover graph*), (*iv*) there is a link between the two clicked urls (*url link graph*) (*v*) there are *l* common terms in the content of the two urls (*link graph*).

Random walks on the click graph. The idea of representing the query log information as a bipartite graph between queries and documents (where the edges are weighted according to the user clicks) has been extensively used in the literature. Craswell and Szummer [23] study a random-walk model on the click graph, and they suggest using the resulting probability distribution of the model for ranking documents to queries. As mentioned in [23], query-document pairs can be considered as "soft" (positive) relevance judgments. These are however are noisy and sparse. The noise is due to the fact that users judge from short summaries and might not click on relevant documents. The sparsity problem is due to the fact that the users may not click on relevant documents. When a large number of documents are relevant, users may click on only a small fraction of them. The random-walk model can be used to reduce the amount of noise and it also alleviates the sparseness problem. One of the main benefits of the approach in [23] is that relevant documents to a query can be ranked highly even if no previous user has clicked on them for that query.

The click-graph can be used in many applications. Some of the applications discussed by Craswell and Szummer in [23] are the following:

- **Query-to-document search.** The problem is to rank relevant documents for a given ad-hoc query. The click graph is used to find documents of high quality and relevant documents for a query. Such documents may not necessarily be easy to determine using pure content-based analysis.

- **Query-to-query suggestion.** Given a query of a user, we want to find other queries that the user might be interested in. The role of the click-graph is determine other relevant queries in the "proximity" of the input query. Examples of finding such related queries can be found in [9, 59].

- **Document-to-query annotation.** The idea is that a query can be used as a concise description of the documents that the users click for that query, and thus queries can be used to represent documents. Studies have shown that the use of such a representation can improve web search [60]. It can be used for other web mining applications [51].

- **Document-to-document relevance feedback.** For this application, the task is to find relevant documents for a given target document, and are also relevant for a user.

The random walk on the click graph models a user who issues queries, clicks on documents according to the edge weights of the graph. These documents inspire the user to issue new queries, which in turn lead to new documents and so on. More formally, we define $\mathcal{G} = (Q \cup D, E)$ is the click graph, with Q and D being the set of queries and documents. We define E being the set of edges, the weight C_{jk} of an edge (j, k) is the number of clicks in the query log between nodes j and k. The weights are then normalized to represent the transition probabilities at the t-th step of the walk. The transition probabilities are defined as follows:

$$\Pr\nolimits_{t+1|t}[k \mid j] = \begin{cases} (1-s)\frac{C_{jk}}{\sum_i C_{ji}}, & \text{if } k \neq j, \\ s, & \text{if } k = j. \end{cases}$$

In other words, a self-loop is added at each node. The random walk is performed by traversing the nodes of the click graph according to the probabilities $\Pr_{t+1|t}[k \mid j]$.

Let $\mathbf{A}$ be the adjacency-matrix of the graph, whose (j, k)-th entry is $\Pr_{t+1|t}[k \mid j]$. Then, if $\mathbf{q}_j$ is a unit vector with an entry equal to 1 at the j-th position and all other entries equal to 0, the probability of a transition from node j to node k in t steps is $\Pr_{t|0}[k \mid j] = [\mathbf{q}_j \mathbf{A}^t]_k$. The notation $[\mathbf{u}]_i$ refers to the i-th entry of vector $\mathbf{u}$. The random-walk models that are typically used in the literature, such as PageRank and much more, consider *forward walks*, and exploit the property that the resulting vector of visiting probabilities $[\mathbf{q}\mathbf{A}^t]$ converges to a fixed distribution. This is the stationary distribution of the random walk, as $t \to \infty$, and is independent of the vector of initial probabilities $\mathbf{q}$. The value $[\mathbf{q}\mathbf{A}^t]_k$, i.e., the value of the stationary distribution at the k-th node, is usually interpreted as the importance of node k in the random walk, and it is used as the score for ranking node k.

Craswell and Szummer consider the idea of running the random walk *backwards*. Essentially the question is which is the probability that the walk started at node k given that after t steps is at node j. Bayes' law gives

$\mathrm{Pr}_{0|t}[k \mid j] \propto \mathrm{Pr}_{t|0}[j \mid k] \, \mathrm{Pr}_0[k]$, where $\mathrm{Pr}_0[k]$ is a *prior* of starting at node k and it is usually set to the uniform distribution, i.e., $\mathrm{Pr}_0[k] = 1/N$. To see the difference between forward and backward random walk, notice that since the stationary distribution of the forward walk is independent from the initial distribution, the limiting distribution of the backward random walk is uniform. Nevertheless, according to Craswell and Szummer, running the walk backwards for a small number of steps (before convergence) gives meaningful differentiation among the nodes in the graph. The experiments in [23] confirm that for ad-hoc search in image databases, the backward walk gives superior precision results than the forward random walk.

Random surfer and random querier. While the classic PageRank algorithm simulates a *random surfer* on the web, the random-walk on the click graph simulates the behavior of a *random querier*: moving between queries and documents according to the clicks of the query log. Poblete et al. [52] observe that searching and surfing the web are the two most common actions of web users, and they suggest building a model that combines these two activities by means of a random walk on a *unified* graph: the union of the hyperlink graph with the click graph.

The random walk on the unified graph is described as follows: At each step, the user selects to move at a random query or a random document with probability $1 - \alpha$. With probability α, the user makes a step, which can be one of two types:

- with probability $1 - \beta$ the user follows a link in the hyperlink graph,

- with probability β the user follows a link in the click graph.

The authors in [52] point out that combining the two graphs is beneficial, because the two graph structures are complementary and each of them can be used to alleviate the shortcomings of the other. For example, using clicks is a way to take into account user feedback, and this improves the robustness of the hyperlink graph to the degrading effects of link-spam. On the other hand, considering hyperlinks and browsing patterns increases the density and the connectivity of the click graph, and the model takes into account pages that users might visit *after* issuing particular queries.

The query-flow graph. We will now change the focus of the discussion to a different type of graphs extracted from query logs. In all our previous discussions, the graphs do not take into account the notion of *time*. In other words, the timestamp information from the query logs is completely ignored. However, if one wants to reason about the querying patterns of users, and the ways that user submit queries in order to achieve more complex information retrieval goals, one has to include the temporal aspect in the analysis of query logs.

In order to capture the querying behavior of users, Boldi et al. [13] define the concept of the *query-flow graph*. This is related to the discussion about sessions and chains at the beginning of this section. The query-flow graph G_{qf} is then defined to be directed graph $G_{qf} = (V, E, w)$ where:

- the set of nodes is $V = Q \cup \{s, t\}$, i.e., the distinct set of queries Q submitted to the search engine and two special nodes s and t, representing a *starting state* and a *terminal state*. These can be interpreted as the begin and end of a chain;

- $E \subseteq V \times V$ is the set of directed edges;

- $w : E \to (0, 1]$ is a weighting function that assigns to every pair of queries $(q, q') \in E$ a weight $w(q, q')$ representing the probability that q and q' are part of the same chain.

Boldi et al. suggest a machine learning method for building the query-flow graph. First, given a query log $\mathcal{L}$, it is assumed that it has been split into a set of sessions $\mathcal{S} = \{S_1, \ldots, S_m\}$. Two queries $q, q' \in Q$ are *tentatively* connected with an edge if there is at least one session in $\mathcal{S}$ in which q and q' are consecutive. Then, for the tentative edges, the weights $w(q, q')$ are learned using a machine learning algorithm. If the weight of an edge is estimated to be 0, then the edge is removed. The features used to learn the weights $w(q, q')$ include *textual features* (such as the cosine similarity, the Jaccard coefficient, and size of intersection between the queries q and q', computed on on sets of stemmed words and on character-level 3-grams), *session features* (such as the number of sessions in which the pair (q, q') appears, the average session length, the average number of clicks in the sessions, the average position of the queries in the sessions, etc.), and *time-related features* (such as the average time difference between q and q' in the sessions in which (q, q') appears). Several of those features have been used in the literature for the problem of segmenting a user session into logical sessions [33]. For learning the weights $w(q, q')$, Boldi et al. use a rule-based model and 5 000 labeled pairs of queries as training data. Boldi et al. argue that the query-flow graph is a useful construct that models user querying patterns and can be used in many applications. One such application is that of query recommendations.

Another interesting application of the query-flow graph is segmenting and assembling chains in user sessions. In this particular application, one complication is that there is not necessarily some timeout constraint in the case of chains. Therefore, as an example, all the queries of a user who is interested in planning a trip to a far-away destination and web searches for tickets, hotels, and other tourist information over a period of several weeks should be grouped in the same chain. Additionally, for the queries composing a chain, it is not required to be consecutive. Following the previous example, the user who is planning the far-away trip may search for tickets in one day, then make some other queries related to a newly released movie, and then return to trip planning the next day by searching for a hotel. Thus, a session may contain queries from many chains. Conversely, a chain may contain queries from many sessions.

In [13] the problem of finding chains in query logs is modeled as an *Asymmetric Traveling Salesman Problem* (ATSP) on the query-flow graph. The formal definition of the chain-finding problem is the following: Let $S = \langle q_1, q_2, \ldots, q_k \rangle$ be the supersession of one particular user. We assume that a query-flow graph has been built by processing a query log that includes S. Then, we define a *chain cover* of S to be a partition of the set $\{1, \ldots, k\}$ into subsets $C_1, \ldots, C_h$. Each set $C_u = \{i_1^u < \cdots < i_{\ell_u}^u\}$ can be thought of as a chain $C_u = \langle s, q_{i_1^u}, \ldots, q_{i_{\ell_u}^u}, t \rangle$, which is associated with probability

$$\Pr[C_u] = \Pr[s, q_{i_1^u}] \Pr[q_{i_1^u}, q_{i_2^u}] \ldots \Pr[q_{i_{\ell_u-1}^u}, q_{i_{\ell_u}^u}] \Pr[q_{i_{\ell_u}^u}, t],$$

We would like to find a chain cover maximizing $\Pr[C_1] \ldots \Pr[C_h]$.

The chain-finding problem is then divided into two subproblems: *session reordering* and *session breaking*. The session reordering problem is to ensure that all the queries belonging to the same search session are consecutive. Then, the session breaking problem is much easier as it only needs to deal with non-intertwined chains.

The session reordering problem is formulated as an instance of the ATSP: Given the query-flow graph G_{qf} with edge weights $w(q, q')$, and given the session $S = \langle q_1, q_2, \ldots q_k \rangle$, consider the subgraph of G_{qf} induced by S. This is defined as the induced subgraph $G_S = (V, E, h)$ with nodes $V = \{s, q_1, \ldots, q_k, t\}$, edges E, and edge weights h defined as $h(q_i, q_j) = -\log \max\{w(q_i, q_j), w(q_i, t) w(s, q_j)\}$. The maximum of the previous expression is taken over the options of splitting and not splitting a chain. For more details about the edge weights of G_S, see [13]. An optimal ordering is a permutation π of $\langle 1, 2, \ldots k \rangle$ that maximizes the expression

$$\prod_{i=1}^{k-1} w(q_{\pi(i)}, q_{\pi(i+1)}).$$

This problem is equivalent to that of finding a Hamiltonian path of minimum weight in this graph.

Session breaking is an easier task, once the session has been re-ordered. It corresponds to the determination of a series of cut-off points in the re-ordered session. One way of achieving this is by determining a threshold η in a validation dataset, and then deciding to break a reordered session whenever $w(q_{\pi(i)}, q_{\pi(i+1)}) < \eta$.

4.3   Query Recommendations

As the next topic of graph mining for web applications and query-log analysis, we discuss the problem of *query recommendations*. Even though the problem statement does not involve graphs, many approaches in the literature work by exploring the graph structures induced from query logs. Examples of such graphs were discussed in the previous section.

The application of query recommendation takes place when search engines offer not only document results but also *alternative queries* in response to the queries they receive from their users. The purpose of those query recommendations is to help users locate information more effectively. Indeed, it has been observed over the past years that users are looking for information for which they do not have sufficient knowledge [10], and thus they may not be able to specify their information needs precisely. The recommendations provided by search engines are typically queries similar to the original one, and they are obtained by analyzing the query logs.

Many of the algorithms for making query recommendations are based on defining similarity measures among queries, and then recommending the most popular queries in the query log among the similar ones to a given query. For computing query similarity, Wen et al. [59] suggest using distance functions based on (i) the keywords or phrases of the query, (ii) string matching of keywords, (iii) the common clicked documents, and (iv) the distance of the clicked documents in some pre-defined hierarchy. Another similarity measure based on common clicked documents was proposed by Beeferman et al. [9]. Baeza-Yates et al. [5] argue that the distance measures proposed by the previous methods have practical limitations, because two related queries may output different documents in their answer sets. To overcome these limitations, they propose to represent queries as term-weighted vectors obtained by aggregating the term-weighted vectors of their clicked documents. Association rule mining has also been used to discover related queries in [28]. The query log is viewed as a set of transactions, where each transaction represents a session in which a single user submits a sequence of related queries in a time interval.

Next we review some of the query recommendation methods that are based on graph structures.

Hitting time. Mei et al. [44] propose a query recommendation method, which is based on the proximity of the queries on the *click graph*. Recall that the click graph is the bipartite graph that has queries and documents as two partitions, and the weight of an edge $w(q, u)$ indicates the number of times that document d has been clicked when query q was submitted. The main idea is based on the concept of structural proximity of specific nodes. When the user submits a query, the corresponding node is located in the click graph, and other recommendations are queries that are located in the *proximity* of the query node.

For a meaningful notion of distance between nodes in the click graph, Mei et al. suggest to use the notion of *hitting time*. The hitting time from a node u to a node v in a graph G is the expected number of steps taken when v is visited for a first time in a random walk starting from u. Hitting time captures not only nodes that are connected by short paths in the graph but also nodes that are connected by many paths. Therefore, it is a *robust* distance measure between graph nodes.

In addition, Mei et al. [44] propose an adaptation of their method that can provide personalized query suggestions. The idea is to adjust the weights of the edges of the click graph so that they can better model the preferences of the user for whom we want to provide a recommendation. Mei et al. observe that models for personalized web search provide estimates of a probability that a user clicks on a certain document. Thus, any personalized algorithm for web search can be combined with their hitting-time method in order to provide personalized recommendations.

Topical query decomposition. A different aspect of query recommendation is addressed by Bonchi et al. [14], who try to overcome a common limitation of many query recommendation algorithms. This limitation is that many of the recommendations are very similar to each other. Instead Bonchi et al. formulate a new problem, which they call *topical query decomposition*. In this new framework, the goal is to find a set of queries that cover different aspects of the original query. The intuition is that such a set of diverse queries can be more useful in cases when the query is too short (and thus imprecise and ambiguous), and it is hard to receive good recommendations based on the query-content only.

The problem statement of topical query decomposition is based again on the click graph. In particular, let q be a query and $D(q)$ be the result set of q, i.e., the neighbor nodes of q in the click graph. We denote with $\mathcal{Q}(q)$ the maximal set of queries p_i, where for each p_i, the set $D(p_i)$ has at least one document in common with the documents returned by q. In other words, we have

$$\mathcal{Q}(q) = \{p_i | \langle p_i, D(p_i) \rangle \in \mathcal{L} \land D(p_i) \cap D(q) \neq \emptyset\}.$$

The goal is to compute a *cover*, i.e., selecting a sub-collection $\mathcal{C} \subseteq \mathcal{Q}(q_i)$ such that it covers almost all of $D(q_i)$. As stated before, the queries in $\mathcal{C}$ should represent coherent, conceptually well-separated set of documents: they should have small overlap, and they should not cover too many documents outside $D(q_i)$.

Bonchi et al. propose two different algorithms for the problem of topical query decomposition. The first algorithm is a top-down approach, based on set covering. Starting from the queries in $Q(q)$, this approach tries to handle the problem as a special instance of the weighted set covering problem. The weight of each query in the cover is given by its internal topical coherence, the fraction of documents in $D(q)$, the number of documents it retrieves that are not in $D(q)$, as well as its overlap with other queries in the solution. The second algorithm is a bottom-up approach, based on clustering. Starting with the documents in $D(q)$, this approach tries to build clusters of documents which are compact in the topics space. Since the resulting clusters are not necessarily document sets associated with queries existing in the query log, a second phase is needed. In this phase, the clusters found in the first phase are "matched" to the sets that correspond to queries in the query log.

Query recommendations based on the query-flow graph. Boldi et al. [13] investigate the alternative approach of finding query recommendations using the query-flow graph instead of the click graph. A random walk approach is used in the this case, as in the approach of Mei et al. [44]. However, in this case, the recommended queries are selected on the basis of a PageRank measure instead of hitting time. We also allow *teleportation* (or jumps) to specific nodes during the random walks in order to bias the walk towards these nodes. In particular, given the query q, the method computes the PageRank values of a random walk on the query-flow graph where the teleportation is always at the node of the graph that corresponds to query q. In this way, queries that are close to q in the graph are favored to be selected as recommendations. The advantage of using the query-flow graph instead of the click graph is that the method favors as recommendations for q queries q' that follow q in actual user sessions. Thus, it is likely that q' are natural continuations of q in an information seeking task performed by users.

Boldi et al. [13] explore various alternatives to that of using random walk on the query-flow graph for the query recommendation problem. One interesting idea is to use normalized PageRank. Here, if $s_q(q')$ is the PageRank score for query q' on a random walk with teleportation to the original query q, instead of using the pure random-walk score $s_q(q')$, they consider the ratio $\hat{s}_q(q') = s_q(q')/r(q')$ where $r(q')$ is the absolute random-walk score of q' (i.e., the one computed using a uniform teleportation vector). The intuition behind this normalization is to avoid recommending very popular queries (like "ebay") that may easily get high PageRank scores even though they are not related with the original query. The experiments in [13] showed that in most cases $\hat{s}_q(q')$ produces rankings that are more reasonable, but sometimes tend to boost by too much the scores with low absolute value $r(q')$. To use a bigger denominator, they also tried dividing with $\sqrt{r(q')}$, which corresponds to the geometric mean between $s_q(q')$ and $\hat{s}_q(q')$.

Another interesting variant of the query-recommendation framework of Boldi et al. is providing recommendations that depend not only on the last query input by the user, but on some of the last queries in the user's history. This approach may help to alleviate the data sparsity problem. This is because the current query may be rare, but among the previous queries there might be queries for which we have enough information in the query flow graph. Basing the recommendation on the user's query history may also help to solve ambiguous queries, as we have more informative suggestions based on what the user is doing during the current session. To take the recent queries of the user into account, one has to modify the random walk, in order to perform the teleportation into the set of last queries, instead of only the one last query. For more details on the method and various examples of recommendations see [13].

Using both the click graph and session data. Finally, we discuss the query-recommendation approach of Cao et al. [17], which uses both the click graph and session data. As in the previous case of Boldi et al., the algorithm of Cao et al. has the advantage that it provides recommendations that are based on the few last queries of the user. The proposed algorithm has two steps. In the first step, the algorithm uses the click-graph in order to clusters all the queries of the query log. In particular, two queries are represented by the vector of neighbor documents in the click graph, and then the queries are clustered based on the Euclidean distance of their representation vectors. A simple greedy clustering algorithm is proposed that can scale to very large query-log data. In the second step, user sessions are processed and each query is represented by the cluster center that was assigned to during the first clustering step. The intuition of representing queries by their cluster center is to address the problem that two queries might have the same search intent. Thus, the authors in [17] prefer to work with "query concepts" rather than individual queries. Then *frequent sequential patterns* are mined from the user sessions. For each frequent sequence of query concepts $c_s = c_1 \ldots c_l$, the concept c_l is used as a candidate concept for the sequence $c'_s = c_1 \ldots c_{l-1}$. A ranked list of candidate concepts c for c'_s is then built based on the occurrences of the concepts c following c'_s in the same session; the more occurrences c has, the higher c is ranked. In practice, it is only needed to keep the representative queries of the top-k (e.g., $k = 5$) candidate concepts. These representative queries are called the *candidate recommendations* for the sequence c'_s and can be used for query recommendation, when c'_s is observed online.

5.   Conclusions

In this chapter we reviewed elements of mining graphs in the context of web applications. We focused on graphs arising in social networks, social media, and query logs. We discussed modeling issues and we presented specific problems in those areas, such as estimating the reputation and the popularity of items in a network, mining query logs, and performing query recommendations. Understanding the structure of graphs appearing in those applications, modeling the complex interactions between entities, and designing algorithms for leveraging the latent knowledge introduces new challenges in the field of graph mining. Classic graph-mining algorithms such as those involving random walks can provide a starting point. However, they often need to be extended and adapted in order to capture the requirements and complexities of the data models and the applications at hand.

References

[1] Lada A. Adamic, Jun Zhang, Eytan Bakshy, and Mark S. Ackerman. Knowledge sharing and yahoo answers: everyone knows something. In *Proceedings of the 17th international conference on World Wide Web (WWW)*, 2008.

[2] Eugene Agichtein, Carlos Castillo, Debora Donato, Aristides Gionis, and Gilad Mishne. Finding high quality content in social media, with an application to community-based question answering. In *Proceedings of ACM WSDM*, pages 183–194, Stanford, CA, USA, February 2008. ACM Press.

[3] Ricardo Baeza-Yates. Graphs from search engine queries. In *Theory and Practice of Computer Science (SOFSEM)*, 2007.

[4] Ricardo Baeza-Yates and Alessandro Tiberi. Extracting semantic relations from query logs. In *Proceedings of the 13th ACM international conference on Knowledge discovery and data mining (KDD)*, 2007.

[5] Ricardo A. Baeza-Yates, Carlos A. Hurtado, and Marcelo Mendoza. Query recommendation using query logs in search engines. In *Current Trends in Database Technology – EDBT Workshops*, 2004.

[6] A. L. Barabasi and R. Albert. Emergence of scaling in random networks. *Science*, 286:509–512, 1999.

[7] Albert-Laszlo Barabasi. *Linked: How Everything Is Connected to Everything Else and What It Means for Business, Science, and Everyday Life.* Plume Books, April 2002.

[8] L. Becchetti, C. Castillo, D. Donato, R. Baeza-Yates, and S. Leonardi. Link analysis for web spam detection. *ACM Transactions on the Web (TWEB)*, 2(1):1–42, February 2008.

[9] Doug Beeferman and Adam Berger. Agglomerative clustering of a search engine query log. In *Proceedings of the 6th ACM international conference on Knowledge discovery and data mining (KDD)*, 2000.

[10] Nicholas J. Belkin. The human element: helping people find what they don't know. *Communications of the ACM*, 43(8), 2000.

[11] Jiang Bian, Yandong Liu, Ding Zhou, Eugene Agichtein, and Hongyuan Zha. Learning to recognize reliable users and content in social media with coupled mutual reinforcement. In *Proceedings of the 18th international conference on World Wide Web (WWW)*, 2009.

[12] P. Boldi, R. Posenato, M. Santini, and S. Vigna. Traps and pitfalls of topic-biased pagerank. In *Proceedings of the 4th International Workshop on Algorithms and Models for the Web-Graph (WAW)*, 2008.

[13] Paolo Boldi, Francesco Bonchi, Carlos Castillo, Debora Donato, Aristides Gionis, and Sebastiano Vigna. The query-flow graph: model and applications. In *Proceeding of the 17th ACM conference on Information and knowledge management (CIKM)*, 2008.

[14] Francesco Bonchi, Carlos Castillo, Debora Donato, and Aristides Gionis. Topical query decomposition. In *Proceedings of the 14th ACM international conference on Knowledge discovery and data mining (KDD)*, 2008.

[15] S. Brin and L. Page. The anatomy of a large-scale hypertextual web search engines. *Computer Networks and ISDN Systems*, 30(1–7):107–117, 1998.

[16] Guido Caldarelli. *Scale-Free Networks*. Oxford University Press, 2007.

[17] Huanhuan Cao, Daxin Jiang, Jian Pei, Qi He, Zhen Liao, Enhong Chen, and Hang Li. Context-aware query suggestion by mining click-through and session data. In *Proceeding of the 14th ACM international conference on Knowledge discovery and data mining (KDD)*, 2008.

[18] Carlos Castillo, Debora Donato, and Aristides Gionis. Estimating the number of citations of a paper using author reputation. In *String Processing and Information Retrieval Symposium (SPIRE)*, 2007.

[19] L. Catledge and J. Pitkow. Characterizing browsing behaviors on the world wide web. *Computer Networks and ISDN Systems*, 6, 1995.

[20] Hyunwoo Chun, Haewoon Kwak, Young H. Eom, Yong Y. Ahn, Sue Moon, and Hawoong Jeong. Comparison of online social relations in volume vs interaction: a case study of cyworld. In *Proceedings of the 8th ACM SIGCOMM conference on Internet measurement (IMC)*, 2008.

[21] CiteSeer, http://citeseer.com.

[22] Nick Craswell, Rosie Jones, Georges Dupret, and Evelyne Viegas, editors. *Workshop on Web Search Click Data (WSCD), held in conjunction with WSDM*, Barcelona, Spain, 2009.

[23] Nick Craswell and Martin Szummer. Random walks on the click graph. In *Proceedings of the 30th annual international ACM conference on Research and development in information retrieval (SIGIR)*, 2007.

[24] G. M. Del Corso, A. Gulli, and F. Romani. Fast pagerank computation via a sparse linear system. *Internet Mathematics*, 2(3), 2005.

[25] Alex Fabrikant, Elias Koutsoupias, and Christos Papadimitriou. Heuristically optimized trade-offs: A new paradigm for power laws in the internet. In *Proceedings of the 29th International Colloquium on Automata, Languages and Programming (ICALP)*, 2002.

[26] Michalis Faloutsos, Petros Faloutsos, and Christos Faloutsos. On power-law relationships of the internet topology. In *Proceedings of the annual ACM conference on Data Communication (SIGCOMM)*, 1999.

[27] D. Fogaras, B. Racz, K. Csalogany, and T. Sarlos. Towards scaling fully personalized pageRank: algorithms, lower bounds, and experiments. *Internet Mathematics*, 2(3):333–358, 2005.

[28] Bruno M. Fonseca, Paulo Braz Golgher, Edleno Silva de Moura, Bruno Pôssas, and Nivio Ziviani. Discovering search engine related queries using association rules. *Journal of Web Engineering*, 2(4), 2004.

[29] Ko Fujimura and Naoto Tanimoto. The eigenrumor algorithm for calculating contributions in cyberspace communities. *Trusting Agents for Trusting Electronic Societies*, pages 59–74, 2005.

[30] Robert Gunning. *The technique of clear writing*. McGraw-Hill, 1952.

[31] Z. Gyongyi, H. Garcia-Molina, and J. Pedersen. Combating Web spam with TrustRank. In *Proceedings of the 30th International Conference on Very Large Data Bases (VLDB)*, pages 576–587, Toronto, Canada, August 2004. Morgan Kaufmann.

[32] T.H. Haveliwala. Topic-sensitive pagerank. In *Proceedings of the eleventh International World Wide Web Conference (WWW)*, Honolulu, Hawaii, 2002.

[33] Rosie Jones and Kristina L. Klinkner. Beyond the session timeout: automatic hierarchical segmentation of search topics in query logs. In *Proceedings of the 16th ACM conference on Conference on information and knowledge management (CIKM)*, 2008.

[34] Rosie Jones, Ravi Kumar, Bo Pang, and Andrew Tomkins. I know what you did last summer: query logs and user privacy. In *Proceeding of the 16th ACM conference on Information and knowledge management (CIKM)*, 2007.

[35] Rosie Jones, Ravi Kumar, Bo Pang, and Andrew Tomkins. Vanity fair: privacy in querylog bundles. In *Proceeding of the 17th ACM conference on Information and knowledge management (CIKM)*, 2008.

[36] S. Kamvar, T. Haveliwala, C. Manning, and G. Golub. Exploiting the block structure of the web for computing pagerank. Technical report, Stanford University, 2003.

[37] J. Peter Kincaid, Robert P. Fishburn, Richard L. Rogers, and Brad S. Chissom. Derivation of new readability formulas for navy enlisted personnel. Technical Report Research Branch Report 8-75, Millington, Tenn, Naval Air Station, 1975.

[38] Jon Kleinberg. Authoritative sources in a hyperlinked environment. *Journal of ACM*, 46(5), 1999.

[39] Aleksandra Korolova, Krishnaram Kenthapadi, Nina Mishra, and Alexandros Ntoulas. Releasing search queries and clicks privately. In *Proceedings of the 18th international conference on World Wide Web (WWW)*, 2009.

[40] R. Kumar, P. Raghavan, S. Rajagopalan, D. Sivakumar, A. Tomkins, and E. Upfal. Stochastic models for the web graph. In *Proceedings of the 41st Annual Symposium on Foundations of Computer Science (FOCS)*, 2000.

[41] Ravi Kumar, Jasmine Novak, Bo Pang, and Andrew Tomkins. On anonymizing query logs via token-based hashing. In *Proceedings of the 16th international conference on World Wide Web (WWW)*, 2007.

[42] A.N. Langville and C.D. Meyer. Updating pagerank with iterative aggregation. In *Proceedings of the 13th International World Wide Web Conference on Alternate track papers & posters (WWW)*, New York, NY, USA, 2004.

[43] G. Harry McLaughlin. SMOG grading: A new readability formula. *Journal of Reading*, 12(8):639–646, 1969.

[44] Qiaozhu Mei, Dengyong Zhou, and Kenneth Church. Query suggestion using hitting time. In *Proceeding of the 17th ACM conference on Information and knowledge management (CIKM)*, 2008.

[45] Michael Mitzenmacher. A brief history of generative models for power law and lognormal distributions. *Internet Mathematics*, 1(2), 2003.

[46] M. Newman. Power laws, pareto distributions and zipf's law. *Contemporary Physics*, 2005.

[47] M. E. J. Newman and Juyong Park. Why social networks are different from other types of networks. *Physical Review E*, 68(3):036122, Sep 2003.

[48] Lawrence Page, Sergey Brin, Rajeev Motwani, and Terry Winograd. The PageRank citation ranking: bringing order to the Web. Technical report, Stanford Digital Library Technologies Project, 1998.

[49] Romualdo Pastor-Satorras, Alexei Vazquez, and Alessandro Vespignani. Dynamical and correlation properties of the internet. *Physical Review Letters*, 87(25):258701, Nov 2001.

[50] Benjamin Piwowarski and Hugo Zaragoza. Predictive user click models based on click-through history. In *Proceedings of the 16th ACM conference on Conference on information and knowledge management (CIKM)*, 2007.

[51] Barbara Poblete and Ricardo Baeza-Yates. A content and structure website mining model. In *Proceedings of the 15th international conference on World Wide Web (WWW)*, 2006.

[52] Barbara Poblete, Carlos Castillo, and Aristides Gionis. Dr. searcher and mr. browser: a unified hyperlink-click graph. In *Proceeding of the 17th*

ACM conference on Information and knowledge management (CIKM), 2008.

[53] Filip Radlinski and Thorsten Joachims. Query chains: learning to rank from implicit feedback. In *Proceeding of the 11th ACM SIGKDD international conference on Knowledge discovery in data mining*, 2005.

[54] Atish Das Sarma, Sreenivas Gollapudi, and Rina Panigrahy. Estimating pagerank on graph streams. In *Proceedings of the 27th ACM Symposium on Principles of Database Systems (PODS)*, 2008.

[55] Stephan H. Strogatz. Exploring complex networks. *Nature*, 410(6825):268–276, March 2001.

[56] Qi Su, Dmitry Pavlov, Jyh-Herng Chow, and Wendell C. Baker. Internet-scale collection of human-reviewed data. In *Proceedings of the 16th international conference on World Wide Web (WWW)*, pages 231–240, New York, NY, USA, 2007. ACM Press.

[57] Jaime Teevan, Eytan Adar, Rosie Jones, and Michael A. S. Potts. Information re-retrieval: repeat queries in yahoo's logs. In *Proceedings of the 30th annual international ACM conference on Research and development in information retrieval (SIGIR)*, 2007.

[58] Stanley Wasserman and Katherine Faust. *Social Network Analysis: Methods and Applications*. Cambridge University Press, 1994.

[59] Ji-Rong Wen, Jian-Yun Nie, and Hong-Jiang Zhang. Clustering user queries of a search engine. In *Proceedings of the 10th international conference on World Wide Web (WWW)*, 2001.

[60] Gui-Rong Xue, Hua-Jun Zeng, Zheng Chen, Yong Yu, Wei-Ying Ma, WenSi Xi, and WeiGuo Fan. Optimizing web search using web click-through data. In *Proceedings of the 13th ACM international conference on Information and knowledge management (CIKM)*, 2004.

[61] Yahoo! Answers, http://answers.yahoo.com.

[62] Soon-Hyung Yook, Filippo Radicchi, and Hildegard Meyer-Ortmanns. Self-similar scale-free networks and disassortativity, Jul 2005.

[63] Jun Zhang, Mark S. Ackerman, and Lada Adamic. Expertise networks in online communities: structure and algorithms. In *Proceedings of the 16th international conference on World Wide Web (WWW)*, 2007.

Chapter 16

GRAPH MINING APPLICATIONS TO SOCIAL NETWORK ANALYSIS

Lei Tang and Huan Liu
Computer Science & Engineering
Arizona State University
L.Tang@asu.edu, Huan.Liu@asu.edu

Abstract   The prosperity of Web 2.0 and social media brings about many diverse social networks of unprecedented scales, which present new challenges for more effective graph-mining techniques. In this chapter, we present some graph patterns that are commonly observed in large-scale social networks. As most networks demonstrate strong community structures, one basic task in social network analysis is community detection which uncovers the group membership of actors in a network. We categorize and survey representative graph mining approaches and evaluation strategies for community detection. We then present and discuss some research issues for future exploration.

Keywords: Social Network Analysis, Graph Mining, Community Detection,

1.   Introduction

Social Network Analysis (SNA) [61] is the study of relations between individuals including the analysis of social structures, social position, role analysis, and many others. Normally, the relationship between individuals, e.g., kinship, friends, neighbors, etc. are presented as a network. Traditional social science involves the circulation of questionnaires, asking respondents to detail their interaction with others. Then a network can be constructed based on the response, with nodes representing individuals and edges the interaction between them. This type of data collection confines traditional SNA to a limited scale, typically at most hundreds of actors in one study.

C.C. Aggarwal and H. Wang (eds.), *Managing and Mining Graph Data*,
Advances in Database Systems 40, DOI 10.1007/978-1-4419-6045-0_16,
© Springer Science+Business Media, LLC 2010

With the prosperity of Internet and Web 2.0, many social networking and social media sites are emerging, and people can easily connect to each other in the cyber space. This also facilitates SNA to a much larger scale — millions of actors or even more in a network; Examples include email communication networks [18], instant messenger networks [33], mobile call networks [39], friends networks [38]. Other forms of complex network, like coauthorship or citation networks [56], biological networks, metabolic pathways, genetic regulatory networks, food web and neural networks, are also examined and demonstrate similar patterns [44]. These large scale networks of various entities yield patterns that are normally not observed in small networks. In addition, they also pose computational challenges as well as new tasks and problems for the SNA.

Social network analysis involves a variety of tasks. To name a few, we list some that are among the most relevant to the data mining field:

- Centrality analysis aims to identify the "most important" actors in a social network. Centrality is a measure to calibrate the "importance" of an actor. This helps to understand the social influence and power in a network.

- Community detection. Actors in a social network form groups[1]. This task identify these communities through the study of network structures and topology.

- Position/Role analysis identifies the role associated with different actors during network interaction. For instance, what is the role of "husband"? Who serves as the bridge between two groups?

- Network modeling attempts to simulate the real-world network via simple mechanisms such that the patterns presented in large-scale complex networks can be captured.

- Information diffusion studies how the information propagates in a network. Information diffusion also facilitates the understanding the cultural dynamics, and infection blocking.

- Network classification and outlier detection. Some actors are labeled with certain information. For instance, in a network with some terrorists identified, is it possible to infer other people who are likely to be terrorists by leveraging the social network information?

- Viral marketing and link prediction. The modeling of the information diffusion process, in conjunction with centrality analysis and communi-

---

[1]In this chapter, community and group are used interchangeably.

ties, can help achieve more cost-effective viral marketing. That is, only a small set of users are selected for marketing. Hopefully, their adoption can influence other members in the network, so the benefit is maximized.

Normally, a social network is represented as a graph. How to mine the patterns in the graph for the above tasks becomes a hot topic thanks to the availability of enormous social network data. In this chapter, we attempt to present some recent trends of large social networks and discuss graph mining applications for social network analysis. In particular, we discuss graph mining applications to community detection, a basic task in SNA to extract meaningful social structures or positions, which also serves as basis for some other related SNA tasks. Representative approaches for community detection are summarized. Interesting emerging problems and challenges are also presented for future exploration.

For convenience, we define some notations used throughout this chapter. A network is normally represented as a graph $G(V, E)$, where V denotes the vertexes (equivalently nodes or actors) and E denotes edges (ties or connections). The connections are represented via adjacency matrix A, where $A_{ij} \neq 0$ denotes $(v_i, v_j) \in E$, while $A_{ij} = 0$ denotes $(v_i, v_j) \notin E$. The degree of node v_i is d_i. If the edges between nodes are directed, the in-degree and out-degree are denoted as d_i^- and d_i^+ respectively. Number of vertexes and edges of a network are $|V| = n$, and $|E| = m$, respectively. The shortest path between a pair of nodes v_i and v_j is called *geodesic*, and the geodesic distance between the two is denoted as $d(i, j)$. $G_s(V_s, E_s)$ represents a subgraph in G. The neighbors of a node v are denoted as $N(v)$. In a directed graph, the neighbors connecting to and from one node v are denoted as $N^-(v)$ and $N^+(v)$, respectively. Unless specified explicitly, we assume a network is unweighted and undirected.

2. Graph Patterns in Large-Scale Networks

Most large-scale networks share some common patterns that are not noticeable in small networks. Among all the patterns, the most well-known characteristics are: *scale-free distribution, small world effect, and strong community structure*.

2.1 Scale-Free Networks

Many statistics in real-world have a typical "scale", a value around which the sample measurements are centered. For instance, the height of all the people in the United States, the speed of vehicles on a highway, etc. But the node degrees in real-world large scale social networks often follow a power law distribution (a.k.a. Zipfian distribution, Pareto distribution [41]). A random

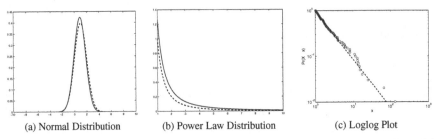

(a) Normal Distribution   (b) Power Law Distribution   (c) Loglog Plot

Figure 16.1. Different Distributions. A dashed curve shows the true distribution and a solid curve is the estimation based on 100 samples generated from the true distribution. (a) Normal distribution with $\mu = 1$, $\sigma = 1$; (b) Power law distribution with $x_{min} = 1$, $\alpha = 2.3$; (c) Loglog plot, generated via the toolkit in [17].

variable X follows a power law distribution if

$$p(x) = Cx^{-\alpha}, \quad x \geq x_{min}, \quad \alpha > 1 \qquad (2.1)$$

here $\alpha > 1$ is to ensure a normalization constant C exists [41]. A power law distribution is also called *scale-free* distribution [8] as the shape of the distribution remains unchanged except for an overall multiplicative constant when the scale of units is increased by a factor. That is,

$$p(ax) = bp(x) \qquad (2.2)$$

where a and b are constants. In other words, there is no characteristic scale with the random variable. The functional form is the same for all the scales. The network with a scale-free distribution for nodal degrees is also called *scale-free network*.

Figures 16.1a and 16.1b demonstrate a normal distribution and a power-law distribution respectively. While the normal distribution has a "center", the power law distribution is highly skewed. For normal distribution, it is extremely rare for an event to occur that are several deviations away from the mean. On the contrary, power law distribution allows the tail to be much longer. That is, it is common that some nodes in a social network have extremely high degrees while the majority have few connections. The reason is that the decay of the tail for a power law distribution is polynomial. It is asymptotically slower than exponential as presented in the decay of normal distribution, resulting in a heavy-tail (or long-tail [6], fat-tail) phenomenon.

The curve of power law distribution becomes a straight line if we plot the degree distribution in a log-log scale, since

$$\log p(x) = -\alpha \log x + \log C$$

This is commonly used by practitioners to rigorously verify whether a distribution follows power law, though some researchers advise more careful statistical examination to fit a power law distribution [17]. It can be verified the cumulative distribution function (cdf) can also be written in the following form:

$$F(X \geq x) \propto x^{-\alpha+1}$$

The samples of rare events (say, extremely high degrees in a network) are scarce, resulting in an unreliable estimation of the density. A more robust estimation is to approximate the cdf. One example of the loglog plot of cdf estimation is shown in Figure 16.1c.

Besides node degrees, some other network statistics are also observed to follow a power law pattern, for example, the largest eigenvalues of the adjacency matrix derived from a network [21], the size of connected components in a network [31], the information cascading size [36], and the densification of a growing network [34]. Scale-free distribution seems common rather than "by chance" for large-scale networks.

2.2 Small-World Effect

Travers and Milgram [58] conducted a famous experiment to examine the average path length for social networks of people in the United States. In the experiments, the subjects involved were asked to send a chain letter to his acquaintances starting from an individual in Omaha, Nebraska or Wichita, Kansas to the target individual in Boston, Massachusetts. Finally, 64 letters arrived and the average path length fell around 5.5 or 6, which later led to the so-called "six degrees of separation". This result is also confirmed recently in a planetary-scale instant messaging network of more than 180 million people, in which the average path length of two messengers is 6.6 [33].

This small world effect is observed in many large scale networks. That is, two actors in a huge network are actually not too far away. To quantify the effect, different network measures are used:

- **Diameter**: a shortest path between two nodes is called a *geodesic*, and *diameter* is the length of the longest geodesic between any pair of nodes in the graph [61]. It might be the case that a network contains more than one connected component. Thus, no path exists between two nodes in different components. In this case, practitioners typically examine the geodesic between nodes of the same component. The diameter is the minimum number of hops required to reach all the connected nodes from any node.

- **Effective Eccentricity**: the minimum number of hops required to reach at least 90% of all connected pairs of nodes in the network [57]. This measure removes the effect of outliers that are connected through a long path.

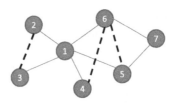

Figure 16.2. A toy example to compute clustering coefficient: $C_1 = 3/10$, $C_2 = C_3 = C_4 = 1$, $C_5 = 2/3$, $C_6 = 3/6$, $C_7 = 1$. The global clustering coefficient following Eqs. (2.5) and (2.6) are 0.7810 and 0.5217, respectively.

- **Characteristic Path Length**: the median of the means of the shortest path lengths connecting each node to all other nodes (excluding unreachable ones) [12]. This measure focuses on the average distance between pairs rather than the maximum one as the diameter.

All the above measures involve the calculation of the shortest path between all pairs of connected nodes. Two simple approaches to compute the diameter are:

- Repeated matrix multiplication. Let A denotes the adjacency matrix of a network, then the non-zero entries in A^k denote those pairs that are connected in k hops. The diameter corresponds to the minimum k so that all entries of A^k are non-zero. It is evident that this process leads to denser and denser matrix, which requires $O(n^2)$ space and $O(n^{2.88})$ time asymptotically for matrix multiplication.

- Breadth-first search can be conducted starting from each node until all or a certain proportion (90% as for effective eccentricity) of the network nodes are reached. This costs $O(n + m)$ space but $O(nm)$ time.

Evidently, both approaches above become problematic when the network scales to millions of nodes. One natural solution is to sample the network, but it often leads to poor approximation. A randomized algorithm achieving better approximation is presented in [48].

2.3   Community Structures

Social networks demonstrate a strong community effect. That is, a group of people tend to interact with each other more than those outside the group. To measure the community effect, one related concept is *transitivity*. In a simple form, friends of a friend are likely to be friends as well. *Clustering coefficient* is proposed specifically to measure the transitivity, the probability of connections between one vertex's neighboring friends.

Definition 2.1 (Clustering Coefficient). *Suppose a node v_i has d_i neighbors, and there are k_i edges among these neighbors, then the clustering coefficient* *is*

$$C_i = \begin{cases} \frac{k_i}{d_i \times (d_i - 1)/2} & d_i > 1 \\ 0 & d_i = 0 \; or \; 1 \end{cases} \tag{2.3}$$

The denominator is essentially the possible number of edges between the neighbors. Take the network in Figure 16.2 as an example. Node v_1 has 5 neighbors v_2, v_3, v_4, v_5, and v_6. Among these neighbors, there are 3 edges (dashed lines) (v_2, v_3), (v_4, v_6) and (v_5, v_6). Hence, the clustering coefficient of v_1 is $3/10$. Alternatively, clustering coefficient can also be equally defined as:

$$C_i = \frac{\text{number of triangles connected to node } v_i}{\text{number of connected triples centered on node } v_i} \tag{2.4}$$

where a triple is a tuple $(v_i, \{v_j, v_k\})$ such that $(v_i, v_j) \in E$, $(v_i, v_k) \in E$, and the flanking nodes v_j and v_k are unordered. For instance, $(v_1, \{v_3, v_6\})$ and $(v_1, \{v_6, v_3\})$ in Figure 16.2 represent the same triple centered on v_1 and there are in total 10 such triples. Triangle denotes an *unordered set* of three vertexes such that each two is connected. The triangles connected to node v_1 are $\{v_1, v_2, v_3\}$, $\{v_1, v_4, v_6\}$ and $\{v_1, v_5, v_6\}$, so $C_1 = 3/10$.

To measure the community structure of a network, two commonly used global clustering coefficients are defined by extending the definition of Eqs. (2.3) and (2.4), respectively.

$$C = \sum_{i=1}^{n} C_i / n \tag{2.5}$$

$$
\begin{aligned}
C &= \frac{\sum_{i=1}^{n} \text{number of triangles connected to node } v_i}{\sum_{i=1}^{n} \text{number of connected triples centered on node } v_i} \\
&= \frac{3 \times \text{number of triangles in the network}}{\text{number of connected triples of nodes}}
\end{aligned}
\tag{2.6}
$$

Eq. (2.5) yields high variance for nodes with less degrees. E.g., for nodes with degree 2, C_i is either 0 or 1. It is commonly used for numerical study [62] whereas Eq. (2.6) is used more for analytical study. In the toy example, the global clustering coefficients based the two formulas are 0.7810 and 0.5217 respectively.

The computation of global clustering coefficient relies on exact counting of triangles in the network which can be computationally expensive [5, 51, 30]. One efficient exact counting method without huge memory requirement is the simple node-iterator (or edge-iterator) algorithm, which essentially traverse all the nodes (edges) to compute the number of triangles connected to each node (edge). Some approximation algorithms are proposed, which require one single pass [13] or multiple passes [9] of the huge edge file. It can be verified that the number of triangles is proportional to the sum of the cube of eigenvalues of the adjacency matrix [59]. Thus, using the few top eigenvalues to approximate the number is also viable.

While clustering coefficient and transitivity concentrate on microscopic view of community effect, communities of macroscopic view also demonstrate intriguing patterns. In real-world networks, a giant component tends to form with the remaining being singletons and minor communities [28]. Even within the giant component, tight but almost trivial communities (connecting to the rest of the network through one or two edges) at very small scales are often observed. Most social networks lack well-defined communities in a large scale [35]. The communities gradually "blend in" the rest of the network as their size expands.

2.4   Graph Generators

As large scale networks demonstrate similar patterns, one interesting question is: what is the innate mechanism of these networks? A variety of graph and network generators have been proposed such that these patterns can be reproduced following some simple rules. The classical model is the random graph model [20], in which the edges connecting nodes are generated probabilistically via flipping a biased coin. It yields beautiful mathematical properties but does not capture the common patterns discussed above. Recently, Watts and Strogatz proposed a model mixing the random graph model and a regular lattice structure, producing small diameter and high clustering effect [62]; a preferential attachment process is presented in [8] to explain the power law distribution exhibited in real-world networks. These two pieces of seminal work stir renewed enthusiasm researching on pursing graph generators to capture some other network patterns. For instance, the availability of dynamic network data enables the possibility to study how a network evolves and how its fundamental network properties vary over time. It is observed that many growing networks are becoming denser with average degrees increasing. Meanwhile, the effective diameter shrinks with the growth of a network [34]. These properties cannot be explained by the aforementioned network models. Thus, a forest-fire model is proposed. While many models focus on global patterns present in networks, the microscopic property of networks is also calling for alternative explanations [32].   Please refer to surveys [40, 14] for more detailed discussion.

3.   Community Detection

As mentioned above, social networks demonstrate strong community effect. The actors in a network tend to form groups of closely-knit connections. The groups are also called communities, clusters, cohesive subgroups or modules in different context. Roughly speaking, individuals interact more frequently within a group than between groups. Detecting cohesive groups in a social network (also termed as *community detection*) remains a core problem in social network analysis. Finding out these groups also helps for other related tasks of social network analysis. Various definitions and approaches are exploited for community detection. Briefly, the criteria of groups fall into four categories: node-centric, group-centric, network-centric, and hierarchy-centric. Below, we elucidate some representative methods in each category.

3.1 Node-Centric Community Detection

Community detection based on node-centric criteria requires *each node* in a group to satisfy certain properties like mutuality, reachability, or degrees.

Groups based on Complete Mutuality. An ideal cohesive group is a *clique*. It is a maximal complete subgraph of three or more nodes all of which are adjacent to each other. For a directed graph, [29] shows that with very high probability, there should exist a complete bipartite in a community. These complete bipartites work as a core for a community. The authors propose to extract an (i, j)-bipartite of which all the i nodes are connected to another j nodes in the graph.

Unfortunately, it is NP-hard to find out the maximum clique in a network. Even an approximate solution can be difficult to find. One brute-force approach to enumerate the cliques is to traverse of all the nodes in the network. For each node, check whether there is any clique of a specified size that contains the node. Then the clique is collected and the node is removed from future consideration. This works for small scale networks, but becomes impractical for large-scale networks. The main strategy to address this challenge is to effectively prune those nodes and edges that are unlikely to be contained in a maximal clique or a complete bipartite.

An algorithm to identify the maximal clique in large social networks is explored in [1]. Each time, a subset of the network is sampled. Based on this smaller set, a clique can be found in a greedy-search manner. The maximal clique found on the subset (say, it contains q nodes) serves as the lower bound for pruning. That is, the maximal clique should contain at least q members, so the nodes with degree less than q can be removed. This pruning process is repeated until the network is reduced to a reasonable size and the maximal clique can be identified.

A similar strategy can be applied to find complete bipartites. A subtle difference of the work in [29] is that it aims to find the complete bipartite of a fixed size, say an (i, j)-bipartite. Iterative pruning is applied to remove those nodes with their out-degree less than j and their in-degree less than i. After this initial pruning, an inclusion-exclusion pruning strategy is applied to either eliminate a node from concentration or discover an (i, j)-bipartite. The authors

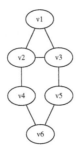

cliques: $\{v_1, v_2, v_3\}$
2-cliques: $\{v_1, v_2, v_3, v_4, v_5\}, \{v_2, v_3, v_4, v_5, v_6\}$
2-clans: $\{v_2, v_3, v_4, v_5, v_6\}$
2-clubs: $\{v_1, v_2, v_3, v_4\}, \{v_1, v_2, v_3, v_5\}, \{v_2, v_3, v_4, v_5, v_6\}$

Figure 16.3. A toy example (reproduced from [61])

proposed to focus first on nodes that are of out-degree j (or of in-degree i) . It is easy to check whether a node belongs to an (i, j)-bipartite by examining whether all its connected nodes have enough connections. So either one node is purged or an (i, j)-bipartite is identified.

Note that clique (or complete bipartite) is a *very* strict definition, and rarely can it be observed in a large size in real-world social networks. This structure is very unstable as the removal of any edge could break this definition. Practitioners typically use identified maximal cliques (or maximal complete bipartites) as cores or seeds for subsequent expansion for a community [47, 29]. Alternatively, other forms of substructures close to a clique are identified as communities as discussed next.

Groups based on Reachability.   This type of community considers the reachability between actors. In the extreme case, two nodes can be considered as belonging to one community if there exists a path between the two nodes. Thus each component[2] is a community. This can be efficiently done in $O(n + m)$ time. However, in real-world networks, a giant component tends to form while many others are singletons and minor communities [28]. For those minorities, it is straightforward to identify them via connected components. More efforts are required to find communities in the giant component.

Conceptually, there should be a short path between any two nodes in a group. Several well studied structures in social science are:

- k-*clique* is a maximal subgraph in which the largest geodesic distance between any two nodes is no greater than k. That is,

$$d(i, j) \leq k \ \ \forall v_i, v_j \in V_s$$

---

[2]Connected nodes form a component.

Note that the geodesic distance is defined on the original network. Thus, the geodesic is not necessarily included in the group structure. So a k-clique may have a diameter greater than k or even become disconnected.

- k-*clan* is a k-clique in which the geodesic distance $d(i,j)$ between all nodes in the subgraph is no greater than k for all paths within the subgraph. A k-clan must be a k-clique, but it is not so vice versa. For instance, $\{v_1, v_2, v_3, v_4, v_5\}$ in Figure 16.3 is a 2-clique, but not 2-clan as the geodesic distance of v_4 and v_5 is 2 in the original network, but 3 in the subgraph.

- k-*club* restricts the geodesic distance within the group to be no greater than k. It is a maximal substructure of diameter k.

All k-clans are k-cliques, and k-clubs are normally contained within k-cliques. These substructures are useful in the study of information diffusion and influence propagation.

Groups based on Nodal Degrees. This requires actors within a group to be adjacent to a relatively large number of group members. Two commonly studied substructures are:

- k-*plex* - It is a maximal subgraph containing n_s nodes, in which each node is adjacent to no fewer than $n_s - k$ nodes in the subgraph. In other words, each node may have no ties up to k group members. A k-plex becomes a clique when $k = 1$.

- k-*core* - It is a substructure that each node (v_i) connects to at least k members within the group, i.e.,

$$d_s(i) \geq k \ \forall v_i \in V_s$$

The definitions of k-plex and k-core are actually complementary. A k-plex with group size equal to n_s, is also a $(n_s - k)$-core. The structures above are normally robust to the removal of edges in the subgraph. Even if we miss one or two edges, the subgraph is still connected. Solving the k-plex and earlier k-clan problems requires involved combinatorial optimization [37]. As mentioned in the previous section, the nodal degree distribution in a social network follows power law, i.e., few nodes with many degrees and many others with few degrees. However, groups based on nodal degrees require all the nodes of a group to have at least a certain number of degrees, which is not very suitable for the analysis of large-scale networks where power law is a norm.

Groups based on Within-Outside Ties. This kind of group forces each node to have more connections to nodes that are within the group than to those outside the group.

- *LS sets*: A set of nodes V_s in a social network is an LS set iff *any of its proper subsets* has more ties to its complement within V_s than to those outside V_s. An important property which distinguishes LS sets from previous cliques, k-cliques and k-plexes, is that any two LS sets are either disjoint or one LS set contains the other [10]. This implies that a hierarchical series of LS sets exist in a network. However, due the strict constraint, large-size LS sets are rarely found in reality, leading to its limited usage for analysis. An alternative generalization is Lambda sets.

- *Lambda sets*: The group should be difficult to disconnect by the removal of edges in the subgraph. Let $\lambda(v_i, v_j)$ denote the number of edges that must be removed from the graph in order to disconnect any two nodes v_i and v_j. A set is called lambda set if

$$\lambda(v_i, v_j) > \lambda(v_k, v_\ell) \quad \forall v_i, v_j, v_k \in V_s, \ \forall v_\ell \in V \setminus V_s$$

  It is a maximal subset of actors who have more edge-independent paths connecting them to each other than to outsiders. The minimum connectivity among the members of a lambda set is denoted as $\lambda(G_s)$.

There are more lambda sets in reality than LS sets, hence it is more practical to use lambda sets in network analysis. Akin to LS sets, lambda sets are also disjoint at an edge-connectivity level λ. To obtain a hierarchical structure of lambda sets, one can adopt a two-step algorithm:

- Compute the edge connectivity between any pair of nodes in the network via "maximum-flow, minimum-cut" algorithms.

- Starting from the highest edge connectivity, gradually join nodes such that $\lambda(v_i, v_j) \geq k$.

Since the lambda sets at each level (k) is disjoint, this generates a hierarchical structure of the nodes. Unfortunately, the first step is computationally prohibitive for large-scale networks as the minimum-cut computation involves each pair of nodes.

3.2   Group-Centric Community Detection

All of the above group definitions are node centric, i.e. each node in the group has to satisfy certain properties. Group-centric criteria, instead, consider the connections inside a group as whole. It is acceptable to have some nodes in the group to have low connectivity as long as the group overall satisfies certain requirements. One such example is *density-based groups*. A subgraph $G_s(V_s, E_s)$ is γ-dense (also called a quasi-clique [1]) if

$$\frac{E_s}{V_s(V_s - 1)/2} \geq \gamma \tag{3.1}$$

Clearly, the quasi-clique becomes a clique when $\gamma = 1$. Note that this density-based group typically does not guarantee the nodal degree or reachbility for each node in the group. It allows the degree of different nodes to vary drastically, thus seems more suitable for large-scale networks.

In [1], the maximum γ-dense quasi-cliques are explored. A greedy algorithm is adopted to find a maximal quasi-clique. The quasi-clique is initialized with a vertex with the largest degree in the network, and then expanded with nodes that are likely to contribute to a large quasi-clique. This expansion continues until no nodes can be added to maintain the γ-density. Evidently, this greedy search for maximal quasi-clique is not optimal. So a subsequent local search procedure (GRASP) is applied to find a larger maximal quasi-clique in the local neighborhood. This procedure is able to detect a close-to-optimal maximal quasi-clique but requires the whole graph to be in main memory. To handle large-scale networks, the authors proposed to utilize the procedure above to find out the lower bound of degrees for pruning. In each iteration, a subset of edges are sampled from the network, and GRASP is applied to find a locally maximal quasi-clique. Suppose the quasi-clique is of size k, it is impossible to include in the maximal quasi-clique a node with degree less than $k\gamma$, all of whose neighbors also have their degree less than $k\gamma$. So the node and its incident edges can be pruned from the graph. This pruning process is repeated until GRASP can be applied directly to the remaining graph to find out the maximal quasi-clique.

For a directed graph like the Web, the work in [19] extends the complete-bipartite core [29] to γ-dense bipartite. (X, Y) is a γ-dense bipartite if

$$\forall x \in X, |N^+(x) \cap Y| \geq \gamma|Y| \qquad (3.2)$$
$$\forall y \in Y, |N^-(y) \cap X| \geq \gamma'|X| \qquad (3.3)$$

where γ and γ' are user provided constants. The authors derive a heuristic to efficiently prune the nodes. Due to the heuristic being used, not all satisfied communities can be enumerated. But it is able to identify some communities for a medium range of community size/density, while [29] favors to detect small communities.

3.3 Network-Centric Community Detection

Network-centric community detection has to consider the connections of the whole network. It aims to partition the actors into a number of disjoint sets. A group in this case is not defined independently. Typically, some quantitative criterion of the network partition is optimized.

Groups based on Vertex Similarity.   Vertex similarity is defined in terms of how similar the actors interact with others. Actors behaving in the same

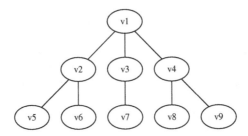

Figure 16.4. Equivalence for Social Position

role during interaction are in the same social position. The position analysis is to identify the social status and roles associated with different actors. For instance, what is the role of "wife"? What is the interaction pattern of "vice president" in a company organization? In position analysis, several concepts with decreasing strictness are studied to define two actors sharing the same social position [25]:

- **Structural Equivalence** Actors i and j are structurally equivalent, if for any actor k that $k \neq i, j$, $(i, k) \in E$ iff $(j, k) \in E$. In other words, actors i and j are connecting to exactly the same set of actors in the network. If the interaction is represented as a matrix, then rows (columns) i and j are the same except for the diagonal entries. For instance, in Figure 16.4, v_5 and v_6 are structurally equivalent. So are v_8 and v_9.

- **Automorphic equivalence** Structural equivalence requires the connections of two actors to be exactly the same, yet it is too restrictive. Automorphic equivalence allows the connections to be isomorphic. Two actors u and v are automorphically equivalent iff all the actors of G can be relabeled to form an isomorphic graph. In the diagram, $\{v_2, v_4\}$, $\{v_5, v_6, v_8, v_9\}$ are automorphically equivalent, respectively.

- **Regular equivalence** Two nodes are regularly equivalent if they have the same profile of ties with other members that are also regularly equivalent. Specifically, u and v are regularly equivalent (denoted as $u \equiv v$) iff

$$(u, a) \in E \Rightarrow \exists b \in V, \; such \; that \; (v, b) \in E \; and \; a \equiv b \qquad (3.4)$$

In the diagram, the regular equivalence results in three equivalence classes $\{v_1\}$, $\{v_2, v_3, v_4\}$, and $\{v_5, v_6, v_7, v_8, v_9\}$.

Structural equivalence is too restrictive for practical use, and no effective approach exists to scale automorphic equivalence or regular equivalence to more than thousands of actors. In addition, in large networks (say, online friends networks), the connection is very noisy. Meaningful equivalence of large scale is difficult to detect. So some simplified similarity measures ignoring the social roles are used in practice, including cosine similarity [27], Jaccard similarity [23], etc. They consider the connections as features for actors, and rely on the fact that actors sharing similar connections tend to reside within the same community. Once the similarity measure is determined, classical k-means or hierarchical clustering algorithm can be applied.

It can be time consuming to compute the similarity between each pair of actors. Thus, Gibson et al. [23] present an efficient two-level shingling algorithm for fast computation of web communities. Generally speaking, the shingling algorithm maps each vector (the connection of actors) into a constant number of "shingles". If two actors are similar, they share many shingles; otherwise, they share few. After initial shingling, each shingle is associated with a group of actors. In a similar vein, the shingling algorithm can be applied to the first-level shingles as well. So similar shingles end up sharing the same meta-shingles. Then all the actors relating to one meta-shingle form one community. This two-level shingling can be efficiently computed even for large-scale networks. Its time complexity is approximately linear to the number of edges. By contrast, normal similarity-based methods have to compute the similarity for each pair of nodes, totaling $O(n^2)$ time at least.

Groups based on Minimum-Cut.   A community is defined as a vertex subset $C \subset V$, such that $\forall v \in C$, v has at least as many edges connecting to vertices in C as it does to vertices in $V \backslash C$ [22]. Flake et al. show that the community can be found via s-t minimum cut given a source node s in the community and a sink node t outside the community as long as both ends satisfy certain degree requirement. Some variants of minimum cut like normalized cut and ratio cut can be applied to SNA as well. Suppose we have a partition of k communities $\pi = (V_1, V_2, \cdots, V_k)$, it follows that

$$\text{Ratio Cut}(\pi) = \sum_{i=1}^{k} \frac{cut(V_i, \bar{V}_i)}{|V_i|} \tag{3.5}$$

$$\text{Normalized Cut}(\pi) = \sum_{i=1}^{k} \frac{cut(V_i, \bar{V}_i)}{vol(V_i)} \tag{3.6}$$

where $vol(V_i) = \sum_{v_j \in V_i} d_j$. Both objectives attempt to minimize the number of edges between communities, yet avoid the bias of trivial-size communities like singletons. Interestingly, both formulas can be recast as an optimization problem of the following type:

$$\min_{S \in \{0,1\}^{n \times k}} Tr(S^T L S) \tag{3.7}$$

where L is the graph Laplacian (normalized Laplacian) for ratio cut (normalized cut), and $S \in \{0,1\}^{n \times k}$ is a community indicator matrix defined below:

$$S_{ij} = \begin{cases} 1 & \text{if vertex } i \text{ belongs to community } j \\ 0 & \text{otherwise} \end{cases}$$

Due to the discreteness property of S, this problem is still NP-hard. A standard way is to adopt a spectral relaxation to allow S to be continuous leading to the following trace minimization problem:

$$\min_{S \in R^{n \times k}} Tr(S^T L S) \quad s.t. \ S^T S = I \qquad (3.8)$$

It follows that S corresponds to the eigenvectors of k smallest eigenvalues (except 0) of Laplacian L. Note that a graph Laplacian always has an eigenvector 1 corresponding to the eigenvalue 0. This vector indicates all nodes belong to the same community, which is useless for community partition, thus is removed from consideration. The obtained S is essentially an approximation to the community structure. In order to obtain a disjoint partition, some local search strategy needs to be applied. An effective and widely used strategy is to apply k-means on the matrix S to find the partitions of actors.

The main computational cost with the above spectral clustering is that an eigenvector problem has to be solved. Since the Laplacian matrix is usually sparse, the eigenvectors correspond to the smallest eigenvalues can be computed in an efficient way. However, the computational cost is still $O(n^2)$, which can be prohibitive for mega-scale networks.

Groups based on Block Model Approximation. Block modeling assumes the interaction between two vertices depends only on the communities they belong to. The actors within the same community are *stochastically equivalent* in the sense that the probabilities of the interaction with all other actors are the same for actors in the same community [46, 4]. Based on this block model, one can apply classical Bayesian inference methods like EM or Gibbs sampling to perform maximum likelihood estimation for the probability of interaction as well as the community membership of each actor.

In a different fashion, one can also use matrix approximation for block models. That is, the actors in the interaction matrix can be reordered in a form such that those actors sharing the same community form a dense interaction block. Based on the stochastic assumption, it follows that the community can be identified based on interaction matrix A via the following optimization [63]:

$$\min_{S,\Sigma} \ell(A; S^T \Sigma S) \qquad (3.9)$$

Ideally, S should be an cluster indicator matrix with entry values being 0 or 1, Σ captures the strength of between-community interaction, and ℓ is the loss function. To solve the problem, spectral relaxation of S can to be adopted. If S is relaxed to be continuous, it is then similar to spectral clustering. If S is constrained to be non-negative, then it shares the same spirit as stochastic block models. This matrix approximation often resorts to numerical optimization techniques like alternating optimization or gradient methods rather than Bayesian inference.

Groups based on Modularity.   Different from other criteria, modularity is a measure which considers the degree distribution while calibrating the community structure. Consider dividing the interaction matrix A of n vertices and m edges into k non-overlapping communities. Let s_i denote the community membership of vertex v_i, d_i represents the degree of vertex i. Modularity is like a statistical test that the null model is a uniform random graph model, in which one actor connects to others with uniform probability. For two nodes with degree d_i and d_j respectively, the expected number of edges between the two in a uniform random graph model is $d_i d_j / 2m$. Modularity measures how far the interaction is deviated from a uniform random graph. It is defined as:

$$Q = \frac{1}{2m} \sum_{ij} \left[A_{ij} - \frac{d_i d_j}{2m} \right] \delta(s_i, s_j) \tag{3.10}$$

where $\delta(s_i, s_j) = 1$ if $s_i = s_j$. A larger modularity indicates denser within-group interaction. Note that Q could be negative if the vertices are split into bad clusters. $Q > 0$ indicates the clustering captures some degree of community structure.

   In general, one aims to find a community structure such that Q is maximized. While maximizing the modularity over hard clustering is proved to be NP hard [11], a spectral relaxation of the problem can be solved efficiently [42]. Let $\mathbf{d} \in Z_+^n$ be the degree vector of all nodes where Z_+^n is the set of positive numbers of n dimensionality, $S \in \{0, 1\}^{n \times k}$ be a community indicator matrix, and the modularity matrix defined as

$$B = A - \frac{\mathbf{d}\mathbf{d}^T}{2m} \tag{3.11}$$

The modularity can be reformulated as

$$Q = \frac{1}{2m} Tr(S^T B S) \tag{3.12}$$

Relaxing S to be continuous, it can be shown that the optimal S is the top-k eigenvectors of the modularity matrix B [42].

Groups based on Latent Space Model.   Latent space model [26, 50, 24] maps the actors into a latent space such that those with dense connections are likely to occupy the latent positions that are not too far away. They assume the interaction between actors depends on the positions of individuals in the latent space. A maximum likelihood estimation can be utilized to estimate the position.

3.4   Hierarchy-Centric Community Detection

Another line of community detection is to build a hierarchical structure of communities based on network topology. This facilitates the examination of communities at different granularity. There are mainly three types of hierarchical clustering: divisive, agglomerative, and structure search.

Divisive hierarchical clustering.   Divisive clustering first partitions the actors into several disjoint sets. Then each set is further divided into smaller ones until the set contains only a small number of actors (say, only 1). The key here is how to split the network into several parts. Some partition methods presented in previous section can be applied recursively to divide a community into smaller sets. One particular divisive clustering proposed for graphs is based on edge betweeness [45]. It progressively removes edges that are likely to be bridges between communities. If two communities are joined by only a few cross-group edges, then all paths through the network from nodes in one community to the other community have to pass along one of these edges. Edge betweenness is a measure to count how many shortest paths between pair of nodes pass along the edge, and this number is expected to be large for those between-group edges. Hence, progressively removing those edges with high betweenness can gradually disconnects the communities, which leads naturally to a hierarchical community structure.

Agglomerative hierarchical clustering.   Agglomerative clustering begins with each node as a separate community and merges them successively into larger communities. Modularity is used as a criterion [15] to perform hierarchical clustering. Basically, a community pair should be merged if doing so results in the largest increase of overall modularity, and the merge continues until no merge can be found to improve the modularity. It is noticed that this algorithm incurs many imbalanced merges (a large community merges with a tiny community), resulting in high computational cost [60]. Hence, the merge criterion is modified accordingly to take into consideration the size of communities. In the new scheme, communities of comparable sizes are joined first, leading to a more balanced hierarchical structure of communities and to improved efficiency.

Structure Search.   Structure search starts from a hierarchy and then searches for hierarchies that are more likely to generate the network. This idea first appears in [55] to maintain a topic taxonomy for group profiling, and then a similar idea is applied for hierarchical construction of communities in social networks. [16] defines a random graph model for hierarchies such that two actors are connected based on the interaction probability of their least common ancestor node in the hierarchy. The authors generate a sequence of hierarchies via local changes of the network and accept it proportional to the likelihood. The final hierarchy is the consensus of a set of comparable hierarchies. The bottleneck with structure search approach is its huge search space. A challenge is how to scale it to large networks.

4.   Community Structure Evaluation

In the previous section, we describe some representative approaches for community detection. Part of the reason that there are so many assorted definitions and methods, is that there is no clear ground truth information about the community structure in a real world network. Therefore, different community detection methods are developed from various applications of specific needs. In this section, we depict strategies commonly adopted to evaluate identified communities in order to facilitate the comparison of different community detection methods.

Depending on network information, different strategies can be taken for comparison:

- Groups with self-consistent definitions. Some groups like cliques, k-cliques, k-clans, k-plexes and k-cores can be examined immediately once a community is identified. If the goal of community detection is to enumerate all the desirable substructures of this sort, the total number of retrieved communities can be compared for evaluation.

- Networks with ground truth. That is, the community membership for each actor is known. This is an ideal case. This scenario hardly presents itself in real-world large-scale networks. It usually occurs for evaluation on synthetic networks (generated based on predefined community structures) [56] or a tiny network [42]. To compare the ground truth with identified community structures, visualization can be intuitive and straightforward [42]. If the number of communities is small (say 2 or 3 communities), it is easy to determine a one-to-one mapping between the identified communities and the ground truth. So conventional classification measures such as error-rate, F1-measure can be used. However, when there are a plurality of communities, it may not be clear what a correct mapping is. Instead, normalized mutual information (NMI) [52] can be adopted to measure the difference of two partitions:

$$
NMI(\pi^a, \pi^b) = \frac{\sum_{h=1}^{k^{(a)}} \sum_{\ell=1}^{k^{(b)}} n_{h,\ell} \log \left(\frac{n \cdot n_{h,l}}{n_h^{(a)} \cdot n_\ell^{(b)}} \right)}{\sqrt{\left(\sum_{h=1}^{k^{(a)}} n_h^{(a)} \log \frac{n_h^a}{n} \right) \left(\sum_{\ell=1}^{k^{(b)}} n_\ell^{(b)} \log \frac{n_\ell^b}{n} \right)}}
\tag{4.1}
$$

where π^a, π^b denotes two different partitions of communities. $n_{h,\ell}$, n_h^a, n_ℓ^b are, respectively, the number of actors simultaneously belonging to the h-th community of π^a and ℓ-th community of π^b, the number of actors in the h-th community of partition π^a, and the number of actors in the ℓ-th community of partition π^b. NMI is a measure between 0 and 1 and equals to 1 when π^a and π^b are the same.

- Networks with semantics. Some networks come with semantic or attribute information of the nodes and connections. In this case, the identified communities can be verified by human subjects to check whether it is consistent with the semantics. For instance, whether the community identified in the Web is coherent to a shared topic [22, 15], whether the clustering of coauthorship network captures the research interests of individuals. This evaluation approach is applicable when the community is reasonably small. Otherwise, selecting the top-ranking actors as representatives of a community is commonly used. This approach is qualitative and hardly can it be applied to all communities in a large network, but it is quite helpful for understanding and interpretation of community patterns.

- Networks without ground truth or semantic information. This is the most common situation, yet it requires objective evaluation most. Normally, one resorts to some quantitative measures for evaluation. One common measure being used is modularity [43]. Once we have a partition, we can compute its modularity. The method with higher modularity wins. Another comparable approach is to use the identified community as a base for link prediction, i.e., two actors are connected if they belong to the same community. Then, the predicted network is compared with the true network, and the deviation is used to calibrate the community structure. Since social network demonstrates strong community effect, a better community structure should predict the connections between actors more accurately. This is essentially checking how far the true network deviates from a block model based on the identified communities.

5.   Research Issues

We have now described some graph mining techniques for community detection, a basic task in social network analysis. It is evident that community detection, though it has been studied for many years, is still in pressing need for effective graph mining techniques for large-scale complex networks. We present some key problems for further research:

- Scalability. One major bottleneck with community detection is scalability. Most existing approaches require a combinatorial optimization formulation for graph mining or eigenvalue problem of the network. Some alternative techniques are being developed to overcome the barrier, including local clustering [49] and multi-level methods [2]. How to find out meaningful communities efficiently and develop scalable methods for mega-scale networks remains a big challenge.

- Community evolution. Most networks tend to evolve over time. How to effectively capture the community evolution in dynamic social networks [56]? Can we find the members which act like the backbone of communities? How does this relate to the influence of an actor? What are the determining factors that result in community evolution [7]? How to profile the characteristics of evolving communities[55]?

- Usage of communities. How to utilize these communities for further social network analysis needs more exploration, especially for those emerging tasks in social media like classification [53], ranking, finding influential actors [3], viral marketing, link prediction, etc. Community structures of a social network can be exploited to accomplish these tasks.

- Utility of patterns. As we have introduced, large-scale social networks demonstrate some distinct patterns that are not usually observable in small networks. However, most existing community detection methods do not take advantage of the patterns in their detection process. How to utilize these patterns with various community detection methods remains unclear. More research should be encouraged in this direction.

- Heterogeneous networks. In reality, multiple relationships can exist between individuals. Two persons can be friends and colleagues at the same time. In online social media, people interact with each other in a variety of forms resulting in a multi-relational (multi-dimensional) network [54]. Some systems also involve multiple types of entities to interact with each other, leading to multi-mode networks [56]. Analysis of these heterogeneous networks involving heterogeneous actors or relations demands further investigation.

The prosperity of social media and emergence of large-scale complex networks poses many challenges and opportunities to graph mining and social network analysis. The development of graph mining techniques can facilitate the analysis of networks in a much larger scale, and help understand human social behaviors. Meanwhile, the common patterns and emerging tasks in social network analysis continually surprise us and stimulate advanced graph mining techniques. In this chapter, we point out the converging trend of the two fields and expect its healthy acceleration in the near future.

References

[1] J. Abello, M. G. C. Resende, and S. Sudarsky. Massive quasi-clique detection. In *LATIN*, pages 598–612, 2002.

[2] A. Abou-Rjeili and G. Karypis. Multilevel algorithms for partitioning power-law graphs. pages 10 pp.–, April 2006.

[3] N. Agarwal, H. Liu, L. Tang, and P. S. Yu. Identifying the influential bloggers in a community. In *WSDM '08: Proceedings of the international conference on Web search and web data mining*, pages 207–218, New York, NY, USA, 2008. ACM.

[4] E. Airodi, D. Blei, S. Fienberg, and E. P. Xing. Mixed membership stochastic blockmodels. *J. Mach. Learn. Res.*, 9:1981–2014, 2008.

[5] N. Alon, R. Yuster, and U. Zwick. Finding and counting given length cycles. *Algorithmica*, 17(3):209–223, 1997.

[6] C. Anderson. *The Long Tail: why the future of business is selling less of more*. 2006.

[7] L. Backstrom, D. Huttenlocher, J. Kleinberg, and X. Lan. Group formation in large social networks: membership, growth, and evolution. In *KDD '06: Proceedings of the 12th ACM SIGKDD international conference on Knowledge discovery and data mining*, pages 44–54, New York, NY, USA, 2006. ACM.

[8] A.-L. Barabasi and R. Albert. Emergence of Scaling in Random Networks. *Science*, 286(5439):509–512, 1999.

[9] L. Becchetti, P. Boldi, C. Castillo, and A. Gionis. Efficient semi-streaming algorithms for local triangle counting in massive graphs. In *KDD '08: Proceeding of the 14th ACM SIGKDD international conference on Knowledge discovery and data mining*, pages 16–24, New York, NY, USA, 2008. ACM.

[10] S. P. Borgatti, M. G. Everett, and P. R. Shirey. Ls sets, lambda sets and other cohesive subsets. *Social Networks*, 12:337–357, 1990.

[11] U. Brandes, D. Delling, M. Gaertler, R. Gorke, M. Hoefer, Z. Nikoloski, and D. Wagner. Maximizing modularity is hard. *Arxiv preprint physics/0608255*, 2006.

[12] T. Bu and D. Towsley. On distinguishing between internet power law topology generators. In *Twenty-First Annual Joint Conference of the IEEE Computer and Communications Societies*, volume 2, pages 638–647 vol.2, 2002.

[13] L. S. Buriol, G. Frahling, S. Leonardi, A. Marchetti-Spaccamela, and C. Sohler. Counting triangles in data streams. In *PODS '06: Proceedings of the twenty-fifth ACM SIGMOD-SIGACT-SIGART symposium on Principles of database systems*, pages 253–262, New York, NY, USA, 2006. ACM.

[14] D. Chakrabarti and C. Faloutsos. Graph mining: Laws, generators, and algorithms. *ACM Comput. Surv.*, 38(1):2, 2006.

[15] A. Clauset, M. Mewman, and C. Moore. Finding community structure in very large networks. *Arxiv preprint cond-mat/0408187*, 2004.

[16] A. Clauset, C. Moore, and M. E. J. Newman. Hierarchical structure and the prediction of missing links in networks. *Nature*, 453:98–101, 2008.

[17] A. Clauset, C. R. Shalizi, and M. E. J. Newman. Power-law distributions in empirical data. *arXiv*, 706, 2007.

[18] J. Diesner, T. L. Frantz, and K. M. Carley. Communication networks from the enron email corpus "it's always about the people. enron is no different". *Comput. Math. Organ. Theory*, 11(3):201–228, 2005.

[19] Y. Dourisboure, F. Geraci, and M. Pellegrini. Extraction and classification of dense communities in the web. In *WWW '07: Proceedings of the 16th international conference on World Wide Web*, pages 461–470, New York, NY, USA, 2007. ACM.

[20] P. Erdøs and A. Renyi. On the evolution of random graphs. *Publ. Math. Inst. Hung. Acad. Sci*, 5:17–61, 1960.

[21] M. Faloutsos, P. Faloutsos, and C. Faloutsos. On power-law relationships of the internet topology. In *SIGCOMM '99: Proceedings of the conference on Applications, technologies, architectures, and protocols for computer communication*, pages 251–262, New York, NY, USA, 1999. ACM.

[22] G. W. Flake, S. Lawrence, and C. L. Giles. Efficient identification of web communities. In *KDD '00: Proceedings of the sixth ACM SIGKDD international conference on Knowledge discovery and data mining*, pages 150–160, New York, NY, USA, 2000. ACM.

[23] D. Gibson, R. Kumar, and A. Tomkins. Discovering large dense subgraphs in massive graphs. In *VLDB '05: Proceedings of the 31st inter-*

national conference on Very large data bases, pages 721–732. VLDB Endowment, 2005.

[24] M. S. Handcock, A. E. Raftery, and J. M. Tantrum. Model-based clustering for social networks. *Journal Of The Royal Statistical Society Series A*, 127(2):301–354, 2007.

[25] R. Hanneman and M. Riddle. *Introduction to Social Network Methods.* http://faculty.ucr.edu/ hanneman/, 2005.

[26] P. D. Hoff and M. S. H. Adrian E. Raftery. Latent space approaches to social network analysis. *Journal of the American Statistical Association*, 97(460):1090–1098, 2002.

[27] J. Hopcroft, O. Khan, B. Kulis, and B. Selman. Natural communities in large linked networks. In *KDD '03: Proceedings of the ninth ACM SIGKDD international conference on Knowledge discovery and data mining*, pages 541–546, New York, NY, USA, 2003. ACM.

[28] R. Kumar, J. Novak, and A. Tomkins. Structure and evolution of online social networks. In *KDD '06: Proceedings of the 12th ACM SIGKDD international conference on Knowledge discovery and data mining*, pages 611–617, New York, NY, USA, 2006. ACM.

[29] R. Kumar, P. Raghavan, S. Rajagopalan, and A. Tomkins. Trawling the web for emerging cyber-communities. *Comput. Netw.*, 31(11-16):1481–1493, 1999.

[30] M. Latapy. Main-memory triangle computations for very large (sparse (power-law)) graphs. *Theor. Comput. Sci.*, 407(1-3):458–473, 2008.

[31] J. Leskovec, L. A. Adamic, and B. A. Huberman. The dynamics of viral marketing. In *EC '06: Proceedings of the 7th ACM conference on Electronic commerce*, pages 228–237, New York, NY, USA, 2006. ACM.

[32] J. Leskovec, L. Backstrom, R. Kumar, and A. Tomkins. Microscopic evolution of social networks. In *KDD '08: Proceeding of the 14th ACM SIGKDD international conference on Knowledge discovery and data mining*, pages 462–470, New York, NY, USA, 2008. ACM.

[33] J. Leskovec and E. Horvitz. Planetary-scale views on a large instant-messaging network. In *WWW '08: Proceeding of the 17th international conference on World Wide Web*, pages 915–924, New York, NY, USA, 2008. ACM.

[34] J. Leskovec, J. Kleinberg, and C. Faloutsos. Graph evolution: Densification and shrinking diameters. *ACM Trans. Knowl. Discov. Data*, 1(1):2, 2007.

[35] J. Leskovec, K. J. Lang, A. Dasgupta, and M. W. Mahoney. Statistical properties of community structure in large social and information net-

works. In *WWW '08: Proceeding of the 17th international conference on World Wide Web*, pages 695–704, New York, NY, USA, 2008. ACM.

[36] J. Leskovec, M. McGlohon, C. Faloutsos, N. Glance, and M. Hurst. Cascading behavior in large blog graphs. In *SIAM International Conference on Data Mining (SDM 2007)*, 2007.

[37] B. McClosky and I. V. Hicks.   Detecting cohesive groups. http://www.caam.rice.edu/ ivhicks/CokplexAlgorithmPaper.pdf, 2009.

[38] A. Mislove, M. Marcon, K. P. Gummadi, P. Druschel, and B. Bhattacharjee. Measurement and analysis of online social networks. In *IMC '07: Proceedings of the 7th ACM SIGCOMM conference on Internet measurement*, pages 29–42, New York, NY, USA, 2007. ACM.

[39] A. A. Nanavati, S. Gurumurthy, G. Das, D. Chakraborty, K. Dasgupta, S. Mukherjea, and A. Joshi. On the structural properties of massive telecom call graphs: findings and implications. In *CIKM '06: Proceedings of the 15th ACM international conference on Information and knowledge management*, pages 435–444, New York, NY, USA, 2006. ACM.

[40] M. Newman. The structure and function of complex networks. *SIAM Review*, 45:167–256, 2003.

[41] M. Newman. Power laws, Pareto distributions and Zipf's law. *Contemporary physics*, 46(5):323–352, 2005.

[42] M. Newman. Finding community structure in networks using the eigenvectors of matrices. *Physical Review E (Statistical, Nonlinear, and Soft Matter Physics)*, 74(3), 2006.

[43] M. Newman. Modularity and community structure in networks. *PNAS*, 103(23):8577–8582, 2006.

[44] M. Newman, A.-L. Barabasi, and D. J. Watts, editors. *The Structure and Dynamics of Networks*. 2006.

[45] M. Newman and M. Girvan. Finding and evaluating community structure in networks. *Physical Review E*, 69:026113, 2004.

[46] K. Nowicki and T. A. B. Snijders. Estimation and prediction for stochastic blockstructures. *Journal of the American Statistical Association*, 96(455):1077–1087, 2001.

[47] G. Palla, I. Derenyi, I. Farkas, and T. Vicsek. Uncovering the overlapping community structure of complex networks in nature and society. *Nature*, 435:814–818, 2005.

[48] C. R. Palmer, P. B. Gibbons, and C. Faloutsos. ANF: a fast and scalable tool for data mining in massive graphs. In *KDD '02: Proceedings of the eighth ACM SIGKDD international conference on Knowledge discovery and data mining*, pages 81–90, New York, NY, USA, 2002. ACM.

[49] S. Papadopoulos, A. Skusa, A. Vakali, Y. Kompatsiaris, and N. Wagner. Bridge bounding: A local approach for efficient community discovery in complex networks. Feb 2009.

[50] P. Sarkar and A. W. Moore. Dynamic social network analysis using latent space models. *SIGKDD Explor. Newsl.*, 7(2):31–40, 2005.

[51] T. Schank and D. Wagner. Finding, counting and listing all triangles in large graphs, an experimental study. In *Workshop on Experimental and Efficient Algorithms*, 2005.

[52] A. Strehl and J. Ghosh. Cluster ensembles — a knowledge reuse framework for combining multiple partitions. *J. Mach. Learn. Res.*, 3:583–617, 2003.

[53] L. Tang and H. Liu. Relational learning via latent social dimensions. In *KDD '09: Proceeding of the 15th ACM SIGKDD international conference on Knowledge discovery and data mining*, 2009.

[54] L. Tang and H. Liu. Uncovering cross-dimension group structures in multi-dimensional networks. In *SDM workshop on Analysis of Dynamic Networks*, 2009.

[55] L. Tang, H. Liu, J. Zhang, N. Agarwal, and J. J. Salerno. Topic taxonomy adaptation for group profiling. *ACM Trans. Knowl. Discov. Data*, 1(4):1–28, 2008.

[56] L. Tang, H. Liu, J. Zhang, and Z. Nazeri. Community evolution in dynamic multi-mode networks. In *KDD '08: Proceeding of the 14th ACM SIGKDD international conference on Knowledge discovery and data mining*, pages 677–685, New York, NY, USA, 2008. ACM.

[57] S. Tauro, C. Palmer, G. Siganos, and M. Faloutsos. A simple conceptual model for the internet topology. In *Global Telecommunications Conference*, volume 3, pages 1667–1671, 2001.

[58] J. Travers and S. Milgram. An experimental study of the small world problem. *Sociometry*, 32(4):425–443, 1969.

[59] C. E. Tsourakakis. Fast counting of triangles in large real networks without counting: Algorithms and laws. *IEEE International Conference on Data Mining*, 0:608–617, 2008.

[60] K. Wakita and T. Tsurumi. Finding community structure in mega-scale social networks: [extended abstract]. In *WWW '07: Proceedings of the 16th international conference on World Wide Web*, pages 1275–1276, New York, NY, USA, 2007. ACM.

[61] S. Wasserman and K. Faust. *Social Network Analysis: Methods and Applications*. Cambridge University Press, 1994.

[62] D. J. Watts and S. H. Strogatz. Collective dynamics of 'small-world' networks. *Nature*, 393:440–442, 1998.

[63] K. Yu, S. Yu, and V. Tresp. Soft clsutering on graphs. In *NIPS*, 2005.

Chapter 17

SOFTWARE-BUG LOCALIZATION WITH GRAPH MINING

Frank Eichinger
Institute for Program Structures and Data Organization (IPD)
Universität Karlsruhe (TH), Germany
eichinger@ipd.uka.de

Klemens Böhm
Institute for Program Structures and Data Organization (IPD)
Universität Karlsruhe (TH), Germany
boehm@ipd.uka.de

Abstract   In the recent past, a number of frequent subgraph mining algorithms has been proposed They allow for analyses in domains where data is naturally graph-structured. However, caused by scalability problems when dealing with large graphs, the application of graph mining has been limited to only a few domains. In software engineering, debugging is an important issue. It is most challenging to localize bugs automatically, as this is expensive to be done manually. Several approaches have been investigated, some of which analyze traces of repeated program executions. These traces can be represented as call graphs. Such graphs describe the invocations of methods during an execution. This chapter is a survey of graph mining approaches for bug localization based on the analysis of dynamic call graphs. In particular, this chapter first introduces the subproblem of reducing the size of call graphs, before the different approaches to localize bugs based on such reduced graphs are discussed. Finally, we compare selected techniques experimentally and provide an outlook on future issues.

Keywords: Software Bug Localization, Program Call Graphs

C.C. Aggarwal and H. Wang (eds.), *Managing and Mining Graph Data*,
Advances in Database Systems 40, DOI 10.1007/978-1-4419-6045-0_17,
© Springer Science+Business Media, LLC 2010

1. Introduction

Software quality is a huge concern in industry. Almost any software contains at least some minor bugs after being released. In order to avoid bugs, which incur significant costs, it is important to find and fix them before the release. In general, this results in devoting more resources to quality assurance. Software developers usually try to find and fix bugs by means of in-depth code reviews, along with testing and classical debugging. Locating bugs is considered to be the most time consuming and challenging activity in this context [6, 20, 24, 26] where the resources available are limited. Therefore, there is a need for semi-automated techniques guiding the debugging process [34]. If a developer obtains some hints where bugs might be localized, debugging becomes more efficient.

Research in the field of software reliability has been extensive, and various techniques have been developed addressing the identification of defect-prone parts of software. This interest is not limited to software-engineering research. In the machine-learning community, automated debugging is considered to be one of the ten most challenging problems for the next years [11]. So far, no bug localization technique is perfect in the sense that it is capable of discovering any kind of bug. In this chapter, we look at a relatively new class of bug localization techniques, the analysis of *call graphs* with *graph-mining* techniques. It can be seen as an approach orthogonal to and complementing existing techniques.

Graph mining, or more specifically *frequent subgraph mining*, is a relatively young discipline in data mining. As described in the other chapters of this book, there are many different techniques as well as numerous applications for graph mining. Probably the most prominent application is the analysis of chemical molecules. As the NP-complete problem of subgraph isomorphism [16] is an inherent part of frequent subgraph mining algorithms, the analysis of molecules benefits from the relatively small size of most of them. Compared to the analysis of molecular data, software-engineering artifacts are typically mapped to graphs that are much larger. Consequently, common graph-mining algorithms do not scale for these graphs. In order to make use of call graphs which reflect the invocation structure of specific program executions, it is key to deploy a suitable *call-graph-reduction* technique. Such techniques help to alleviate the scalability problems to some extent and allow to make use of graph-mining algorithms in a number of cases. As we will demonstrate, such approaches work well in certain cases, but some challenges remain. Besides scalability issues that are still unsolved, some call-graph-reduction techniques lead to another challenge: They introduce edge weights representing call frequencies. As graph-mining research has concentrated on structural and categorical domains, rather than on quantitative weights, we are not aware of any algorithm specialized in mining *weighted graphs*. Though this chapter presents a technique to analyze graphs with weighted edges, the technique is a composition of established algorithms rather than a universal weighted graph mining algorithm. Thus, besides mining large graphs, weighted graph mining is a further challenge for graph-mining research driven by the field of software engineering.

The remainder of this chapter is structured as follows: Section 2 introduces some basic principles of call graphs, bugs, graph mining and bug localization with such graphs. Section 3 gives an overview of related work in software engineering employing data-analysis techniques. Section 4 discusses different call-graph-reduction techniques. The different bug-localization approaches are presented and compared in Section 5 and Section 6 concludes.

2. Basics of Call Graph Based Bug Localization

This section introduces the concept of dynamic call graphs in Subsection 2.1. It presents some classes of bugs in Subsection 2.2 and Subsection 2.3 explains how bug localization with call graphs works in principle. A brief overview of key aspects of graph and tree mining in the context of this chapter is given in Subsection 2.4.

2.1 Dynamic Call Graphs

Call graphs are either *static* or *dynamic* [17]. A *static call graph* [1] can be obtained from the source code. It represents all methods[1] of a program as nodes and all possible method invocations as edges. *Dynamic call graphs* are of importance in this chapter. They represent an execution of a particular program and reflect the actual invocation structure of the execution. Without any further treatment, a call graph is a *rooted ordered tree*. The main-method of a program usually is the root, and the methods invoked directly are its children. Figure 17.1a is an abstract example of such a call graph where the root Node a represents the main-method.

Unreduced call graphs typically become very large. The reason is that, in modern software development, dedicated methods typically encapsulate every single functionality. These methods call each other frequently. Furthermore, iterative programming is very common, and methods calling other methods occur within loops, executed thousands of times. Therefore, the execution of even a small program lasting some seconds often results in call graphs consisting of millions of edges.

The size of call graphs prohibits a straightforward mining with state-of-the-art graph-mining algorithms. Hence, a reduction of the graphs which com-

---

[1]In this chapter, we use *method* interchangeably with *function*.

presses the graphs significantly but keeps the essential properties of an individual execution is necessary. Section 4 describes different reduction techniques.

2.2   Bugs in Software

In the software-engineering literature, there is a number of different definitions of *bugs, defects, errors, failures, faults* and the like. For the purpose of this chapter, we do not differentiate between them. It is enough to know that a bug in a program execution manifests itself by producing some other results than specified or by leading to some unexpected runtime behavior such as crashes or non-terminating runs. In the following, we introduce some types of bugs which are particularly interesting in the context of call graph based bug localization.

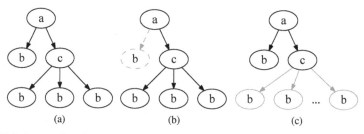

Figure 17.1. An unreduced call graph, a call graph with a structure affecting bug, and a call graph with a frequency affecting bug.

- **Crashing and non-crashing bugs:** *Crashing bugs* lead to an unexpected termination of the program. Prominent examples include null pointer exceptions and divisions by zero. In many cases, e.g., depending on the programming language, such bugs are not hard to find: A stack trace is usually shown which gives hints where the bug occurred. Harder to cope with are *non-crashing bugs*, i.e., failures which lead to faulty results without any hint that something went wrong during the execution. As non-crashing bugs are hard to find, all approaches to discover bugs with call-graph mining focus on them and leave aside crashing bugs.

- **Occasional and non-occasional bugs:** *Occasional bugs* are bugs which occur with some but not with any input data. Finding occasional bugs is particularly difficult, as they are harder to reproduce, and more test cases are necessary for debugging. Furthermore, they occur more frequently, as *non-occasional bugs* are usually detected early, and occasional bugs might only be found by means of extensive testing. As all bug-localization techniques presented in this chapter rely on comparing call graphs of failing and correct program executions, they deal with occasional bugs only. In other words, besides examples of failing program executions, there needs to be a certain number of correct executions.

- **Structure and call frequency affecting bugs**: This distinction is particularly useful when designing call graph based bug-localization techniques. *Structure affecting bugs* are bugs resulting in different shapes of the call graph where some parts are missing or occur additionally in faulty executions. An example is presented in Figure 17.1b, where Node *b* called from *a* is missing, compared to the original graph in Figure 17.1a. In this example, a faulty `if`-condition in Node *a* could have caused the bug. In contrast, *call frequency affecting bugs* are bugs which lead to a change in the number of calls of a certain subtree in faulty executions, rather than to completely missing or new substructures. In the example in Figure 17.1c, a faulty loop condition or a faulty `if`-condition inside a loop in Method *c* are typical causes for the increased number of calls of Method *b*.

As probably any bug-localization technique, call graph based bug localization is certainly not able to find all kinds of software bugs. For example, it is possible that bugs do not affect the call graph at all. For instance, if some mathematical expression calculates faulty results, this does not necessarily affect subsequent method calls and call graph mining can not detect this. Therefore, call graph based bug localization should be seen as a technique which complements other techniques, as the ones we will describe in Section 3. In this chapter we concentrate on deterministic bugs of single-threaded programs and leave aside bugs which are specific for such situations. However, the techniques described in the following might locate such bugs as well.

2.3 Bug Localization with Call Graphs

So far, several approaches have been proposed to localize bugs by means of call-graph mining [9, 13, 14, 25]. We will present them in detail in the following sections. In a nutshell, the approaches consist of three steps:

1 Deduction of call graphs from program executions, assignment of labels *correct* or *failing*.

2 Reduction of call graphs.

3 Mining of call graphs, analysis of the resulting frequent subgraphs.

Step 1: Deriving call graphs is relatively simple. They can be obtained by tracing program executions while testing, which is assumed to be done anyway. Furthermore, a classification of program executions as *correct* or *failing* is needed to find discriminative patterns in the last step. Obtaining the necessary information can be done easily, as quality assurance widely uses test suites which provide the correct results [18].

Step 2: Call-graph reduction is necessary to overcome the huge sizes of call graphs. This is much more challenging. It involves the decision how much information lost is tolerable when compressing the graphs. However, even if reduction techniques can facilitate mining in many cases, they currently do not allow for mining of arbitrary software projects. Details on call-graph reduction are presented in Section 4.

Step 3: This step includes frequent subgraph mining and the analysis of the resulting frequent subgraphs. The intuition is to search for patterns typical for faulty executions. This often results in a ranking of methods suspected to contain a bug. The rationale is that such a ranking is given to a software developer who can do a code review of the suspicious methods. The specifics of this step vary widely and highly depend on the graph-reduction scheme used. Section 5 discusses the different approaches in detail.

2.4   Graph and Tree Mining

Frequent subgraph mining has been introduced in earlier chapters of this book. As such techniques are of importance in this chapter, we briefly recapitulate those which are used in the context of bug localization based on call graph mining:

- **Frequent subgraph mining:** Frequent subgraph mining searches for the complete set of subgraphs which are frequent within a database of graphs, with respect to a user defined minimum support. Respective algorithms can mine connected graphs containing labeled nodes and edges. Most implementations also handle directed graphs and pseudo graphs which might contain self-loops and multiple edges. In general, the graphs analyzed can contain cycles. A prominent mining algorithm is *gSpan* [32].

- **Closed frequent subgraph mining:** Closed mining algorithms differ from regular frequent subgraph mining in the sense that only closed subgraphs are contained in the result set. A subgraph sg is called closed if no other graph is contained in the result set which is a supergraph of sg and has exactly the same support. Closed mining algorithms therefore produce more concise result sets and benefit from pruning opportunities which may speed up the algorithms. In the context of this chapter, the *CloseGraph* algorithm [33] is used, as closed subgraphs proved to be well suited for bug localization [13, 14, 25].

- **Rooted ordered tree mining:** Tree mining algorithms (a survey with more details can be found in [5]) work on databases of trees and exploit their characteristics. Rooted ordered tree mining algorithms work on *rooted ordered trees*, which have the following characteristics: In contrast to *free trees*, *rooted trees* have a dedicated root node, the `main`-method in call trees. *Ordered trees* preserve the order of outgoing edges of a node, which is not encoded in arbitrary graphs. Thus, call trees can keep the information that a certain node is called before another one from the same parent. Rooted ordered tree mining algorithms produce result sets of rooted ordered trees. They can be embedded in the trees from the original tree database, preserving the order. Such algorithms have the advantage that they benefit from the order, which speeds up mining significantly. Techniques in the context of bug localization sometimes use the *FREQT* rooted ordered tree mining algorithm [2]. Obviously, this can only be done when call trees are not reduced to graphs containing cycles.

3. Related Work

This chapter of the book surveys bug localization based on graph mining and dynamic call graphs. As many approaches orthogonal to call-graph mining have been proposed, this section on related work provides an overview of such approaches.

The most important distinction for bug localization techniques is if they are *static* or *dynamic*. Dynamic techniques rely on the analysis of program runs while static techniques do not require any execution. An example for a static technique is source code analysis which can be based on code metrics or different graphs representing the source code, e.g., static call graphs, control-flow graphs or program-dependence graphs. Dynamic techniques usually trace some information during a program execution which is then analyzed. This can be information on the values of variables, branches taken during execution or code segments executed.

In the remainder of this section we briefly discuss the different static and dynamic bug localization techniques. At the end of this section we present recent work in *mining of static program-dependence graphs* in a little more detail, as this approach makes use of graph mining. However, it is static in nature as it does not involve any program executions. It is therefore not similar to the mining schemes based on dynamic call graphs described in the remainder of this chapter.

Mining of Source Code. Software-complexity metrics are measures derived from the source code describing the complexity of a program or its methods. In many cases, complexity metrics correlate with defects in software [26, 34]. A standard technique in the field of 'mining software repositories' is to map post-release failures from a bug database to defects in static source code. Such a mapping is done in [26]. The authors derive standard complexity metrics from source code and build regression models based on them and the information if the software entities considered contain bugs. The regression models can then predict post-release failures for new pieces of software. A similar study uses decision trees to predict failure probabilities [21]. The approach in [30] uses regression techniques to predict the likelihood of bugs based on static usage relationships between software components. All approaches mentioned require a large collection of bugs and version history.

Dynamic Program Slicing.   Dynamic program slicing [22] can be very useful for debugging although it is not exactly a bug localization technique. It helps searching for the exact cause of a bug if the programmer already has some clue or knows where the bug appears, e.g., if a stack trace is available. Program slicing gives hints which parts of a program might have contributed to a faulty execution. This is done by exploring data dependencies and revealing which statements might have affected the data used at the location where the bug appeared.

Statistical Bug Localization.   Statistical bug localization is a family of dynamic, mostly data focused analysis techniques. It is based on instrumentation of the source code, which allows to capture the values of variables during an execution, so that patterns can be detected among the variable values. In [15], this approach is used to discover program invariants. The authors claim that bugs can be detected when unexpected invariants appear in failing executions or when expected invariants do not appear. In [23], variable values gained by instrumentation are used as features describing a program execution. These are then analyzed with regression techniques, which leads to potentially faulty pieces of code. A similar approach, but with a focus on the control flow, is [24]. It instruments variables in condition statements. It then calculates a ranking which yields high values when the evaluation of these statements differs significantly in correct and failing executions.

The instrumentation-based approaches mentioned either have a large memory footprint [6] or do not capture all bugs. The latter is caused by the usual practice not to instrument every part of a program and therefore not to watch every value, but to instrument sampled parts only. [23] overcomes this problem by collecting small sampled parts of information from productive code on large numbers of machines via the Internet. However, this does not facilitate the discovery of bugs before the software is shipped.

Analysis of Execution Traces.   A technique using tracing and visualization is presented in [20]. It relies on a ranking of program components based on the information which components are executed more often in failing program executions. Though this technique is rather simple, it produces good bug-localization results. In [6], the authors go a step further and analyze sequences of method calls. They demonstrate that the temporal order of calls is more promising to analyze than considering frequencies only. Both techniques can be seen as a basis for the more sophisticated call graph based techniques this chapter focuses on. The usage of call sequences instead of call frequencies is a generalization which takes more structural information into account. Call graph based techniques then generalize from sequence-based techniques. They do so by using more complex structural information encoded in the graphs.

Mining of Static Program-Dependence Graphs.   Recent work of Chang et al. [4] focuses on discovering *neglected conditions*, which are also known as *missing paths*, *missing conditions* and *missing cases*. They are a class of bugs which are in many cases *non-crashing occasional bugs* (cf. Subsection 2.2) – dynamic call graph based techniques target such bugs as well. An example of a neglected condition is a forgotten case in a switch-statement. This could lead to wrong behavior, faulty results in some occasions and is in general non-crashing.

Chang et al. work with static *program-dependence graphs* (*PDGs*) [28] and utilize graph-mining techniques. PDGs are graphs describing both control and data dependencies (edges) between elements (nodes) of a method or of an entire program. Figure 17.2a provides an example PDG representing the method $add(a, b)$ which returns the sum of its two parameters. Control dependencies are displayed by solid lines, data dependencies by dashed lines. As PDGs are static, only the number of instructions and dependencies within a method limit their size. Therefore, they are usually smaller than dynamic call graphs (see Sections 2 and 4). However, they typically become quite large as well, as methods often contain many dependencies. This is the reason why they cannot be mined directly with standard graph-mining algorithms. PDGs can be derived from source code. Therefore, like other static techniques, PDG analysis does not involve any execution of a program.

The idea behind [4] is to first determine *conditional rules* in a software project. These are rules (derived from PDGs, as we will see) occurring frequently within a project, representing fault-free patterns. Then, rule violations are searched, which are considered to be *neglected conditions*. This is based on the assumption that the more a certain pattern is used, the more likely it is to be a valid rule. The conditional rules are generated from PDGs by deriving *(topo-*

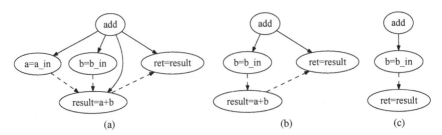

Figure 17.2. An example PDG, a subgraph and a topological graph minor.

logical) graph minors[2]. Such graph minors represent transitive intraprocedural dependencies. They can be seen – like subgraphs – as a set of smaller graphs describing the characteristics of a PDG. The PDG minors are obtained by employing a *heuristic maximal frequent subgraph-mining algorithm* developed by the authors. Then, an expert has to confirm and possibly edit the graph minors (also called *programming rules*) found by the algorithm. Finally, a *heuristic graph-matching algorithm*, which is developed by the authors as well, searches the PDGs to find the rule violations in question.

From a technical point of view, besides the PDG representation, the approach relies on the two new heuristic algorithms for maximal frequent subgraph mining and graph matching. Both techniques are not investigated from a graph theoretic point of view nor evaluated with standard data sets for graph mining. Most importantly, there are no guarantees for the heuristic algorithms: It remains unclear in which cases graphs are not found by the algorithms. Furthermore, the approach requires an expert to examine the rules, typically hundreds, by hand. However, the algorithms do work well in the evaluation of the authors.

The evaluation on four open source programs demonstrates that the approach finds most neglected conditions in real software projects. More precisely, 82% of all rules are found, compared to a manual investigation. A drawback of the approach is the relatively high false-positive rate which leads to a bug-detection precision of 27% on average.

Though graph-mining techniques similar to dynamic call graph mining (as presented in the following) are used in [4], the approaches are not related. The work of Chang et al. relies on static PDGs. They do not require any program execution, as dynamic call graphs do.

---

[2]A *graph minor* is a graph obtained by repeated deletions and edge contractions from a graph [10]. For *topological graph minors* as used in [4], in addition, paths between two nodes can be replaced with edges between both nodes. Figure 17.2 provides (a) an example PDG along with (b) a subgraph and (c) a topological graph minor. The latter is a minor of both, the PDG and the subgraph. Note that in general any subgraph of a graph is a minor as well.

4.   Call-Graph Reduction

As motivated earlier, reduction techniques are essential for call graph based bug localization: Call graphs are usually very large, and graph-mining algorithms do not scale for such sizes. Call-graph reduction is usually done by a lossy compression of the graphs. Therefore, it involves the tradeoff between keeping as much information as possible and a strong compression. As some bug localization techniques rely on the temporal order of method executions, the corresponding reduction techniques encode this information in the reduced graphs.

In Subsection 4.1 we describe the possibly easiest reduction technique, which we call *total reduction*. In Subsection 4.2 we introduce various techniques for the reduction of iteratively executed structures. As some techniques make use of the temporal order of method calls during reduction, we describe these aspects in Subsection 4.3. We provide some ideas on the reduction of recursion in Subsection 4.4 and conclude the section with a brief comparison in Subsection 4.5.

4.1   Total Reduction

The *total reduction* technique is probably the easiest technique and yields good compression. In the following, we introduce two variants:

- **Total reduction** (R_{total}). Total reduction maps every node representing the same method in the call graph to a single node in the reduced graph. This may give way to the existence of loops (i.e., the output is a regular graph, not a tree), and it limits the size of the graph (in terms of nodes) to the number of methods of the program. In bug localization, [25] has introduced this technique, along with a temporal extension (see Subsection 4.3).

- **Total reduction with edge weights** ($R_{\text{total\_w}}$). [14] has extended the plain total reduction scheme (R_{total}) to include call frequencies: Every edge in the graph representing a method call is annotated with an edge weight. It represents the total number of calls of the callee method from the caller method in the original graph. These weights allow for more detailed analyses.

Figure 17.3 contains examples of the total reduction techniques: (a) is an unreduced call graph, (b) its total reduction (R_{total}) and (c) its total reduction with edge weights ($R_{\text{total\_w}}$).

In general, total reduction (R_{total} and $R_{\text{total\_w}}$) reduces the graphs quite significantly. Therefore, it allows graph mining based bug localization with software projects larger than other reduction techniques. On the other hand, much

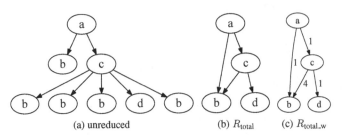

(a) unreduced   (b) R_{total} (c) $R_{\text{total\_w}}$

Figure 17.3. Total reduction techniques.

information on the program execution is lost. This concerns frequencies of the executions of methods (R_{total} only) as well as information on different structural patterns within the graphs (R_{total} and $R_{\text{total\_w}}$). In particular, the information is lost in which context (at which position within a graph) a certain substructure is executed.

4.2 Iterations

Next to total reduction, reduction based on the compression of iteratively executed structures (i.e., caused by loops) is promising. This is due to the frequent usage of iterations in today's software. In the following, we introduce two variants:

- **Unordered zero-one-many reduction** ($R_{\text{01m\_unord}}$). This reduction technique omits equal substructures of executions which are invoked more than twice from the same node. This ensures that many equal substructures called within a loop do not lead to call graphs of an extreme size. In contrast, the information that some substructure is executed several times is still encoded in the graph structure, but without exact numbers. This is done by doubling substructures within the call graph. Compared to total reduction (R_{total}), more information on a program execution is kept. The downside is that the call graph generally is much larger.

  This reduction technique is inspired by Di Fatta et al. [9] (cf. $R_{\text{01m\_ord}}$ in Subsection 4.3), but does not take the temporal order of the method executions into account. [13, 14] have used it for comparisons with other techniques which do not make use of temporal information.

- **Subtree reduction** (R_{subtree}). This reduction technique, proposed in [13, 14], reduces subtrees executed iteratively by deleting all but the first subtree and inserting the call frequencies as edge weights. In general, it therefore leads to smaller graphs than $R_{\text{01m\_unord}}$. The edge weights allow for a detailed analysis; they serve as the basis of the analysis technique described in Subsection 5.2. Details of the reduction technique are given in the remainder of this subsection.

Note that with R_{total}, and with $R_{01\text{m\_unord}}$ in most cases as well, the graphs of a correct and a failing execution with a *call frequency affecting bug* (cf. Subsection 2.2) are reduced to exactly the same graph. With R_{subtree} (and with $R_{\text{total\_w}}$ as well), the edge weights would be different when call frequency affecting bugs occur. Analysis techniques can discover this (cf. Subsection 5.2).

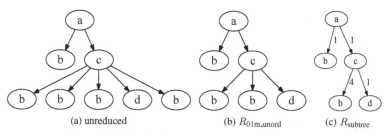

(a) unreduced   (b) $R_{01\text{m\_unord}}$ (c) R_{subtree}

Figure 17.4. Reduction techniques based on iterations.

Figure 17.4 illustrates the two iteration-based reduction techniques: (a) is an unreduced call graph, (b) its zero-one-many reduction without temporal order ($R_{01\text{m\_unord}}$) and (c) its subtree reduction (R_{subtree}). Note that the four calls of b from c are reduced to two calls with $R_{01\text{m\_unord}}$ and to one edge with weight 4 with R_{subtree}. Further, the graph resulting from R_{subtree} has one node more than the one obtained from $R_{\text{total\_w}}$ in Figure 17.3c, but the same number of edges.

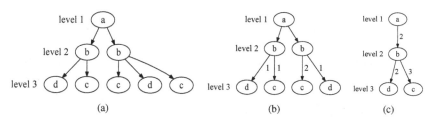

(a)   (b)   (c)

Figure 17.5. A raw call tree, its first and second transformation step.

For the subtree reduction (R_{subtree}), [14] organizes the call tree into n horizontal levels. The root node is at *level* 1. All other nodes are in levels numbered with the distance to the root. A na"ve approach to reduce the example call tree in Figure 17.5a would be to start at *level* 1 with Node a. There, one would find two child subtrees with a different structure – one could not merge anything. Therefore, one proceeds level by level, starting from *level* $n-1$, as described in Algorithm 22. In the example in Figure 17.5a, one starts in *level* 2.

The left Node b has two different children. Thus, nothing can be merged there. In the right b, the two children c are merged by adding the edge weights of the merged edges, yielding the tree in Figure 17.5b. In the next level, *level* 1, one processes the root Node a. Here, the structure of the two successor subtrees is the same. Therefore, they are merged, resulting in the tree in Figure 17.5c.

Algorithm 22 Subtree reduction algorithm.

1: **Input:** a call tree organized in n levels
2: **for** $level = n - 1$ **to** 1 **do**
3: **for each** *node* **in** *level* **do**
4:    merge all isomorph child-subtrees of *node*,
      sum up corresponding edge weights
5: **end for**
6: **end for**

4.3   Temporal Order

So far, the call graphs described just represent the occurrence of method calls. Even though, say, Figures 17.3a and 17.4a might suggest that b is called before c in the root Node a, this information is not encoded in the graphs. As this might be relevant for discriminating faulty and correct program executions, the bug-localization techniques proposed in [9, 25] take the temporal order of method calls within one call graph into account. In Figure 17.6a, increasing integers attached to the nodes represent the order. In the following, we present the corresponding reduction techniques:

- **Total reduction with temporal edges** ($R_{\text{total\_tmp}}$). In addition to the total reduction (R_{total}), [25] uses so called *temporal edges*. The authors insert them between all methods which are executed consecutively and are invoked from the same method. They call the resulting graphs *software-behavior graphs*. This reduction technique includes the temporal order from the raw ordered call trees in the reduced graph representations. Technically, temporal edges are directed edges with another label, e.g., 'temporal', compared to other edges which are labeled, say, 'call'.

  As the graph-mining algorithms used for further analysis can handle edges labeled differently, the analysis of such graphs does not give way to any special challenges, except for an increased number of edges. In consequence, the totally reduced graphs loose their main advantage, their small size. However, taking the temporal order into account might help discovering certain bugs.

- **Ordered zero-one-many reduction** (R_{01m_ord}). This reduction technique proposed by Di Fatta et al. [9] makes use of the temporal order. This is done by representing the graph as a *rooted ordered tree*, which can be analyzed with an order aware mining algorithm. To include the temporal order, the reduction technique is changed as follows: While R_{01m_unord} omits any equal substructure which is invoked more than twice from the same node, here only substructures are removed which are executed more than twice in direct sequence. This facilitates that all temporal relationships are retained. E.g., in the reduction of the sequence b, b, b, d, b (see Figure 17.6) only the third b is removed, and it is still encoded that b is called after d once.

Depending on the actual execution, this technique might lead to extreme sizes of call trees. For example, if within a loop a Method a is called followed by two calls of b, the reduction leads to the repeated sequence a, b, b, which is not reduced at all. The rooted ordered tree miner in [9] partly compensates the additional effort for mining algorithms caused by such sizes, which are huge compared to R_{01m_unord}. Rooted ordered tree mining algorithms scale significantly better than usual graph mining algorithms [5], as they make use of the order.

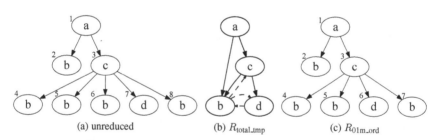

(a) unreduced $\qquad$ (b) R_{total_tmp} $\qquad$ (c) R_{01m_ord}

Figure 17.6. Temporal information in call graph reductions.

Figure 17.6 illustrates the two graph reductions which are aware of the temporal order. (The integers attached to the nodes represent the invocation order.) (a) is an unreduced call graph, (b) its total reduction with temporal edges (dashed, R_{total_tmp}) and (c) its ordered zero-one-many reduction (R_{01m_ord}). Note that, compared to R_{01m_unord}, R_{01m_ord} keeps a third Node b called from c, as the direct sequence of nodes labeled b is interrupted.

4.4 Recursion

Another challenge with the potential to reduce the size of call graphs is recursion. The total reductions (R_{total}, R_{total_w} and R_{total_tmp}) implicitly handle recursion as they reduce both iteration and recursion. E.g., when every method is collapsed to a single node, (self-)loops implicitly represent recursion. Besides that, recursion has not been investigated much in the context of call-graph reduction and in particular not as a starting point for reductions in addition to iterations. The reason for that is, as we will see in the following, that the reduction of recursion is less obvious than reducing iterations and might finally result in the same graphs as with a total reduction. Furthermore, in compute-intensive applications, programmers frequently replace recursions with iterations, as this avoids costly method calls. Nevertheless, we have investigated recursion-based reduction of call graphs to a certain extent and present some approaches in the following. Two types of recursion can be distinguished:

- **Direct recursion.** When a method calls itself directly, such a method call is called a *direct recursion*. An example is given in Figure 17.7a where Method b calls itself. Figure 17.7b presents a possible reduction represented with a self-loop at Node b. In Figure 17.7b, edge weights as in R_{subtree} represent both frequencies of iterations and the depth of direct recursion.

- **Indirect recursion.** It may happen that some method calls another method which in turn calls the first one again. This leads to a chain of method calls as in the example in Figure 17.7c where b calls c which again calls b etc. Such chains can be of arbitrary length. Obviously, such *indirect recursions* can be reduced as shown in Figures 17.7c–(d). This leads to the existence of loops.

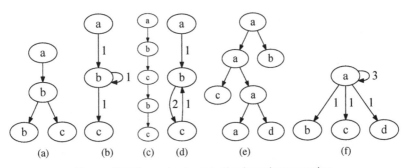

Figure 17.7. Examples for reduction based on recursion.

Both types of recursion are challenging when it comes to reduction. Figures 17.7e–(f) illustrate one way of reducing direct recursions. While the subsequent reflexive calls of a are merged into a single node with a weighted self-loop, b, c and d become siblings. As with total reductions, this leads to new structures which do not occur in the original graph. In bug localization, one might want to avoid such artifacts. E.g., d called from exactly the same method as b could be a structure-affecting bug which is not found when such artifacts occur. The problem with indirect recursion is that it can be hard to detect and becomes expensive to detect all occurrences of long-chained recursion. To conclude, when reducing recursions, one has to be aware that, as with total reduction, some artifacts may occur.

4.5   Comparison

To compare reduction techniques, we must look at the level of compression they achieve on call graphs. Table 17.1 contains the sizes of the resulting graphs (increasing in the number of edges) when different reduction techniques are applied to the same call graph. The call graph used here is obtained from an execution of the Java diff tool taken from [8] used in the evaluation in [13, 14]. Clearly, the effect of the reduction techniques varies extremely depending on the kind of program and the data processed. However, the small program used illustrates the effect of the various techniques. Furthermore it can be expected that the differences in call-graph compressions become more significant with increasing call graph sizes. This is because larger graphs tend to offer more possibilities for reductions.

Reduction	Nodes	Edges
R_{total}, $R_{\text{total\_w}}$	22	30
R_{subtree}	36	35
$R_{\text{total\_tmp}}$	22	47
$R_{\text{01m\_unord}}$	62	61
$R_{\text{01m\_ord}}$	117	116
unreduced	2199	2198

Table 17.1. Examples for the effect of call graph reduction techniques.

Obviously, the total reduction (R_{total} and $R_{\text{total\_w}}$) achieves the strongest compression and yields a reduction by two orders of magnitude. As 22 nodes remain, the program has executed exactly this number of different methods. The subtree reduction (R_{subtree}) has significantly more nodes but only five more edges. As – roughly speaking – graph-mining algorithms scale with the number of edges, this seems to be tolerable. We expect the small increase in the number of edges to be compensated by the increase in structural information encoded. The unordered zero-one-many reduction technique ($R_{\text{01m\_unord}}$) again yields somewhat larger graphs. This is because repetitions are represented as doubled substructures instead of edge weights. With the total reduction with temporal edges ($R_{\text{total\_tmp}}$), the number of edges increases by roughly 50% due to the temporal information, while the ordered zero-one-many reduction ($R_{\text{01m\_ord}}$) almost doubles this number. Subsection 5.4 assesses the effectiveness of bug localization with the different reduction techniques along with the localization methods.

Clearly, some call graph reduction techniques also are expensive in terms of runtime. However, we do not compare the runtimes, as the subsequent graph mining step usually is significantly more expensive.

To summarize, different authors have proposed different reduction techniques, each one together with a localization technique (cf. Section 5): the total reduction ($R_{\text{total\_tmp}}$) in [25], the zero-one-many reduction (R_{01m_ord}) in [9] and the subtree reduction (R_{subtree}) in [13, 14]. Some of the reductions can be used or at least be varied in order to work together with a bug localization technique different from the original one. In Subsection 5.4, we present original and varied combinations.

5.   Call Graph Based Bug Localization

This section focuses on the third and last step of the generic bug-localization process from Subsection 2.3, namely frequent subgraph mining and bug localization based on the mining results. In this chapter, we distinguish between structural approaches [9, 25] and the frequency-based approach used in [13, 14].

In Subsections 5.1 and 5.2 we describe the two kinds of approaches. In Subsection 5.3 we introduce several techniques to integrate the results of structural and frequency-based approaches. We present some comparisons in Subsection 5.4.

5.1   Structural Approaches

Structural approaches for bug localization can locate *structure affecting bugs* (cf. Subsection 2.2) in particular. Approaches following this idea do so either in isolation or as a complement to a frequency-based approach. In most cases, a likelihood $P(m)$ that Method m contains a bug is calculated, for every method. This likelihood is then used to rank the methods. In the following, we refer to it as *score*. In the remainder of this subsection, we introduce and discuss the different structural scoring approaches.

The Approach by Di Fatta et al.   In [9], the R_{01m_ord} call-graph reduction is used (cf. Section 4), and the rooted ordered tree miner *FREQT* [2] is employed to find frequent subtrees. The call trees analyzed are large and lead to scalability problems. Hence, the authors limit the size of the subtrees searched to a maximum of four nodes. Based on the results of frequent subtree mining, they define the *specific neighborhood* (SN). It is the set of all subgraphs contained in all call graphs of failing executions which are not frequent in call graphs of correct executions:

$$SN := \{sg \mid (supp(sg, D_{\text{fail}}) = 100\%) \wedge \neg(supp(sg, D_{\text{corr}}) \geq minSup)\}$$

where $supp(g, D)$ denotes the support of a graph g, i.e., the fraction of graphs in a graph database D containing g. D_{fail} and D_{corr} denote the sets of call graphs of failing and correct executions. [9] uses a minimum support $minSup$ of 85%.

Based on the *specific neighborhood*, a *structural score* P_{SN} is defined:

$$P_{\text{SN}}(m) := \frac{supp(g_m, SN)}{supp(g_m, SN) + supp(g_m, D_{\text{corr}})}$$

where g_m denotes all graphs containing Method m. Note that P_{SN} assigns the value 0 to methods which do not occur within SN and the value 1 to methods which occur in SN but not in correct program executions D_{corr}.

The Approach by Eichinger et al. The notion of *specific neighborhood* (SN) has the problem that no support can be calculated when the SN is empty.[3] Furthermore, experiments of ours have revealed that the P_{SN}-scoring only works well if a significant number of graphs is contained in SN. This depends on the graph reduction and mining techniques and has not always been the case in the experiments. Thus, to complement the frequency-based scoring (cf. Subsection 5.2), another structural score is defined in [14]. It is based on the set of frequent subgraphs which occur in failing executions only, SG_{fail}. The structural score P_{fail} is calculated as the support of m in SG_{fail}:

$$P_{\text{fail}}(m) := supp(g_m, SG_{\text{fail}})$$

Further Support-based Approaches. Both the P_{SN}-score [9] and the P_{fail}-score [14] have their weaknesses. Both approaches consider structure affecting bugs which lead to additional substructures in call graphs corresponding to failing executions. In the SN, only substructures occurring in *all* failing executions (D_{fail}) are considered – they are ignored if a single failing execution does not contain the structure. The P_{fail}-score concentrates on subgraphs occurring in failing executions only (SG_{fail}), although they do not need to be contained in *all* failing executions. Therefore, both approaches might not find structure affecting bugs leading not to additional structures but to fewer structures. The weaknesses mentioned have not been a problem so far, as they have rarely affected the respective evaluation, or the combination with another ranking method has compensated it.

---

[3] [9] uses a simplistic fall-back approach to deal with this effect.

One possible solution for a broader structural score is to define a score based on two support values: The support of every subgraph sg in the set of call graphs of correct executions $supp(sg, D_{corr})$ and the respective support in the set of failing executions $supp(sg, D_{fail})$. As we are interested in the support of methods and not of subgraphs, the maximum support values of all subgraphs sg in the set of subgraphs SG containing a certain Method m can be derived:

$$s_{fail}(m) := \max_{\{sg \mid sg \in SG,\, m \in sg\}} supp(sg, D_{fail})$$

$$s_{corr}(m) := \max_{\{sg \mid sg \in SG,\, m \in sg\}} supp(sg, D_{corr})$$

Example 17.1. *Think of Method* a, *called from the* main*-method and containing a bug. Let us assume there is a subgraph* main $\rightarrow$ a *(where '$\rightarrow$' denotes an edge between two nodes) which has a support of 100% in failing executions and 40% in correct ones. At the same time there is the subgraph* main $\rightarrow$ a $\rightarrow$ b *where* a *calls* b *afterwards. Let us say that the bug occurs exactly in this constellation. In this situation,* main $\rightarrow$ a $\rightarrow$ b *has a support of 0% in D_{corr} while it has a support of 100% in D_{fail}. Let us further assume that there also is a much larger subgraph* sg *which contains* a *and occurs in 10% of all failing executions. The value s_{fail}(a) therefore is 100%, the maximum of 100% (based on subgraph* main $\rightarrow$ a*), 100% (based on* main $\rightarrow$ a $\rightarrow$ b*) and 10% (based on* sg*).*

With the two relative support values s_{corr} and s_{fail} as a basis, new structural scores can be defined. One possibility would be the absolute difference of s_{fail} and s_{corr}:

$$P_{fail\text{-}corr}(m) = |s_{fail}(m) - s_{corr}(m)|$$

Example 17.2. *To continue Example 17.1, $P_{fail\text{-}corr}$(a) is 60%, the absolute difference of 40% (s_{corr}(a)) and 100% (s_{fail}(a)). We do not achieve a higher value than 60%, as Method* a *also occurs in bug-free subgraphs.*

The intuition behind $P_{fail\text{-}corr}$ is that both kinds of structure affecting bugs are covered: (1) those which lead to additional structures (high s_{fail} and low to moderate s_{corr} values like in Example 17.2) and (2) those leading to missing structures (low s_{fail} and moderate to high s_{corr}). In cases where the support in both sets is equal, e.g., both are 100% for the main-method, $P_{fail\text{-}corr}$ is zero. We have not yet evaluated $P_{fail\text{-}corr}$ with real data. It might turn out that different but similar scoring methods are better.

The Approach by Liu et al.   Although [25] is the first study which applies graph mining techniques to dynamic call graphs to localize non-crashing bugs, this work is not directly compatible to the other approaches described so far. In [25], bug localization is achieved by a rather complex classification process, and it does not generate a ranking of methods suspected to contain a bug, but a set of such methods.

The work is based on the $R_{\text{total\_tmp}}$ reduction technique and works with total reduced graphs with temporal edges (cf. Section 4). The call graphs are mined with a variant of the *CloseGraph* algorithm [33]. This step results in frequent subgraphs which are turned into binary features characterizing a program execution: A boolean feature vector represents every execution. In this vector, every element indicates if a certain subgraph is included in the corresponding call graph. Using those feature vectors, a support-vector machine (SVM) is learned which decides if a program execution is correct or failing. More precisely, for every method, two classifiers are learned: one based on call graphs including the respective method, one based on graphs without this method. If the precision rises significantly when adding graphs containing a certain method, this method is deemed more likely to contain a bug. Such methods are added to the so-called *bug-relevant function set*. Its functions usually line up in a form similar to a stack trace which is presented to a user when a program crashes. Therefore, the bug-relevant function set serves as the output of the whole approach. This set is given to a software developer who can use it to locate bugs more easily.

5.2 Frequency-based Approach

The frequency-based approach for bug localization by Eichinger et al. [13, 14] is in particular suited to locate *frequency affecting bugs* (cf. Subsection 2.2), in contrast to the structural approaches. It calculates a score as well, i.e., the likelihood to contain a bug, for every method.

After having performed frequent subgraph mining with the *CloseGraph* algorithm [33] on call graphs reduced with the R_{subtree} technique, Eichinger et al. analyze the edge weights. As an example, a call-frequency affecting bug increases the frequency of a certain method invocation and therefore the weight of the corresponding edge. To find the bug, one has to search for edge weights which are increased in failing executions. To do so, they focus on frequent subgraphs which occur in both correct and failing executions. The goal is to develop an approach which automatically discovers which edge weights of call graphs from a program are most significant to discriminate between *correct* and *failing*. To do so, one possibility is to consider different *edge types*, e.g., edges having the same calling Method m_s (start) and the same method called m_e (end). However, edges of one type can appear more than once within one subgraph and, of course, in several different subgraphs. Therefore, the authors analyze every edge in every such location, which is referred to as a *context*. To specify the exact location of an edge in its context within a certain subgraph, they do not use the method names, as they may occur more than once. Instead, they use a unique id for the calling node (id_s) and another one for the method called (id_e). All ids are valid within their subgraph. To sum up, edges in its context in a certain subgraph sg are referenced with the following tuple: (sg, id_s, id_e). A certain bug does not affect all method calls (edges) of the same type, but method calls of the same type in the same context. Therefore, the authors assemble a feature table with every edge in every context as columns and all program executions in the rows. The table cells contain the respective edge weights. Table 17.2 serves as an example.

	$a \to b$ (sg_1, id_1, id_2)	$a \to b$ (sg_1, id_1, id_3)	$a \to b$ (sg_2, id_1, id_2)	$a \to c$ (sg_2, id_1, id_3)	$\cdots$	Class
g_1	0	0	13	6513	$\cdots$	*correct*
g_2	512	41	8	12479	$\cdots$	*failing*
$\cdots$	$\cdots$	$\cdots$	$\cdots$	$\cdots$	$\cdots$	$\cdots$

Table 17.2. Example table used as input for feature-selection algorithms.

The first column contains a reference to the program execution or, more precisely, to its reduced call graph $g_i \in G$. The second column corresponds to the first subgraph (sg_1) and the edge from id_1 (Method a) to id_2 (Method b). The third column corresponds to the same subgraph (sg_1) but to the edge from id_1 to id_3. Note that both id_2 and id_3 represent Method b. The fourth column represents an edge from id_1 to id_2 in the second subgraph (sg_2). The fifth column represents another edge in sg_2. Note that ids have different meanings in different subgraphs. The last column contains the class *correct* or *failing*. If a certain subgraph is not contained in a call graph, the corresponding cells have value 0, like g_1, which does not contain sg_1. Graphs (rows) can contain a certain subgraph not just once, but several times at different locations. In this case, averages are used in the corresponding cells of the table.

The table structure described allows for a detailed analysis of edge weights in different contexts within a subgraph. Algorithm 23 describes all subsequent steps in this subsection. After putting together the table, Eichinger et al. deploy a standard feature-selection algorithm to score the columns of the table and thus the different edges. They use an entropy-based algorithm from the *Weka* data-mining suite [31]. It calculates the information gain $InfoGain$ [29] (with respect to the class of the executions, *correct* or *failing*) for every column (Line 23 in Algorithm 23). The information gain is a value between 0 and 1, interpreted as a likelihood of being responsible for bugs. Columns with an information gain of 0, i.e., the edges always have the same weights in both classes, are discarded immediately (Line 23 in Algorithm 23).

Call graphs of failing executions frequently contain bug-like patterns which are caused by a preceding bug. Eichinger et al. call such artifacts *follow-up*

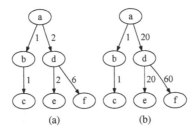

Figure 17.8. Follow-up bugs.

bugs. Figure 17.8 illustrates a follow-up bug: (a) represents a bug free version, (b) contains a call frequency affecting bug in Method a which affects the invocations of d. Here, this method is called 20 times instead of twice. Following the R_{subtree} reduction, this leads to a proportional increase in the number of calls in Method d. [14] contains more details how follow-up bugs are detected and removed from the set of edges E (Line 23 of Algorithm 23).

Algorithm 23 Procedure to calculate $P_{\text{freq}}(m_s, m_e)$ and $P_{\text{freq}}(m)$.

1: **Input:** a set of edges $e \in E$, $e = (sg, id_s, id_e)$
2: assign every $e \in E$ its information gain $Info\,Gain$
3: $E = E \setminus \{e \mid e.Info\,Gain = 0\}$
4: remove follow-up bugs from E
5: $E_{(m_s,m_e)} = \{e \mid e \in E \wedge e.id_s.label = m_s \wedge e.id_e.label = m_e\}$
6: $P_{\text{freq}}(m_s, m_e) = \max_{e \in E_{(m_s,m_e)}} (e.Info\,Gain)$
7: $E_m = \{e \mid e \in E \wedge e.id_s.label = m\}$
8: $P_{\text{freq}}(m) = \max_{e \in E_m} (e.Info\,Gain)$

At this point, Eichinger et al. calculate likelihoods of method invocations containing a bug, for every invocation (described by a calling Method m_s and a method called m_e). They call this score $P_{\text{freq}}(m_s, m_e)$, as it is based on the call frequencies. To do the calculation, they first determine sets $E_{(m_s,m_e)}$ of edges $e \in E$ for every method invocation in Line 23 of Algorithm 23. In Line 23, they use the $\max()$ function to calculate $P_{\text{freq}}(m_s, m_e)$, the maximum $Info\,Gain$ of all edges (method invocations) in E. In general, there are many edges in E with the same method invocation, as an invocation can occur in different contexts. With the $\max()$ function, the authors assign every invocation the score from the context ranked highest.

Example 17.3. *An edge from* a *to* b *is contained in two subgraphs. In one subgraph, this edge* a $\rightarrow$ b *has a low InfoGain value of 0.1. In the other subgraph, and therefore in another context, the same edge has a high InfoGain* *value of 0.8, i.e., a bug is relatively likely. As one is interested in these cases, lower scores for the same invocation are less important, and only the maximum is considered.*

At the moment, the ranking does not only provide the score for a method invocation, $P_{\text{freq}}(m_s, m_e)$, but also the subgraphs where it occurs and the exact embeddings. This information might be important for a software developer. The authors report this information additionally. To ease comparison with other approaches not providing this information, they also calculate $P_{\text{freq}}(m)$ for every calling Method m in Lines 23 and 23 of Algorithm 23. The explanation is analogous to the one of the calculation of $P_{\text{freq}}(m_s, m_e)$ in Lines 23 and 23.

5.3   Combined Approaches

As discussed before, structural approaches are well-suited to locate *structure affecting bugs*, while frequency-based approaches focus on *call frequency affecting bugs*. Therefore, it seems to be promising to combine both approaches. [13] and [14] have investigated such strategies.

In [13], Eichinger et al. have combined the frequency-based approach with the P_{SN}-score [9]. In order to calculate the resulting score, the authors use the approach by Di Fatta et al. [9] without temporal order: They use the $R_{\text{01m\_unord}}$ reduction with a general graph miner, *gSpan* [32], in order to calculate the structural P_{SN}-score. They derived the frequency-based P_{freq}-score as described before after mining the same call graphs but with the R_{subtree} reduction and the *CloseGraph* algorithm [33] and different mining parameters. In order to combine the two scores derived from the results of two graph mining runs, they calculated the arithmetic mean of the normalized scores:

$$P_{\text{comb}[13]}(m) = \frac{P_{\text{freq}}(m)}{2 \max_{n \in sg \in D} (P_{\text{freq}}(n))} + \frac{P_{\text{SN}}(m)}{2 \max_{n \in sg \in D} (P_{\text{SN}}(n))}$$

where n is a method in a subgraph sg in the database of all call graphs D.

As the combined approach in [13] leads to good results but requires two costly graph-mining executions, the authors have developed a technique in [14] which requires only one graph-mining execution: They combine the frequency-based score with the simple structural score P_{fail}, both based on the results from one *CloseGraph* [33] execution. They combine the results with the arithmetic mean, as before:

$$P_{\text{comb}[14]}(m) = \frac{P_{\text{freq}}(m)}{2 \max_{n \in sg \in D} (P_{\text{freq}}(n))} + \frac{P_{\text{fail}}(m)}{2 \max_{n \in sg \in D} (P_{\text{fail}}(n))}$$

5.4 Comparison

We now present the results of our experimental comparison of the bug localization and reduction techniques introduced in this chapter. The results are based on the (slightly revised) experiments in [13, 14].

Most bug localization techniques as described in this chapter produce ordered lists of methods. Someone doing a code review would start with the first method in such a list. The maximum number of methods to be checked to find the bug therefore is the position of the faulty method in the list. This position is our measure of result accuracy. Under the assumption that all methods have the same size and that the same effort is needed to locate a bug within a method, this measure linearly quantifies the intellectual effort to find a bug. Sometimes two or more subsequent positions have the same score. As the intuition is to count the maximum number of methods to be checked, all positions with the same score have the number of the last position with this score. If the first bug is, say, reported at the third position, this is a fairly good result, depending on the total number of methods. A software developer only has to do a code review of maximally three methods of the target program.

Our experiments feature a well known Java diff tool taken from [8], consisting of 25 methods. We instrumented this program with fourteen different bugs which are artificial, but mimic bugs which occur in reality and are similar to the bugs used in related work. Each version contains one – and in two cases two – bugs. See [14] for more details on these bugs. We have executed each version of the program 100 times with different input data. Then we have classified the executions as correct or failing with a test oracle based on a bug free reference program.

The experiments are designed to answer the following questions:

1 How do *frequency-based approaches* perform compared to *structural* ones? How can *combined approaches* improve the results?

2 In Subsection 4.5 we have compared *reduction techniques* based on the compression ratio achieved. How do the different reduction techniques perform in terms of bug localization precision?

3 Some approaches make use of the *temporal order* of method calls. The call graph representations tend to be much larger than without. Do such graph representations improve precision?

In concrete terms, we compare the following five alternatives:

- E_{01m}: The structural P_{SN}-scoring approach similar to [9] (cf. Subsection 5.1), but based on the unordered R_{01m_unord} reduction.

- $E_{subtree}$: The frequency-based P_{freq}-scoring approach as in [13, 14] (cf. Subsection 5.2) based on the $R_{subtree}$ reduction.

- $E_{comb[13]}$: The combined approach from [13] (cf. Subsection 5.3) based on the R_{01m_unord} and $R_{subtree}$ reductions.

- $E_{comb[14]}$: The combined approach from [14] (cf. Subsection 5.3) based on the $R_{subtree}$ reduction.

- E_{total}: The combined approach as in [14] (cf. Subsection 5.3) but with the R_{total_w} reduction like in [25] (but with weights and without temporal edges, cf. Subsection 5.1).

We present the results (the number of the first position in which a bug is found) of the five experiments for all fourteen bugs in Table 17.3. We represent a bug which is not discovered with the respective approach with '25', the total number of methods of the program. Note that with the frequency-based and the combined method rankings, there usually is information available where a bug is located within a method, and in the context of which subgraph it appears. The following comparisons leave aside this additional information.

Exp.\Bug	1	2	3	4	5	6	7	8	9	10	11	12	13	14
E_{01m}	25	3	1	3	2	4	3	1	1	6	4	4	25	4
$E_{subtree}$	3	3	1	1	1	3	3	1	25	2	3	3	3	3
$E_{comb[13]}$	1	3	1	2	2	1	2	1	3	1	2	4	8	5
$E_{comb[14]}$	3	2	1	1	1	2	2	1	18	2	2	3	3	3
E_{total}	1	5	1	4	3	5	5	2	25	2	5	4	6	3

Table 17.3. Experimental results.

Structural, Frequency-Based and Combined Approaches.   Comparing the results from E_{01m} and $E_{subtree}$, the frequency-based approach ($E_{subtree}$) performs almost always as good or better than the structural one (E_{01m}). This demonstrates that analyzing numerical call frequencies is adequate to locate bugs. Bugs 1, 9 and 13 illustrate that both approaches alone cannot find certain bugs. Bug 9 cannot be found by comparing call frequencies ($E_{subtree}$). This is because Bug 9 is a modified condition which always leads to the invocation of a certain method. In consequence, the call frequency is always the same. Bugs 1 and 13 are not found with the purely structural approach (E_{01m}). Both are typical call frequency affecting bugs: Bug 1 is in an if-condition inside a loop and leads to more invocations of a certain method. In Bug 13, a modified for-condition slightly changes the call frequency of a method inside the loop. With the R_{01m_unord} reduction technique used in E_{01m}, Bug 2 and 13 have the same graph structure both with correct and with failing executions. Thus, it is difficult to impossible to identify structural differences.

The combined approaches in $E_{comb[13]}$ and $E_{comb[14]}$ are intended to take structural information into account as well to improve the results from $E_{subtree}$. We do achieve this goal: When comparing $E_{subtree}$ and $E_{comb[14]}$, we retain the already good results from $E_{subtree}$ in nine cases and improve them in five.

When looking at the two combination strategies, it is hard to say which one is better. $E_{comb[13]}$ turns out to be better in four cases while $E_{comb[14]}$ is better in six ones. Thus, the technique in $E_{comb[14]}$ is slightly better, but not with every bug. Furthermore, the technique in $E_{comb[13]}$ is less efficient as it requires two graph-mining runs.

Reduction Techniques. Looking at the call-graph-reduction techniques, the results from the experiments discussed so far reveal that the subtree-reduction technique with edge weights ($R_{subtree}$) used in $E_{subtree}$ as well as in both combined approaches is superior to the zero-one-many reduction (R_{01m_unord}). Besides the increased precision of the localization techniques based on the reduction, $R_{subtree}$ also produces smaller graphs than R_{01m_unord} (cf. Subsection 4.5).

E_{total} evaluates the total reduction technique. We use R_{total_w} as an instance of the total reduction family. The rationale is that this one can be used with $E_{comb[14]}$. In most cases, the total reduction (E_{total}) performs worse than the subtree reduction ($E_{comb[14]}$). This confirms that the subtree-reduction technique is reasonable, and that it is worth to keep more structural information than the total reduction does. However, in cases where the subtree reduction produces graphs which are too large for efficient mining, and the total reduction produces sufficiently small graphs, R_{total_w} can be an alternative to $R_{subtree}$.

Temporal Order. The experimental results listed in Table 17.3 do not shed any light on the influence of the temporal order. When applied to the buggy programs used in our comparisons, the total reduction with temporal edges (R_{total_tmp}) produces graphs of a size which cannot be mined in a reasonable time. This already shows that the representation of the temporal order with additional edges might lead to graphs whose size is not manageable any more. In preliminary experiments of ours, we have repeated E_{01m} with the R_{01m_ord} reduction and the *FREQT* [2] rooted ordered tree miner in order to evaluate the usefulness of the temporal order. Although we systematically varied the different mining parameters, the results of these experiments in general are not better than those in E_{01m}. Only in two of the 14 bugs the temporal-aware approach has performed better than E_{01m}, in the other cases it has performed worse. In a comparison with the $R_{subtree}$ reduction and the *gSpan* algorithm [32], the R_{01m_ord} reduction with the ordered tree miner displayed a significantly increased runtime by a factor of 4.8 on average.[4] Therefore, our preliminary result is that the incorporation of the temporal order does not increase the precision of bug localizations. This is based on the bugs considered so far, and more comprehensive experiments would be needed for a more reliable statement.

Threats to Validity.   The experiments carried out in this subsection, as well as in the respective publications [9, 13, 14, 25], illustrate the ability to locate bugs based on dynamic call graphs using graph mining techniques. From a software engineering point of view, three issues remain for further evaluations: (1) All experiments are based on artificially seeded bugs. Although these bugs mimic typical bugs as they occur in reality, a further investigation with real bugs, e.g., from a real software project, would prove the validity of the proposed techniques. (2) All experiments feature rather small programs containing the bugs. The programs rarely consist of more than one class and represent situations where bugs could be found relatively easy by a manual investigation as well. When solutions for the current scalability issues are found, localization techniques should be validated with larger software projects. (3) None of the techniques considered has been directly compared to other techniques such as those discussed in Section 3. Such a comparison, based on a large number of bugs, would reveal the advantages and disadvantages of the different techniques. The *iBUGS* project [7] provides real bug datasets from large software projects such as *AspectJ*. It might serve as a basis to tackle the issues mentioned.

6.   Conclusions and Future Directions

This chapter has dealt with the problem of localizing software bugs, as a use case of graph mining. This localization is important as bugs are hard to detect manually. Graph mining based techniques identify structural patterns in trace data which are typical for failing executions but rare in correct. They serve as hints for bug localization. Respective techniques based on call graph mining first need to solve the subproblem of call graph reduction. In this chapter we have discussed both reduction techniques for dynamic call graphs and approaches analyzing such graphs. Experiments have demonstrated the usefulness of our techniques and have compared different approaches.

---

[4]In this comparison, *FREQT* was restricted as in [9] to find subtrees of a maximum size of four nodes. Such a restriction was not set in *gSpan*. Furthermore, we expect a further significant speedup when *CloseGraph* [33] is used instead of *gSpan*.

All techniques surveyed in this chapter work well when applied to relatively small software projects. Due to the NP-hard problem of subgraph isomorphism inherent to frequent subgraph mining, none of the techniques presented is directly applicable to large projects. One future challenge is to overcome this problem, be it with more sophisticated graph-mining algorithms, e.g., scalable approximate mining or discriminative techniques, or smarter bug-localization frameworks, e.g., different graph representations or constraint based mining. One starting point could be the granularity of call graphs. So far, call graphs represent method invocations. One can think of smaller graphs representing interactions at a coarser level, i.e., classes or packages. [12] presents encouraging results regarding the localization of bugs based on class-level call graphs. As future research, we will investigate how to turn these results into a scalable framework for locating bugs. Such a framework would first do bug localization on a coarse level before 'zooming in' and investigating more detailed call graphs.

Call graph reduction techniques introducing edge weights trigger another challenge for graph mining: weighted graphs. We have shown that the analysis of such weights is crucial to detect certain bugs. Graph-mining research has focused on structural issues so far, and we are not aware of any algorithm for explicit mining of weighted graphs. Next to reduced call graphs, such algorithms could mine other real world graphs as well [3], e.g., in logistics [19] and image analysis [27].

Acknowledgments

We are indebted to Matthias Huber for his contributions. We further thank Andreas Zeller for fruitful discussions and Valentin Dallmeier for his comments on early versions of this chapter.

References

[1] F. E. Allen. Interprocedural Data Flow Analysis. In *Proc. of the IFIP Congress*, 1974.

[2] T. Asai, K. Abe, S. Kawasoe, H. Arimura, H. Sakamoto, and S. Arikawa. Efficient Substructure Discovery from Large Semi-structured Data. In *Proc. of the 2nd SIAM Int. Conf. on Data Mining (SDM)*, 2002.

[3] D. Chakrabarti and C. Faloutsos. Graph Mining: Laws, Generators, and Algorithms. *ACM Computing Surveys (CSUR)*, 38(1):2, 2006.

[4] R.-Y. Chang, A. Podgurski, and J. Yang. Discovering Neglected Conditions in Software by Mining Dependence Graphs. *IEEE Transactions on Software Engineering*, 34(5):579–596, 2008.

[5] Y. Chi, R. Muntz, S. Nijssen, and J. Kok. Frequent Subtree Mining – An Overview. *Fundamenta Informaticae*, 66(1–2):161–198, 2005.

[6] V. Dallmeier, C. Lindig, and A. Zeller. Lightweight Defect Localization for Java. In *Proc. of the 19th European Conf. on Object-Oriented Programming (ECOOP)*, 2005.

[7] V. Dallmeier and T. Zimmermann. Extraction of Bug Localization Benchmarks from History. In *Proc. of the 22nd IEEE/ACM Int. Conf. on Automated Software Engineering (ASE)*, 2007.

[8] I. F. Darwin. *Java Cookbook*. O'Reilly, 2004.

[9] G. Di Fatta, S. Leue, and E. Stegantova. Discriminative Pattern Mining in Software Fault Detection. In *Proc. of the 3rd Int. Workshop on Software Quality Assurance (SOQUA)*, 2006.

[10] R. Diestel. *Graph Theory*. Springer, 2006.

[11] T. G. Dietterich, P. Domingos, L. Getoor, S. Muggleton, and P. Tadepalli. Structured Machine Learning: The Next Ten Years. *Machine Learning*, 73(1):3–23, 2008.

[12] F. Eichinger and K. Böhm. Towards Scalability of Graph-Mining Based Bug Localisation. In *Proc. of the 7th Int. Workshop on Mining and Learning with Graphs (MLG)*, 2009.

[13] F. Eichinger, K. Böhm, and M. Huber. Improved Software Fault Detection with Graph Mining. In *Proc. of the 6th Int. Workshop on Mining and Learning with Graphs (MLG)*, 2008.

[14] F. Eichinger, K. Böhm, and M. Huber. Mining Edge-Weighted Call Graphs to Localise Software Bugs. In *Proc. of the European Conf. on Machine Learning and Principles and Practice of Knowledge Discovery in Databases (ECML PKDD)*, 2008.

[15] M. D. Ernst, J. Cockrell, W. G. Griswold, and D. Notkin. Dynamically Discovering Likely Program Invariants to Support Program Evolution. *IEEE Transactions on Software Engineering*, 27(2):99–123, 2001.

[16] M. R. Garey and D. S. Johnson. *Computers and Intractability: A Guide to the Theory of NP-Completeness*. W. H. Freeman, 1979.

[17] S. L. Graham, P. B. Kessler, and M. K. Mckusick. gprof: A Call Graph Execution Profiler. In *Proc. of the ACM SIGPLAN Symposium on Compiler Construction*, 1982.

[18] M. J. Harrold, R. Gupta, and M. L. Soffa. A Methodology for Controlling the Size of a Test Suite. *ACM Transactions on Software Engineering and Methodology (TOSEM)*, 2(3):270–285, 1993.

[19] W. Jiang, J. Vaidya, Z. Balaporia, C. Clifton, and B. Banich. Knowledge Discovery from Transportation Network Data. In *Proc. of the 21st Int. Conf. on Data Engineering (ICDE)*, 2005.

[20] J. A. Jones, M. J. Harrold, and J. Stasko. Visualization of Test Information to Assist Fault Localization. In *Proc. of the 24th Int. Conf. on Software Engineering (ICSE)*, 2002.

[21] P. Knab, M. Pinzger, and A. Bernstein. Predicting Defect Densities in Source Code Files with Decision Tree Learners. In *Proc. of the Int. Workshop on Mining Software Repositories (MSR)*, 2006.

[22] B. Korel and J. Laski. Dynamic Program Slicing. *Information Processing Letters*, 29(3):155–163, 1988.

[23] B. Liblit, A. Aiken, A. X. Zheng, and M. I. Jordan. Bug Isolation via Remote Program Sampling. *ACM SIGPLAN Notices*, 38(5):141–154, 2003.

[24] C. Liu, X. Yan, L. Fei, J. Han, and S. P. Midkiff. SOBER: Statistical Model-Based Bug Localization. *SIGSOFT Software Engineering Notes*, 30(5):286–295, 2005.

[25] C. Liu, X. Yan, H. Yu, J. Han, and P. S. Yu. Mining Behavior Graphs for "Backtrace" of Noncrashing Bugs. In *Proc. of the 5th SIAM Int. Conf. on Data Mining (SDM)*, 2005.

[26] N. Nagappan, T. Ball, and A. Zeller. Mining Metrics to Predict Component Failures. In *Proc. of the 28th Int. Conf. on Software Engineering (ICSE)*, 2006.

[27] S. Nowozin, K. Tsuda, T. Uno, T. Kudo, and G. Bakir. Weighted Substructure Mining for Image Analysis. In *Proc. of the Conf. on Computer Vision and Pattern Recognition (CVPR)*, 2007.

[28] K. J. Ottenstein and L. M. Ottenstein. The Program Dependence Graph in a Software Development Environment. *SIGSOFT Software Engineering Notes*, 9(3):177–184, 1984.

[29] J. R. Quinlan. *C4.5: Programs for Machine Learning*. Morgan Kaufmann Publishers, 1993.

[30] A. Schröter, T. Zimmermann, and A. Zeller. Predicting Component Failures at Design Time. In *Proc. of the 5th Int. Symposium on Empirical Software Engineering*, 2006.

[31] I. H. Witten and E. Frank. *Data Mining: Practical Machine Learning Tools and Techniques with Java Implementations*. Morgan Kaufmann Publishers, 2005.

[32] X. Yan and J. Han. gSpan: Graph-Based Substructure Pattern Mining. In *Proc. of the 2nd IEEE Int. Conf. on Data Mining (ICDM)*, 2002.

[33] X. Yan and J. Han. CloseGraph: Mining Closed Frequent Graph Patterns. In *Proc. of the 9th ACM Int. Conf. on Knowledge Discovery and Data Mining (KDD)*, 2003.

[34] T. Zimmermann, N. Nagappan, and A. Zeller. Predicting Bugs from History. In T. Mens and S. Demeyer, editors, *Software Evolution*, pages 69–88. Springer, 2008.

Chapter 18

A SURVEY OF GRAPH MINING TECHNIQUES FOR BIOLOGICAL DATASETS

S. Parthasarathy

The Ohio State University
2015 Neil Ave, DL395, Columbus, OH
srini@cse.ohio-state.edu

S. Tatikonda

The Ohio State University
2015 Neil Ave, DL395, Columbus, OH
tatikond@cse.ohio-state.edu

D. Ucar

The Ohio State University
2015 Neil Ave, DL395, Columbus, OH
ucar@cse.ohio-state.edu

Abstract

Mining structured information has been the source of much research in the data mining community over the last decade. The field of bioinformatics has emerged as important application area in this context. Examples abound ranging from the analysis of protein interaction networks to the analysis of phylogenetic data. In this article we survey the principal results in the field examining them both from the algorithmic contributions and applicability in the domain in question. We conclude this article with a discussion of the key results and identify some interesting directions for future research.

Keywords: Graph Mining, Tree Mining, Biological Networks, Community Discovery

C.C. Aggarwal and H. Wang (eds.), *Managing and Mining Graph Data,*
Advances in Database Systems 40, DOI 10.1007/978-1-4419-6045-0_18,
© Springer Science+Business Media, LLC 2010

1.   Introduction

Advances in data collection and storage technology have led to a proliferation of structured information available to organizations and individuals. This information is often also available to the user in a myriad of formats and across multiple media. This is especially true in the vibrant field of bioinformatics where an increasing large number of problems are represented in structured or semi-structured format. Examples abound ranging from protein interaction networks (graphs) to phylogenetic datasets (trees), and from XML repositories of proteomic data (trees) to regulatory networks (graphs). The size and number of such data stores is growing rapidly.

Such data may arise directly out of experimental observations (e.g. PPI network complexes from mass spectrometry) or may be a convenient abstraction for housing relational information (e.g. Protein Data Bank). Other examples include mRNA measurements from microarray studies can be used to infer pairwise gene relations that imply co-expression of two genes. Regulatory relations between DNA binding proteins and genes can also be identified via various experimental technologies such as ChIP-chip, ChIP-seq, or DamID. Learning a biological network structure from experimental data that reflects the real world relations is a challenge in itself. Where data mining, in particular graph mining, can help is in the analysis of such structure data for the discovery of useful information. such as identification of common or useful substructures and detecting anomalous or unusual structures.

In this article we survey the use of graph mining for bioinformatics problems. This topic has been heavily researched over the last decade and we review the relevant material. We take a broad view of the term *graph mining* here. Since trees are simply connected acyclic graphs we include approaches that leverage tree mining algorithms as well. Additionally within the domain of graph mining there are approaches that focus on harvesting patterns from a single large graph or network and those that focus on extracting patterns from multiple graphs. We also cover other variants of graphs in our discussion including different tree variants, directed and bi-partite graphs.

The rest of this article is broadly divided into four sections. Section 2 discusses the use of tree mining algorithms for bioinformatics problems. For example, RNA secondary structures can be represented in the form of a tree. A forest of such RNA structure trees can be employed to characterize a newly sequenced novel RNA structure by identification of common topological patterns [93]. In particular we survey the role played by frequent tree mining algorithms, tree alignment, and statistical methods in this context.

In Section 3 we discuss algorithms that target the identification of frequent sub-patterns across multiple networks. For example in a recent study [53] it was shown how 39 co-expression networks of Budding Yeast can be analyzed for coherent dense subgraphs across many of these networks. The discovered subgraphs then used to predict functionality of unknown genes. In particular we survey the role played by frequent graph mining algorithms and motif discovery algorithms in this context.

In Section 4 we discuss approaches that mine single and large biological networks for the identification of important subnetwork structures, such as identification of densely interacting communities from PPI networks or gene co-expression networks. In particular we discuss the role played by community discovery and graph clustering algorithms in the presence of uncertainty and noise in this context.

Finally in Section 5 we conclude this survey with a discussion of some open problems in the field.

2.   Mining Trees

Trees are widely used to represent various biological structures like glycans, RNAs, and phylogenies.

Glycans are carbohydrate sugar chains attached to some lipids or proteins, and they are considered the third class of information-encoding biological macromolecules subsequent to DNA and proteins. The field of characterizing and studying is known as *glycomics*, akin to genomics and proteomics. Glycans play a critical role in many biological processes including embryonic development, cell to cell communication, coordination of immune functions, tumor progression, and protein regulations and interactions. Glycans are composed of monosaccharides (sugars) that are linked by glycosidic bonds. Unlike DNA and proteins which are simple strings of nucleotides and amino acids, monosaccharides may be linked to one or more other sugars, thereby forming a branched tree structure – they are often represented as rooted ordered labeled trees. In some cases, though rare, glycans may contain cycles due to rare cyclization of carbohydrate structures (e.g., cyclodextrins) [48]. There exist a number of representation schemes (KCF [5], LINUCS [13], GLYDE [87], GlycoCT [48], and GLYDE-II [83]) and database systems (CarbBank [1], SWEET-DB [75], KEGG/GLYCAN [45], EuroCarbDB [2], GlycoSuiteDB [26]) to store glycan data.

Ribonucleic acid (RNA) is a type of molecule that consists of a long chain of nucleotide units. RNA molecules play an important role in several key functionalities which include translation, splicing, gene regulation, and synthesis of proteins. As with all biomolecules, the function of RNAs is intimately related to their structure. The secondary structure of RNAs is a list of base

---

[1] http://bssv01.lancs.ac.uk/gig/pages/gag/carbbank.htm
[2] http://www.eurocarbdb.org/

pairs satisfying certain constraints. It is formed by folding the single-stranded RNA molecule back onto itself, and it provides a scaffold for the tertiary structure [82, 107]. The secondary structure is often modeled (with some approximations) as trees [11, 34, 35, 74, 93]. Since the exact experimental determination of RNA structure is difficult [33], scientists often employ computational methods for predicting the structure of various biological molecules. These methods provide a deeper understanding of RNAs structural repertoire, and thereby help in identifying new functional RNAs.

In *Phylogenetics*, trees are used as a fundamental data structure to represent and study evolutionary connections among different organisms as understood by ancestor–descendant relationships. The Tree of Life [3] is an example of such a tree illustrating the phylogeny of life on Earth that is based on the collective evidence from many different fields of biology and bioscience. The organisms over which a phylogenetic tree is induced are referred to as *taxa*, and they form the leaf nodes in the tree. The internal nodes denote the *speciation* and *duplication* events which result in *orthologs* and *paralogs*, respectively. Speciation is the origin of a new species capable of making a living in a new way from the species from which it arose. Paralogs are genes related by duplication within a genome. While traditional Phylogenetics relied on morphological data obtained by measuring and quantifying the phenotypic properties of representative organisms, more recent studies use gene or amino acid sequences encoding encoding proteins as the basis for classification. There exist a number of different approaches to construct these trees from input data [4] – distance matrix based methods, maximum parsimony, maximum likelihood, Bayesian inference etc. The trees produced by these methods can either be rooted or unrooted. Sometimes it is possible to force them to produce rooted trees by supplying an *outgroup*, which is an organism that is clearly less related to rest of the organisms. Such an outgroup is likely to be present near the root node. We now describe different techniques to analyze such tree structured biological data.

2.1   Frequent Subtree Mining

Frequent pattern mining is one of the fundamental data mining task that asks for a set of all substructures which appear more than a (user specified) threshold number of times in a given database. The subtree patterns obtained from tree databases are extremely useful in a variety of tasks such as structure prediction, identification of functional modules, consensus substructure discovery etc. We briefly describe some of these applications below.

The common techniques that are used to infer the phylogenies such as maximum parsimony [32] usually produce multiple trees for a given set of input sequences or genes. When the number of these output trees is too large to suggest meaningful evolutionary relations, Biologists use *consensus trees* or *supertrees* in order to summarize the output trees [77, 101]. One may also use such trees to infer common relations among trees produced from multiple different tree induction methods. Shasha and Zhang have studied the quality of consensus trees by extracting frequent *cousin pairs* from a set of phylogenetic trees modeled as rooted unordered trees [95]. A cousin pair defined as a pair of nodes which share the same ancestor node. The kinship in a cousin pair is captured via a distance measure that is measured using the depth of involved nodes. Given two parameters d and θ, their algorithm extracts all cousin pairs whose distance is at most d and whose frequency is at least θ. The discovered frequent pairs are also shown to be useful in discovering co-occurring patterns in multiple phylogenies, in evaluating the quality of consensus trees, and in finding kernel trees from a group of phylogenies.

The idea of frequent cousin pairs can be extended to more complex substructures, and they can be discovered by using traditional frequent subtree mining algorithms [117, 120]. From a biological standpoint, these agreement subtrees identify the set of species that are evolutionarily related according to a majority of trees under inspection. Zhang and Wang showed that these subtrees capture more important relationships when compared to consensus trees [120]. Hadzic *et al.* have applied similar methods on the 'Prions' database that describes protein instances stored for human Prion proteins [42].

Due to common evolutionary origins, there are often common substructures among multiple structurally similar RNAs. For instance, the occurrence of smaller snoRNA motifs within the larger hTR RNA structure, indicating a functional relation between these RNAs [79]. Uncovering such structural similarities is believed to help in discovering novel functional and evolutionary relationships among RNAs, which are not easily achieved by methods like sequence alignment [34]. Algorithms to extract common RNA substructures have been applied for the purpose of predicting RNA folding [69] and in functional studies of RNA processing mechanisms [93].

More recently, frequent subtree mining have been applied on glycan databases. Hashimoto *et al.* have developed an α-closed frequent subtree mining algorithm [46]. A frequent subtree S is considered α-closed unless support(S') $\geq$ max($\alpha \cdot$ support(T), $minsup$) for any supertree S' of S, where $0 \leq \alpha \leq 1$ and $minsup$ is the user defined support threshold. It mines maximal subtrees when α is set to 0 and closed subtrees when $\alpha = 1$. Instead of ranking the resulting subtrees based on their frequency, they rank them based on statistical hypothesis testing. This is because the frequencies of subtrees are easily biased by the frequencies of constituent monosaccharides. Based on their statistical ranking method, they developed a glycan classification method that is similar to a well known linear soft margin SVMs [90]. Such a method essentially makes use of frequent subtrees obtained from a class of glycans in predicting whether or not a new glycan belongs to the given class.

2.2   Tree Alignment and Comparison

Comparison of two or more tree structures is a fundamental problem in many fields including RNA secondary structures comparison, syntactic pattern recognition, image clustering, genetics, chemical structure analysis, and Glycan structure analysis. The comparison among RNA secondary structures are known to be useful in identifying conserved structural motifs in folding process [93] and in constructing taxonomy trees [69]. The unordered tree comparisons can help in morphological problems arising in genetics – for example, in determining genetic diseases based on ancestry tree patterns [97].

Early research has focused on extending sequence matching algorithms to tree structures. The concepts related to longest common subsequence, shortest common supersequence, and string edit distance have been extended to largest common subtree (LCT) [1, 64, 118], smallest common supertree (SCS) [37, 41, 88, 110], and tree edit distance (TED) [12, 104, 119], respectively. In Phylogenetics, the longest common subtree problem is commonly referred to as Maximum Agreement Subtree (MAST) problem [36]. Biologists use MASTs to reconcile different evolutionary trees built over same taxa, and thereby to discover compatible relationships among those trees [63]. A number of efficient algorithms have been proposed for this purpose [31, 41, 64]. Aoki *et al.* studied the application of these techniques to index and query carbohydrate databases like KEGG [4].

Supertrees, on the other hand, can not only retain all or most of the information from the source trees but they can also find novel relationships which do not co-occur on any one source tree [88]. Supertrees in Phylogenetics can be built over source trees which share some but not necessarily all taxa. There are primarily two ways to build these supertrees. The first class of methods convert the topology of each source tree into a data matrix [85]. These matrices are then combined into a single large matrix, which is then used to construct the most parsimonious tree. When the given source trees are compatible, more direct methods can be used [25, 37]. In such a case, a backbone tree made up of taxa that common to given taxa is first constructed. By analyzing and thereby projecting each branch in backbone tree onto source trees, a combined supertree is constructed. The resulting supertrees are often referred to as *strict* since they do not conflict with any phylogenetic relationships in any source tree.

The tree edit distance between two trees refers to the number of minimum number of basic edit operations (relabel, insert, and delete) required to transform one tree into the other. This notion was first explored by Selkow [92], which was later generalized by Tai [104]. This conventional definition of edit distance has been extended to include more complex operations such as subtree insertions, subtree moves etc. [18, 17]. There has been a tremendous amount of work being done in developing fast algorithms to compute tree edit distance for both ordered and unordered trees. Most of the algorithms, similar to methods which compute string edit distance, follow dynamic programming based approaches. Bille has recently surveys several important algorithms that solve this problem [12]. These concepts have further been extended to RNA structures by taking their primary, secondary, and tertiary structures into account [40, 57].

Jiang et al. introduced the idea of *tree alignment* [58], which is in spirit similar to sequence alignment. An alignment between two trees is obtained by first inserting special nodes (labeled with spaces) into both trees such that the resulting trees have same structure. A cost model is defined over the set of opposing labels. The problem then is to find an optimal alignment which minimizes the sum of the costs of all opposing pairs [112]. Hochsmann *et al* designed a method for computing multiple alignments of RNA secondary structures, which was then used used to cluster RNA molecules purely based on their structure [50]. Bafna and Muthukrishnan presented a method to align a given RNA sequence with some unknown secondary structure to one with known sequence and structure. Such a method helps in RNA structure prediction in the case when the structure of a closely related sequence is known [9].

Glycan structure alignment techniques have been proposed by using traditional tree alignment algorithms and glycosidic linkage score matrices. These alignment techniques, just like popular sequence alignment methods, are useful when analyzing newly discovered glycans. Aoki *et al.* have proposed KCaM [5], an extension of popular Smith-Waterman sequence alignment technique [98], to perform exact and approximate glycan alignment. The approximate algorithm aligns monosaccharides while allowing gaps in the alignment, and the exact matching algorithm aligns linkages while disallowing any gaps, thus resulting in a stricter criterion for alignments. In a similar spirit, Aoki *et al.* have developed a glycan substitution matrix [2] to measure the similarity between monosaccharides, as in amino acid similarity represented by amino acid substitution matrices like BLOSUM [47]. Such a matrix can be used to discover those links that are positioned similarly, and thus potentially denote similar functionality. Thus, it is can be used to improve the alignment algorithms like KCaM to produce more biologically meaningful results. Kawano *et al.* have developed techniques to predict glycan structures from incomplete or noisy data such as DNA microarray data by making use of knowledge about known glycan structures from KEGG GLYCAN database [62].

There is also an interesting notion of tree alignment, when the problem is discussed with respect to phylogenetic trees. While the traditional tree induction methods act upon sequence data to estimate the tree structure, tree alignment methods operate in reverse direction. More precisely, given a set of sequences from different species and a phylogenetic tree depicting the ancestral relationship among these species, compute an optimal alignment of the sequences by the means of constructing a minimum-cost evolutionary tree. Such methods are useful in determining the possible ancestral molecular sequences (which correspond to internal nodes in the tree) that gave rise to the extant sequences through a series of mutational events [56, 113].

2.3   Statistical Models

While analyzing glycan structures, unlike in phylogenies and RNA structures, it is often important to capture dependencies that are not bounded simply by the edges of the tree structure. In order to learn such patterns, a tree structured probabilistic model called as the Probabilistic Sibling-dependent Tree Markov Model (PSTMM) was developed [3, 108, 109]. It incorporates not only the dependency between a node and its its parent but also between a node and its eldest sibling. EM based learning algorithms were also proposed to learn parameters of the model. Hashimoto *et al.* have improved this for computational complexity by proposing ordered tree Markov model (OTMM) [44]. Instead of incorporating dependencies to both elder sibling and parent from each node, it uses only one dependency – where the eldest sibling depended only on the parent, and each younger sibling only depended on its older sibling. These methods have been applied to align multiple glycan trees, and thereby to detect biologically significant common subtrees in these alignments, where the trees are automatically classified into subtypes already known in glycobiology.

Ohtsubo and Marth showed that many motifs are involved in a variety of diseases including cancer i.e., these motifs act as *biomarkers* [81]. They also showed that the methods to predict characteristic glycan substructures (motifs) from a set of known glycans may be useful in predicting biomarkers of interest. Several research works have developed kernel methods for glycan biomarker classification and prediction. Hizukuri *et al.* developed a similarity measure known as *trimer kernel* for comparing glycan structures that takes the biological properties of involved glycans into account [49]. They have subsequently used this method in the framework of Support Vector Machines (SVMs) to extract characteristic functional units (motifs) specific to leukemia. This method was further extended by Koboyama *et al.* who developed a kernel that measures the similarity between two between two labeled trees by counting the number of common q-length substrings known as *tree q-grams* [68]. Recently, Yamanishi *et al.* have developed a class of kernel functions which can be used for classifying glycans and detecting discriminative glycan motifs with SVMs [114]. The hierarchical model that they proposed handles the issue of large number of features required by the q-gram kernel. A kernel for each q was first developed, upon which another kernel was trained to extract the best feature from the best kernel.

3. Mining Graphs for the Discovery of Frequent Substructures

Graphs are important tools to model complex structures from various domains. Further characterization of these complex structures can be accomplished through the discovery of basic substructures that are frequently occurring. Identification of such repeating patterns might be useful for diverse biological applications such as classification of protein structural families, investigation of large and frequent sub-pathways in metabolic networks, and decomposition of Protein Protein Interaction (PPI) graphs into motifs. In this section, we focus on mining frequent subgraphs from biological networks. First, we look at various methods to identify subgraphs that occur frequently in a large collection of graphs. Next, we discuss substructures that occur significantly more often than expected by chance in a single and large graph, which are known as motifs. We cover different strategies for identification of such structures and their applications on diverse biological networks.

3.1   Frequent Subgraph Mining

Frequent subgraph mining (FSM) aims to find all (connected) frequent subgraphs from a graph database. More formally, given a set of graphs G, and a support threshold $minSup$, FSM finds all subgraphs (s_G) such that fraction of graphs in G of which s_G is a subgraph is greater than the $minSup$. There are two major challenges that are associated with FSM analysis: subgraph isomorphism and efficient enumeration of all frequent subgraphs. Subgraph isomorphism problem, which is an NP-complete problem, detects whether two given networks have the same structure. Therefore, time and space requirements for the existing FSM algorithms increase exponentially with the increasing pattern size and number of graphs. To design algorithms that scale to large biological graphs, techniques that simplify the problem by alternative graph modeling or graph summarization have been proposed. These algorithms are successfully utilized on diverse biological graphs for various purposes, including the identification of recurrently co-expressed gene groups and detection of frequently occurring subgraphs in a collection of metabolic pathways.

Koyuturk *et al.* developed a scalable algorithm for mining pathway substructures that are frequently encountered over different metabolic pathways [66]. A metabolic pathway is defined as a collection of metabolites M, enzymes Z, and reactions R. Each reaction $r \in R$, is associated with a set of enzymes ($Z(r) \in Z$) and a set of substrates and products which are metabolites. The algorithm aims to discover common motifs of enzyme interactions. Therefore, they re-model the metabolic pathways as directed graphs which emphasize enzyme interactions. In their representation, nodes represent enzymes, and a directed edge from an enzyme to another implies that the product of the first enzyme is consumed by a reaction catalyzed by the second. After constructing a collection of these graphs, they mine this collection to identify the maximal connected subgraphs that are contained in at least a pre-defined number of these graphs, where this number is determined by the support threshold. This model enforces unique node labeling to eliminate the subgraph isomorphism problem. This enforcement also enables the use of frequent itemset mining algorithms for the problem at hand by specifying edge-sets as the itemsets. In frequent itemset mining problem, each transaction is a collection of items, and the problem is to identify all frequent sets of items that occur in more than a specified number of these transactions. Koyuturk *et al,* reduced their problem into a frequent itemset mining problem by enforcing a connectivity constraint on edge-sets. They proposed an extension to a previously suggested frequent-itemset mining algorithm based on backtracking [38] which grows candidate subgraphs by only considering edges from a candidate edge set. Using their algorithm pathway graphs of 155 organisms collected from the KEGG database have been analyzed. They extracted considerably large sub-pathways that are frequent across these organism-specific pathway graphs. An example discovered sub-pathway of glutamate includes 4 nodes and 6 edges and it occurs in 45 of the 155 organisms. In a latter work, You *et al* applied SUBDUE system to obtain meaningful patterns from metabolic pathways [116]. SUBDUE is a system that identifies interesting and repetitive substructures based on graph compression and the minimum description length principles [51]. The best graphical pattern S that minimize the description length (MDL) of itself and that of the original input graph G when it is compressed with pattern S is identified with this system. First they identify the best pattern in G, which minimizes the MDL based criteria. Next, S is included into a hierarchy, where G is compressed with S. All such patterns in the input graph G are obtained, until no more compression is possible. The SUBDUE system is successfully applied on metabolic pathways to find unique and common patterns among a collection of pathways [116].

Another major application of FSM in biological domain is the identification of recurrent patterns from many gene co-expression networks. Gene co-expression networks are built on the basis of mRNA abundance measured by microarray technologies. In a gene co-expression network, nodes represent genes, and two nodes are linked if the corresponding genes have significantly similar expression patterns over different microarray samples. Similarity between two genes is typically measured by the absolute value of the correlation coefficient between their expression profiles [52]. Next, based on a thresholding procedure, co-expression similarities are transformed into a measure of interaction strength. Different gene association networks can be constructed using different thresholding principles, i.e., hard or soft thresholding [52]. Although a gene co-expression network derived from a single microarray study can include many spurious edges, a recent study pointed out that genes co-expressed across multiple studies are more likely to be real and to correspond to functional groups [70]. Therefore, mining frequent gene groups across many gene co-expression networks has drawn recent attention. However, extant FSM algorithms do not scale to large gene co-expression graphs. In addition, as pointed by Hu *et al.*, frequency concept may not be enough to capture biologically interesting substructures. For this purpose, they proposed an algorithm, named CODENSE [53], that identifies frequent, coherent, and dense subgraphs across large collection of co-expression networks. According to their definition, all edges of a coherent subgraph frequently co-occur (and not co-occur) in the whole set of graphs. On the other hand, in a dense subgraph, the number of edges is close to the maximal possible number. Thus, coherent and dense structures better represent biological modules. Their algorithm starts with building a summary graph by eliminating infrequent edges from the input graphs. Another algorithm developed by the same group, MODES algorithm, is employed to extract dense subgraphs of the summary graph. For each of these dense summary subgraphs, edge occurrence profiles which is a binary matrix that indicates occurrence of dense summary graph edges in the original set of graphs are constructed. Using these profiles, a second-order graph is built to indicate the co-occurrence of edges across all graphs. In this representation, each edge is transformed into a node, and two nodes are connected if their corresponding edge occurrence profiles show high similarity. They shoved that coherent graphs across input graphs will be dense in the second-order graph. Therefore, at the final step of the CODENSE, dense subgraphs of the second-order graph are identified. CODENSE algorithm is scalable as it operates on two metagraphs, namely summary graph and second order graph, instead of operating on individual networks. Dense patterns of these meta structures are identified, instead of patterns from individual graphs. It is also adjustable for exact or approximate pattern matching. CODENSE is applied on 39 co-expression networks of Budding Yeast organism to obtain functionally homogeneous gene clusters. These clusters are further employed in order to predict functionality of 169 unknown yeast genes. They showed that a significant portion of their predictions are supported by the literature [53].

CODENSE assumes that frequent subgraphs will be coherent across all graphs, on the other hand, it is possible to have subgraphs that are coherent only in a subset of these graphs. In order to take this fact into consideration, Huang *et al.* proposed an algorithm based on biclustering [55]. They start by identifying bi-cluster seeds from edge occurrence profiles. First, sub-matrices that are all 1s are identified from the edge co-occurrence matrix. Then, based on a Simulated Annealing methodology these initial structures are expanded. Connected components among these expanded seeds are identified and returned by their algorithm as recurring frequent subgraphs. They employed their algorithm on 65 co-expression datasets obtained from 65 different microarray studies. In a follow-up work conducted to identify frequently occurring gene subgraphs across many co-expression graphs, Yan *et al.* [115] studied a stepwise algorithm which constructs a neighbor association summary graph by clustering co-expression networks into groups. A neighbor association summary graph measures the association of two vertices based on their connections with their neighbors across input graphs. Two vertices that co-occur in many small frequent dense vertex sets have a high weight in the neighbor association graph. Once they build the neighbor association graph, they decompose it into (overlapping) dense subgraphs and then eliminate discovered dense subgraphs if their corresponding vertex-sets are not frequently dense enough. They named their algorithm NeMo for Network Module Mining. NeMo is applied on 105 human microarray datasets and recurrent co-expression clusters are identified. Functional homogeneity of these clusters are validated based on ChIP-chip data and conserved motif data [115].

For the automatic identification of common motifs in most any scientific molecular dataset, MotifMiner, a general and scalable toolkit has been proposed [23]. MotifMiner represents the information between a pair of nodes (atoms), A_i and A_j, as a mining bond. The mining bond $M(A_i, A_j)$ is a triplet of $< type(A_i), type(A_j), attr(A_i, A_j) >$ form. The information contained in $attr(A_i, A_j)$ vary depending on the resolution of the structure. As an example, if the structure is at the atomic level, $attr(A_i, A_j)$ can contain the distances between atoms A_i and A_j. This enables the flexibility to analyze several disparate domains, including protein, drug, and MD simulation datasets. Using mining bond definition, a k size structure is defined as $str_k = S, A_1, ..., A_k$, where A_i is the ith atom and S is the set of mining bonds describing this structure. MotifMiner employs a Range pruning methodology to limit the search for viable strongly connected sub-structures and a Candidate pruning methodology to prune the search space of possible frequent structures. In addition, Recursive Fuzzy Hashing is used for rapid matching of structures while determining the frequency of occurrence. Distance Binning and Resolution principle is also proposed to work in conjunction with Recursive Fuzzy Hashing to handle noise in the input data. MotifMiner has been evaluated on various datasets, including pharmaceutical data, tRNA data, protein data, molecular dynamics simulations [24]. In a follow-up study, Li *et al.* proposed several extensions, i.e., sliding resolution, handling boundary conditions, and enforcing local structure linkage, to the MotifMiner algorithm [72] in order to improve both the running time and the quality of the results. They also incorporated the domain constraints into the original MotifMiner algorithm for mining and aligning protein 3D structures. To evaluate the efficacy of the revised algorithm they used it to align the proteins Rad53 and Chk2, both of which contain FHA domain. FHA domains have very few conserved residues, which limits the use of sequence alignment algorithms for their alignment. The aligned result (depicted in Figure 18.1) is similar to structure-aided sequence alignment done manually [29], particularly at structurally similar regions. In a more recent work, a parallel implementation of this toolkit has been proposed [111]. The parallelized version demonstrate good speedup on real-world datasets.

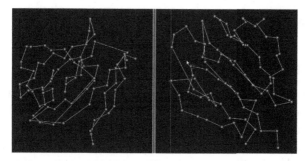

Figure 18.1. Structural alignment of two FHA domains. FHA1 of Rad53 (left) and FHA of Chk2 (right)

Jin *et al.* generalized the problem of frequent subgraph mining to mine frequent large-scale structures from graphs [59]. They developed a framework, Topological Structure Miner (TSMiner), that is based on a well-established mathematical concept known as topological minor. A topological minor of a given graph can be obtained by contracting the independent paths of one of its subgraphs into edges. Topological structures of a graph are derived from topological minors. Frequent subgraphs of a graph can be mined as a special case of frequent topological structures, but their framework is able to capture structures missed by standard algorithms. They proposed a scalable incremental algorithm to enumerate frequent topological structures. The concept of occurrence lists in order to efficiently count the support of a potential frequent topological structure is introduced. They employed this tool to search for potential protein-lipid binding sites in membrane proteins. Six membrane proteins, that are known to bind with cardiolipins (CL), are first represented in the form of graphs. In these graphs, amino acids represent nodes (20 different labels) and links exist between nodes if two amino acids are close enough to each other.

Two of the topological structures discovered with their toolkit are depicted in Figure 18.2. Such large structures cannot be obtained by using standard motif mining algorithms. As noted by the authors, the identified topological structures are mainly composed of polar (N, T, S), charged (K), and aromatic (W) residues, which is in agreement with biophysics literature.

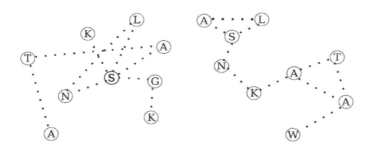

Figure 18.2. Frequent Topological Structures Discovered by TSMiner

3.2 Motif Discovery in Biological Networks

In addition to subgraphs that are frequent across many networks, substructures that are repeated frequently within a single and large network can be useful for knowledge discovery. A motif of a graph refers to a substructure, which is repeated considerably inside the graph. There are two main approaches, frequency-based and statistical, to determine the significance of this repetition. The frequency-based approach considers a subgraph as a motif if it is occurring more than a threshold number of times. On the other hand, statistical approach labels a subgraph as motif if it is occurring more than the expected number of times with respect to random networks. Network motifs can be particularly effective in understanding the modularity and the global structure of biological networks. For example in the case of PPI networks, motifs can be useful for the identification of protein complexes and other protein groupings that are related to the mechanics of the living organism. In the case of regulatory networks, motifs enable understanding gene regulation mechanisms and it also enables researchers to develop models and experiments to understand these mechanics.

Milo *et al.* is the first to define network motifs and find them in networks from biochemistry, neurobiology, ecology, and engineering [78]. They defined network motifs as *patterns of interconnections occurring in complex networks at numbers that are significantly higher than those in randomized networks*. Their analysis revealed some common (and diverse) motifs across fields. As an example, they shoved that the directed triangle motif, known as the feed-forward loop, exists in both transcription-regulatory and neural networks, whereas four-node feedback loops only emerge in electric circuits but not in biological systems. To identify such motifs, Milo *et al.* exhaustively enumerated all subgraphs of n nodes in the studied networks, where n is limited to 3 and 4. They then generated random networks while keeping the number of nodes, links and the degree distribution unchanged. Subgraphs of these random networks are counted and these counts are used to determine motifs. As an alternative to exact counting, in a follow-up work they proposed a *sampling method for subgraph counting* [61]. Instead of enumerating subgraphs exhaustively, subgraphs are sampled to estimate their relative frequency. The method starts by picking a random edge from the network and then expanding the corresponding subgraph iteratively by picking random neighboring edges. At each iteration, a list of all candidate edges are generated for the next random pick. The subgraph is expanded until it reaches a pre-defined size. Although being an extension over the exhaustive search, this algorithm is also limited to finding small-size motifs. In the transcription network of E. coli, subgraph samples of sizes 3 to 8 have been reported. Higher order motifs composed of five and six nodes in this network are tabulated in their study [61].

Protein-protein interaction networks accumulate pairwise or group-wise physical interactions of proteins into a network structure. Motifs of these networks can be utilized to characterize and better understand the group-level relations. For identification of large size motifs in Protein-Protein Interaction (PPI) networks, a scalable algorithm, NEtwork MOtif FINDER [19] has been proposed as an extension to subgraph mining algorithms. This algorithm is based on formation of frequent trees of varying size from 2 to k, which are then used to partition the graph into a set of graphs such that each graph embeds a size-k tree. In the next step, frequent size-k graphs are generated by performing graph join operations. Frequency of these size-k graphs can be counted in randomized networks. NEMOFINDER describes frequent subgraphs that are also unique as *Network Motifs*. Uniqueness of a subgraph is determined by the number of times a subgraph is more frequent in the real graph than randomized graphs. Existing Apriori-based algorithm are not able to capture interesting network motifs that are repeated and unique. Uniqueness of these size-k graphs are calculated based on their number of occurrences in real input graph and the randomized graphs. They build their algorithm as an extension to the SPIN [54] algorithm with the possibility of overlapping subgraphs. The input to the NEMOFINDER algorithm is a PPI network, and user defined thresholds for frequency, uniqueness, and maximal network size. The algorithm outputs Network Motifs that are frequent and unique with respect to the defined thresholds. Employing their algorithm on the PPI network of budding yeast, they discovered motifs up to size 12. They later proposed an extension to the NEMOFINDER, named LaMoFinder, which takes into consideration labels of nodes [20]. While applying LaMoFinder to discover PPI network motifs, they used Gene Ontology terms as node labels [20]. They first mine an unannotated network for motifs. Next, motifs are labeled with Gene Ontology functions. Their analysis showed that by incorporating labels they are not able to capture only the topological shapes but also biological context of motifs. Labeled motifs extracted from a real world PPI network are employed for protein function prediction.

In a more recent work, Grochow and Kellis [39] proposed an algorithm to avoid the limitations of exact counting and subgraph sampling based motif mining algorithms. Their algorithm works by exhaustively searching for instances of a single query graph in a network. They proposed a motif-centric alternative to existing methods which is based on an improved isomorphism test, i.e., symmetry breaking. The algorithm identifies all instances of a query graph H, given a target network G. They extended isomorphism test based on the most constrained neighbor concept. They defined the most constrained neighbor of the already-mapped nodes which is the least possible nodes to be mapped to. They also introduced and enforced several symmetry-braking conditions, to make sure that there is a unique map from the query graph H to each instance of H in G. They utilized their algorithm to find motifs in two biological networks: PPI network of S. cerevisiae and Transcriptional network of S. cerevisiae. The former is composed of 1379 nodes and 2473 edges, where motifs of 15 and 20 nodes can be identified with the proposed algorithm. From the latter one, which has 685 nodes and 1052 edges, a 29-node motif that corresponds to the cellular transcription machinery has been identified. In addition to being scalable for finding larger motifs, this algorithm also enables exploring motif clustering and querying a particular subgraph. Moreover, the algorithm is very easy to parallelize by counting each subgraph on a separate processor.

4.   Mining Graphs for the Discovery of Modules

Different forms of real-life associations between biological entities have been detected by various technologies and these associations have been accumulated in the public databases in the form of complex networks. Understanding these complex structures often require breaking them into small components and identifying the interactions between these components. These components are composed of nodes which are more relevant to each other than with outsiders and they are commonly referred as communities or modules. Decomposition of a given graph into its modules can also be very effective in the analysis of biological networks. Some biological networks are naturally decomposed into such components, which are commonly referred as modular networks. Some examples of biological modules are transcriptional modules, protein complexes, gene functional groups, and signaling pathways.

The most well-known biological modular networks is the Protein-Protein Interaction(PPI) Network. The number and coverage of public databases that collect experimental data on protein physical bindings of diverse organisms have been increasing with the advancements in high-throughput techniques. Although there is no established standard database of PPIs today, there have been efforts to integrate existing interactions in publicly available databases. As of today, Human Protein Reference Database (HPRD) footnote(http: //www.hprd.org) includes 34,624 Protein-protein interactions between Human proteins that are derived from a number of platforms such as Mass Spectrometric Analysis, Yeast two-hybrid based protein-protein interaction, and Co-immunoprecipitation and mass spectrometry-based protein-protein interaction. Similarly, another freely accessible database BIOGRID [100] includes more than 238,634 raw interactions from various organisms including Saccharomyces cerevisiae, Caenorhabditis elegans, Drosophil melanogaster and Homo sapiens. These large collections of protein interactions are naturally represented in the form of networks to facilitate the process of knowledge discovery. Modular nature of these networks has been investigated by different algorithms and the identified modules have been utilized for a better characterization of the unknown proteins.

Gene co-expression networks are another example of biological networks that exhibit modular structure [15, 102]. In these network structures, nodes represent genes and edges between nodes refer to genes that are expressed similarly over studied conditions. Gene groups that indicate a similar expression pattern can be defined as a gene module, where a functionality between the elements of this module is likely to be shared [91, 102]. Another modular biological network that have been excessively studied is the Regulatory networks. They model activation (or suppression) of a gene by specific DNA binding proteins in the form of a directed graph. Modules that can be deduced from regulatory networks correspond to a set of co-regulated genes as well as their common regulators. Given all these application areas, effective identification of modules from diverse biological networks has great potential for a better understanding of studied organisms.

In this section we discuss different methodologies that are proposed for the detection of network modules or communities in biological graphs. Here, a community can be defined as a densely connected group of nodes, where only a few connections exist between different communities [80]. First, we look at algorithms that extract community structures from networks. Next, we discuss clustering algorithms that have been proposed to decompose the whole structure into subgroups, where similarity within group elements is maximized, and between groups is minimized.

4.1   Extracting Communities

In the analysis of PPI networks, of particular interest to many scientists is to study protein interaction networks to isolate densely interacting regions, also known as communities, since they are presumed to be protein complexes or functional modules. A protein complex can be defined as a set of proteins that bind to each other in order to accomplish a cellular level task. Identification of these structures is useful to understand cell functioning, to predict functionality of unknown proteins. The interest in their identification is motivated by the fact that proteins heavily interacting within themselves, usually participate into the same biological processes. Thus, discovery of dense subgraphs from PPI networks is recognized as an important task for the identification of protein complexes. Based on this underlying principle, a set of algorithms that employ local dense regions of PPI networks to discover putative complexes have been proposed.

Bader et al [8] proposed a three-step algorithm; Molecular COmplex DEtection (MCODE) to identify clusters of proteins that are heavily interacting. MCODE starts with weighting each node of the network based on the density of its local neighborhood. Next, nodes with high weights are assigned as seeds and starting from these seed nodes initial clusters are obtained by iteratively including neighboring nodes to the cluster. Finally an optional third step is proposed to filter proteins according to a connectivity criteria. They evaluated MCODE on an integrated dataset of Budding Yeast that is composed of 9088 protein-protein interactions among 4379 proteins from the MIPS, YPD, and PreBIND databases. They predicted 166 complexes from this network. 52 of these complexes matched with known protein complexes in the MIPS database. MCODE bases on the observation that proteins share functions with their immediate neighbors. In a more recent work, Chua et al utilized another observation based on level-2 interactions in PPI networks [22]. They derived a topological weighting schema, namely the Functional Similarity Weight (FS-Weight) that enables weighting both direct and indirect (i.e., 'level-2') interactions. FS-Weight makes use of estimated reliability of each interaction to reduce the impact of noise. The reliability of each experimental source is estimated by the fraction of unique interactions in which at least one level-4 Gene Ontology term is shared. FS-Weight also favors two proteins that share many common neighbors from a reliable source. Number of non-common neighbors are also included into the calculation in order to reduce potential false positive inferences. Based on FS-weights, the studied PPI network is expanded with 'level-2' interactions and filtered by eliminating interactions with small FS-weights. After this preprocessing step, they identify cliques in the modified PPI network and iteratively merged cliques to form larger subgraphs that are still dense. More recently, Li et al [73] proposed an algorithm named DE-

CAFF (Dense Neighborhood Extraction using Connectivity and conFidence measures) which employs the Hub Removals algorithm [86]. DECAFF initially identifies local dense neighborhoods of each protein by iteratively removing nodes with low degrees from the local neighborhoods. These local cliques are merged with the dense subgraphs detected by the Hub Removal algorithm [86] based on a Neighborhood Affinity criteria. Neighborhood Affinity of two subgraphs is calculated based on their size and the number of their common neighbors. Finally DECAFF improves the quality of final clusters by removing subgraphs with low reliability scores. The reliability of a subgraph is defined as the average reliability of all interactions of that subgraph, where interaction reliability is deduced from functional relevance of its two interacting proteins.

In addition to PPI networks, scientists are also interested in identifying community structures from gene co-expression networks. Expression profiles obtained through microarray studies can be transformed into gene co-expression networks, where nodes represent genes and two nodes are linked if the corresponding genes behave significantly similar across different samples (i.e., co-expression). Scientists are particularly interested in the problem of identifying gene subnetworks that have similar expression patterns under different conditions [103] since they have been theorized to have the same cellular function [30]. To find gene groups that have similar expression patterns, Hartuv and Shamir proposed an algorithm that recursively splits the weighted co-expression graph into its highly connected components [43]. A highly connected component is defined as a subnetwork which includes at least two nodes, i.e., $n > 1$, and which can only be disconnected after the removal of more than $n/2$ edges. Their algorithm, namely the *Highly Connected Subgraphs*(HCS), at each iteration splits the network into subgraphs until a highly connected component is identified. Shamir and Sharan [94] proposed an extension of the HCS algorithm, CLICK - CLuster Identification via Connectivity Kernels. In each step of their algorithm, a minimum cut of the input graph is computed, which outputs two subgraphs. Subgraphs which satisfied certain criterion are labeled as kernels. Each kernel is attributed with a fingerprint similarity that is calculated based on its elements. After all the kernels are identified, nodes that are not part of any kernels are further analyzed and the ones that are similar to any of the kernels are included into the kernel and the kernel's fingerprint is re-calculated - adoption step in the algorithm. Next, kernels that are similar enough are merged and the adoption operation is repeated. Adoption and kernel merging steps are repeated until there are no more changes in the kernel structures. Final kernels are outputted as gene clusters obtained by the CLICK algorithm. They have shown that their algorithm outperform existing clustering algorithms when applied on various gene expression datasets, originating from various studies, such as the yeast cell cycle dataset, or the response of human fibroblasts to serum.

Regulatory modules can be inferred from diverse datasets including ChIP-chip, motif, and gene expression datasets. A regulatory module is composed of a set of genes that are co-regulated by a common set of regulators. In order to identify such modules from ChIP-chip data and gene expression profiles, GRAM algorithm is proposed [10]. A set of genes that are bind with the same regulator set is obtained from the ChIP-chip binding p-values with an exhaustive search. Subsequently, a subset of this set that are similarly expressed is selected to serve as a seed. Then, the algorithm identifies genes that are similarly expressed with the seed genes and that are connected to the same set of transcription factors based on a relaxed binding criteria. Lemmens *et al.* improved the GRAM algorithm by incorporating motif data as an additional source [71]. In the seed discovery step, they discover seeds composed of genes that are co-expressed (deduced from mRNA measurements), that bind to the same regulators (deduced from ChIP-chip data), and that have the same motifs in their intergenic regions (deduced from Motif data). they employed an Apriori-like algorithm in order to identify such seeds. And a p-value is assigned to asses the quality of each seed. In the second seed extension step, gene content of the seeds are extended. For this purpose, each gene is ranked according to their correlation with the mean expression profile of the seed genes, and the ones that are similar enough (according to a cut-off) are included into the module. They employed their algorithm for the discovery of Budding Yeast regulatory modules by integrating ChIP-chip, motif, and gene expression datasets.

4.2 Clustering

Clustering algorithms can also be effective in identifying the modules of biological networks. In contrast to community discovery approaches, clustering (or graph partitioning) decompose the whole network structure into groups. A clustering algorithm locates every node of the graph into a community or a module.

To elucidate gene functions at a global scale, clustering of gene co-expression networks have been investigated. Since genes that are on the same pathways or belong to the same functional complexes are often co-regulated, they often exhibit similar expression patterns under diverse conditions. Thus, identifying and studying groups of highly-interacting genes in co-expression networks is an important step towards characterizing genes at a global scale. For this purpose, a variety of existing graph partitioning algorithms can be leveraged. Spectral methods that target weighted cuts [96] form an important class of such algorithms. Multi-level graph partitioning algorithms such as Metis [60] and Graclus[27] are well known to scale well for large networks.

Divisive/agglomerative approaches have also been popular in network analysis [80], but they are expensive and do not scale well [16]. Markov Clustering (MCL) [28], a graph clustering algorithm based on (stochastic) flow simulation, has proved to be highly effective at clustering biological networks [14]. A variant of this algorithm known as MLR-MCL [89] have been proposed recently to address the scalability of MCL algorithm.

In addition to these diverse graph partitioning algorithms, other classical clustering algorithms have also been employed – e.g., the hierarchical clustering [99], the k-means clustering [76], and the self-organizing maps [65]. Besides the application of standard clustering algorithms, clustering algorithms that are more suitable for the specific task have been studied. Among these are the biclustering algorithms which identify a group of genes that behave similarly only for a subset of all conditions. Given a gene expression matrix of samples and genes, biclustering algorithms perform clustering in two dimensions simultaneously [21]. Statistically significant sub-matrices of a subset of genes and a subset of samples are the identified biclusters. Cheng and Church proposed a greedy approach in order to find maximal sized biclusters that satisfy a certain condition on the residue scores [21]. Their algorithm identifies each biclusters separately by iteratively removing rows and columns until the mean squared residue score for the sub-matrix (an assessment for the quality of bi-cluster) is smaller than a threshold and by iteratively adding rows and columns while the quality assessment score does not exceed threshold. Each run of the algorithm identifies a sub-matrix (bi-cluster) separately, and the next bi-cluster is identified after the found sub-matrix is masked by randomization. using this algorithm, they identified biclusters from gene expression datasets of Human and Yeast. Later, Koyuturk *et al* proposed a work which associates statistical significance to the extracted biclusters. To discover binary biclusters from a quantized gene expression matrix, they formulate this problem as an optimization problem based on the statistical significance objective. Fast heuristics are proposed so solve this optimization problem in a scalable manner. The algorithm is tested on quantized breast tumor gene expression matrix [67]. Tanay *et al.* converted bi-clustering problem into a graph theory problem using bi-partite modeling [106]. Initially the expression data is converted into a bi-partite of genes and samples. More formally a graph $G(V, S, E)$ is constructed where V is set of genes, S is set of conditions, and there exists and edge between v and s, $(v, s) \in E$ if, g is expressionally responsive in sample s. This modeling reduces the biclustering problem into the problem of finding the densest subgraphs in G. Since the identification of heaviest bi-clique is an NP-complete problem, authors restricted the search space by assuming a degree bound on one side of the bipartite graph. Later Tanay applied SAMBA algorithm on the gene expression dataset of 96 human tissue samples [105]. In that work, they compared their work against, Cheng

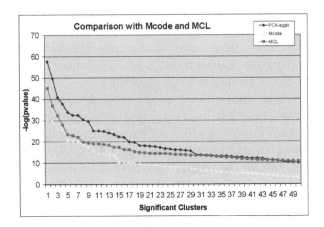

Figure 18.3. Benefits of Ensemble Strategy for Community Discovery in PPI networks in comparison to community detection algorithm MCODE and clustering algorithm MCL. The Y-axis represents -log(p-value).

and Church's algorithm [21] and observed that biclusters from SAMBA are better in terms of their statistical significance.

An ensemble clustering algorithm is also studied on biological networks to generate a more robust clustering compared to individual clustering algorithms [6]. Cluster ensembles can be defined as a mapping from a set of clusterings generated by a variety of sources into a single consensus clustering arrangement. Asur *et al.* proposed an ensemble clustering for the PPI decomposition problem. First different topological weighting schemes are proposed to generate different views of the unweighted PPI network. Next, these different views are clustered with different algorithms to obtain a set of base clusterings of the network. These clusterings are integrated into a Cluster Membership Matrix which is reduced in size to eliminate redundancy and to scale the consensus determination problem based on PCA. Subsequently standard hierarchical clustering algorithms are utilized for computing the consensus clustering (recursive bisections (PCA-rbr) and agglomerative clustering (PCA-agglo)). When compared with existing community detection and clustering algorithms, they observed that their algorithm is able to produce topologically and biologically more significant clusters (as shown in Figure 18.3). The Y-axis represents distribution of Gene Ontology enrichment p-values. Smaller p-values represent more significantly enriched groups with a particular Gene Ontology term.

In addition to biclustering and ensemble clustering strategies, scientists also studied soft clustering algorithms for biological networks, which enables assigning multiple-cluster membership to multi-faceted biological entities. To enable multiple cluster membership for proteins while identifying PPI clusters, Asur et al [6] proposed a soft ensemble clustering technique that is a step further from their PCA based consensus clustering. This adapted algorithm, after obtaining the initial consensus clustering, iteratively calculates the strength of each protein's membership to each consensus cluster based on shortest path distances. Proteins that have high propensity towards multiple membership are then assigned to their alternate clusters. To test the efficacy of this soft clustering algorithm, the compared their algorithm with the original ensemble clustering. As can be seen in Figure 18.4, they observed that, allowing multiple membership to proteins, improves the overall accuracy of the clustering, as evident from the smaller p-values of GO enrichment analysis.

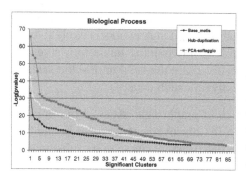

Figure 18.4. Soft Ensemble Clustering improves the quality of extracted clusters. The Y-axis represents -log(p-value).

A soft bi-clustering algorithm (MF-PINCoC), an extension to the algorithm PINCoC, has been proposed to identify overlapping dense subgraphs by using a local search technique has been proposed recently [84]. The PINCoC algorithm applies a greedy search strategy in order to find the local optimal sub-matrices in terms of a quality function. More recently, Avogadri *et al.* proposed an ensemble fuzzy clustering for decomposing gene expression datasets into its overlapping clusters [7]. They first generate multiple views of the data by using random projections. A random projection maps data from a high-dimensional space to a lower dimensional space. On these views, they applied fuzzy k-means algorithm and these fuzzy clustering arrangements are combined into a similarity matrix. They again employed fuzzy k-means on this similarity matrix to identify fuzzy consensus clustering [7]. This algorithm is applied on four different microarray datasets and compared against different ensemble strategies.

5.   Discussion

In this article we surveyed the principal results in the field of graph mining that relate to the application domain of bioinformatics. We examined these results along three directions: i) from the perspective of mining tree-structured data; ii) from the perspective of mining multiple graphs or networks; and iii) from the perspective mining of mining a single (large) network in the presence of noise and uncertainty.

Both data mining and the field of bioinformatics are young and vibrant and thus there are ample opportunities for interesting lines of future research at their intersection. Sticking to the theme of this article – graph mining in bioinformatics – below we list several such opportunities. This list is by no means a comprehensive list but highlight some of the potential opportunities researchers may avail of.

- Scalable algorithms for analyzing time varying networks: A large majority of the work to date in this field has focused on the analysis of static networks. While there have been some recent efforts to analyze dynamic biological networks, research in this arena is at its infancy. With anticipated advances in technology where much more temporal data is likely to become available temporal analysis of such networks is likely to be an important arena of future research. Underpinning this effort, given the size and dynamics of the data involved are the need to develop scalable algorithms for processing and analyzing such data.

- Discovering anomalous structures in graph data: Again while most of the work to date has focused on the discovery of frequent or modular structure within such data – the discovery of anomalous substructures often has a crucial role to play in such domains. Defining what constitutes an anomaly, how to compute it efficiently while leveraging the ambient knowledge in the domain in question are some of the challenges to be addressed.

- Integrating data from multiple, possibly conflicting sources: A fundamental challenge in bioinformatics in general is that of data integration. Data is available in many formats and often times are in conflict. For example protein interaction data produced by various experimental methods (mass spectrometry, Yeast2Hybrid, in-silico) are often in conflict. Research into methods that are capable of resolving such conflicts while still discovering useful patterns are needed.

- Incorporating domain information: It has been our observation that often we as data mining researchers tend to under-utilize available domain information. This may arise out of ignorance (the field of bioinformatics is very vast) or simply omitted from the training phase as a means to confirm the utility of the proposed methods (to maintain the sanctity of the validation procedure). We believe a fresh look at how domain knowledge can be embedded in existing approaches and better validation methodologies in close conjunction with domain experts must be looked into.

- Uncertainty-aware and noise-tolerant methods: While this has certainly been an active area of research in the bioinformatics community in general, and in the field of graph mining in bioinformatics in particular, there are still many open problems here. Incorporating uncertainty is necessarily a domain-dependent issue and probabilistic approaches offer exciting possibilities. Additionally leveraging topological, relational and other semantic characteristics of the data effectively is an interesting topic for future research. A related challenge here is to model trust and provenance related information.

- Ranking and summarizing patterns harvested: While ranking and summarizing patterns has been the subject of much research in the data mining and network science community the role of such methods in bioinformatics has been much less researched. We expect this to be a very important and active area of research especially since often times evaluating and validating patterns discovered can be an expensive and time consuming process. In this context research into ranking algorithms for bioinformatics that leverage domain knowledge and mechanisms for summarizing patterns harvested is an exciting opportunity for future research.

References

[1] Akutsu, T. (1992). An RNC algorithm for finding a largest common subtree of two trees. *IEICE Transactions on Information and Systems*, 75(1):95–101.

[2] Aoki, K., Mamitsuka, H., Akutsu, T., and Kanehisa, M. (2005). A score matrix to reveal the hidden links in glycans. *Bioinformatics*, 21(8):1457–1463.

[3] Aoki, K., Ueda, N., Yamaguchi, A., Kanehisa, M., Akutsu, T., and Mamitsuka, H. (2004a). Application of a new probabilistic model for recognizing complex patterns in glycans.

[4] Aoki, K., Yamaguchi, A., Okuno, Y., Akutsu, T., Ueda, N., Kanehisa, M., and Mamitsuka, H. (2003). Efficient tree-matching methods for accurate carbohydrate database queries. *Genome Informatics Sl*, pages 134–143.

[5] Aoki, K., Yamaguchi, A., Ueda, N., Akutsu, T., Mamitsuka, H., Goto, S., and Kanehisa, M. (2004b). KCaM (KEGG Carbohydrate Matcher): a software tool for analyzing the structures of carbohydrate sugar chains. *Nucleic acids research*, 32(Web Server Issue):W267.

[6] Asur, S., Ucar, D., and Parthasarathy, S. (2007). An ensemble framework for clustering protein protein interaction networks. *Bioinformatics*, 23(13):i29.

[7] Avogadri, R. and Valentini, G. (2009). Fuzzy ensemble clustering based on random projections for DNA microarray data analysis. *Artificial Intelligence in Medicine*, 45(2-3):173–183.

[8] Bader, G. and Hogue, C. (2003). An automated method for finding molecular complexes in large protein interaction networks. *BMC Bioinfomatics*, 4:2.

[9] Bafna, V., Muthukrishnan, S., and Ravi, R. (1995). Computing similarity between RNA strings. *In Combinatorial Pattern Matching (CPM), volume 937 of LNCS.*

[10] Bar-Joseph, Z., Gerber, G., Lee, T., Rinaldi, N., Yoo, J., Robert, F., Gordon, D., Fraenkel, E., Jaakkola, T., Young, R., et al. (2003). Computational discovery of gene modules and regulatory networks. *Nature Biotechnology*, 21(11):1337–1342.

[11] Benedetti, G. and Morosetti, S. (1996). A graph-topological approach to recognition of pattern and similarity in RNA secondary structures. *Biophysical chemistry*, 59(1-2):179–184.

[12] Bille, P. (2005). A survey on tree edit distance and related problems. *Theoretical computer science*, 337(1-3):217–239.

[13] Bohne-Lang, A., Lang, E., Ferster, T., and von der Lieth, C. (2001). LINUCS: linear notation for unique description of carbohydrate sequences. *Carbohydrate research*, 336(1):1–11.

[14] Brohee, S. and van Helden, J. (2006). Evaluation of clustering algorithms for protein-protein interaction networks. *BMC bioinformatics*, 7(1):488.

[15] Butte, A. and Kohane, I. (2000). Mutual information relevance networks: functional genomic clustering using pairwise entropy measurements. In *Pac Symp Biocomput*, volume 5, pages 418–429.

[16] Chakrabarti, D. and Faloutsos, C. (2006). Graph mining: Laws, generators, and algorithms. *ACM Computing Surveys (CSUR)*, 38(1).

[17] Chawathe, S. and Garcia-Molina, H. (1997). Meaningful change detection in structured data. *ACM SIGMOD Record*, 26(2):26–37.

[18] Chawathe, S., Rajaraman, A., Garcia-Molina, H., and Widom, J. (1996). Change detection in hierarchically structured information. In *Proceedings of the 1996 ACM SIGMOD international conference on Management of data*, pages 493–504. ACM New York, NY, USA.

[19] Chen, J., Hsu, W., Lee, M., and Ng, S. (2006). NeMoFinder: Dissecting genome-wide protein-protein interactions with meso-scale network motifs. In *Proceedings of the 12th ACM SIGKDD international conference on Knowledge discovery and data mining*, pages 106–115. ACM New York, NY, USA.

[20] Chen, J., Hsu, W., Lee, M. L., and Ng, S.-K. (2007). Labeling network motifs in protein interactomes for protein function prediction. *Data Engineering, International Conference on*, 0:546–555.

[21] Cheng, Y. and Church, G. (2000). Biclustering of expression data. In *Proceedings of the Eighth International Conference on Intelligent Systems for Molecular Biology table of contents*, pages 93–103. AAAI Press.

[22] Chua, H., Ning, K., Sung, W., Leong, H., and Wong, L. (2007). Using indirect protein-protein interactions for protein complex prediction. In *Computational Systems Bioinformatics: Proceedings of the CSB 2007 Conference*, page 97. Imperial College Press.

[23] Coatney, M. and Parthasarathy, S. (2005a). MotifMiner: Efficient discovery of common substructures in biochemical molecules. *Knowledge and Information Systems*, 7(2):202–223.

[24] Coatney, M. and Parthasarathy, S. (2005b). Motifminer: Efficient discovery of common substructures in biochemical molecules. *Knowl. Inf. Syst.*, 7(2):202–223.

[25] Constantinescu, M. and Sankoff, D. (1995). An efficient algorithm for supertrees. *Journal of Classification*, 12(1):101–112.

[26] Cooper, C., Harrison, M., Wilkins, M., and Packer, N. (2001). GlycoSuiteDB: a new curated relational database of glycoprotein glycan structures and their biological sources. *Nucleic Acids Research*, 29(1):332.

[27] Dhillon, I., Guan, Y., and Kulis, B. (2005). A fast kernel-based multilevel algorithm for graph clustering. *Proceedings of the 11th ACM SIGKDD*, pages 629–634.

[28] Dongen, S. (2000). *Graph clustering by flow simulation*. PhD thesis, PhD Thesis, University of Utrecht, The Netherlands.

[29] Durocher, D., Taylor, I., Sarbassova, D., Haire, L., Westcott, S., Jackson, S., Smerdon, S., and Yaffe, M. (2000). The molecular basis of FHA domain: phosphopeptide binding specificity and implications for phospho-dependent signaling mechanisms. *Molecular Cell*, 6(5):1169–1182.

[30] Eisen, M., Spellman, P., Brown, P., and Botstein, D. (1998). Cluster analysis and display of genome-wide expression patterns. *Proceedings of the National Academy of Sciences*, 95(25):14863–14868.

[31] Farach, M. and Thorup, M. (1994). Fast comparison of evolutionary trees. In *Proceedings of the fifth annual ACM-SIAM symposium on Discrete algorithms*, pages 481–488. Society for Industrial and Applied Mathematics Philadelphia, PA, USA.

[32] Fitch, W. (1971). Toward defining the course of evolution: minimum change for a specific tree topology. *Systematic zoology*, 20(4):406–416.

[33] Furtig, B., Richter, C., Wohnert, J., and Schwalbe, H. (2003). NMR spectroscopy of RNA. *ChemBioChem*, 4(10):936–962.

[34] Gan, H., Pasquali, S., and Schlick, T. (2003). Exploring the repertoire of RNA secondary motifs using graph theory; implications for RNA design. *Nucleic acids research*, 31(11):2926.

[35] Gardner, P. and Giegerich, R. (2004). A comprehensive comparison of comparative RNA structure prediction approaches. *BMC bioinformatics*, 5(1):140.

[36] Gordon, A. (1979). A measure of the agreement between rankings. *Biometrika*, 66(1):7–15.

[37] Gordon, A. (1986). Consensus supertrees: the synthesis of rooted trees containing overlapping sets of labeled leaves. *Journal of Classification*, 3(2):335–348.

[38] Gouda, K. and Zaki, M. (2001). Efficiently mining maximal frequent itemsets. In *Proceedings of the 2001 IEEE International Conference on Data Mining*, pages 163–170.

[39] Grochow, J. and Kellis, M. (2007). Network motif discovery using subgraph enumeration and symmetry-breaking. *Lecture Notes in Computer Science*, 4453:92.

[40] Guignon, V., Chauve, C., and Hamel, S. (2005). An edit distance between RNA stem-loops. *Lecture notes in computer science*, 3772:333.

[41] Gupta, A. and Nishimura, N. (1998). Finding largest subtrees and smallest supertrees. *Algorithmica*, 21(2):183–210.

[42] Hadzic, F., Dillon, T., Sidhu, A., Chang, E., and Tan, H. (2006). Mining substructures in protein data. In *IEEE ICDM 2006 Workshop on Data Mining in Bioinformatics (DMB 2006)*, pages 18–22.

[43] Hartuv, E. and Shamir, R. (2000). A clustering algorithm based on graph connectivity. *Information processing letters*, 76(4-6):175–181.

[44] Hashimoto, K., Aoki-Kinoshita, K., Ueda, N., Kanehisa, M., and Mamitsuka, H. (2006a). A new efficient probabilistic model for mining labeled ordered trees. In *Proceedings of the 12th ACM SIGKDD international conference on Knowledge discovery and data mining*, pages 177–186.

[45] Hashimoto, K., Goto, S., Kawano, S., Aoki-Kinoshita, K., Ueda, N., Hamajima, M., Kawasaki, T., and Kanehisa, M. (2006b). KEGG as a glycome informatics resource. *Glycobiology*, 16(5):63–70.

[46] Hashimoto, K., Takigawa, I., Shiga, M., Kanehisa, M., and Mamitsuka, H. (2008). Mining significant tree patterns in carbohydrate sugar chains. *Bioinformatics*, 24(16):i167.

[47] Henikoff, S. and Henikoff, J. (1992). Amino acid substitution matrices from protein blocks. *Proceedings of the National Academy of Sciences*, 89(22):10915–10919.

[48] Herget, S., Ranzinger, R., Maass, K., and Lieth, C. (2008). GlycoCT: a unifying sequence format for carbohydrates. *Carbohydrate Research*, 343(12):2162–2171.

[49] Hizukuri, Y., Yamanishi, Y., Nakamura, O., Yagi, F., Goto, S., and Kanehisa, M. (2005). Extraction of leukemia specific glycan motifs in humans by computational glycomics. *Carbohydrate research*, 340(14):2270–2278.

[50] Hochsmann, M., Voss, B., and Giegerich, R. (2004). Pure multiple RNA secondary structure alignments: a progressive profile approach. *IEEE/ACM Transactions on Computational Biology and Bioinformatics (TCBB)*, 1(1):53–62.

[51] Holder, L., Cook, D., and Djoko, S. (1994). Substructure discovery in the subdue system. In *Proc. of the AAAI Workshop on Knowledge Discovery in Databases*, pages 169–180.

[52] Horvath, S. and Dong, J. (2008). Geometric interpretation of gene coexpression network analysis. *PLoS Computational Biology*, 4(8).

[53] Hu, H., Yan, X., Huang, Y., Han, J., and Zhou, X. (2005). Mining coherent dense subgraphs across massive biological networks for functional discovery. *Bioinformatics*, 21(1):213–221.

[54] Huan, J., Wang, W., Prins, J., and Yang, J. (2004). Spin: mining maximal frequent subgraphs from graph databases. In *Proceedings of the tenth ACM SIGKDD international conference on Knowledge discovery and data mining*, pages 581–586. ACM New York, NY, USA.

[55] Huang, Y., Li, H., Hu, H., Yan, X., Waterman, M., Huang, H., and Zhou, X. (2007). Systematic discovery of functional modules and context-specific functional annotation of human genome. *Bioinformatics*, 23(13):i222.

[56] Jiang, T., Lawler, E., and Wang, L. (1994). Aligning sequences via an evolutionary tree: complexity and approximation. In *Proceedings of the twenty-sixth annual ACM symposium on Theory of computing*, pages 760–769. ACM New York, NY, USA.

[57] Jiang, T., Lin, G., Ma, B., and Zhang, K. (2002). A general edit distance between RNA structures. *Journal of Computational Biology*, 9(2):371–388.

[58] Jiang, T., Wang, L., and Zhang, K. (1995). Alignment of trees: an alternative to tree edit. *Theoretical Computer Science*, 143(1):137–148.

[59] Jin, R., Wang, C., Polshakov, D., Parthasarathy, S., and Agrawal, G. (2005). Discovering frequent topological structures from graph datasets.

In *Proceedings of the eleventh ACM SIGKDD international conference on Knowledge discovery in data mining*, pages 606–611. ACM New York, NY, USA.

[60] Karypis, G. and Kumar, V. (1999). A fast and high quality multilevel scheme for partitioning irregular graphs. *SIAM Journal on Scientific Computing*, 20(1):359.

[61] Kashtan, N., Itzkovitz, S., Milo, R., and Alon, U. (2004). Efficient sampling algorithm for estimating subgraph concentrations and detecting network motifs. *Bioinformatics*, 20(11):1746–1758.

[62] Kawano, S., Hashimoto, K., Miyama, T., Goto, S., and Kanehisa, M. (2005). Prediction of glycan structures from gene expression data based on glycosyltransferase reactions. *Bioinformatics*, 21(21):3976–3982.

[63] Keselman, D. and Amir, A. (1994). Maximum agreement subtree in a set of evolutionary trees-metrics and efficient algorithms. In *Annual Symposium on Foundations of Computer Science*, volume 35, pages 758–758. IEEE Computer Society Press.

[64] Khanna, S., Motwani, R., and Yao, F. (1995). Approximation algorithms for the largest common subtree problem.

[65] Kohonen, T. (1995). Self-organizing maps. *Springer, Berlin*.

[66] Koyuturk, M., Grama, A., and Szpankowski, W. (2004a). An efficient algorithm for detecting frequent subgraphs in biological networks. *Bioinformatics*, 20(90001).

[67] Koyuturk, M., Szpankowski, W., and Grama, A. (2004b). Biclustering gene-feature matrices for statistically significant dense patterns. In *2004 IEEE Computational Systems Bioinformatics Conference, 2004. CSB 2004. Proceedings*, pages 480–484.

[68] Kuboyama, T., Hirata, K., Aoki-Kinoshita, K., Kashima, H., and Yasuda, H. (2006). A gram distribution kernel applied to glycan classification and motif extraction. *Genome Informatics Series*, 17(2):25.

[69] Le, S., Owens, J., Nussinov, R., Chen, J., Shapiro, B., and Maizel, J. (1989). RNA secondary structures: comparison and determination of frequently recurring substructures by consensus. *Bioinformatics*, 5(3):205–210.

[70] Lee, H., Hsu, A., Sajdak, J., Qin, J., and Pavlidis, P. (2004). Coexpression analysis of human genes across many microarray data sets. *Genome Research*, 14(6):1085–1094.

[71] Lemmens, K., Dhollander, T., De Bie, T., Monsieurs, P., Engelen, K., Smets, B., Winderickx, J., De Moor, B., and Marchal, K. (2006). Inferring transcriptional modules from ChIP-chip, motif and microarray data. *Genome biology*, 7(5):R37.

[72] Li, H., Marsolo, K., Parthasarathy, S., and Polshakov, D. (2004). A new approach to protein structure mining and alignment. *Proceedings of the ACM SIGKDD Workshop on Data Mining and Bioinformatics (BIOKDD)*, pages 1–10.

[73] Li, X., Foo, C., and Ng, S. (2007). Discovering protein complexes in dense reliable neighborhoods of protein interaction networks. In *Computational Systems Bioinformatics: Proceedings of the CSB 2007 Conference*, page 157. Imperial College Press.

[74] Liu, N. and Wang, T. (2006). A method for rapid similarity analysis of RNA secondary structures. *BMC bioinformatics*, 7(1):493.

[75] Loß, A., Bunsmann, P., Bohne, A., Loß, A., Schwarzer, E., Lang, E., and Von der Lieth, C. (2002). SWEET-DB: an attempt to create annotated data collections for carbohydrates. *Nucleic acids research*, 30(1):405–408.

[76] MacQueen, J. (1967). Some methods for classification and analysis of multivariate observations. *Proceedings of 5th Berkeley Symposium on Mathematical Statistics and Probability*, pages 281–297.

[77] Margush, T. and McMorris, F. (1981). Consensusn-trees. *Bulletin of Mathematical Biology*, 43(2):239–244.

[78] Milo, R., Shen-Orr, S., Itzkovitz, S., Kashtan, N., Chklovskii, D., and Alon, U. (2002). Network motifs: simple building blocks of complex networks. *Science*, 298(5594):824–827.

[79] Mitchell, J., Cheng, J., and Collins, K. (1999). A box H/ACA small nucleolar RNA-like domain at the human telomerase RNA 3'end. *Molecular and cellular biology*, 19(1):567–576.

[80] Newman, M. and Girvan, M. (2004). Finding and evaluating community structure in networks. *Physical Review E*, 69:026113.

[81] Ohtsubo, K. and Marth, J. (2006). Glycosylation in cellular mechanisms of health and disease. *Cell*, 126(5):855–867.

[82] Onoa, B. and Tinoco, I. (2004). RNA folding and unfolding. *Current Opinion in Structural Biology*, 14(3):374–379.

[83] Packer, N., von der Lieth, C., Aoki-Kinoshita, K., Lebrilla, C., Paulson, J., Raman, R., Rudd, P., Sasisekharan, R., Taniguchi, N., and York, W. (2008). Frontiers in glycomics: Bioinformatics and biomarkers in disease. *Proteomics*, 8(1).

[84] Pizzuti, C. and Rombo, S. (2008). Multi-functional protein clustering in ppi networks. In *Bioinformatics Research and Development*, pages 318–330.

[85] Ragan, M. (1992). Phylogenetic inference based on matrix representation of trees. *Molecular Phylogenetics and Evolution*, 1(1):53.

[86] Ravasz, E., Somera, A., Mongru, D., Oltvai, Z., and Barabasi, A. (2002). Hierarchical organization of modularity in metabolic networks.

[87] Sahoo, S., Thomas, C., Sheth, A., Henson, C., and York, W. (2005). GLYDE-an expressive XML standard for the representation of glycan structure. *Carbohydrate research*, 340(18):2802–2807.

[88] Sanderson, M., Purvis, A., and Henze, C. (1998). Phylogenetic supertrees: assembling the trees of life. *Trends in Ecology & Evolution*, 13(3):105–109.

[89] Satuluri, V. and Parthasarathy, S. (2009). Scalable Graph Clustering using Stochastic Flows: Applications to Community Discovery. In *Proceedings of the 15th ACM SIGKDD international conference on Knowledge discovery and data mining*, pages 737–746.

[90] Schelkopf, B. and Smola, A. (2002). *Learning with kernels: Support vector machines, regularization, optimization, and beyond.* MIT press.

[91] Segal, E., Shapira, M., Regev, A., Pe'er, D., Botstein, D., Koller, D., and Friedman, N. (2003). Module networks: identifying regulatory modules and their condition-specific regulators from gene expression data. *Nature genetics*, 34(2):166–176.

[92] Selkow, S. (1977). The tree-to-tree editing problem. *Information processing letters*, 6(6):184–186.

[93] Shapiro, B. and Zhang, K. (1990). Comparing multiple RNA secondary structures using tree comparisons. *Bioinformatics*, 6(4):309–318.

[94] Sharan, R. and Shamir, R. (2000). CLICK: A clustering algorithm with applications to gene expression analysis. 8:307–316.

[95] Shasha, D., Wang, J., and Zhang, S. (2004). Unordered tree mining with applications to phylogeny. In *in Proceedings of International Conference on Data Engineering*, pages 708–719.

[96] Shi, J. and Malik, J. (2000). Normalized cuts and image segmentation. *IEEE Transactions on pattern analysis and machine intelligence*, 22(8):888–905.

[97] Shih, F. and Mitchell, O. (1989). Threshold decomposition of gray-scale morphology into binarymorphology. *IEEE Transactions on Pattern Analysis and Machine Intelligence*, 11(1):31–42.

[98] Smith, T. and Waterman, M. (1981). Identification of common molecular subsequences. *J. Mol. Bwl*, 147:195–197.

[99] Sneath, S. (1973). Hierarchical clustering.

[100] Stark, C., Breitkreutz, B., Reguly, T., Boucher, L., Breitkreutz, A., and Tyers, M. (2006). BioGRID: a general repository for interaction datasets. *Nucleic acids research*, 34(Database Issue):D535.

[101] Stockham, C., Wang, L., and Warnow, T. (2002). Statistically based postprocessing of phylogenetic analysis by clustering. *Bioinformatics*, 18(3):465–469.

[102] Stuart, J., Segal, E., Koller, D., and Kim, S. (2003a). A gene-coexpression network for global discovery of conserved genetic modules. *Science*, 302(5643):249–255.

[103] Stuart, J., Segal, E., Koller, D., and Kim, S. (2003b). A gene-coexpression network for global discovery of conserved genetic modules. *Science*, 302(5643):249–255.

[104] Tai, K. (1979). The tree-to-tree correction problem. *Journal of the Association for Computing Machm© ry*, 26(3):422–433.

[105] Tanay, A., Sharan, R., Kupiec, M., and Shamir, R. (2004). Revealing modularity and organization in the yeast molecular network by integrated analysis of highly heterogeneous genomewide data. *Proceedings of the National Academy of Sciences*, 101(9):2981–2986.

[106] Tanay, A., Sharan, R., and Shamir, R. (2002). Discovering statistically significant biclusters in gene expression data. *Bioinformatics*, 18(Suppl 1):S136–S144.

[107] Tinoco, I. and Bustamante, C. (1999). How RNA folds. *Journal of molecular biology*, 293(2):271–281.

[108] Ueda, N., Aoki, K., and Mamitsuka, H. (2004). A general probabilistic framework for mining labeled ordered trees. In *Proceedings of the Fourth SIAM International Conference on Data Mining*, pages 357–368.

[109] Ueda, N., Aoki-Kinoshita, K., Yamaguchi, A., Akutsu, T., and Mamitsuka, H. (2005). A probabilistic model for mining labeled ordered trees: Capturing patterns in carbohydrate sugar chains. *IEEE Transactions on Knowledge and Data Engineering*, 17(8):1051–1064.

[110] Valiente, G. (2002). *Algorithms on trees and graphs*. Springer.

[111] Wang, C. and Parthasarathy, S. (2004). Parallel algorithms for mining frequent structural motifs in scientific data. In *Proceedings of the 18th annual international conference on Supercomputing*, pages 31–40. ACM New York, NY, USA.

[112] Wang, L., Jiang, T., and Gusfield, D. (1997). A more efficient approximation scheme for tree alignment. In *Proceedings of the first annual international conference on Computational molecular biology*, pages 310–319. ACM New York, NY, USA.

[113] Wang, L., Jiang, T., and Lawler, E. (1996). Approximation algorithms for tree alignment with a given phylogeny. *Algorithmica*, 16(3):302–315.

[114] Yamanishi, Y., Bach, F., and Vert, J. (2007). Glycan classification with tree kernels. *Bioinformatics*, 23(10):1211.

[115] Yan, X., Mehan, M., Huang, Y., Waterman, M., Yu, P., and Zhou, X. (2007). A graph-based approach to systematically reconstruct human transcriptional regulatory modules. *Bioinformatics*, 23(13):i577.

[116] You, C. H., Holder, L. B., and Cook, D. J. (2006). Application of graph-based data mining to metabolic pathways. *Data Mining Workshops, International Conference on*, 0:169–173.

[117] Zaki, M. (2005). Efficiently mining frequent trees in a forest: Algorithms and applications. *IEEE Transactions on Knowledge and Data Engineering*, 17(8):1021–1035.

[118] Zhang, K. and Jiang, T. (1994). Some MAX SNP-hard results concerning unordered labeled trees. *Information Processing Letters*, 49(5):249–254.

[119] Zhang, K. and Shasha, D. (1989). Simple fast algorithms for the editing distance between trees and related problems. *SIAM journal on computing*, 18:1245.

[120] Zhang, S. and Wang, T. (2008). Discovering Frequent Agreement Subtrees from Phylogenetic Data. *IEEE Transactions on Knowledge and Data Engineering*, 20(1):68–82.

Chapter 19

TRENDS IN CHEMICAL GRAPH DATA MINING

Nikil Wale
Computer Science & Engineering
University of Minnesota, Twin Cities, US
nwale@cs.umn.edu

Xia Ning
Computer Science & Engineering
University of Minnesota, Twin Cities, US
xning@cs.umn.edu

George Karypis
Computer Science & Engineering
University of Minnesota, Twin Cities, US
karypis@cs.umn.edu

Abstract

Mining chemical compounds *in silico* has drawn increasing attention from both academia and pharmaceutical industry due to its effectiveness in aiding the drug discovery process. Since graphs are the natural representation for chemical compounds, most of the mining algorithms focus on mining chemical graphs. Chemical graph mining approaches have many applications in the drug discovery process that include structure-activity-relationship (SAR) model construction and bioactivity classification, similar compound search and retrieval from chemical compound database, target identification from phenotypic assays, *etc.* Solving such problems *in silico* through studying and mining chemical graphs can provide novel perspective to medicinal chemists, biologist and toxicologist. Moreover, since the large scale chemical graph mining is usually employed at the early stages of drug discovery, it has the potential to speed up the entire drug discovery process. In this chapter, we discuss various problems and algorithms related to mining chemical graphs and describe some of the state-of-the-art chemical graph mining methodologies and their applications.

C.C. Aggarwal and H. Wang (eds.), *Managing and Mining Graph Data*,
Advances in Database Systems 40, DOI 10.1007/978-1-4419-6045-0_19,
© Springer Science+Business Media, LLC 2010

Keywords: Chemical Graph, Descriptor Spaces, Classification, Ranked Retrieval, Scaffold Hopping, Target Fishing.

1.   Introduction

Labeled graphs (either topological or geometric) have been a promising abstraction to capture the characteristics of datasets arising in many fields such as the world wide web, social networks, biology, and chemistry ([9], [13], [30], [49]). The vertices of these graphs correspond to the entities in the objects and the edges correspond to the relations between them. This graph-based representation can directly capture many of the sequential, topological, geometric, and other relational characteristics of such datasets. For example, in the domain of the world wide web and social networks the entire set of objects and their relations are represented via a single large graph ([13]). In biology, objects to be mined are represented either as a single large graph (e.g., metabolic and signaling pathways) or via separate graphs (e.g., protein structures) ([65], [30], [33]). In chemistry, each object to be mined is represented via a separate graph (e.g., molecular graphs) ([49]).

Graph mining over the above representations has found applications in the domain of web data analysis such as the analysis of XML documents and weblogs, web searches, web document analysis *etc*([9]). Graph mining is also being used in social sciences for the analysis of social networks that help understand social phenomenon and group behavior([13]). In the domain of traditional sciences like biology and chemistry, graph mining has found numerous important applications. For example, in biology graphs can be used to directly model the key topological and geometric characteristics of protein molecules. Vertices in these graphs will correspond to different amino acids. The edges will correspond to the connections of amino acids in the protein's backbone or the non-covalent bonds(i.e., contact points) in the 3D structure. Mining these graph patterns provides important insights into protein structure and function ([22], [3]).

In chemistry, graphs can be used to directly model the key topological and geometric characteristics of chemical structures. Vertices in these graphs correspond to different atoms and the edges correspond to bonds that connect atoms ([29]). Mining on a set of chemical compounds or molecules helps in understanding the key characteristics of a set molecules for a given process (such as toxicity and biological activity) and has become the primary application area of chemical graph mining ([49], [40]). The typical applications performed on chemical structures include mining sub-structures in a given set of ligands ([40]), mining databases to retrieve other relevant compounds, clustering of chemical compounds based on common sub-structures, and predicting compound bioactivity by classification, regression and ranking techniques ([2], [28]).

Most of the mining algorithms operate on the assumption that the properties and biological activity of a chemical compound are related to its structure ([2], [28]). This assumption is widely referred to as the structure-activity-relationship principle or simply SAR. Hansch ([17]) demonstrated that the biological activity of a chemical compound can be mathematically expressed as a function of its physiochemical properties, which led to the development of quantitative methods for modeling structure-activity relationships (QSAR). Since that work, many different approaches have been developed for building such structure-activity-relationship (SAR) models. All of these models are derived using some notion of structural similarity between chemical compounds. The similarity is determined using a similarity function over a descriptor-space representation, and the descriptor-space is most commonly generated from chemical graphs. These models have become an essential tool for predicting biological activity from the structural properties of a molecule.

The rest of this chapter will review some of the current trends in chemical graph mining and modeling. It will highlight some of the techniques that exist and that were recently developed for representing chemical compounds, building classification models, retrieving compounds from databases, and identifying the proteins that the compounds will bind to. The chapter concludes by outlining some of the future research directions in this field.

2. Topological Descriptors for Chemical Compounds

Descriptor-based representations of chemical compounds are used extensively in cheminformatics, as they represent a convenient and computationally efficient way to capture key characteristics of the compounds' structures ([2], [28]). Such representations have extensive applications to similarity search and various structure-driven prediction problems for activity, toxicity, absorption, distribution, metabolism and excretion ([2]). Many of these descriptors are derived by mining structural patterns from a set of molecular graphs of the chemical compounds. Such descriptors include topological descriptors derived directly from the topology of molecular graphs and 2D/3D pharmacophore descriptors that describe the critical atoms/atom groups that are highly likely to be involved in protein-ligand binding ([7], [32], [55], [28]). In the rest of this section we review some of the topological descriptors that are used extensively to represent chemical compounds and analyze their different properties. This includes both a set of time-tested descriptors as well as recently developed descriptors that have shown promising results.

2.1   Hashed Fingerprints (FP)

Hash fingerprints are generally used to encode the 2D structural characteristics of a chemical compound into a fixed bit vector and are used extensively for various tasks in chemical informatics. These fingerprints are typically generated by enumerating all cycles and linear paths up to a given number of bonds and hashing each of these cycles and paths into a fixed bit-string ([7], [4], [51], [20]). The specific bit-string that is generated depends on the number of bonds, the number of bits that are set, the hashing function, and the length of the bit-string. The key property of these fingerprint descriptors is that they encode a very large number of sub-structures into a compact representation. Many variants of these fingerprints exist, some use predefined structural fragments in conjunction with the fingerprints, for example, Unity fingerprints ([51]), others count the number of times a bit position is set, for example, hologram ([20]). However, a recent study has shown that the performance of most of these fingerprints is comparable ([26]).

2.2   Maccs Keys (MK)

Molecular Design Limited (MDL) has created the key based fingerprints Maccs Keys ([32]) based on pattern matching of a chemical compound structure to a pre-defined set of structural fragments. These fragments have been identified by domain experts ([10]) to be important for bioactivity of chemical compounds. The original set of descriptors consists of 166 structural fragments and each such fragment becomes a key and occupies a fixed position in the descriptor space. This approach relies on pre-defined rules to encapsulate the essential molecular descriptors a-priori and does not learn them from the chemical dataset. This descriptor space is notably different from fingerprint based descriptor space. Unlike fingerprints, no *folding* (hashing) is performed on the sub-structures.

2.3   Extended Connectivity Fingerprints (ECFP)

Molecular descriptors and fingerprints based on the extended connectivity concept have been described by several authors ([42], [19]). The earliest concept of such a descriptor-space was described in [59]. Recently, these fingerprints have been popularized by their implementation within Pipeline Pilot ([11]). These fingerprints are generated by first assigning some initial label to each atom and then applying a Morgan type algorithm ([34]) to generate the fingerprints. Morgan's algorithm consists of l iterations. In each iteration, a new label is generated and assigned to each atom by combining the current labels of the neighboring atoms (i.e, connected via a bond). The union of the labels assigned to all the atoms over all the l iterations are used as the descriptors to represent each compound. The key idea behind this descriptor generation algorithm is to capture the topology around each atom in the form of shells whose radius ranges from 1 to l. Thus, these descriptors can capture rather complex topologies. The value for l is a user supplied parameter and typically ranges from two to six.

2.4   Frequent Subgraphs (FS)

A number of methods have been proposed in recent years to mine frequently occurring subgraphs (sub-structures) in a chemical graph database ([37], [61], [27]). Frequent subgraphs of a chemical graph database D are defined as all subgraphs that are present in at least σ ($\sigma \leq |D|$) of compounds of the database, where σ is the absolute minimum frequency requirement (also called absolute minimum support constraint). These frequent subgraphs can be used as descriptors for the compounds in that database. A descriptor space formed out of frequently occurring subgraphs depends on the value of σ. Therefore, the descriptor space can change for a particular problem instance if the value of σ is changed. An advantage of such a descriptor space is that it can create descriptors suitable for a given dataset. Moreover, the substructures mined consist of arbitrary sizes and topologies. A potential disadvantage of this method is that it is unclear how to select a suitable value of σ for a given problem. A very high value will fail to discover important subgraphs whereas a very low value will result in combinatorial explosion of frequent subgraphs.

2.5   Bounded-Size Graph Fragments (GF)

Recently, a new descriptor space, Graph Fragments (GF), has been developed consisting of sub-structures or fragments that exist in a compound library ([55]). Graph Fragments of a chemical graph database D are defined as all connected subgraphs present in every chemical graph of D that has a size of less than or equal to the user supplied parameter l. Therefore, GF descriptor space is a subset of the FS descriptor space generated using a absolute minimum support threshold of 1. However, instead of the minimum support threshold used in generating FS, the user supplied parameter l is used to control the combinatorial complexity of the fragment generation process for GF and put an upper bound on the size of fragments generated. An efficient algorithm to generate the GF descriptors for a library of compounds is described in [55].

2.6   Comparison of Descriptors

A careful analysis of the descriptor spaces described in the previous section illustrate four dimensions along which these schemes compare with each other and represent some of the choices that have been explored in designing fragment-based or fragment-derived descriptors for chemical compounds. Ta-

Table 19.1. Design choices made by the descriptor spaces.

| | | Previously developed descriptors | | |
	Generation	Topological Complexity	Precise	Complete Coverage
FP	dynamic	Low	No	Yes
MK	static	Low to High	Yes	Maybe
ECFP	dynamic	Low to High	Maybe	Yes
FS	dynamic	Low to High	Yes	Maybe
GF	dynamic	Low to High	Yes	Yes

FP refers to the hashed fingerprints, MK to Maccs keys, ECFP to extended connectivity fingerprints, FS to frequent subgraphs, and GF to graph fragments.

ble 19.1 summarizes the characteristics of these descriptor spaces along the four dimensions. The first dimension is associated with whether the fragments are determined directly from the dataset at hand or they have been pre-identified by domain experts. The fragments of Maccs keys have been determined a priori whereas all other descriptors are determined directly from the dataset. The advantage of a priori approach is that it can capture domain knowledge. However, due to the fixed set of fragments identified a priori it might not adapt to the characteristics for a particular dataset. The second dimension is associated with the topological complexity of the actual fragments. Schemes like fingerprints use simple topologies consisting of paths and cycles. Descriptors such as extended connectivity fingerprints, frequent subgraphs and graph fragments allow topologies with arbitrary complexity. Topologically complex fragments along with simple ones might enrich the descriptor space. The third dimension is associated with whether or not the fragments are being precisely represented in the descriptor space. Most schemes generate descriptors that are precise in the sense that there is a one-to-one mapping between the fragments and the dimensions of the descriptor space. In contrast, due to the hashing approach, descriptors such as fingerprints and extended connectivity fingerprints lead to imprecise representations (i.e., many fragments can map to the same dimension of the descriptor space). Depending on the number of these many-to-one mappings, these descriptors can lead to representations with varying degree of information loss. Finally, the fourth dimension is associated with the ability of the descriptor space to cover all or nearly all of the dataset. Descriptor spaces created from fingerprints, extended connectivity fingerprints, and graph fragments are guaranteed to contain fragments or hashed fragments from each one of the compounds. On the other hand, descriptor spaces corresponding to Maccs keys and frequent sub-structures may lead to a descriptor-based representation of the dataset in which some of the compounds have no or a very small number of descriptors. A descriptor space that covers all the compounds

Table 19.2. SAR performance of different descriptors.

Datasets	fp	ECFP	MK	FS	GF
NCI1	0.30	0.32	0.29	0.27	**0.33**
NCI109	0.27	**0.32**	0.24	0.26	**0.32**
NCI123	0.25	**0.27**	0.24	0.23	**0.27**
NCI145	0.30	0.35	0.28	0.30	**0.37**
NCI167	0.06	0.06	0.04	0.06	**0.07**
NCI220	**0.33**	0.28	0.26	0.21	0.29
NCI33	0.26	0.31	0.26	0.25	**0.33**
NCI330	0.34	**0.36**	0.31	0.24	**0.36**
NCI41	0.25	**0.36**	0.28	0.30	**0.36**
NCI47	0.26	**0.31**	0.26	0.24	**0.31**
NCI81	0.27	**0.28**	0.25	0.24	**0.28**
NCI83	0.26	**0.31**	0.26	0.25	**0.31**

The numbers correspond to the ROC_{50} values of SVM-based SAR models for twelve screening assays obtained from NCI. The ROC_{50} value is the area under the receiver operating characteristic curve (ROC) up to the first 50 false positives. These values were computed using a 5-fold cross-validation approach. The descriptors being evaluated are: graph fragments (GF) ([55]), extended connectivity fingerprints (ECFP) ([28]), Chemaxon's fingerprints (fp) (Chemaxon Inc.) ([4]), Maccs keys (MK) (MDL Information Systems Inc.) ([32]), and frequent subgraphs (FS) ([8]).

of a dataset has the advantage of encoding some amount of information for every compound.

The qualitative comparison of the descriptors along the lines discussed above is shown in Table 19.1. This table shows that unlike other descriptors, GF descriptors satisfy all the key properties described earlier such as dynamic generation, complex topology, precise representation, and complete coverage. For example, unlike path-based structural descriptors (fp) and extended-connectivity fingerprints, they are guaranteed to have a one-to-one mapping between a fragment and a dimension in the descriptor space. Moreover, unlike fingerprints, they impose no limit on the complexity of the descriptor's structures ([55]) and unlike Maccs Keys, the descriptors are dynamically generated from the dataset at hand. Lastly, unlike FS, which may suffer from partial coverage, this descriptor space is ensured to have 100% coverage by eliminating the minimum support criterion and generating all fragments. Therefore, GF descriptors allow for better representation of the underlying compounds and they are expected to show better performance in the context of SAR based classification and retrieval approaches.

A quantitative comparison in Table 19.2 shows classification results from a recent study ([55]) using the NCI datasets obtained from the PubChem Project ([39]). These results empirically show that the GF descriptor space achieves a performance that is either better or comparable to that achieved by currently used descriptors, indicating that the above mentioned properties are important to capture the compounds' structural characteristics.

3.   Classification Algorithms for Chemical Compounds

Numerous approaches have been developed for building classifying models for various classes of interest (e.g., active/inactive, toxic/non-toxic, *etc*). Depending on the class of interest, these models are often called structure-activity-relationship (SAR) or structure-property-relationship (SPR) models. Over the years, these approaches have evolved from the initial regression-based techniques used by Hansch ([17]), to methods that utilize complex statistical model estimation procedures ([24], [28], [42], [2]). Among them, methods based on Support Vector Machines (SVM) ([52]) have recently become very popular as they have been shown to produce highly accurate SAR and SPR models for a wide-range of problems ([14], [57], [25], [24], [55], [15]). Two broad classes of SVM-based methods have been developed. The first operate on the descriptor-space representation of the chemical compounds, whereas the second use various graph kernels that operate directly on the compounds' molecular graphs. However, despite their differences, the absolute performance achieved by these methods is often comparable, and no winning methodology has emerged.

3.1   Approaches based on Descriptors

The descriptor-space based approaches first represent each chemical compound as a high-dimensional (frequency) vector based on the set of descriptors that they contain (e.g., hashed fingerprints, graph fragments, etc) and then utilize various vector-space-based kernel functions to determine the similarity between the various compounds ([8], [49], [55], [57], [14]). Such functions include linear, radial basis function, Tanimoto coefficient, and Min-Max kernel ([49], [55]). The performance of these kernels has been extensively evaluated with each other and the results have showed that the Tanimoto coefficient (also known as the extended Jacquard similarity) and the Min-Max kernels are often among the best performing schemes ([49], [55]). The Tanimoto coefficient is defined as

$$\mathcal{K}_{TC}(X, Y) = \frac{\sum\limits_{i=1}^{M} x_i y_i}{\sum\limits_{i=1}^{M} (x_i^2 + y_i^2 - x_i y_i)}, \tag{3.1}$$

and the Min-Max kernel is defined as

$$\mathcal{K}_{MM}(X,Y) = \frac{\sum\limits_{i=1}^{M} min(x_i, y_i)}{\sum\limits_{i=1}^{M} max(x_i, y_i)}, \tag{3.2}$$

where the terms x_i and y_i are the values along the ith dimension of the M dimensional X and Y vectors, respectively.

A number of variations of these descriptor-based approaches have also been developed. One of them, which is applicable when the descriptor spaces contain a very large number of dimensions, involves the use of various feature selection techniques to reduce the effective dimensionality of the descriptor space by retaining only those descriptors that are over-represented in some classes ([8], [31], [58]). Another variation, which is designed for descriptor spaces that contain descriptors of different sizes, calculates a different similarity value for the descriptors belonging to each of the different sizes and then combines them to yield a single similarity value ([55]). This approach ensures that each individual size contributes equally to the overall similarity score and that the score is not unnecessarily dominated by the large-size descriptors, which are often more abundant.

3.2   Approaches based on Graph Kernels

The approaches based on graph kernels determine the similarity of two chemical compounds by directly comparing their molecular graphs without having to generate an intermediate descriptor-based representation ([47], [49], [40], [33]). A number of graph kernels have been developed and used in the context of building SAR and SPR models. This includes approaches that measure the similarity between two molecular graphs as the size of their maximum common subgraph ([41]), by using powers of adjacency matrices ([40]), by calculating Markov random walks on the underlying graphs ([40]), and by using weighted substructure matching between two graphs ([33]). For instance, the kernels based on powers of adjacency matrices count shared labelled sequences (paths) between two chemical graphs. Markov random walk kernels also compute the matches generated by walks (paths) on the two chemical compounds. However, as the name suggests, the match is derived by markov random walks on the two graphs. Note that the above two kernels are similar in flavor to path-based descriptor-space similarity described earlier. Weighted substructure matching kernel assigns weights based on the number of embeddings of a common substructure found in the two chemical graphs. In this approach, a substructure of size l is centered around an atom and consists of all atoms and bonds that can be reached by a path of length l via this atom. This kernel is similar in flavor to the extended connectivity fingerprints (ECFP) described earlier. However, in the case of this kernel function, no explicit descriptor-space is generated.

4.   Searching Compound Libraries

Searching large databases of chemical compounds, often referred to as *compound libraries*, in order to identify compounds that share the same bioactivity (i.e., they bind to the same protein or class of proteins) with a certain *query* compound is arguably the most widely used operation involving chemical compounds and an essential step towards the iterative optimization of a compound's binding affinity, selectivity, and other pharmaceutically relevant properties. This search is usually performed against different libraries (e.g., corporate library, libraries of commercially available compounds, libraries of patented compounds, etc) and provide key information that can be used to identify other more potent compounds and to guide the synthesis of small-scale libraries around the initial query compounds.

Depending on the initial properties of the query compound and the goal of the iterative optimization process, there are two distinct types of operations that the database search mechanisms needs to support. The first is the standard *rank-retrieval* operation whose goal is to identify compounds that are similar to the query in terms of their bioactivity. The second is the *scaffold-hopping* operation whose goal is to identify compounds that are similar to the query in terms of their bioactivity but their structures are different from that of the query (different scaffolds). This latter operation is used when the query compound has some undesirable properties such as toxicity, bad ADME (absorption, distribution, metabolism and excretion), or may be promiscuous ([18], [45]). Since these properties are often shared by the compounds that have very similar structures, it is important to identify as many chemical compounds as possible that not only show the desired activity for the biomolecular target but also have different structures (come from diverse chemical classes or chemo-types) ([64], [18], [48]). Furthermore, scaffold-hopping is also important from the point of view of un-patented chemical space. Many important lead compounds and drug candidates have already been patented. In order to find new therapies and offer alternative treatments it is important for a pharmaceutical company to discover novel leads significantly different from the existing patented chemical space.

The solution to the ranked-retrieval operation relies on the well known fact that the chemical structure of a compound relates to its activity (SAR). As such, effective solutions can be devised that rank the compounds in the database based on how structurally similar they are to the query. However, for scaffold-hopping, the compounds retrieved must be structurally *sufficiently* similar to possess similar bioactivity but at the same time must be structurally *dissimilar* enough to be a novel chemotype. This is a much harder operation than simple ranked-retrieval as it has the additional constraint of maximizing dissimilarity that runs counter to the relationship between the structure of a compound and its activity.

The rest of this section describes two sets of techniques for performing the ranked-retrieval and scaffold-hopping operations. The first are inspired by advances in automatic relevance feedback mechanism and use techniques such as the automatic query expansion to identify structurally different compounds from the query. The second measure the similarity between the query and a compound by taking into account additional information beyond their structure-based similarities. This *indirect* way of measuring similarity enables the retrieval of compounds that are structurally different from the query but at the same time possess the desired bioactivity. The indirect similarities are derived by analyzing the similarity network formed by the query and the database compounds. These indirect similarity based techniques operate on the descriptor-space representation of the compounds and are independent of the selected descriptor-space.

4.1 Methods Based on Direct Similarity

Many methods have been proposed for ranked-retrieval and scaffold-hopping that directly operate on the underlying descriptor space representation. These *direct similarity* based methods can be divided into two groups. The first contains methods that rely on better designed descriptor-space representations, whereas the second contains methods that are not specific to any descriptor-space representation but utilize different retrieval strategies to improve the overall performance.

Among the first set of methods, 2D descriptors described in Section 2 such as path-based fingerprints (fp), dictionary based keys (MACCS) and more recently Extended Connectivity fingerprints (ECFP) as well as Graph Fragments (GF) have all been successfully applied for the retrieval problem([55]). However, for scaffold-hopping, pharmacophore based descriptors such as ErG ([48]) have been shown to outperform 2D topology based descriptors ([48], [64]). Lastly, descriptors based on 3D structure or conformations of the molecule have also been applied successfully for scaffold-hopping ([64], [45]).

The second set of methods include the turbo search based schemes ([18]) which utilize ideas from automatic relevance feedback mechanism ([1]). The turbo search techniques operate as follows. Given a query q, they start by retrieving the top-k compounds from the database. Let A be the $(k + 1)$-size set that contains q and the top-k compounds. For each compound $c \in A$, all the compounds in the database are ranked in decreasing order based on their similarity to c, leading to $k + 1$ ranked lists. These lists are combined to obtain the final similarity of each compound with respect to the initial query. Similar methods based on consensus scoring, rank averaging, and voting have also been investigated ([64]).

4.2 Methods Based on Indirect Similarity

Recently, a set of techniques to improve the scaffold-hopping performance have been introduced that are based on measuring the similarity between the query and a compound by taking into account additional information beyond their descriptor-space-based representation ([54], [56]). These methods are motivated by the observation that if a query compound q is structurally similar to a database compound c_i and c_i is structurally similar to another database compound c_j, then q and c_j could be considered as being similar or related even though they may have zero or very low direct similarity. This *indirect* way of measuring similarity can enable the retrieval of compounds that are structurally different from the query but at the same time, due to associativity, possess the same bioactivity properties with the query.

The set of techniques developed to capture such indirect similarities are inspired by research in the fields of information retrieval and social network analysis. These techniques derive the indirect similarities by analyzing the network formed by a k-nearest-neighbor graph representation of the query and the database compounds. The network linking the database compounds with each other *and* with the query is determined by using a *k-nearest-neighbor* (NG) and a *k-mutual-nearest-neighbor* (MG) graph. Both of these graphs contain a node for each of the compounds as well as a node for the query. However, they differ on the set of edges that they contain. In the k-nearest-neighbor graph there is an edge between a pair of nodes corresponding to compounds c_i and c_j, if c_i is in the k-nearest-neighbor list of c_j or vice-versa. In the k-mutual-nearest-neighbor graph, an edge exists only when c_i is in the k-nearest-neighbor list of c_j *and* c_j is in the k-nearest-neighbor list of c_i. As a result of these definitions, each node in NG will be connected to at least k other nodes (assuming that each compound has a non-zero similarity to at least k other compounds), whereas in MG, each node will be connected to at most k other nodes.

Since the neighbors of each compound in these graphs correspond to some of its most structurally similar compounds and due to the relation between structure and activity (SAR), each pair of adjacent compounds will tend to have similar activity. Thus, these graphs can be considered as network structures for capturing bioactivity relations.

A number of different approaches have been developed for determining the similarity between nodes in social networks that take into account various topological characteristics of the underlying graphs ([50], [13]).For the problem of scaffold-hopping, the similarity between a pair of nodes is determined as a function of the intersection of their adjacency lists ([54], [56]), which takes into account all two-edge paths connecting these nodes. Specifically, the similarity between c_i and c_j with respect to graph G is given by

$$\mathrm{isim}_G(c_i, c_j) = \frac{\mathrm{adj}_G(c_i) \cap \mathrm{adj}_G(c_j)}{\mathrm{adj}_G(c_i) \cup \mathrm{adj}_G(c_j)}, \tag{4.1}$$

where $\mathrm{adj}_G(c_i)$ and $\mathrm{adj}_G(c_j)$ are the adjacency lists of c_i and c_j in G, respectively.

This measure assigns a high similarity value to a pair of compounds if both are very similar to a large set of common compounds. Thus, compounds that are part of reasonably tight clusters (i.e., a set of compounds whose structural similarity is high) will tend to have high indirect similarities as they will most likely have a large number of common neighbors. In such cases, the indirect similarity measure re-enforces the existing high direct similarities between compounds. However, the indirect similarity between a pair of compounds c_i and c_j can also be high even if their direct similarity is low. This can happen when the compounds in $\mathrm{adj}_G(c_i) \cap \mathrm{adj}_G(c_j)$ match different structural descriptors of c_i and c_j. In such cases, the indirect similarity measure is capable of identifying relatively weak structural similarities, making it possible to identify scaffold-hopping compounds.

Given the above graph-based indirect similarity measures, various strategies can be employed to retrieve compounds from the database. Three such strategies are discussed below. The first corresponds to that used by the standard ranked-retrieval method, whereas the other two are inspired by information retrieval methods used for automatic relevance feedback ([1]) and are specifically designed to improve the scaffold-hopping performance.

Best-Sim Retrieval Strategy. This is the most widely used retrieval strategy and it simply returns the compounds that are the most similar to the query. Specifically, if A is the set of compounds that have been retrieved thus far, then the next compound c_{next} that is selected is given by

$$c_{next} = \arg \max_{c_i \in D - A} \{\mathrm{isim}(c_i, q)\}. \tag{4.2}$$

This compound is added to A, removed from the database, and the overall process is repeated until the desired number of compounds has been retrieved ([56]).

Best-Sum Retrieval Strategy. This retrieval strategy incorporates additional information from the set of compounds retrieved thus far (set A). Specifically, the compound selected, c_{next}, is the one that has the highest average similarity to the set $A \cup \{q\}$. That is,

$$c_{next} = \arg\max_{c_i \in D-A}\{\text{isim}(c_i, A \cup \{q\})\}. \tag{4.3}$$

The motivation behind this approach is that due to SAR, the set A will contain a relatively large number of active compounds. Thus, by modifying the similarity between q and a compound c to also include how similar c is to the compounds in the set A, a similarity measure that is re-enforced by A's active compounds is obtained ([56]). This enables the retrieval of active compounds that are similar to the compounds present in A even if their similarity to the query is not very high; thus, enabling scaffold-hopping.

Best-Max Retrieval Strategy.   A key characteristic of the retrieval strategy described above is that the final ranking of each compound is computed by taking into account *all* the similarities between the compound and the compounds in the set A. Since the compounds in A will tend to be structurally similar to the query compound, this approach is rather conservative in its attempt to identify active compounds that are structurally different from the query (i.e., scaffold-hops).

To overcome this problem, a retrieval strategy was developed ([56]) that is based on the best-sum approach but instead of selecting the next compound based on its average similarity to the set $A \cup \{q\}$, it selects the compound that is the most similar to *one* of the compounds in $A \cup \{q\}$. That is, the next compound is given by

$$c_{next} = \arg\max_{c_i \in D-A}\{ \max_{c_j \in A \cup \{q\}} \text{isim}(c_i, c_j)\}. \tag{4.4}$$

In this approach, if a compound c_j other than q has the highest similarity to some compound c_i in the database, c_i is chosen as c_{next} and added to A irrespective of its similarity to q. Thus, the query-to-compound similarity is not necessarily included in every iteration as in the other schemes, allowing this strategy to identify compounds that are structurally different from the query.

4.3   Performance of Indirect Similarity Methods

The performance of indirect similarity-based retrieval strategies based on the NG as well as MG graph was compared to direct similarity based on Tanimoto coefficient ([56]). The compounds were represented using different descriptor-spaces (GF, ECFP, and ErG). The quantitative results showed that indirect similarity is consistently, and in many cases substantially, better than direct similarity. Figure 19.1 shows a part of the results in [56] which compare MG based indirect similarity to direct Tanimoto coefficient (TM) similarity searching using ECFP descriptors. It can be observed from the figure

Figure 19.1. Performance of indirect similarity measures (MG) as compared to similarity searching using the Tanimoto coefficient (TM).

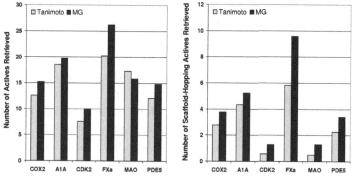

Tanimoto indicates the performance of similarity searching using the Tanimoto coefficient with extended connectivity descriptors; MG indicates the performance of similarity searching using the indirect similarity approach on the mutual neighbors graph formed using extended connectivity fingerprints.

that indirect similarity outperforms direct similarity for scaffold-hopping active retrieval in all of six datasets that were tested. It can also be observed that indirect similarity outperforms direct similarity for active compound retrieval in all datasets except MAO. Moreover, the relative gains achieved by indirect similarity for the task of identifying active compounds with different scaffolds is much higher, indicating that it performs well in identifying compounds that have similar biomolecule activity even when their direct similarity is low.

5. Identifying Potential Targets for Compounds

Target-based drug discovery, which involves selection of an appropriate target (typically a single protein) implicated in a disease state as the first step, has become the primary approach of drug discovery in pharmaceutical industry ([2], [46]). This was made possible by the advent of High Throughput Screening (HTS) technology in the late 1980s that enabled rapid experimental testing of a large number of chemical compounds against the target of interest. HTS is now routinely utilized to identify the most promising compounds (*hits*) that show desired binding/activity against a given target. Some of these compounds then go through the long and expensive process of optimization, and eventually one of them may go to clinical trials. If clinical trails are successful then the compound becomes a drug. HTS technology ushered in a new era of drug discovery by reducing the time and money taken to find hits that will have a high chance of eventually becoming a drug.

However, the increased number of candidate hits from HTS did not increase the number of actual drugs coming out of the drug discovery pipeline. One of the principal reasons for this failure is that the above approach only focuses on the target of interest, taking a very narrow view of the disease. As such, it may lead to unsatisfactory phenotypic effects such as toxicity, promiscuity, and low efficacy in the later stages of drug discovery ([46]). More recently, research focus is shifting to directly screen molecules to identify desirable phenotypic effects using cell-based assays. This screening evaluates properties such as toxicity, promiscuity and efficacy from the onset rather than in later stages of drug discovery ([23], [46]). Moreover, toxicity and off-target effects are also a focus of early stages of conventional target-based drug discovery ([5]). But from the drug discovery perspective, target identification and subsequent validation has become the rate limiting step in order to tackle the above issues ([12]). Targets must be identified for the hits in phenotypic assay experiments and for secondary pharmacology as the activity of hits against all of its potential targets sheds light on the toxicity and promiscuity of these hits ([5]). Therefore, the identification of all likely targets for a given chemical compound, also called *Target Fishing* ([23]), has become an important problem in drug discovery.

Computational techniques are becoming increasingly popular for target fishing due to large amounts of data from high-throughput screening (HTS), microarrays, and other experiments ([23]). Given a compound, these techniques initially assign a score to each potential target based on some measure of likelihood that the compound binds to the target. These techniques then select as the compound's targets either those targets whose score is above a certain cut-off or a small number of the highest scoring targets. Some of the early target fishing methods utilized approaches based on reverse docking ([5]) and nearest-neighbor classification ([35]). Reverse docking approaches dock a compound against all the targets of interest and identify as the most likely targets those that achieve the best binding affinity score. Note that these approaches are applicable only for proteins with resolved 3D structure and as such their applicability is somewhat limited. The nearest-neighbor approaches rely on the structure-activity-relationship (SAR) principle and identify as the most likely targets for a compound the targets whose nearest neighbors show activity against. In these approaches the solution to the target fishing problem only depends on the underlying descriptor-space representation, the similarity function employed, and the definition of nearest neighbors. However, the performance of these approaches has been recently surpassed by a new set of *model-based* methods that solve the target fishing problem using various machine-learning approaches to learn models for each one of the potential targets based on their known ligands ([36], [25], [53]). These methods are further discussed in the subsequent sections.

5.1   Model-based Methods For Target Fishing

Two different approaches have been employed to build models suitable for target fishing. In the first approach, a separate SAR model is built for every target. For a given test compound, these models are used to obtain a score for each target against this compound. The highest scoring targets are then considered as the most likely targets that this compound will bind to ([36], [53], [23]). This approach is similar to the reverse docking approach described earlier. However, the target scores for a compound are obtained from the models built for each target instead of the docking procedure. The second approach treats target fishing problem as an instance of the multilabel prediction problem and uses category ranking algorithms([6]) to solve this problem ([53]).

Bayesian Models for Target Fishing (Bayesian). This approach utilizes multi-category bayesian models ([36]) wherein a model is built for every target in the database using SAR data available for each target. Compounds that show activity against a target are used as positives for that target and the rest of the compounds are treated as negatives. The input to the algorithm is a training set consisting of a set of chemical compounds and a set of targets. A model is learned for every target given a descriptor-space representation of training chemical compounds ([36]). For a new chemical compound whose targets have to be predicted, an estimator score is computed for each target reflecting the likelihood of activity against this target using the learned models. The target can be ranked according to their estimator scores and the targets that get high scores can be considered as the most likely targets for this compound.

SVM-based Method (SVM rank). This approach for solving the ranking problem builds for each target a one-versus-rest binary SVM classifier ([53]). Given a test chemical compound c, the classifier for each target will then be applied to obtain a prediction score. The ranking of the targets will be obtained by simply sorting the targets based on their prediction scores. If there are N targets in the set of targets $\mathcal{T}$ and $f_i(c)$ is the score obtained for the ith target, then the final ranking $\mathcal{T}^*$ is obtained by

$$\mathcal{T}^* = \underset{\tau_i \in \mathcal{T}}{\mathrm{argsort}} \left\{ f_i(c) \right\}, \tag{5.1}$$

where argsort returns an ordering of the targets in decreasing order of their prediction scores $f_i(c)$. Note that this approach assumes that the prediction scores obtained from the N binary classifiers are directly comparable, which may not necessarily be valid. This is because different classes may be of different sizes and/or less separable from the rest of the dataset, indirectly affecting the nature of the binary model that was learned, and consequently its prediction scores. This SVM-based sorting method is similar to the approach proposed by Kawai and co-workers ([25]).

Cascaded SVM-based Method (Cascade SVM). A limitation of the previous approach is that by building a series of one-vs-rest binary classifiers,

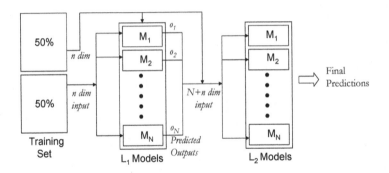

Figure 19.2. Cascaded SVM Classifiers.

it does not explicitly couple the information on the multiple categories that each compound belongs to during model training. As such it cannot capture dependencies that might exist between the different categories. A promising approach that has been explored to capture such dependencies is to formulate it as a cascaded learning problem ([53], [16]). In these approaches, two sets of binary one-vs-rest classification models for each category, referred to as L_1 and L_2, are connected together in a cascaded fashion. The L_1 models are trained on the initial inputs and their outputs are used as input, either by themselves or in conjunction with the initial inputs, to train the L_2 models. This cascaded process is illustrated in Figure 19.2. During prediction time, the L_1 models are first used to obtain predictions which are used as input to the L_2 models which produces the final predictions. Since the L_2 models incorporate information about the predictions produced by the L_1 models, they can potentially capture inter-category dependencies.

A two level SVM based method inspired by the above approach is described in [53]. In this method, both the L_1 and L_2 models consist of N binary one-vs-rest SVM classifiers, one for each target in the set of targets $\mathcal{T}$. The L_1 models correspond exactly to the set of models built by the one-vs-rest method discussed in the previous approach. The representation of each compound in the training set for the L_2 models consists of its descriptor-space based representation and its output from each of the N L_1 models. Thus, each compound c corresponds to an $n + N$ dimensional vector, where n is the dimensionality of the descriptor space. The final ranking $\mathcal{T}^*$ of the targets for a test compound will be obtained by sorting the targets based on their prediction scores from the L_2 models ($f_i^{L_2}(c)$). That is,

$$\mathcal{T}^* = \underset{\tau_i \in \mathcal{T}}{\operatorname{argsort}} \left\{ f_i^{L_2}(c) \right\}, \qquad (5.2)$$

Ranking Perceptron Based Method (RP).   This approach is based on the online version of the ranking perceptron algorithm proposed to learn a ranking

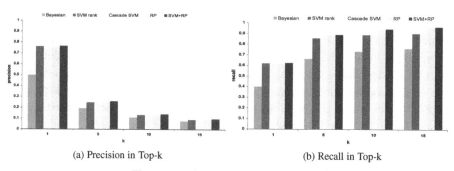

(a) Precision in Top-k   (b) Recall in Top-k

Figure 19.3. Precision and Recall results

function on a set of categories developed by Crammer and Singer ([6], [53]). This algorithm takes as input a set of objects and the categories that they belong to and learns a function that for a given object c it ranks the different categories based on the likelihood that c binds to the corresponding targets. During the learning phase, the distinction between categories is made only via a binary decision function that takes into account whether a category is part of the object's categories (relevant set) or not (non-relevant set). As a result, even though the output of this algorithm is a total ordering of the categories, the learning is only dependent on the partial orderings induced by the set of relevant and non-relevant categories.

The algorithm employed for target fishing extends the work of Crammer and Singer by introducing margin based updates and extending the online version to a batch setting([53]). It learns a linear model W that corresponds to a $N \times n$ matrix, where N is the number of targets and n is the dimensionality of the descriptor space. Thus, the above method can be directly applied on the descriptor-space representation of the training set of chemical compounds.

Finally, the prediction score for compound c_i and target τ_j is given by $\langle W_j, c_i \rangle$, where W_j is the jth row of W, c_i is the descriptor-space representation of the compound, and $\langle \cdot, \cdot \rangle$ denotes a dot-product operation. Therefore, the predicted ranking for a test chemical compound c is given by

$$\mathcal{T}^* = \operatorname*{argsort}_{\tau_j \in \mathcal{T}} \{\langle W_j, c \rangle\}. \tag{5.3}$$

SVM+Ranking Perceptron-based Method (SVM+RP). A limitation of the above ranking perceptron method over the SVM-based methods is that it is a weaker learner as (i) it learns a linear model, and (ii) it does not provide any guarantees that it will converge to a good solution when the dataset is not linearly separable. In order to partially overcome these limitations a scheme that is similar in nature to the cascaded SVM-based approach previously described was developed in which the L_2 models are replaced by a ranking perceptron ([53]). Specifically, N binary one-vs-rest SVM models are trained, which form the set of L_1 models. Similar to the cascade SVM method, the representation of each compound in the training set for the L_2 models consists of its descriptor-space based representation and its output from each of the N L_1 models. Finally, a ranking model W learned using the ranking perceptron described in the previous section. Since the L_2 model is based on the descriptor-space based representation and the outputs of the L_1 models, the size of W is $N \times (n + N)$.

5.2 Performance of Target Fishing Strategies

An extensive evaluation of the different Target Fishing methods was performed recently ([53]) which primarily used the PubChem ([39]) database to extract target-specific dose-response confirmatory assays. Specifically, the ability of the five methods to identify relevant categories in the top-k ranked categories was assessed in this work. The results were analyzed along this direction because this directly corresponds to the use case scenario where a user may want to look at top-k predicted targets for a test compound and further study or analyze them for toxicity, promiscuity, off-target effects, pathway analysis *etc*([53]). The comparisons utilized precision and recall metric in top-k for each of the five schemes. as shown in Figures 19.3a) and 19.3b). These figures show the actual precision and recall values in top-k by varying k from one to fifteen.

These figures indicate that for identifying one of the correct categories or targets in the top 1 predictions, cascade SVM outperforms all the other schemes in terms of both precision and recall. However, as k increases from one to fifteen, the precision and recall results indicate that the best performing scheme is the SVM+Ranking Perceptron and it outperforms all other schemes for both precision as well as recall. Moreover, these values in figure 19.3b) show that as k increases from one to fifteen, both the ranking perceptron based schemes (RP and SVM+RP) start performing consistently better that others in identifying all the correct categories. The two ranking perceptron based schemes also achieve average precision values that are better than other schemes in the top fifteen (Figure 19.3a)).

6. Future Research Directions

Mining and retrieving chemical data for a single biomolecular target and building SAR models on it has been traditionally used to predict as well as analyze the bioactivity and other properties of chemical compounds and plays a key role in drug discovery. However, in recent years the wide-spread use of High-Throughput Screening (HTS) technologies by the pharmaceutical industry has generated a wealth of protein-ligand activity data for large compound libraries against many biomolecular targets. The data has been systematically collected and stored in centralized databases ([38]). At the same time, the completion of the human genome sequencing project has provided a large number of "druggable" protein targets ([44]) that can be used for therapeutic purposes. Additionally, a large fraction of the protein targets that have or are currently been investigated for therapeutic purposes are confirmed to belong to a small number of gene families ([62]). The combination of these three factors has led to the development of methods that utilize information that goes beyond the traditional single biomolecular target's chemical data analysis. In recent years, the trend has been to integrate chemical data with protein and genetic data (bioinformatics data) and analyze the problem over multiple proteins or different protein families. Consequently, Chemogenomics ([43]), Poly-Pharmacology ([38])and Target Fishing ([23]) have emerged as important problems in drug discovery.

Another new direction that utilizes graph mining is network pharmacology. A fundamental assumption in drug discovery that has been applied widely in the past decades is the "one gene, one drug, on disease" assumption. However, the increasing failure in translating drug candidates into effective therapies raises the challenges to this assumption. Recent studies show that the modulating or effecting an individual gene or gene product has little effects on disease network. For example, under laboratory conditions, many single-gene knockouts by themselves exhibit little or no effects on phenotype and only 19% of genes were found to be essential across a number of model organisms ([63]). This robustness of phenotype can be understood in terms of redundant functions and alternative compensatory signalling routes. In addition, large scale functional genomics studies reveal the importance of polypharmacology, which suggests that is, instead of focusing on drugs that are maximally selective against a single drug target, the focus should be to select the drug candidates that interact with multiple proteins that are essential in the biological network. This new paradigm is refereed to as network pharmacology ([21]).

Graph mining has also been utilized to study the drug-target interaction network. Such networks provide topological information between drug and target interactions that once explored may suggest novel perspective in terms of drug discovery that is not possible by looking at drugs and targets in isolation. Learning from drug-target interaction networks has been focused on predicting drugs for targets that are novel, or that have only a few drugs known (*Target Hopping*). These methods tend to leverage the knowledge of both targets and the drug simultaneously to obtain characteristics of drug-target interaction networks. Many of the learning methods utilize Support Vector Machine (SVM). In this approach, novel kernels have been developed that relate drugs and targets explicitly. For example, Yamanish *et al.*([60]), developed profiles to represent interactions between drugs and targets, and then used kernel regression to the relationship among the interactions. Their framework enables predictions of unknown drug-target interactions.

With the improvement in high throughput technologies in chemistry, genomics, proteomics, and chemical genetics, graph mining is set to play an important role in the understanding of human disease and pursuit of novel therapies for these diseases.

References

[1] Ricardo Baeza-Yates and Berthier Ribeiro-Neto. *Modern Information Retrieval.* Addison Wesley, first edition, 1999.

[2] H.J. Bohm and G. Schneider. *Virtual Screening for Bioactive Molecules.* Wiley-VCH, 2000.

[3] K. M. Borgwardt, C. S. Ong, S. Schonauer, S. V. Vishwanathan, A. Smola, and H. P. Kriegel. Protein function prediction via graph kernels. *BMC Bioinformatics*, 21:47–56, 2005.

[4] Chemaxon. *Screen, Chemaxon Inc.*, 2005.

[5] Y. Z. Chen and C. Y. Ung. Prediction of potential toxicity and side effect protein targets of a small molecule by a ligand-protein inverse docking approach. *J Mol Graph Model*, 20(3):199–218, 2001.

[6] K. Crammer and Y. Singer. A new family of online algorithms for category ranking. *Journal of Machine Learning Research.*, 3:1025–1058, 2003.

[7] Daylight. *Daylight Toolkit, Daylight Inc, Mission Viejo, CA, USA*, 2008.

[8] M. Deshpande, M. Kuramochi, N. Wale, and G. Karypis. Frequent substructure-based approaches for classifying chemical compounds. *IEEE TKDE.*, 17(8):1036–1050, 2005.

[9] Inderjit S. Dhillon. Co-clustering documents and words using bipartite spectral graph partitioning. In *Knowledge Discovery and Data Mining*, pages 269–274, 2001.

[10] J. L. Durant, B. A. Leland, D. R. Henry, and J. G. Nourse. Reoptimization of mdl keys for use in drug discovery. *J. Chem. Info. Model.*, 42(6):1273–1280, 2002.

[11] ECFP. *Pipeline Pilot, Accelrys Inc: San Diego CA 2008.*, 2006.

[12] Ulrike S Eggert and Timothy J Mitchison. Small molecule screening by imaging. *Curr Opin Chem Biol*, 10(3):232–237, Jun 2006.

[13] F. Fouss, A. Pirotte, J. Renders, and M. Saerens. Random walk computation of similarities between nodes of a graph with application to collaborative filtering. *IEEE TKDE*, 19(3):355–369, 2007.

[14] H. Geppert, T. Horvath, T. Gartner, S. Wrobel, and J. Bajorath. Support-vector-machine-based ranking significantly improves the effectiveness of similarity searching using 2d fingerprints and multiple reference compounds. *J. Chem. Inf. Model.*, 48:742–746, 2008.

[15] M. Glick, J. L. Jenkins, J. H. Nettles, H. Hitchings, and J. H. Davies. Enrichment of high-throughput screening data with increasing levels of noise using support vector machines, recursive partitioning, and laplacian-modified naive bayesian classifiers. *J. Chem. Inf. Model.*, 46:193–200, 2006.

[16] S. Godbole and S. Sarawagi. Discriminative methods for multi-labeled classification. *PAKDD.*, pages 22–30, 2004.

[17] C. Hansch, P. P. Maolney, T. Fujita, and R. M. Muir. Correlation of biological activity of phenoxyacetic acids with hammett substituent constants and partition coefficients. *Nature*, 194:178–180, 1962.

[18] J. Hert, P. Willet, and D. Wilton. New methods for ligand based virtual screening: Use of data fusion and machine learning to enchance the effectiveness of similarity searching. *J. Chem. Info. Model.*, 46:462–470, 2006.

[19] J. Hert, P. Willett, D. J. Wilton, P. Acklin, K. Azzaoui, E. Jacoby, and A. Schuffenhauer. Comparison of topological descriptors for similarity-based virtual screening using multiple bioactive reference structures. *Org Biomol Chem*, 2(22):3256–66, 2004.

[20] Hologram. *Hologram Fingerprints, Tripos Inc. 1699 South Hanley Road, St Louis, MO 63144-2913, USA.* http://www.tripos.com, 2003.

[21] Andrew L. Hopkins. Network pharmacology: the next paradigm in drug discovery. *Nat Chem Biol*, 4(11):682–690, November 2008.

[22] J. Huan, D. Bandyopadhyay, W. Wang, J. Snoeyink, J. Prins, and A. Tropsha. Comparing graph representations of protein structure for mining family-specific residue-based packing motifs. *J. Comput. Biol.*, 12(6):657–671, 2005.

[23] J. L. Jenkins, A. Bender, and J. W. Davies. In silico target fishing: Predicting biological targets from chemical structure. *Drug Discovery Today*, 3(4):413–421, 2006.

[24] R. N. Jorissen and M. K. Gibson. Virtual screening of molecular databases using support vector machines. *J. Chem. Info. Model.*, 45(3):549–561, 2005.

[25] K. Kawai, S. Fujishima, and Y. Takahashi. Predictive activity profiling of drugs by topological-fragment-spectra-based support vector machines. *J. Chem. Info. Model.*, 48(6):1152–1160, 2008.

[26] T. Kogej, O. Engkvist, N. Blomberg, and S. Moresan. Multifingerprint based similarity searches for targeted class compound selection. *J. Chem. Info. Model.*, 46(3):1201–1213, 2006.

[27] M. Kuramochi and G. Karypis. An efficient algorithm for discovering frequent subgraphs. *IEEE TKDE.*, 16(9):1038–1051, 2004.

[28] A. R. Leach and V. J. Gillet. *An Introduction to Chemoinformatics.* Springer, 2003.

[29] Andrew R. Leach. *Molecular Modeling: Principles and Applications.* Prentice Hall, Englewood Cliffs, NJ, second edition, 2001.

[30] W. Liu, W. Lin, A. Davis, F. Jordan, H. Yang, and M. Hwang. A network perspective on the topological importance of enzymes and their phylogenetic conservation. *BMC Bioinformatics*, 8:121, 2007.

[31] Y. Liu. A comparative study on feature selection methods for drug discovery. *J. Chem. Inf. Comput. Sci.*, 44:1823–1828, 2004.

[32] MDL. *MDL Information Systems Inc., San Leandro, CA, USA.* http://www.mdl.com, 2004.

[33] S. Menchetti, F. Costa, and P. Frasconi. Weighted decomposition kernels. *Proceedings of the 22nd International Conference in Machine Learning.*, 119:585–592, 2005.

[34] H. L. Morgan. The generation of unique machine description for chemical structures: a technique developed at chemical abstract services. *Journal of Chemical Documentation*, 5:107–113, 1965.

[35] J. Nettles, J. Jenkins, A. Bender, Z. Deng, J. Davies, and M. Glick. Bridging chemical and biological space: "target fishing" using 2d and 3d molecular descriptors. *J Med Chem*, 49:6802–6810, Nov 2006.

[36] Nidhi, M. Glick, J. Davies, and J. Jenkins. Prediction of biological targets for compounds using multiple-category bayesian models trained on chemogenomics databases. *J Chem Inf Model*, 46:1124–1133, 2006.

[37] S. Nijssen and J. Kok. A quickstart in frequent structure mining can make a difference. *Proceedings of SIGKDD*, pages 647–652, 2004.

[38] G. V. Paolini, R. H. Shapland, W. P. Van Hoorn, J. S. Mason, and A. Hopkins. Global mapping of pharmacological space. *Nature biotechnology*, 24:805–815, 2006.

[39] Pubchem. *The PubChem Project*, 2007.

[40] L. Ralaivola, S. J. Swamidassa, H. Saigo, and P. Baldi. Graph kernels for chemical informatics. *Neural Networks*, 18(8):1093–1110, 2005.

[41] J. W. Raymond and P. Willett. Maximum common subgraph isomorphism algorithms for the matching of chemical structures. *J. Comp. Aided Mol. Des.*, 16(7):521–533, 2002.

[42] D. Rogers, R. Brown, and M. Hahn. Using extended-connectivity fingerprints with laplacian-modified bayesian analysis in high-throughput screening. *J. Biomolecular Screening*, 10(7):682–686, 2005.

[43] D. Rognan. Chemogenomic approaches to rational drug design. *Br J Pharmacol*, 152(1):38–52, Sep 2007.

[44] A. P. Russ and S. Lampel. The druggable genome: an update. *Drug Discov Today*, 10(23-24):1607–10, 2005.

[45] Jamal C. Saeh, Paul D. Lyne, Bryan K. Takasaki, and David A. Cosgrove. Lead hopping using svm and 3d pharmacophore fingerprints. *J. Chem. Info. Model.*, 45:1122–113, 2005.

[46] Frank Sams-Dodd. Target-based drug discovery: is something wrong? *Drug Discov Today*, 10(2):139–147, Jan 2005.

[47] A.J. Smola and R. Kondor. Kernels and regularization on graphs. In *Proceedings COLT and Kernels Workshop*, pages 144–158. M.Warmuth and B. Schelkopf, 2003.

[48] Nikolaus Stiefl, Ian A. Watson, Kunt Baumann, and Andrea Zaliani. Erg: 2d pharmacophore descriptor for scaffold hopping. *J. Chem. Info. Model.*, 46:208–220, 2006.

[49] S. J. Swamidass, J. Chen, J. Bruand, P. Phung, L. Ralaivola, and P. Baldi. Kernels for small molecules and the prediction of mutagenicity, toxicity and anti-cancer activity. *Bioinformatics*, 21(1):359–368, 2005.

[50] B. Teufel and S. Schmidt. Full text retrieval based on syntactic similarities. *Information Systems*, 31(1), 1988.

[51] Unity. *Unity Fingerprints, Tripos Inc. 1699 South Hanley Road, St Louis, MO 63144-2913, USA.* http://www.tripos.com, 2003.

[52] V. Vapnik. *Statistical Learning Theory*. John Wiley, New York, 1998.

[53] N. Wale and G. Karypis. Target identification for chemical compounds using target-ligand activity data and ranking based methods. Technical Report TR-08-035, University of Minnesota, 2008. Accepted: Jour. Chem. Inf. Model, Published on the web, September 18, 2009.

[54] N. Wale, G. Karypis, and I. A. Watson. Method for effective virtual screening and scaffold-hopping in chemical compounds. *Comput Syst Bioinformatics Conf*, 6:403–414, 2007.

[55] N. Wale, I. A. Watson, and G. Karypis. Comparison of descriptor spaces for chemical compound retrieval and classification. *Knowledge and Information Systems*, 14:347–375, 2008.

[56] N. Wale, I. A. Watson, and G. Karypis. Indirect similarity based methods for effective scaffold-hopping in chemical compounds. *J. Chem. Info. Model.*, 48(4):730–741, 2008.

[57] A. M. Wassermann, H. Geppert, and J. Bajorath. Searching for target-selective compounds using different combinations of multiclass support vector machine ranking methods, kernel functions, and fingerprint descriptors. *J. Chem. Inf. Model.*, 49:582–592, 2009.

[58] J. Wegner, H. Frohlich, and Andreas Zell. Feature selection for descriptor based classification models. 1. theory and ga-sec algorithm. *J. Chem. Inf. Comput. Sci.*, 44:921–930, 2004.

[59] P. Willett. A screen set generation algorithm. *J. Chem. Inf. Comput. Sci.*, 19:159–162, 1979.

[60] Y. Yamanishi, M. Araki, A. Gutteridge, W. Hondau, and M. Kanehisa. Prediction of drug-target interaction networks from the integration of chemical and genomic spaces. *Bioinformatics*, 24:232–240, 2008.

[61] Xifeng Yan and Jiawei Han. gspan: Graph-based substructure pattern mining. *ICDM*, pages 721–724, 2002.

[62] M. Yildirim, K. Goh, M. Cusick, A. Barabasi, and M. Vidal. Drug-target network. *Nat Biotechnol*, 25(10):1119–1126, Oct 2007.

[63] Brian P. Zambrowicz and Arthur T. Sands. Modeling drug action in the mouse with knockouts and rna interference. *Drug Discovery Today: TARGETS*, 3(5):198 – 207, 2004.

[64] Qiang Zhang and Ingo Muegge. Scaffold hopping through virtual screening using 2d and 3d similarity descriptors: Ranking, voting and consensus scoring. *J. Chem. Info. Model.*, 49:1536–1548, 2006.

[65] Ziding Zhang and Martin G Grigorov. Similarity networks of protein binding sites. *Proteins*, 62(2):470–478, Feb 2006.

Index